AF360698

Intelligent Transportation Systems for Electric Vehicles

Intelligent Transportation Systems for Electric Vehicles

Editor

João Carlos Amaro Ferreira

MDPI • Basel • Beijing • Wuhan • Barcelona • Belgrade • Manchester • Tokyo • Cluj • Tianjin

João Carlos Amaro Ferreira
ISCTE-IUL
Portugal

Editorial Office
MDPI
St. Alban-Anlage 66
4052 Basel, Switzerland

This is a reprint of articles from the Special Issue published online in the open access journal *Energies* (ISSN 1996-1073) (available at: https://www.mdpi.com/journal/energies/special_issues/ Intelligent_Systems_Electric_Vehicles).

For citation purposes, cite each article independently as indicated on the article page online and as indicated below:

LastName, A.A.; LastName, B.B.; LastName, C.C. Article Title. *Journal Name* **Year**, *Article Number*, Page Range.

ISBN 978-3-03943-048-2 (Hbk)
ISBN 978-3-03943-049-9 (PDF)

Contents

About the Editor

João C. Ferreira is an Assistant Professor, with habilitation, from ISCTE-IUL in Lisbon, Portugal. He graduated in Physics at the Technical University of Lisbon (UTL/IST), Portugal, received an MSC in Telecommunication and a PhD degree in Computer Science Engineering from UTL/IST and a second PhD in Industrial Engineering from Minho University. His professional and research interests are in data science, text mining, IoT, AI, intelligent transportation systems (ITS), blockchain, network security, and electric vehicles. He has authored over 200 scientific papers in different areas of computer science and served as general chair of several international conferences such, as OAIR 2013, INTSYS18, INTSYS19, INTSYS20. He has received five best paper awards at conferences. Recently, he has served as an expert/reviewer for several top scientific journals with more than 150 revisions. He participates in the evaluation process of more than 20 projects, such as EU Commission, ANI (Agência Nacional de Inovação) and FCT (Portuguese agency that supports science, technology, and innovation). He participates in more than 30 EU projects (FP7/H2020 and PT2020), in six of as PI. Since 2015, he has been an IEEE senior member. He has served as a Guest Editor and Topic Editor in MDPI for Energies, Electronics, and Sensors and been the editor of three books. He was previously a CIS chapter president (2017–2018) and the Coordinator of the Master in Decision Support Systems (MSIAD), and summer school of smart cities and winter school IoT and blockchain in 2019 and 2020. He has one patent in edge computing for a vessel monitor system. His current projects include H2020 Infrastress, Sparta, ENSUREC and MARISA, Interreg Block4Coop, PT2020 Monitoriçao persistente Pista and Digital Demo.

Preface to "Intelligent Transportation Systems for Electric Vehicles"

We are delighted to announce that the Special Issue of *Energies* on "Intelligent Transportation Systems (ITS) for Electric Vehicles (EV)" received 25 submissions of which 14 from Asia, Europe, and America have been selected. In less than a year, 30 citations have already been achieved. Published work reflects major investigation topics from the charging process, regulation of urban complexes, a real case study on EV charging on a condo, EV charging stations and dispersed generation, parking guidance, and change of user behavior in the city. Two other papers cover the topics "Vehicle Electrification: New Challenges and Opportunities for Smart Grids" and "Optimal Charging Navigation Strategy Design for Rapid Charging Electric Vehicles". Another paper explores train operation and bus fleets. The other published papers are as follows: "Vehicle Extreme Fast Charging Station"; "Optimal Design for a Shared Swap Charging System Considering the Electric Vehicle Battery Charging Rate"; "A Method for the Optimization of Daily Activity Chains Including Electric Vehicles and On-Road Object Detection System Using Monovision and Radar Fusion"; "Energy Management Strategies for Hybrid Electric Vehicles: Review, Classification, Comparison, and Outlook".

The EV and charging station (CS) markets have been growing exponentially, and a forecast from the International Energy Agency estimates an increase of EV sales from the current 3 million to 125 million by 2030. The CS market is growing by 40% a year and is currently worth $300 billion. Electromobility and ITS are essential components in decarbonizing road transportation and play an essential role in the mobility process of smart cities. ITS also play an essential role in this transformation, owing to the flexibility of the EV charging process and EV operation, which operates as an energy storage device; it also helps to facilitate the market penetration of renewable energy resources.

We would like to thank the extensive list of external reviewers from several areas of expertise and from numerous countries around the world and the editor Addison Su, who initiated this topic and provided support during the entire process.

Energies provides an excellent forum for all researchers, developers, and practitioners to discuss all science and technology aspects that are relevant to energy and in this case related to intelligent systems and electric vehicles.

João Carlos Amaro Ferreira
Editor

Article

Vehicle Electrification: New Challenges and Opportunities for Smart Grids

Vitor Monteiro [1,*] , **Jose A. Afonso** [2] , **Joao C. Ferreira** [3] **and Joao L. Afonso** [1]

1   Centro ALGORITMI, University of Minho, 4800-058 Guimarães, Portugal; jla@dei.uminho.pt
2   CMEMS-UMinho Center, University of Minho, 4800-058 Guimarães, Portugal; jose.afonso@dei.uminho.pt
3   ISTAR-IUL, Instituto Universitário de Lisboa (ISCTE-IUL), 1649-026 Lisboa, Portugal;
    joao.carlos.ferreira@iscte-iul.pt
*   Correspondence: vmonteiro@dei.uminho.pt; Tel.: +351-253-510-392

Received: 3 December 2018; Accepted: 24 December 2018; Published: 29 December 2018

Abstract: Nowadays, concerns about climate change have contributed significantly to changing the paradigm in the urban transportation sector towards vehicle electrification, where purely electric or hybrid vehicles are increasingly a new reality, supported by all major automotive brands. Nevertheless, new challenges are imposed on the current electrical power grids in terms of a synergistic, progressive, dynamic and stable integration of electric mobility. Besides the traditional unidirectional charging, more and more, the adoption of a bidirectional interconnection is expected to be a reality. In addition, whenever the vehicle is plugged-in, the on-board power electronics can also be used for other purposes, such as in the event of a power failure, regardless if the vehicle is in charging mode or not. Other new opportunities, from the electrical grid point of view, are even more relevant in the context of off-board power electronics systems, which can be enhanced with new features as, for example, compensation of power quality problems or interface with renewable energy sources. In this sense, this paper aims to present, in a comprehensive way, the new challenges and opportunities that smart grids are facing, including the new technologies in the vehicle electrification, towards a sustainable future. A theoretical analysis is also presented and supported by experimental validation based on developed laboratory prototypes.

Keywords: vehicle electrification; smart grids; renewable energy sources; energy storage systems; power quality; bidirectional; power electronics.

---

1. Introduction

Nowadays, modern societies are facing the well-known problems of environmental air pollution, forcing the adoption of new strategies for mitigating greenhouse gas emissions [1,2]. Some of the actions for alleviating such emissions are mainly offered by emerging smart grids, and are sustained by: (a) Renewable energy sources (RES), on small- and large-scale; (b) energy storage systems (ESS), as a support of RES adoption; and (c) vehicle electrification encompassing advanced functionalities [3–7]. This is even more evident considering that the technologies in the field of industrial and power electronics have evolved in recent years, contributing towards a profound and motivating change of paradigm [8,9]. As a positive consequence, new electronics applications encompassing communication technologies, supported by the Internet of Thing (IoT) concept, will transform the electrical power grid into a dynamic, autonomous, secure and flexible infrastructure [10–13].

Concerning RES, in recent decades, the production of electricity from this type of source (mainly supported by wind and solar) has grown significantly as a contribution for optimizing the energy management in macro- and micro-scenarios. In this perspective, the operation and optimization aspects regarding the introduction of RES in microgrids is envisaged in [14], whereas an ample perspective of the RES contribution for disseminating the new paradigm of smart grids is presented in [15]. In order to optimize the power generation from RES, especially considering the intermittency associated with their production, it will also be fundamental, in the near future, to combine the inclusion of flexible ESS, allowing the establishment of an efficient harmonization between power production, storage, and consumption. The present status and the perspectives for the inclusion of RES with intermittent and unpredictable production is presented in [16], the balancing strategy for power usage from RES, regarding the user demand, is presented in [17], and a review about the role of ESS for mitigating the inconsistency of energy production from RES is offered in [18].

Alongside with RES and ESS, the large-scale adoption of vehicle electrification, principally the electric vehicle (EV), will also be vital for smart grids and smart homes dissemination, as well as for reducing energy costs and greenhouse gas emissions [19,20]. A synergistic use of RES with the charging infrastructure of EVS charging toward opportunities related to the RES and EVs power optimization is offered in [21]. A complete survey concerning the electrification of transportation contextualized in smart grids is present in [22]. The collaboration of EVs and RES toward cost and emission reductions is introduced in [23]. The particular case of the power coordination between EVs and RES in a smart home level is presented in [24]. Concerning this scenario, several perspectives can be adopted. For example, smart charging approaches for EVs conceived to maximize the usage of energy from RES are introduced in [25]. Designed to enhance the grid performance, a scheduling strategy considering the uncertainties from RES and EVs is proposed in [26]. A solar docking charging station for EVs is described in [27]. The impact of EVs and solar photovoltaic panels (PV) prospecting the future enegy generation portfolio is investigated in [28]. A cost minimization for reducing the effect of intermittency in a solar docking charging station with EVs is proposed in [29]. An innovative integrated topology for RES and EVs is proposed and experimentally validated in [30]. A harmonized scheduling of distributed energy assets, optimizing the energy management of a smart home is offered in [31]. The optimization of a smart home prospecting demand response strategies is presented in [32]. A smart charging management for EVs in smart homes is proposed in [33], a control methodology for the EV charging, considering RES and uncertainties as for the energy price, is proposed in [34], and a demand-side energy management including EVs, ESS, RES is presented in [35].

From a global point of view, a complete outline about the status and issues toward the vehicle electrification is offered in [36], whereas an economic investigation of consumers' lookout for the electric mobility supremacy is presented in [37]. The impact that vehicle electrification can cause in the electrical grid is presented in [38], and a survey concerning the vehicle electrification encompassed in a smart grid background is presented in [39]. On the other hand, as the title indicates, this paper focuses on the challenges and opportunities that arise from vehicle electrification, concretely in terms of the utilization of the on-board and off-board EV battery chargers (EVBCs) for innovative operation modes. Thus, besides the traditional operation modes, grid-to-vehicle (G2V) and vehicle-to-grid (V2G), this paper focuses on the possibility to integrate power quality features in the on-board and off-board EVBCs, as a contribution for the grid-side, as well as on the framework with unified technologies with RES and ESS.

Contextualizing the aforementioned aspects, harmonized for vehicle electrification, the main contributions of this paper encompass proposals in the following areas: (a) New opportunities of operation toward on-board EVBCs in a future perspective of smart homes; (b) new opportunities of operation toward single- and three-phase off-board EVBCs in a future perspective of smart grids; (c) new operation modes of off-board EVBCs considering the perspective of improving power quality aspects for the grid-side; (d) new operation modes for single- and three-phase off-board EVBCs considering a unified integration with RES.

After this brief introduction, Section 2 introduces the operating principle of an EVBC, highlighting the different configurations of on-board and off-board strands. Section 3 comprehensively presents the challenges and opportunities that on-board and off-board EVBCs represent for smart grids and smart homes (considering single- and three-phase interfaces). Section 4 presents three laboratory prototypes of EVBCs encompassing innovative features, as well as a brief experimental validation. Finally, Section 5 highlights the main conclusions that can be drawn from this paper.

2. EV Battery Chargers: Principle of Operation

This section introduces the principle of operation of EV battery chargers (EVBCs), as well as the future perspectives in terms of on-board and off-board systems, wired and wireless systems, and integrated coordination towards smart grids.

2.1. On-Board and Off-Board Systems

The principle of operation of an on-board and an off-board EVBC, also highlighting its internal constitution, is presented in this section. Internally, an EVBC is composed of power electronics converters and their control systems, responsible for controlling the EV battery charging and, in conjunction with the other elements of the EV, for establishing a communication with the energy management system of the smart grid or smart home with the concrete objective of defining set points of operation. Figure 1 shows the basic and classical structure of an EVBC, composed by two power converters (a grid-side one interfacing with the electrical grid and a battery-side one interfacing with the EV battery) and by the digital control system common to both power converters. Since the control is done with a closed-loop algorithm, this figure also shows the main control variables that are necessary to acquire, as well as the output control signals for the semiconductors of the power converters.

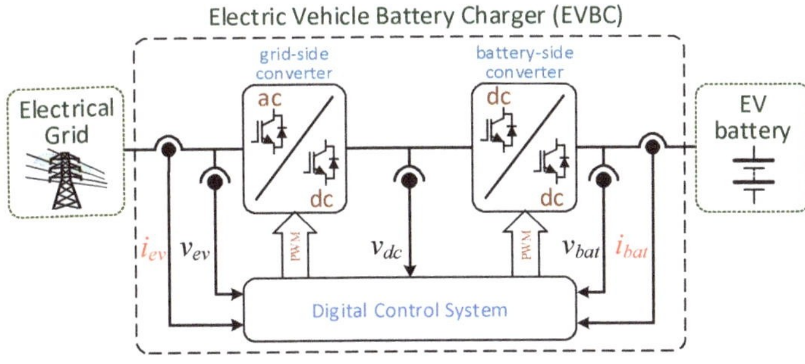

Figure 1. Structure of an electric vehicle battery charger (EVBC) composed of two power converters (a grid-side one interfacing the electrical grid and a battery-side one interfacing the EV battery) and the digital control system.

This is the customary organization of an EVBC, however, it can be classified according to its arrangement with respect to the EV, i.e., on-board and off-board. An EVBC is classified as on-board when the power electronics required to charge the EV battery are inside the EV, i.e., the converter responsible for controlling the stages of battery charging is inside the EV (usually more than a single controlled stage of voltage and current). Figure 2 shows the interface of an EV with the power grid through an on-board EVBC and an off-board EVBC. The power converters of the on-board EVBC are responsible for the bidirectional power flow between the electrical grid and the EV battery. For the grid-side, the ac-dc converter can be controlled by current or voltage according to the operating mode and for the battery-side, the dc-dc converter can also be controlled by current or voltage according to the intended operating mode for the charging system (c.f. Section 3). As shown in the figure, the operating mode is defined by specific control algorithms, whose management is in accordance with the information from the battery management system (BMS), i.e., the BMS establishes the limits of voltage and current during the charging or discharging processes. On the other hand, an EVBC is classified as off-board when the power electronics required to charge the EV battery are outside the EV, i.e., the converters that are responsible for controlling the battery charging stages (usually a single current controlled stage) are outside the EV. An off-board EVBC is composed by a grid-side converter and by a battery-side converter, both allowing bidirectional power flow between the electrical grid (with current control) and the EV battery (also with current control), i.e., in both cases, it is similar to the operation presented previously for the on-board EVBC. Moreover, for the off-board structure, the control of the operating mode is also defined in accordance with the information provided by the BMS.

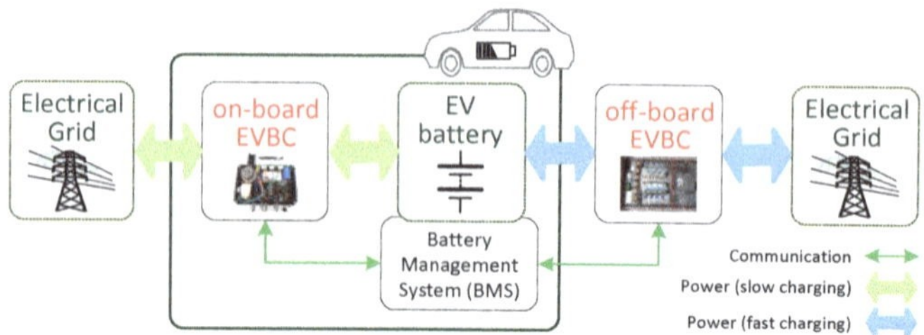

Figure 2. Interface of an EV with the power grid through an on-board EVBC and an off-board EVBC.

2.2. Wireless Charging Systems

In the previous section the on-board and off-board EV battery chargers (EVBCs) were introduced. Due to weight and volume restrictions from the EV perspective, normally on-board EVBCs are designed for far lower power ratings than off-board EVBCs. However, it is important to distinguish that, from the electrical grid point of view, in both cases, the EVBCs can have a galvanic isolation or not, either for safety reasons or for convenience, in order to reduce operating voltage levels (i.e., the levels between the grid and the EV battery). In addition to the galvanic isolation that can be implemented for on-board and off-board systems, EVBCs can also be classified as wired or wireless, depending on whether there is a physical link between the electrical grid and the EVBC. Usually, in a wireless system, a part of the power electronics converter is outside the EV (off-board) and the other part is inside the EV (on-board).

Wireless charging systems are becoming popular for different appliances and for EVs; therefore, the main automakers are also developing realistic solutions for their EVs, including a significant range of charging levels. These wireless charging systems, which have been explored by different companies over the last decades, are also seen as a key opportunity to disseminate the electric mobility market, since it is a new exciting experience for the user. Complete overviews about wireless charging technologies for applications in electric mobility are presented in [40–43]. A basic wireless charging system consists in a fixed ground pad that stays below the EV during the charging and a receiving system that stays embedded in the inferior part of the EV. In addition to the need to increase the efficiency of the power transfer between the ground pad and the EV, which will involve the use of innovative technologies of power converters, the full adoption of wireless charging systems will rely on industry standards, universal communication with any EV and the charging pad, and safety issues for human beings and animals.

2.3. EV in Smart Grids: Coordination and Power Quality

As demonstrated in [44] and [45], the EV dissemination signifies a vast contribution for electrical grids, both in terms of future trends and control coordinating strategies. For example, the collaborative operation between EVs and RES is introduced in [46], and the contextualization with smart homes and microgrids is presented in [47] and [48]. Taking into account the EV operation in G2V and V2G modes framed in smart grids, several key points can be addressed. For instance, the impact on distribution systems is analyzed in [49], the coordination with RES scenarios is explored in [50], the contribution to reducing operating costs and to regulating the grid voltage frequency is explored in [51], and the dynamic operation as a function of other appliances is investigated in [52]. Besides the G2V and V2G modes, the EV can also contribute to improving power quality issues. For instance, the EVBC operation as an active filter is introduced in [53] and [54], and the EVBC contribution for compensating reactive power in the electrical grid is investigated in [55] and [56]. In the context of power quality, an overview about power quality in smart grids is established in [57], a collaborative support between RES and EVs for enhancing power grid support is analyzed in [58], the key aspects of the EVs integration into smart grids are discussed in [59], and innovative operations for the EVs connected into power grids toward mitigating issues of power quality are proposed in [60].

3. Opportunities for Smart Grids

Section 2 introduced the different structures that can be considered for an EVBC. As the analysis of the structures in terms of power electronics is not the objective of this paper, but the challenges and opportunities of vehicle electrification in smart grids and smart homes, the particular details of the topologies of the converters, either hardware or software, are not presented in this section.

3.1. On-Board EV Battery Charger

This section presents the main operation modes of an on-board EVBC, taking into account its limitations and the opportunities that they can offer for the operation in smart grids and smart homes, concretely, in terms of power controllability and new functionalities obtained for the installation where the EV is plugged-in. As an example case, Figure 3 illustrates the integration of an EV (including the on-board EVBC) into a smart home. As shown, the EV battery is charged through an on-board EVBC, which is connected to the electrical grid in parallel with the home loads, i.e., when present, the EV is treated as an additional home load. As illustrated, bidirectional communication is considered between the smart home and the electrical grid toward a smart grid perspective in terms of controllability.

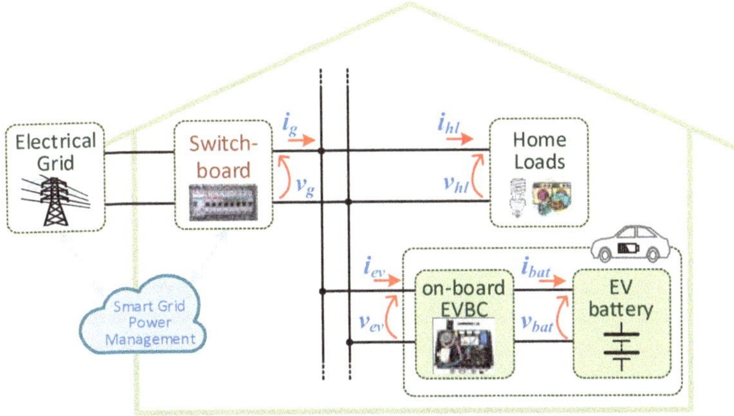

Figure 3. Illustration of the integration of an on-board EVBC into a smart home.

3.1.1. Grid-to-Vehicle (G2V)

The G2V operation mode is exclusively concerned with the EV battery charging directly from the grid, and it is usually the only mode of operation available on EVs. As exemplified in Figure 4, the on-board EVBC is connected to the electrical grid through a smart home with unidirectional power flow and with bidirectional communication between the smart home, the electrical grid and the on-board EVBC. With this operation mode, the value of the EVBC grid-side current (i_{ev}) does not take into account the other loads connected in the same electrical installation (e.g., in the case of a home, the total current is limited by the main circuit breaker, which may be triggered if the limit is surpassed). The principle of operation representative of the G2V mode in a smart home is presented in Figure 5, where the grid voltage (v_g), the grid current (i_g), the home loads current (i_{hl}) and the EVBC grid-side current (i_{ev}) are represented. In order to avoid deteriorating the power quality indices in the electrical grid, the EVBC current is sinusoidal and in phase with the grid voltage. As shown, realistic conditions are considered in terms of distorted grid voltage (v_g) and home loads current (i_{hl}).

Figure 4. On-board EVBC: Grid-to-vehicle (G2V) operation mode.

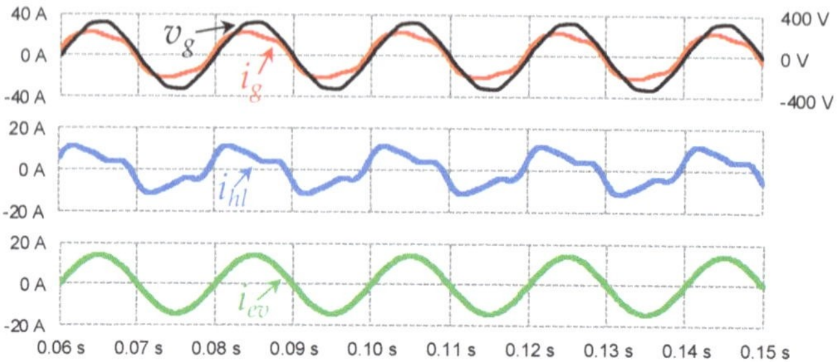

Figure 5. Principle of operation representative of the G2V mode.

Similar to the (basic) G2V mode, the controlled G2V mode refers to the EV battery charging directly from the grid, but with adjustment of the operating power value according to the other connected loads [52]. Besides, with this operation mode, for example, the EVBC operating power may be adjusted according to the power injected by RES, aiming to balance the power production and consumption from the smart home perspective, and without harming the power quality on the grid-side (e.g., frequency and amplitude deviations on the grid voltage). In order to implement this operation mode, it is necessary to establish a communication between the EVBC and the grid (or the home energy management system, i.e., when considering the EV integration into a smart home). The principle of operation representative of the controlled G2V mode is exemplified in Figure 6. Similarly to the aforementioned G2V mode, the EVBC operates with a sinusoidal grid-side current; however, its amplitude is adjusted in real-time according to the other loads of the home. In the transition from case #1 to case #2, a home load was turned-off (the current consumption, i_{hl}, decreases), so the EVBC increases its operating power (increases the current consumption, i_{ev}). Nevertheless, the maximum operating power of the EVBC, which is internally controlled, cannot be exceeded in any circumstance. Applying this control strategy to the EV battery charging, the maximum operating power of the smart home is never exceeded, maintaining the same value. This can be observable in the amplitude of the grid current (i_g).

3.1.2. Vehicle-to-Grid (V2G)

The V2G operation mode refers to the return of part of the energy stored in the EV battery to the grid conferring to the convenience of the grid management system and the EV user, representing a benefit for the electrical grid, because it allows using the EV as an ESS for supporting the grid stability. Contrary to the G2V, in this operation mode, the grid-side and the battery-side converters must be used in bidirectional mode, representing a perspective for the EVBCs of the future EVs. Moreover, this mode requires communication with a grid aggregator, in order to define in which schedules the EVBC operates in this mode, as well as the amount of power that is necessary to return to the grid. This operation mode, illustrated in Figure 7, is controlled according to the power injected into the grid, but it can also be controlled based on the loads connected in the same electrical installation. Figure 8 presents some results illustrating the V2G mode. Initially, in case #1, the EVBC is operating in V2G mode by injecting power into the grid without any control over the other loads, and then, in case #2, the EVBC injects power into the grid as a function of the other loads. In this specific case, a load was turned off; therefore, the power injected increases proportionally. As can be observed, in both cases, the EVBC grid-side current is in phase opposition with the voltage, meaning that power is injected into the grid.

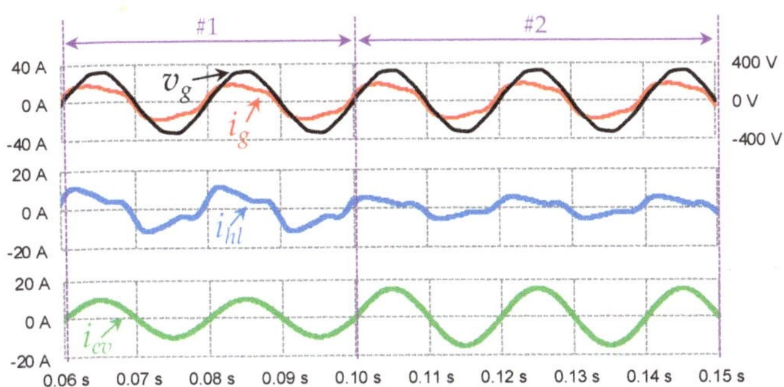

Figure 6. Principle of operation representative of the controlled G2V mode.

Figure 7. On-board EVBC: vehicle-to-grid (V2G) operation mode.

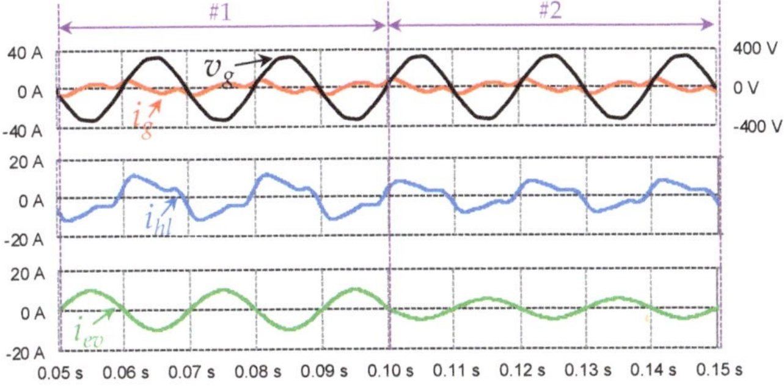

Figure 8. The principle of operation representative of the V2G mode.

3.1.3. Vehicle-to-Load (V2L)–Voltage Source

In the previously presented operation modes, the EVBC is controlled in order to absorb or inject power into the grid, where the grid-side converter operates with a current feedback control, i.e., the voltage is imposed by the grid and the EVBC defines the current waveform. In the operation mode as a voltage source, the EVBC operates independently from the grid, i.e., it can be used as a voltage source to power loads according to the user's convenience. The principle of operation representative of the vehicle-to-load (V2L) mode, i.e., as a voltage source, is presented in Figure 9. This operation mode is useful, for example, in remote locations where a voltage source is only necessary for short periods. It may also be useful in campsites, or in extreme situations of catastrophic events where the grid may be unavailable. Thus, in this operation mode, the grid-side converter operates with a voltage feedback control, i.e., the voltage is imposed by the EVBC and the current waveform is defined by the linear or nonlinear loads connected to the EVBC. As operation mode uses the energy from the EV battery, the EV owner is responsible for the management of the battery state-of-charge, e.g., regarding the minimum acceptable state-of-charge for the next travel. Internally, the EV battery is protected by the BMS. Since this operation mode can be used in multiple locations and for various purposes (e.g., smart homes, remote locations, or in islanding mode), it represents a new contribution for the future smart grids. It is meaningful to note that Nissan already has a system entitle "LEAF-to-Home", where the "EV Power Station" interfaces an EV and a house [61]. However, the key drawback of the Nissan system is that it can only be used where it is installed, i.e., it cannot be used generically with the EV in any place other than the home.

In Figure 10 the operating principle of this mode is presented, where i_g represents the home current, i_{ev} represents the EVBC grid-side current, and i_{hl} the loads current. As can be seen, when the EVBC is functioning as a voltage source, the EVBC current is the same as the load current, and the voltage applied to the loads is the voltage produced by the EVBC, whose value is equal to the nominal value of the grid voltage. This figure is divided into three distinct cases. In case #1, the EVBC is not operating in any mode. In case #2, the EVBC is not connected to the grid and starts to produce a sinusoidal voltage, but no load has yet been connected to the EVBC. In case #3, the EVBC is producing a sinusoidal voltage and a load is connected to the EVBC. In this case, since a nonlinear load was connected, it results in a consumed current with a high harmonic content.

Figure 9. On-board EVBC: Vehicle-to-load (V2L) operation mode (as a voltage source).

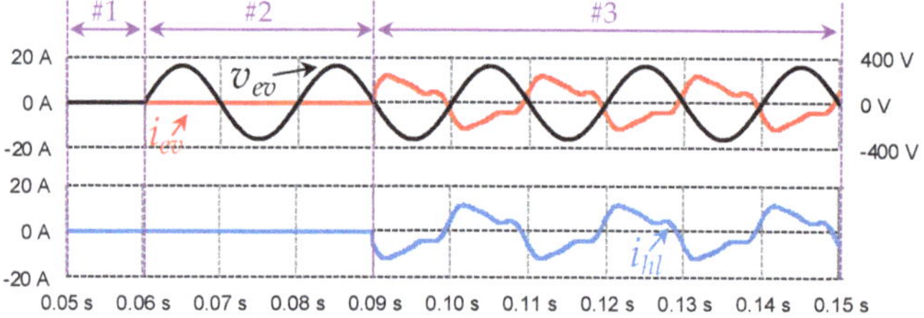

Figure 10. The principle of operation representative of the V2L mode (as a voltage source).

3.1.4. Vehicle-to-Home (V2H)–Uninterruptible Power Supply

In addition to the operation mode presented earlier, the EVBC can also operate as a voltage source, but with the characteristics of an off-line uninterruptible power supply (UPS). This mode represents a new opportunity for smart homes because, in the event of a power failure, the EVBC can operate almost instantly as a voltage source for the smart home. In this operation mode, communication is required between the EVBC and the smart home, in order to identify a power outage and even to some selected priority loads. The principle of operation representative of the vehicle-to-home (V2H) mode as a UPS is presented in Figure 11, which clearly identifies that the EVBC operates in unidirectional mode and disconnected from the electrical grid.

Like the previously presented operation mode, the grid-side converter of the EVBC operates with a voltage control feedback and the current is defined by the loads. However, unlike the previous mode, it is necessary to measure the grid voltage in order to detect when a voltage failure happens. Whenever this occurs, a control signal is sent to the general circuit breaker, to isolate the loads from the grid, and, almost instantaneously, the EVBC starts its operation as a voltage source. Later, when the grid voltage is restored, the EVBC recognizes this situation. Then, after some cycles of the grid voltage, it begins a synchronization with the phase of the voltage and, as soon as the control system is synchronized, it makes the transition to the normal mode, i.e., the loads are fed again by the grid. After this process, the EVBC may either go to an idle state or start another operation mode, such as G2V or V2G. Figure 12 illustrates the operating principle of this mode, showing the grid voltage (v_g), the grid current (i_g), the loads voltage (v_{hl}), the loads current (i_{hl}), the voltage produced by EVBC (v_{ev}), and the EVBC current (i_{ev}). This operation mode is divided into four cases. In case #1 the EVBC is connected to the grid to charge the batteries through the G2V operation mode (i.e., with a sinusoidal current and unitary power factor). In case #2, there is a fault in the grid voltage, detected by the EVBC, which starts operating in UPS mode, feeding the loads. In case #3, the grid voltage is restored and the EVBC stops operating in the UPS mode and the loads are fed back through the electrical grid again, as in case #1. As shown, even with a distorted grid voltage in cases #1 and #3, during the outage (case #2), the EVBC produces a sinusoidal voltage.

Figure 11. On-board EVBC: Vehicle-to-home (V2H) operation mode (as an off-line uninterruptible power supply).

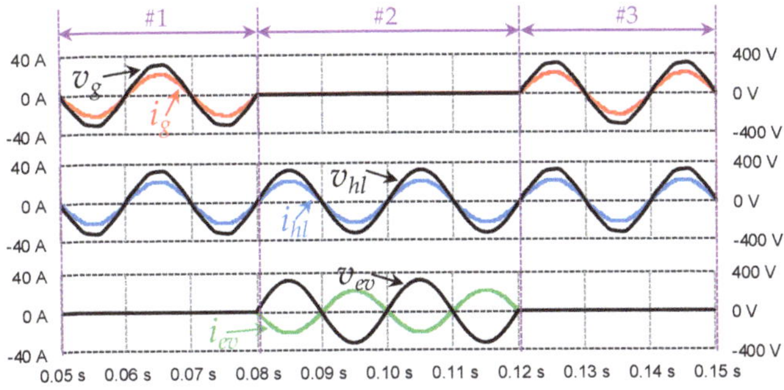

Figure 12. The principle of operation representative of the V2H mode (as an off-line uninterruptible power supply).

3.2. Off-Board EV Battery Charger

The main operation modes of an off-board EVBC are presented in this section, addressing the opportunities that they can offer for a contextualized operation with smart grids and smart homes, both in terms of controllability and new features that can be obtained for the installation where the EV is plugged-in. It is relevant to note that an off-board EVBC can be classified as slow, semi-fast, fast or ultra-fast; however, the modes of operation presented in this section are independent of this classification. Moreover, an off-board EVBC can be installed into the electrical grid through a single- or a three-phase interface. As an example case, Figure 13 illustrates the integration of an off-board EVBC into an industry. As shown, the EV battery is charged through the off-board EVBC, which is connected to the electrical grid in parallel with the home loads. Therefore, the off-board EVBC is continuously linked to the electrical grid independently of the EV presence. As illustrated, bidirectional communication is considered between the industry and the electrical grid towards a smart grid perspective in terms of controllability.

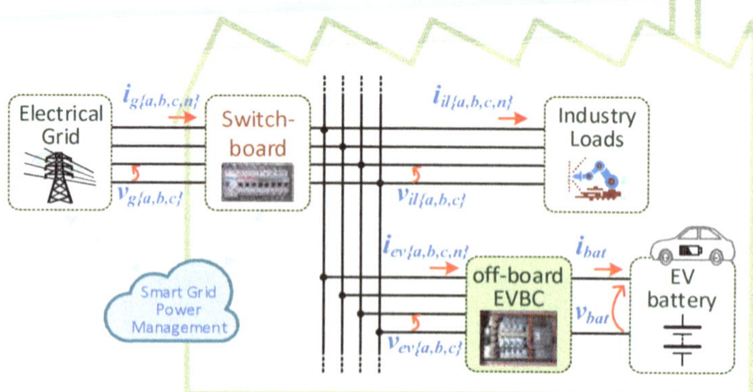

Figure 13. Illustration of the integration of an off-board EVBC into an industry.

3.2.1. Grid-to-Vehicle and Vehicle-to-Grid

As with an on-board EVBC, an off-board EVBC also enables the G2V mode, regardless of the grid operation in terms of power management. Therefore, the central difference between an on-board and an off-board EVBC is the operating power, where, customarily, the off-board EVBC operates with a significantly higher power. In addition to the G2V operation mode, the V2G operation mode is also possible in an off-board EVBC, however, this mode is presented as a new opportunity for off-board EVBCs, since it is important to note that, nowadays, off-board EVBCs are unidirectional, but for operation in V2G mode they need to be bidirectional.

The principle of operation representative of an EVBC in G2V and V2G modes is presented in Figure 14, where it is clearly identified that the EVBC operates in a bidirectional mode in terms of active power and in terms of a communication with the electrical grid. During the charging of the EV battery using an off-board EVBC, the intention is to carry out the process as quickly as possible. However, as the EV is connected to an off-board EVBC only for brief periods, the EV can return a minor part of the stored energy (e.g., for a power management strategy). For example, if the EV is plugged-in for performing a fast charging of about 30 minutes operating with a power of 30 kW, but for 30 s, it returns a power of 10 kW back to the electrical grid, then the total time will be exceeded in just 1 minute and 40 s. This represents a small amount of power for the electrical grid, but considering an EV fleet with this functionality and an electrical grid with prediction and management algorithms, this operation mode is quite attractive for off-board EVBCs.

Figure 14. Integration of an off-board EVBC into the electrical power grid for G2V and V2G operation modes.

3.2.2. Power Quality Compensator

As noted earlier, off-board EVBCs are installed externally to the EV and are only used when it is necessary to charge the EV battery, a task that happens only for brief minutes. Thus, it is foreseeable that these systems may be out of operation during various periods along the day. In this sense, since an off-board EVBC is installed in a specific electrical installation, there is a new opportunity of operation for the off-board EVBC that is related to the possibility to compensate power quality problems for the electrical grid (caused by the nonlinear loads consuming distorted currents). This operation mode is even more significant knowing that it can be accomplished whether or not the EV is present, i.e., the power quality compensation can be performed at the same time that the batteries are being charged (G2V mode) without a detrimental effect on the EV battery or the EVBC operation. Moreover, it should be noted that this operation mode may also be used with the V2G mode. As the compensation is made to the grid-side, it is not necessary to transfer active power between the off-board EVBC and the electrical grid, which is a pertinent benefit to this operation mode. Moreover, no extra hardware is required for the operation in this mode, and it is only necessary to obtain the instantaneous current value of the installation.

Figure 15 presents the framework of an off-board EVBC considering this new opportunity of operation, in which it is possible to compensate power quality problems related to power factor, current harmonics, and current imbalances. These power quality problems are caused by the loads presented in the industry, where the off-board EVBC is also installed. Given the offered benefit, this operation mode is especially suitable for EVBCs installed in industrial zones, where it is intended to minimize problems of power quality and to charge the EV battery as fast as possible. The operating principle of this mode is shown in Figure 16. Therefore, a three-phase installation is considered as a case example. In case #1, the off-board EVBC is connected to the grid without compensating power quality problems and without an EV in G2V or V2G modes. Therefore, the currents of the electrical installation are the consumed currents of the industry loads. In case #2, the process of power quality compensation (e.g., power factor and harmonics) is started without any EV plugged-in. In this case, as can be seen, on the grid-side, the power factor became unitary and the currents became sinusoidal and balanced. In case #3, in addition to the previous compensation, the off-board EVBC also starts the EV battery charging. As can be seen, the currents on the grid-side remain sinusoidal and balanced, only increasing its amplitude, corresponding to the active power required to charge the EV battery. Once again, it is imperative to remember that the operating active power of the EVBC is only used to charge the EV battery.

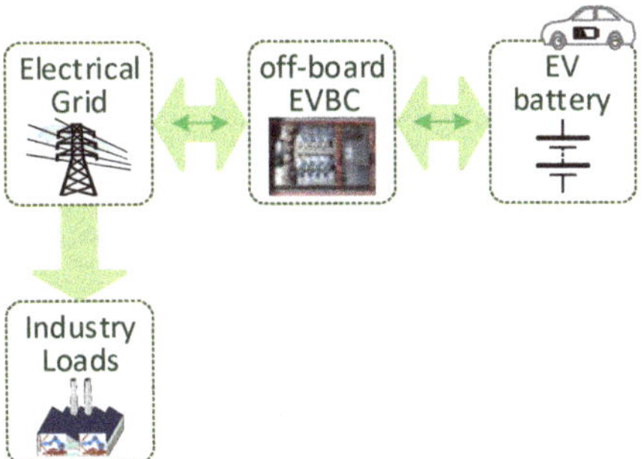

Figure 15. Integration of an off-board EVBC into the electrical power grid for operation as a power quality compensator.

3.2.3. Unified Operation of Power Quality Compensator with Renewables

In the previous sections, new opportunities were presented for the EVBC operation in the context of smart grids. Knowing the influence that RES represent for the evolution and maturation of smart grids, as well as to minimize the impact that the EV battery charging represents in terms of the necessary power, their integration as close as possible to the off-board EVBC is of extreme relevance. For instance, solar photovoltaic panels, as an example of RES, can be used as a solar rooftop in EV charging stations or industries in order to accomplish with the opportunity mentioned in this section. In addition, since both off-board EVBCs and RES interface with the electrical grid using a grid-side converter, and each system has its own dc-link, a new opportunity is proposed in this section for smart grids, which consists of unifying both systems with only one interface with the electrical grid. Moreover, bearing in mind the opportunity identified in the previous section, the new opportunity identified a single interface with the electrical grid and the unification (through the dc-link) of the EVBC and the RES (which can be photovoltaic panels or wind turbines) is presented in Figure 17. Furthermore, with this unified strategy, it is conceivable to increase the efficiency compared to the customary solution in which two interfaces are used with the electrical grid, since the power coming from the RES can be directly used by the battery-side converter to charge the EV battery, without the need to use the grid-side converter.

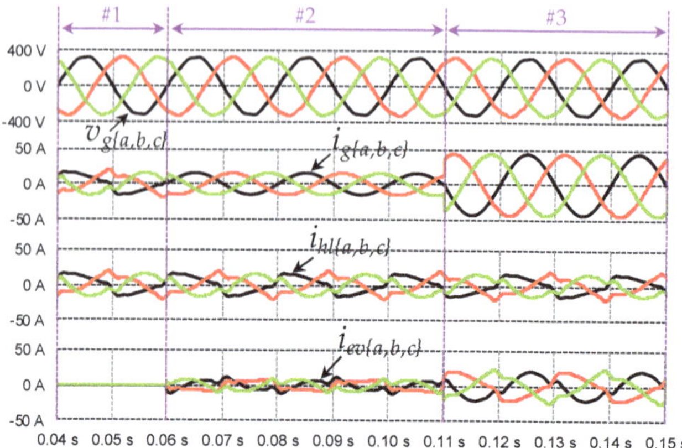

Figure 16. The principle of operation representative of the off-board EVBC operation as a power quality compensator.

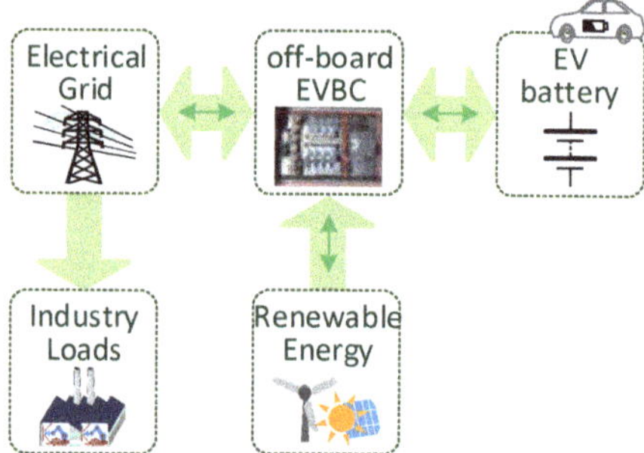

Figure 17. Off-board EVBC: Unified operation as a power quality compensator with renewables.

Figure 18 presents the principle of operation of this mode, in part, supported by operations presented in the previous sections. In case #1, the off-board EVBC operates in the same way as in Figure 16, whereas in case #2 the off-board EVBC is used to inject power into the grid from the RES. As can be seen in this case, the effective values of the grid currents decrease due to the injected power from the RES, but the EV battery continues with the same charging power, meaning that the EV battery is being charged with energy from the RES.

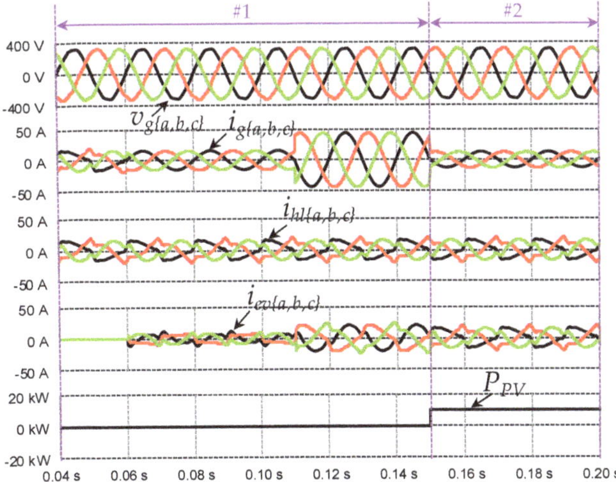

Figure 18. Principle of operation representative of the off-board EVBC: Unified operation as a power quality compensator with renewables.

3.2.4. Unified Operation of Power Quality Compensator, Renewables and Energy Storage Systems

Following the opportunity presented in the previous section, providing the off-board EVBC with another interface on the dc-link, for an ESS, results in a new opportunity that includes the main aspects of smart grids in terms of power electronics. For instance, it should be noted that solar photovoltaic panels, as an example of RES, can be used as a solar rooftop in industries. In this way, it is possible to fit, in a single off-board EVBC equipment, the aspects of electric mobility, power quality, RES and ESS. Figure 19 shows this new opportunity, in which it is possible to recognize a single interface with the electrical grid and the unification (through the dc-link) of the converters for the EV battery charging, RES and ESS. Figure 20 shows the operation principle based on this proposal. In case #1, the off-board EVBC operates in the same way as shown in Figure 18. In case #2, after the EV charging process is completed, the energy produced by the RES is stored in the ESS. As shown in this case, the off-board EVBC compensates the currents of the industry loads and stores the power from the RES in the ESS at the same time.

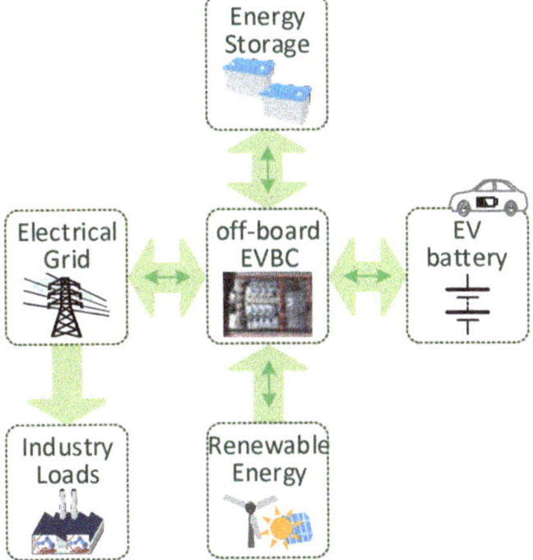

Figure 19. Off-board EVBC: Unified operation of power quality compensator, renewables and energy storage systems.

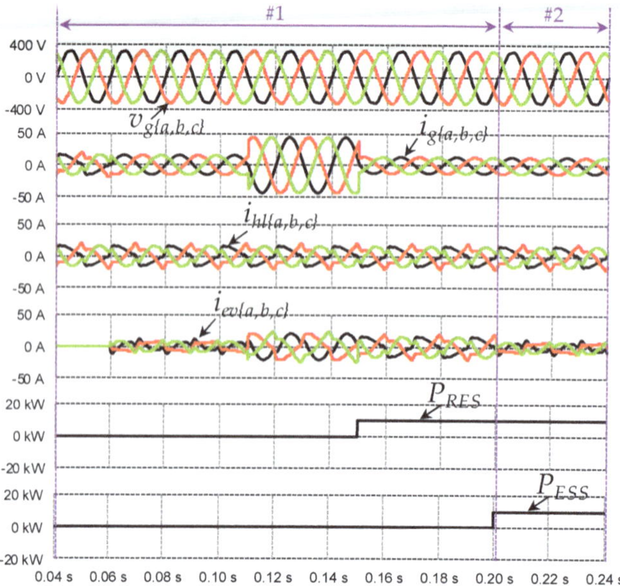

Figure 20. Principle of operation representative of the off-board EVBC: Unified operation of power quality compensator, renewables and energy storage systems.

4. Laboratory Prototypes

This section presents three examples of laboratory prototypes employing some of the operation modes referred to in the previous section. Figure 21 shows a photograph of a laboratory workbench where an on-board system and two off-board systems are presented. These prototypes were developed considering real-scale applications.

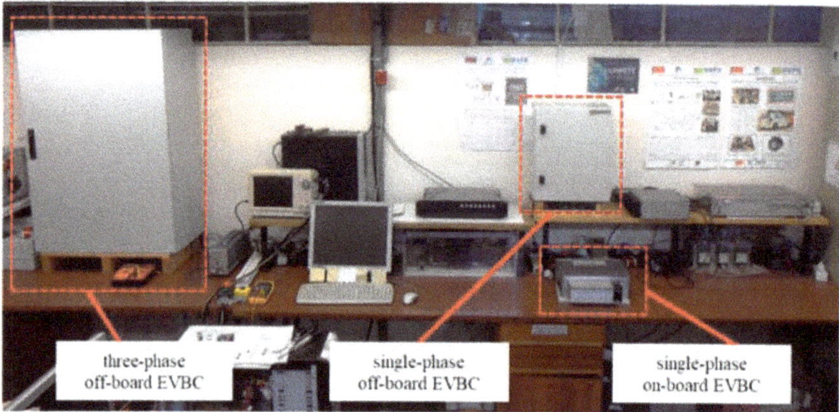

Figure 21. Photography of the laboratory workbench with the on-board and off-board EVBCs.

The on-board EVBC interfaces the electrical grid through a single-phase connection and allows operating in controlled G2V and V2G modes, as well as in voltage source mode or UPS mode. It is composed of two power stages, a grid-side converter and a battery-side converter, joined by a common dc-link.

One off-board EVBC interfaces with the electrical grid through a three-phase connection. It is constituted by two power converters (three-phase grid-side and battery-side) and allows the simultaneous and controlled G2V operation, as well as the compensation of power quality problems for the grid (e.g., current harmonics and reactive power), which is a feature that can be accomplished independently of the EV presence.

The other off-board EVBC interfaces the electrical grid through a single-phase connection. It is also constituted by two power stages, but with three accessible interface ports. A bidirectional ac-dc grid-side converter is employed to interface the electrical grid. In the dc-side, two separate interfaces are possible: A bidirectional dc-dc battery-side converter for the EV battery interface (empowering the G2V and V2G modes) and a unidirectional dc-dc converter for the interface of a RES (in this case a set of photovoltaic panels were considered). The simultaneous compensation of power quality problems is also possible.

It is fundamental to note that it is not the objective of this paper, and in particular this section, to establish a careful description about the internal constitution of the converters presented in the various prototypes, as well as the various control algorithms. In counterpart, the evident objective is to present a set of EVBCs that are capable of operating in several modes, representing new challenges and opportunities for smart grids and smart homes.

Figure 22 presents some experimental results for the single-phase on-board EVBC shown in Figure 21. Figure 22a presents the grid-side current (i_{ev}) and voltage (v_g), evidencing a sinusoidal current and a distorted voltage, meaning that the control algorithm employs a phase-locked loop, contributing to reduce power quality problems. A unitary power factor, voltage, and current in phase, is also verified. Figure 22b shows an experimental result with the on-board EVBC operating as a UPS. This result is divided into two distinctive circumstances: In case #1, the EVBC operates in G2V mode and in case #2 in UPS mode. At the beginning of case #2, a power outage occurs in the electrical grid, interrupting the G2V mode for transition to the UPS mode. This transition was performed automatically, and the power outage was detected by measuring the root mean square (rms) grid voltage (v_g). As shown, the battery-side current (i_{bat}) is positive during case #1 (meaning the EV battery charging) and negative during case #2 (meaning the EV battery is discharging for the UPS operation). The dc-link voltage (v_{dc}) is always positive, controlled by the grid-side converter during case #1 and controlled by the battery-side during case #2. As shown, a sudden variation occurs in the dc-link voltage aiming to compensate as fast as possible the power outage, i.e., the transition for the UPS mode. Lastly, Figure 22c shows the on-board EVBC operation as a UPS. As can be observed, the voltage produced by the on-board EVBC (v_{hl}) is sinusoidal even when it supplies a nonlinear load, which is characterized by a high-distorted current consumption (i_{hl}).

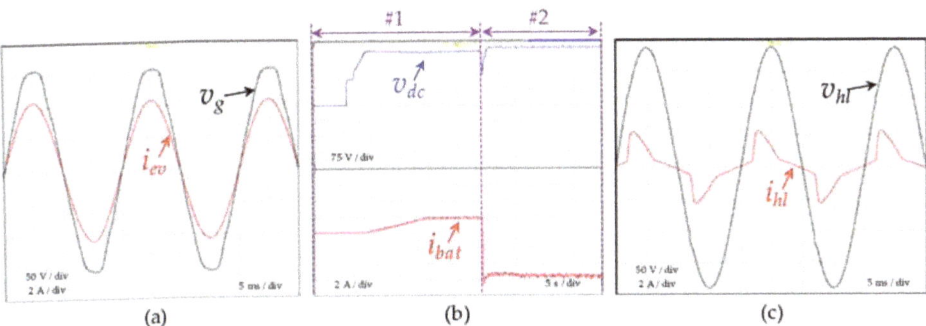

Figure 22. Experimental results of the single-phase on-board EVBC: (**a**) During 50 ms in the grid-to-vehicle (G2V) operation mode-grid-side current (i_{ev}) and voltage (v_g); (**b**) during 50 s in the operation as a UPS-dc-link voltage (v_{dc}) and battery-side current (i_{bat}); (**c**) During 50 ms in the operation as a UPS-produced voltage (v_{hl}) and consumed current (i_{hl}).

Figure 23 presents some experimental results for the single-phase off-board EVBC. Figure 23a shows the power in the three interfaces of the single-phase off-board EVBC, namely: The electrical grid interface (P_G); the RES interface, in this case, photovoltaic panels (P_{RES}); and the EV battery interface (P_{EV}). This experimental result is divided into two distinct cases. During the case #1, the EV battery is charged from the grid and the RES is not producing energy, meaning that the grid-side power (P_G) is equal to the battery-side power (P_{EV}) and the RES power (P_{RES}) is zero. In this case, the single-phase off-board EVBC operates in G2V mode. As can be seen, the operating power is increased up to its nominal value aiming to avoid sudden variations for the grid and for the EV battery. During case #2, the single-phase off-board EVBC operates in V2G mode and the PV panels start their production, meaning that the power injected into the electrical grid (P_G) is the sum of the power from the EV battery (P_{EV}) and the power from the RES (P_{RES}). Figure 23b shows the grid-side current (i_{ev}) and voltage (v_g) of the off-board EVBC, as well as the dc-link voltage (v_{dc}), the EV battery-side current (i_{bat}), and the RES (photovoltaic panels) current (i_{pv}). In this mode (G2V), the grid-side current has a sinusoidal waveform, it is in phase with the voltage, and the dc variables are controlled according to the operation of the single-phase off-board EVBC. Lastly, Figure 23c shows the grid-side current (i_{ev}) with a sinusoidal waveform, but in opposition with the voltage waveform (v_g), signifying that the single-phase off-board EVBC is operating in V2G mode, i.e., it is injecting power into the grid.

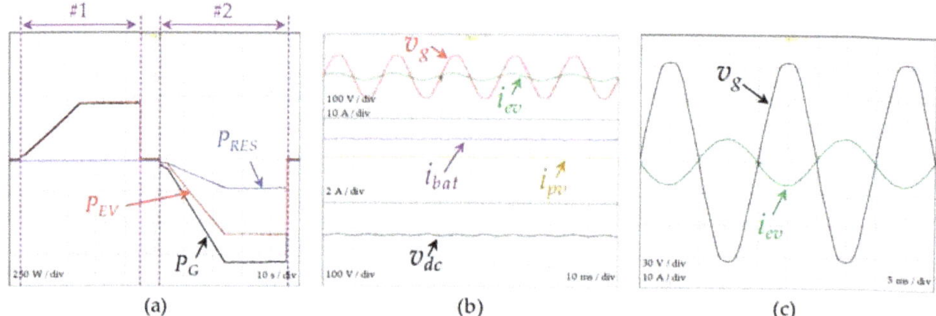

Figure 23. Experimental results of the off-board EVBC considering the integration of renewables (photovoltaic panels): (**a**) During 100 s in the grid-to-vehicle (G2V) and vehicle-to-grid (V2G) operation modes with renewables—power in the grid side (P_G), power in the EV side (P_{EV}) and power in the RES side (P_{RES}); (**b**) during 100 ms in the G2V mode-grid-side current (i_{ev}) and voltage (v_g), dc-link voltage (v_{dc}), EV battery current (i_{bat}), and photovoltaic panels current (i_{pv}); (**c**) during 50 ms in the V2G mode-grid-side current (i_{ev}) and voltage (v_g).

Figure 24 presents some experimental results of the three-phase off-board EVBC. Figure 24a shows the grid-side currents (i_{ga}, i_{gb}, i_{gc}) during the G2V mode, the dc-link voltage (v_{dc}), and the EV battery current (i_{bat}). As can be observed, the currents are balanced and sinusoidal waveforms are presented. The dc-side variables (v_{dc} and i_{bat}) are controlled in conformity. Figure 24b presents the grid-side voltages (v_{ga}, v_{gb}, v_{gc}) and the consuming lagged currents (i_{ga}, i_{gb}, i_{gc}) in order to produce reactive power for the grid, confirming the operation of the off-board EVBC for compensating power quality problems associated with reactive power caused by the connected loads. Lastly, Figure 24c presents the aforementioned variables, but in two distinct cases. In case #1, the off-board EVBC is operating in G2V mode (grid-side currents and voltages in phase) and, in case #2, it is operating with distorted grid-side currents to compensate the current harmonics presented in the installation (i.e., from the grid point of view, the electrical installation operates with sinusoidal currents), validating the operation of the off-board EVBC for compensating power quality problems related with harmonics.

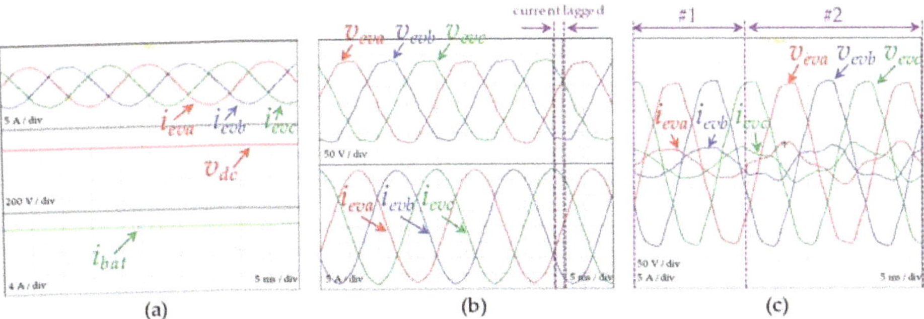

Figure 24. Experimental results of the off-board EVBC compensating power quality problems: (a) During 50 ms in the G2V mode-grid-side currents (i_{eva}, i_{evb}, i_{evc}), dc-link voltage (v_{dc}), and EV battery current (i_{bat}); (b) during 50 ms compensating power quality problems related with power factor-grid-side currents (i_{eva}, i_{evb}, i_{evc}) and voltages (v_{ga}, v_{gb}, v_{gc}); (c) during 50 ms compensating power quality problems related with current harmonics-grid-side currents (i_{eva}, i_{evb}, i_{evc}) and voltages (v_{ga}, v_{gb}, v_{gc}).

5. Conclusions

Aiming towards an effective change of paradigm in the urban transportation sector, vehicle electrification already represents a reality, with a large predictable expansion and related opportunities ahead. However, new challenges are introduced in order to diminish undesirable effects that this revolution can cause to the electrical power grid. In this sense, this paper introduces a comprehensive discussion of new challenges and opportunities in terms of operation modes that can be addressed by on-board and off-board electric vehicle battery chargers (EVBCs) toward smart grids.

Throughout the paper, it has been demonstrated that on-board EVBCs can be used for other purposes than just grid-to-vehicle (G2V) and vehicle-to-grid (V2G) operation. On the other hand, as far as off-board EVBCs are concerned, it has been demonstrated that they can be unified with other features of smart grids, such as renewable energy sources (RES) and energy storage systems (ESS), and, at the same time, compensating power quality problems in the places where they are installed. In this sense, the different opportunities that on-board and off-board EVBCs represent for smart grids have been widely presented and discussed, evidencing the resultant advantages of new technologies toward a sustainable future. As a case example, an experimental validation employing laboratory prototypes of on-board and off-board EVBCs was presented, abstracting the topologies of the power converters.

Author Contributions: V.M., J.A.A., J.C.F. and J.L.A. contributed equally to the conceptualization and writing of the paper.

Funding: This work was supported by COMPETE: POCI-01-0145-FEDER-007043 and FCT—Fundação para a Ciência e Tecnologia within the Project Scope: UID/CEC/00319/2013. This work was financed by the ERDF —European Regional Development Fund, through the Operational Programme for Competitiveness and Internationalisation—COMPETE 2020 Programme, and by National Funds through the Portuguese funding agency, FCT—Fundação para a Ciência e a Tecnologia, within project SAICTPAC/0004/2015-POCI-01-0145-FEDER-016434. This work is part of the FCT project 0302836 NORTE-01-0145-FEDER-030283.

Conflicts of Interest: The authors declare no conflict of interest.

References

1. Bose, B.K. Global Warming—Energy, Environmental Pollution, and the Impact of Power Electronics. *IEEE Ind. Electron. Mag.* **2010**, *4*, 6–17. [CrossRef]
2. Bozchalui, M.C.; Cañizares, C.A.; Bhattacharya, K. Optimal Energy Management of Greenhouses in Smart Grids. *IEEE Trans. Smart Grid* **2015**, *6*, 827–835. [CrossRef]
3. Amin, S.M.; Giacomoni, A.M. Giacomoni, Smart Grid—Safe, Secure, Self-Healing. *IEEE Power Energy Mag.* **2012**, *10*, 33–40. [CrossRef]

4. Gungor, V.C.; Sahin, D.; Kocak, T.; Ergut, S.; Buccella, C.; Cecati, C.; Hancke, G.P. Smart Grid and Smart Homes—Key Players and Pilot Projects. *IEEE Ind. Electron. Mag.* **2012**, *6*, 18–34. [CrossRef]

5. Moslehi, K.; Kumar, R. A Reliability Perspective of the Smart Grid. *IEEE Trans. Smart Grid* **2010**, *1*, 57–64. [CrossRef]

6. Galus, M.D.; Vayá, M.G.; Krause, T.; Andersson, G. The Role of Electric Vehicles in Smart Grids. *WIREs Energy Environ.* **2013**, *2*, 384–400. [CrossRef]

7. Li, D.; Jayaweera, S.K. Distributed Smart-Home Decision-Making in a Hierarchical Interactive Smart Grid Architecture. *IEEE Trans. Parallel Distrib. Syst.* **2015**, *26*, 75–84. [CrossRef]

8. Hashmi, M.H.; Hänninen, S.; Mäki, K. Survey of Smart Grid Concepts, Architectures, and Technological Demonstrations Worldwide. In Proceedings of the IEEE PES Conference on Innovative Smart Grid Technologies Latin America, Medellin, Colombia, 19–21 October 2011; pp. 1–7.

9. Yu, X.; Cecati, C.; Dillon, T.; Simoes, M.G. The New Frontier of Smart Grids: An Industrial Electronics Perspective. *IEEE Ind. Electron. Mag.* **2011**, *5*, 49–63. [CrossRef]

10. Vojdani, A. Smart Integration: The Smart Grid Needs Infrastructure That is Dynamic and Flexible. *IEEE Power Energy Mag.* **2008**, *6*, 71–79. [CrossRef]

11. Gungor, V.C.; Sahin, D.; Kocak, T.; Ergut, S.; Buccella, C.; Cecati, C.; Hancke, G.P. Smart Grid Technologies: Communication Technologies and Standards. *IEEE Trans. Ind. Inform.* **2011**, *7*, 529–539. [CrossRef]

12. Liu, N.; Chen, J.; Zhu, L.; Zhang, J.; He, Y. A Key Management Scheme for Secure Communications of Advanced Metering Infrastructure in Smart Grid. *IEEE Trans. Ind. Electron.* **2013**, *60*, 4746–4756. [CrossRef]

13. Meliopoulos, A.S.; Cokkinides, G.; Huang, R.; Farantatos, E.; Choi, S.; Lee, Y.; Yu, X. Smart Grid Technologies for Autonomous Operation and Control. *IEEE Trans. Smart Grid* **2011**, *2*, 1–10. [CrossRef]

14. Yan, B.; Luh, P.B.; Warner, G.; Zhang, P. Operation and Design Optimization of Microgrids With Renewables. *IEEE Trans. Autom. Sci. Eng.* **2017**, *14*, 573–585. [CrossRef]

15. Blaabjerg, F.; Guerrero, J.M. Smart Grid and Renewable Energy Systems. In Proceedings of the ICEMS International Conference on Electrical Machines and Systems, Beijing, China, 20–23 August 2011; pp. 1–10.

16. Ackermann, T.; Carlini, E.M.; Ernst, B.; Groome, F.; Orths, A.; O'Sullivan, J.; de la Torre Rodriguez, M.; Silva, V. Integrating Variable Renewables in Europe: Current Status and Recent Extreme Events. *IEEE Power Energy Mag.* **2015**, *13*, 67–77. [CrossRef]

17. Bragard, M.; Soltau, N.; Thomas, S.; De Doncker, R.W. The Balance of Renewable Sources and User Demands in Grids: Power Electronics for Modular Battery Energy Storage Systems. *IEEE Trans. Power Electron.* **2010**, *25*, 3049–3056. [CrossRef]

18. Beaudin, M.; Zareipour, H.; Schellenberglabe, A.; Rosehart, W. Energy Storage for Mitigating the Variability of Renewable Electricity Sources: An Updated Review. *Energy Sustain. Dev.* **2010**, *14*, 302–314. [CrossRef]

19. Hernandez, J.E.; Kreikebaum, F.; Divan, D. Flexible Electric Vehicle (EV) Charging to Meet Renewable Portfolio Standard (RPS) Mandates and Minimize Green House Gas Emissions. In Proceedings of the IEEE ECCE Energy Conversion Congress and Exposition, Atlanta, GA, USA, 12–16 September 2010; pp. 4270–4277.

20. Alam, M.R.; Reaz, M.B.I.; Ali, M.A.M. A Review of Smart Homes—Past, Present, and Future. *IEEE Trans. Syst. Man Cybern. C Appl. Rev.* **2012**, *42*, 1190–1203. [CrossRef]

21. Erickson, L.E.; Robinson, J.; Brase, G.; Cutsor, J. *Solar Powered Charging Infrastructure for Electric Vehicles: A Sustainable Development*, 1st ed.; CRC Press: Boca Raton, FL, USA, 2017.

22. Raghavan, S.S.; Khaligh, A. Electrification Potential Factor: Energy-Based Value Proposition Analysis of Plug-In Hybrid Electric Vehicles. *IEEE Trans. Veh. Technol.* **2012**, *61*, 1052–1059. [CrossRef]

23. Saber, A.Y.; Venayagamoorthy, G.K. Plug-in Vehicles and Renewable Energy Sources for Cost and Emission Reductions. *IEEE Trans. Ind. Electron.* **2011**, *58*, 1229–1238. [CrossRef]

24. Martínezij, I.J.; Garcia-Villalobos, J.; Zamora, I.; Eguia, P. Energy Management of Micro Renewable Energy Source and Electric Vehicles at Home Level. *J. Mod. Power Syst. Clean Energy* **2017**, *5*, 979–990.

25. Lopes, J.P.; Almeida, P.M.; Silva, A.M.; Soares, F.J. Smart Charging Strategies for Electric Vehicles: Enhancing Grid Performance and Maximizing the Use of Variable Renewable Energy Resources. In Proceedings of the EVS24 International Battery, Hybrid and Fuel Cell Electric Vehicle Symposium, Stavanger, Norway, 13–16 May 2009; pp. 1–11.

26. Saber, A.Y.; Venayagamoorthy, G.K. Resource Scheduling Under Uncertainty in a Smart Grid with Renewables and Plug-in Vehicles. *IEEE Syst. J.* **2012**, *6*, 103–109. [CrossRef]

27. Robalino, D.M.; Kumar, G.; Uzoechi, L.O.; Chukwu, U.C.; Mahajan, S.M. Design of a Docking Station for Solar Charged Electric and Fuel Cell Vehicles. In Proceedings of the IEEE International Conference on Clean Electrical Power, Capri, Italy, 9–11 June 2009; pp. 655–660.

28. Vithayasrichareon, P.; Mills, G.; MacGill, I.F. Impact of Electric Vehicles and Solar PV on Future Generation Portfolio Investment. *IEEE Trans. Sustain. Energy* **2015**, *6*, 899–908. [CrossRef]

29. Tushar, W.; Yuen, C.; Huang, S.; Smith, D.B.; Poor, H.V. Vincent Poor, Cost Minimization of Charging Stations With Photovoltaics: An Approach with EV Classification. *IEEE Trans. Intell. Transp. Syst.* **2016**, *17*, 156–169. [CrossRef]

30. Monteiro, V.; Pinto, J.G.; Afonso, J.L. Experimental Validation of a Three-Port Integrated Topology to Interface Electric Vehicles and Renewables with the Electrical Grid. *IEEE Trans. Ind. Inform.* **2018**, *14*, 2364–2374. [CrossRef]

31. Pedrasa, M.A.; Spooner, T.D.; MacGill, I.F. Coordinated Scheduling of Residential Distributed Energy Resources to Optimize Smart Home Energy Services. *IEEE Trans. Smart Grid* **2010**, *1*, 134–143. [CrossRef]

32. Tsui, K.M.; Chan, S.C. Demand Response Optimization for Smart Home Scheduling Under Real-Time Pricing. *IEEE Trans. Smart Grid* **2012**, *3*, 1812–1821. [CrossRef]

33. Monteiro, V.D.; Pinto, J.G.; Exposto, B.F.; Ferreira, J.C.; Afonso, J.L. Smart Charging Management for Electric Vehicle Battery Chargers. In Proceedings of the IEEE VPPC Vehicle Power and Propulsion Conference, Coimbra, Portugal, 27–30 October 2014; pp. 1–5.

34. Zhang, T.; Chen, W.; Han, Z.; Cao, Z. Charging Scheduling of Electric VehiclesWith Local Renewable Energy Under Uncertain Electric Vehicle Arrival and Grid Power Price. *IEEE Trans. Veh. Technol.* **2014**, *63*, 2600–2612. [CrossRef]

35. Tushar, M.H.; Zeineddine, A.W.; Assi, C. Demand-Side Management by Regulating Charging and Discharging of the EV, ESS, and Utilizing Renewable Energy. *IEEE Trans. Ind. Inform.* **2018**, *14*, 117–126. [CrossRef]

36. Boulanger, A.G.; Chu, A.C.; Maxx, S.; Waltz, D.L. Vehicle Electrification: Status and Issues. *Proc. IEEE* **2011**, *99*, 1116–1138. [CrossRef]

37. Fan, Z.; Oviedo, R.M.; Gormus, S.; Kulkarni, P. The Reign of EVs? An Economic Analysis from Consumer's Perspective. *IEEE Electrif. Mag.* **2014**, *2*, 61–71. [CrossRef]

38. Dyke, K.J.; Schofield, N.; Barnes, M. The Impact of Transport Electrification on Electrical Networks. *IEEE Trans. Ind. Electron.* **2010**, *57*, 3917–3926. [CrossRef]

39. Su, W.; Eichi, H.; Zeng, W.; Chow, M.Y. A Survey on the Electrification of Transportation in a Smart Grid Environment. *IEEE Trans. Ind. Electron.* **2012**, *8*, 1–10. [CrossRef]

40. Li, S.; Mi, C.C. Wireless Power Transfer for Electric Vehicle Applications. *IEEE J. Emerg. Sel. Top. Power Electron.* **2015**, *3*, 4–17.

41. Musavi, F.; Edington, M.; Eberle, W. Wireless Power Transfer: A Survey of EV Battery Charging Technologies. In Proceedings of the IEEE ECCE Energy Conversion Congress and Exposition, Raleigh, NC, USA, 15–20 September 2012; pp. 1804–1810.

42. Wang, S.; Dorrell, D. Review of Wireless Charging Coupler for Electric Vehicles. In Proceedings of the IEEE IECON Annual Conference of the Industrial Electronics Society, Vienna, Austria, 10–13 November 2013; pp. 7272–7277.

43. Ching, T.W.; Wong, Y.S. Review of Wireless Charging Technologies for Electric Vehicles. In Proceedings of the IEEE PESA International Conference on Power Electronics Systems and Applications, Hong Kong, China, 11–13 December 2013; pp. 1–4.

44. Salmasi, F.R. Control Strategies for Hybrid Electric Vehicles: Evolution, Classification, Comparison, and Future Trends. *IEEE Trans. Veh. Technol.* **2007**, *56*, 2393–2404. [CrossRef]

45. Lopes, J.A.; Soares, F.J.; Almeida, P.M. Rocha Almeida, Integration of Electric Vehicles in the Electric Power Systems. *Proc. IEEE* **2011**, *99*, 168–183. [CrossRef]

46. Lam, A.Y.; Leung, K.C.; Li, V.O. Capacity Estimation for Vehicle-to-Grid Frequency Regulation Services With Smart Charging Mechanism. *IEEE Trans. Smart Grid* **2016**, *7*, 156–166. [CrossRef]

47. Zipperer, A.; Aloise-Young, P.A.; Suryanarayanan, S.; Roche, R.; Earle, L.; Christensen, D.; Bauleo, P.; Zimmerle, D. Electric Energy Management in the Smart Home: Perspectives on Enabling Technologies and Consumer Behavior. *Proc. IEEE* **2013**, *101*, 2397–2408. [CrossRef]

48. Tuohar, M.H.; Abel, C.; Maier, M.; Uddin, M.F. Smart Microgrids: Optimal Joint Scheduling for Electric Vehicles and Home Appliances. *IEEE Trans. Smart Grid* **2014**, *5*, 239–250. [CrossRef]
49. Yilmaz, M.; Krein, P.T. Review of the Impact of Vehicle-to-Grid Technologies on Distribution Systems and Utility Interfaces. *IEEE Trans. Power Electron.* **2013**, *28*, 5673–5689. [CrossRef]
50. Lund, H.; Kempton, W. Integration of renewable energy into the transport and electricity sectors through V2G. *Energy Policy* **2008**, *36*, 3578–3587. [CrossRef]
51. Gao, S.; Chau, K.T.; Liu, C.; Wu, D.; Chan, C.C. Chan, Integrated Energy Management of Plug-in Electric Vehicles in Power Grid With Renewables. *IEEE Trans. Veh. Technol.* **2014**, *63*, 3019–3027. [CrossRef]
52. Monteiro, V.; Carmo, J.P.; Pinto, J.G.; Afonso, J.L. A Flexible Infrastructure for Dynamic Power Control of Electric Vehicle Battery Chargers. *IEEE Trans. Veh. Technol.* **2016**, *65*, 4535–4547. [CrossRef]
53. Rodrigues, M.C.; Souza, I.D.; Ferreira, A.A.; Barbosa, P.G.; Braga, H.A. Simultaneous Active Power Filter and G2V (or V2G) Operation of EV On-Board Power Electronics. In Proceedings of the IEEE IECON Industrial Electronics Conference, Vienna, Austria, 10–13 November 2013; pp. 4684–4689.
54. Monteiro, V.; Pinto, J.G.; Afonso, J.L. Operation Modes for the Electric Vehicle in Smart Grids and Smart Homes: Present and Proposed Modes. *IEEE Trans. Veh. Tech.* **2016**, *65*, 1007–1020. [CrossRef]
55. Sun, Y.; Liu, W.; Su, M.; Li, X.; Wang, H.; Yang, J. A Unified Modeling and Control of a Multi-Functional Current Source-Typed Converter for V2G Application. *Electr. Power Syst. Res.* **2014**, *106*, 12–20. [CrossRef]
56. Buja, G.; Bertoluzzo, M.; Fontana, C. Reactive Power Compensation Capabilities of V2G-Enabled Electric Vehicles. *IEEE Trans. Power Electon.* **2017**, *32*, 9447–9459. [CrossRef]
57. An, L.U.; Qianming, X.U.; Fujun, M.A.; Yandong, C.H. Overview of Power Quality Analysis and Control Technology for the Smart Grid. *J. Mod. Power Syst. Clean Energy* **2016**, *4*, 1–9.
58. Traube, J.; Lu, F.; Maksimovic, D. Photovoltaic Power System with Integrated Electric Vehicle DC Charger and Enhanced Grid Support. In Proceedings of the EPE/PEMC International Power Electronics and Motion Control Conference, Novi Sad, Serbia, 4–6 September 2012; pp. 1–5.
59. Masoum, A.S.; Deilami, S.; Moses, P.S.; Abu-Siada, A. Ahmed Abu-Siada, Impacts of Battery Charging Rates of Plug-in Electric Vehicle on Smart Grid Distribution Systems. In Proceedings of the IEEE ISGT Innovative Smart Grid Technologies Conference Europe, Gothenberg, Sweden, 11–13 October 2010; pp. 1–6.
60. Boynuegri, A.R.; Uzunoglu, M.; Erdinc, O.; Gokalp, E. A new perspective in grid connection of electric vehicles: Different operating modes for elimination of energy quality problems. *Appl. Energy* **2014**, *132*, 435–451. [CrossRef]
61. Green Car Congress. Nissan to launch the 'LEAF to Home' V2Hpower Supply System with Nichicon 'EV Power Station' in June. Available online: http://www.greencarcongress.com/2012/05/leafvsh-20120530.html (accessed on 30 May 2012).

Article

Optimal Distribution Grid Operation Using DLMP-Based Pricing for Electric Vehicle Charging Infrastructure in a Smart City

Bruno Canizes [1],* , João Soares [1] , Zita Vale [1] and Juan M. Corchado [2,3,4]

1   GECAD—Knowledge Engineering and Decision Support Research Center—Polytechnic of Porto (IPP), R. Dr. António Bernardino de Almeida, 431, 4200-072 Porto, Portugal; jan@isep.ipp.pt (J.S.); zav@isep.ipp.pt (Z.V.)
2   University of Salamanca, 37008 Salamanca, Spain; corchado@usal.es
3   Osaka Institute of Technology, 5 Chome-16-1 Omiya, Asahi Ward, Osaka 535-8585, Japan
4   University of Technology Malaysia, Pusat Pentadbiran Universiti Teknologi Malaysia, Skudai 81310, Johor, Malaysia; corchado@usal.es
*   Correspondence: brmrc@isep.ipp.pt

Received: 22 January 2019; Accepted: 14 February 2019; Published: 20 February 2019

Abstract: The use of electric vehicles (EVs) is growing in popularity each year, and as a result, considerable demand increase is expected in the distribution network (DN). Additionally, the uncertainty of EV user behavior is high, making it urgent to understand its impact on the network. Thus, this paper proposes an EV user behavior simulator, which operates in conjunction with an innovative smart distribution locational marginal pricing based on operation/reconfiguration, for the purpose of understanding the impact of the dynamic energy pricing on both sides: the grid and the user. The main goal, besides the distribution system operator (DSO) expenditure minimization, is to understand how and to what extent dynamic pricing of energy for EV charging can positively affect the operation of the smart grid and the EV charging cost. A smart city with a 13-bus DN and a high penetration of distributed energy resources is used to demonstrate the application of the proposed models. The results demonstrate that dynamic energy pricing for EV charging is an efficient approach that increases monetary savings considerably for both the DSO and EV users.

Keywords: charging behaviors; distribution locational marginal pricing; distribution networks; electric mobility; electric vehicle; operation; reconfiguration; renewable energy sources; smart city; smart grid

---

1. Introduction

The efforts to minimize the carbon footprint using a large-scale integration of renewable energy sources (RES), such as wind and solar energy, have led to innovative developments in power distribution systems around the world. Moreover, a new agreement in the European Union (EU) aims to achieve 27% penetration of RES by 2030 [1], as one-third of EU countries have already achieved the 2020 target [2].

Currently, many people move to cities in search of a better quality of life, and this contributes to the continuous expansion of urban areas, which play a major role in modern economies. However, the urban population is responsible for most greenhouse gas emissions, and the United Nations estimates that the urban population will reach 70% of the world's total population by 2050 [3]. Consequently, it is necessary to make intelligent use of resources in urban environments, contributing to the development of smart cities [3]. The energy infrastructure of a smart city (SC), the so-called smart grid (SG), is one of the most important urban infrastructures that allows creating a sustainable city [4]. To this end, it is necessary to modernize grid functionalities through the implementation of innovative technologies, concretely, SG-enabling technologies for information and communication, sensing and measurement, automation, control, renewable generation integration, and storage [5,6].

One of the primary sources of CO_2 emissions is transportation [7,8]. Several authors have been analyzing the benefit of changing from traditional transportation (internal combustion engines) to EVs, in minimizing the transport sector's greenhouse gas emissions. It is widely acknowledged that the shift from internal combustion engines to EVs has many environmental and economic advantages. However, the increasing number of EVs makes it necessary to develop new infrastructure continually for EV charging, and this, in turn, leads to a growing energy demand [9,10]. These charging infrastructures are going to burden the distribution power grid [11–13], namely the high charging loads of fast EV charging stations. Furthermore, some distribution network operating parameters are going to degrade. Several published works describe the negative impacts of EV charging on the following distribution network parameters:

- Voltage profile [14–20];
- Peak load increase [21–24];
- Harmonic distortions [25–30].

Thus, a high EV penetration level may congest the distribution network. Congestion problems can be managed by the DSO, who reinforces the system through long-term planning or market-based congestion control methods [31]. The transmission systems concept of locational marginal pricing (LMP) can be extended to distribution systems [32]; it uses distributed generation (DG) units to handle congestion in distribution networks [33–37] and is usually referred to as distribution locational marginal pricing (DLMP). To deal with EV demand congestion in DN, the work in [38] proposed a step-wise congestion management whereby the DSO predicts congestion for the next day and publishes day-ahead tariff prior to the clearing of the day-ahead market, while [39] solved the social welfare optimization of the distribution system considering EV aggregators as price takers in the local DSO market and demand price elasticity. Liu et al. presented in [40] a market-based mechanism taken from the DLMP concept to alleviate possible distribution system congestion caused by the integration of EVs and heat pumps. Similarly, the authors in [41] proposed a DLMP based on quadratic programming to deal with the congestion in distribution networks with a high penetration of EVs and heat pumps.

As is known, the EVs are additional electric loads and represent mobile energy storage, usually with long resting times. Several mathematical models presented in [42–47] also studied the impact of EV charging in the distribution networks. The works in [48–53] assessed several possibilities for demand-side management, as well as better coordination of charging processes through price incentives that mitigate the impact of EV charging during peak-loads. The works in [54–58] proposed an increase in EV charging flexibility, contributing to increased utilization of the highly-variable renewable energy. Moreover, one of the main challenges in facilitating integrated EV charging in the distribution network is EV user behavior modeling and prediction [59]. Optimal control for allocating EV charging time and energy optimally has been proposed by Gan et al. [60]. However, the model requires that users frequently provide the charging schedule, requiring significant effort on the part of the customer. The algorithms developed in [61] used an EV random user behavior model with renewable generation for EV scheduling, while [62] provided a smart charging strategy according to time-of-use price from the day-ahead forecast. The authors in [63–66] examined EV users' charging behavior and measured psychological variables, an analysis that can help develop new charging strategies.

SCs feature an active power architecture with a high penetration of distributed energy resources (DER), challenging the conventional control and operation framework designed for passive distribution networks. In this context, the loads can be supplied not only by traditional generation units at the upstream power systems, but also by the DER [67]. Thus, a distribution network reconfiguration (DNR) will be a very important and significant strategy for the DSO. DNR is a process that changes the network topology using the remote switches such that all the network constraints are considered. Traditionally, the DNR is associated with system power loss minimization [68,69]; however, in the SG, context the DNR must not only meet the classic objectives, such as power loss, minimization of power not supplied, and improvement of the voltage profile, but also the problems related to the high DER integration and the intelligent reconfiguration related to the SG paradigm [70–72]. Several works considering mathematical [73–75], heuristics and metaheuristics [76,77], and hybrid models [78,79] were developed to deal with DNR and DER penetration.

The above-cited literature has not addressed distribution network operation and reconfiguration simultaneously in an SG context; neither have they considered the high penetration of DER and EV user behavior, nor dynamic EV charging price through DLMPs. Thus, to the best of the authors' knowledge, the answer to the question "Can dynamic EV charging price, have a positive impact on both the operation of the smart distribution network and on EV user behavior?" has not yet been answered. To answer this question, the authors combined an EV behavior simulator with a proposed innovative smart DLMP-based distribution network operation/reconfiguration. This kind of problem is classified as mixed-integer nonlinear programming (MINLP) due to its nonlinear features requiring significant computer resources. To deal with the issue of computational burden and at the same time improve the tractability of the model, the Benders decomposition method is used. This method uses duality theory [80,81] in linear and nonlinear mathematical programming, and it deals with complex problems by splitting them into subproblems. The main goal is to minimize all the DSO expenditures. To this end, the proposed methodology seeks the following:

- Minimize power loss;
- Minimize power not supplied;
- Minimize line congestion;
- Minimize the power generation curtailment;
- Minimize the power from external suppliers;
- Distribution network radial topology.

Considering the research gaps in previous works, this paper presents the following major contributions:

1. The use of an EV user behavior simulator. This simulator is used to simulate the EV user behavior aspects, such as: stochastic EV user aspects, importance of EV charging price, importance of comfort, choosing slow or fast charge, and the user sensibility of the the state of the battery;
2. Present a distribution network operation/reconfiguration optimization problem in an SG context with high DER penetration concerning the behavior aspects of the EV users and the dynamic EV charging price considering DLMPs using the Benders decomposition method;
3. Analyze how and to what extent dynamic EV charging prices contribute positively to smart distribution network operation;
4. Understand how and to what extent dynamic EV charging prices can contribute to a positive impact on the electric vehicles' charging cost.

To demonstrate the application of the proposed methodology, the BISITE (https://bisite.usal.es/en) laboratory's SC mockup model has been used with a 13-bus distribution network and high DER penetration. This paper is organized as follows: after this Introduction, Section 2 presents the proposed methodology and the details of the DLMP-based network operation, as well as the simulation of urban mobility. To verify the performance of the proposed methodology, a case study has been conducted and described in Section 3. The results and its discussion are presented in Section 4. Finally, Section 5 presents the most relevant conclusions.

2. Proposed Methodology

This section presents a detailed description of the adopted methodology (depicted in the Figure 1). Section 2.1 provides information about the EV user behavior simulator, while Section 2.2 describes the DLMP-based network operation model using the Benders decomposition method.

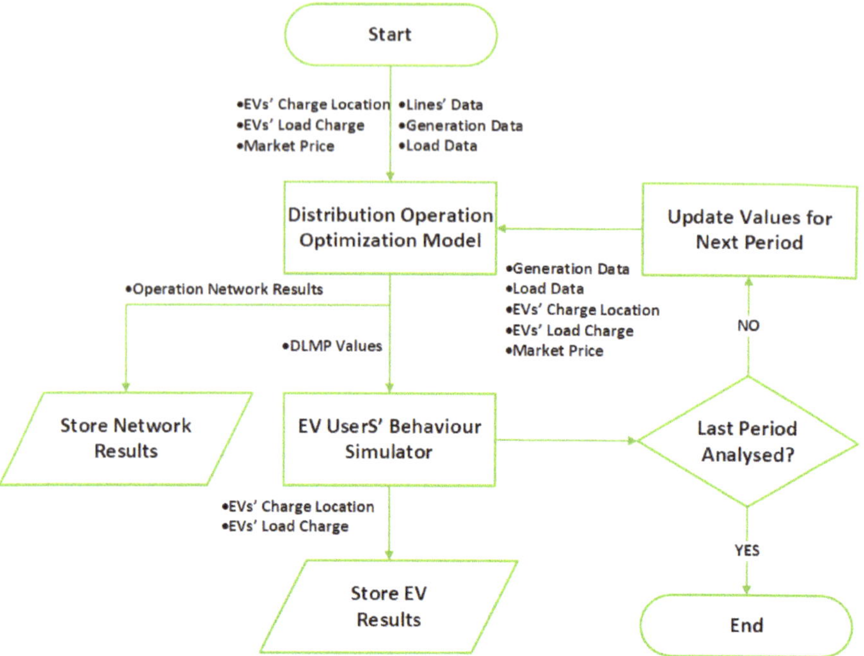

Figure 1. The proposed methodology's flowchart.

2.1. Simulator of Urban Mobility

The simulator module is able to generate a realistic population, considering the size of the network and the parking lots. It has several global and behavior-related parameters (user profiles) discussed later in this section. Figure 2 shows the flowchart of the simulator. After receiving the necessary information from the optimization model, i.e., the DLMP price in each bus of the network, the simulator loops for every individual car to perform the next period's decision (i.e., 15 min). There are only two possible types of decisions: the decision to travel to a destination or a charging decision. Indeed, some trips take more than 15 min, so the car can just keep traveling for a certain number of periods. According to each user preference and behavior, decisions will be affected by the price and distance to the parking lot. Since, in a realistic scenario, some prefer extra comfort even if they pay more, e.g., choosing a fast charge or closer parking lot, the simulator allows defining this range of preferences for each car. These preferences will affect the efficacy of the dynamic EV charging prices, since individual behaviors may neglect lower prices. Nevertheless, our case study provides a different range of behavioral aspects to provide an accurate research outcome in this work.

To determine the dynamic EV charging price, the simulator uses the following Equation (1):

$$DEP = (DLMP + Tariff MV + ACNR) \cdot PLG \cdot VAT \tag{1}$$

where:
DEP: Dynamic EV charging price for each period (€/kWh)
DLMP: Distribution locational marginal pricing (€/kWh)

TariffMV: Energy tariff price for each period (in the Case Study Section, the reader can find the energy tariff price for each considered period) (€/kWh)

PLG: Additional profit margin of the parking owner

VAT: Value-added tax

ACNR: Additional cost related to the fixed term of network price rate to be charged to the customer (€/kWh) and given by (2):

$$ACNR = \frac{\left(\frac{0.397 \cdot CP}{720}\right)}{OPR} \tag{2}$$

The contracted power cost is 0.397 €/kW/month, to be paid to the DSO monthly (www.erse.pt); the *CP* is the charging power of the parking lot; 720 are the hours per month; and *OPR* is the parking occupation rate. With the *ACNR* term added to Equation (1), the contracted power cost is transferred to the final consumer. Moreover, it is important to note that *OPR* is introduced to approximate the real occupation rate of each parking lot and thus affects the *ACNR* cost, which decreases for each customer as the *OPR* rate increases.

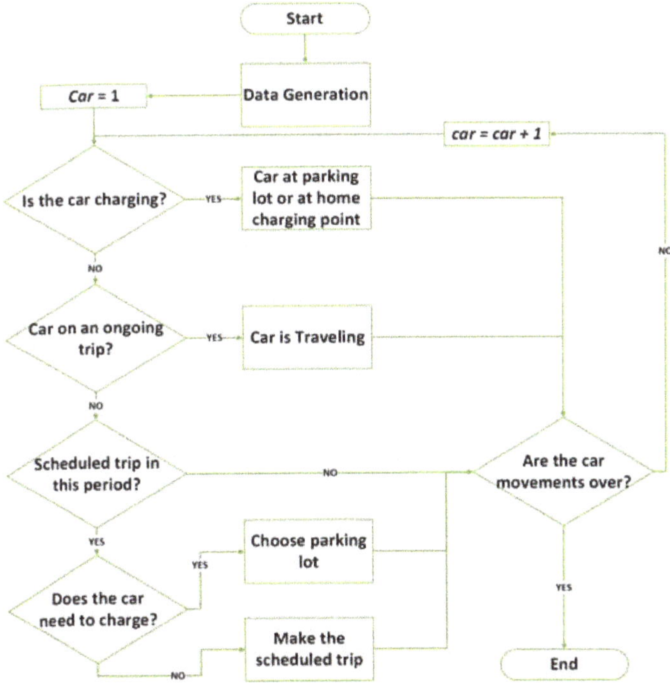

Figure 2. Flowchart for the EV users' behavior simulator.

The global parameters of the simulator are described in Table 1. These are permanent parameters in the simulation; however, their values can be modified according to the needs of each study. Since car travel is simulated using simplified mathematics for vehicle movement, parameter *cdist* represents the penalty on a given distance between two points, e.g., Origin A and Destination B, that the vehicle has to travel (trips). Ideally, the minimum distance to reach Destination B (e.g., work) would be the Euclidean distance; however, in a real-world scenario, the road network is not optimal in this sense. Parameter *sf* can be used to easily change the scale of the map and increase or decrease distances regarding a reference scenario. This allows easily studying the effects on the travel times and charging needs when the urban distance is varied. Parameter *hcpower* enables setting the amount of charge power available when users decide to charge at home. Parameter *chargineff* is the charging efficiency considered for the energy transactions with the electrical grid.

Table 1. Global simulator parameters.

Parameter	Description
ncars	Number of EVs
cdist	Distance increase between two Euclidean points
sf	Scale factor of the map
hcpower	Home charging power
chargineff	Charging mode efficiency

Table 2 describes the parameters related to the user profile, namely regarding the initial location of the car when the simulation begins and its location in subsequent steps. Each car in the simulation replicates the parameters depicted in this table. Among the defined parameters, the weights of *w1*, *w2*, and *ti* are significant. The weights correspond to the importance attributed to distance and price, respectively, while *ti* is used to prioritize trips, for instance going to work cannot be postponed. The weights allow the simulator to compute the behavioral score formula and in this way to decide where to charge the vehicle if needed. For users that give more preference to price while driving long distances in the quest for parking lots with lower charging prices, these prices are dynamic in time and space depending on the DLMP status of the grid. Users with *hc = 0* cannot charge at home, but can charge at parking lots (street charging).

Table 2. User profile parameters.

Parameter	Description
Ilocation	Initial location of the car (usually home)
Clocation	Current location at period *j*
ISoC	Initial state of charge
CSoC	Current state of charge
ae	Car average economy
aeppkm	Car average economy percentage per kilometer
arp	Available range preference
times	Table with the times in which the scheduled trips will be made
as	Average speed
nd	Number of destinations each car has
dest1	Table with the coordinates of the places of the trips to be carried out
i	Boolean variable that determines whether the car will have more than one destination
w1, w2	Weights used in the calculation of the score to determine the best place for charging
ti	Table with the importance of each trip (1 being the least important and 3 the most)
hc	Boolean variable that determines whether the car has a home charger or not

2.2. DLMP-Based Network Operation

DLMP has been studied to provide electricity players with the effective economic signals for optimizing their assets. It is known that the resistance of the distribution network lines is higher than that of transmission lines. Thus, the distribution system losses can be considered one of the main factors that affect the DLMP.

bus voltage regulation is a critical issue, especially with DER proliferation, that is faced by the DSO. Therefore, the DLMP could reflect the voltage impact on the distribution system's economical operation.

In the proposed methodology, DLMP is defined through Lagrangian multipliers of the corresponding constraints (power balance) of the optimization problem, whose goal is to minimize the DSO expenditures.

The distribution network operation and reconfiguration problem in an SG context with high DER penetration concerning the behavior aspects of the EV users and dynamic EV charging price considering DLMPs is classified as MINLP due to the nonlinearity features. To solve complex problems like this, the Benders decomposition is an adequate technique [80,81]. This technique was presented in 1962 by Jacobus Franciscus Benders to solve mixed integer problems [82]. This method is based on the principle that the main problem can be decomposed into sub-problems. The Benders decomposition technique uses duality theory in linear and nonlinear mathematical programming to split a problem whose resolution is difficult into sub-problems [80]. These sub-problems consider specific variables that are solved iteratively until the optimal solution is reached [83].

The problem can be divided into subproblems (a master problem and one or more slave problems). The master subproblem is usually a linear or mixed integer problem including fewer technical constraints. On the other hand, slave subproblems are linear or nonlinear and attempt to validate if the solution of the master problem is technically feasible or not. At this level, the network's technical constraints are considered. A flowchart of the Benders decomposition technique used in this proposed research work is presented in Figure 3, and the diagram of the DLMP-based distribution network operation/reconfiguration model is presented in Figure 4. In Sections 2.2.1 and 2.2.2, the explanation of the Benders decomposition procedure is discussed.

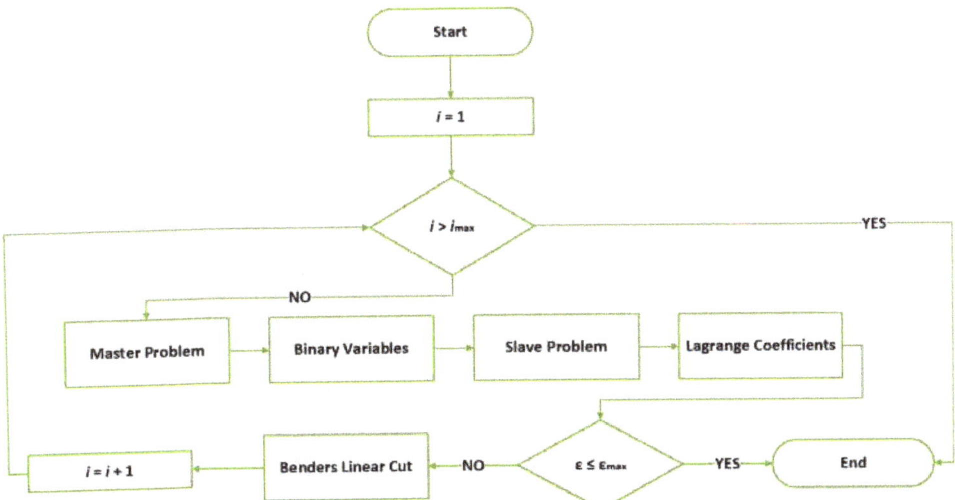

Figure 3. The Benders decomposition flowchart.

This work deals with a non-convex and non-linear slave subproblem (namely in the power flow equations) in which the zero-duality gap is not guaranteed. Thus, the Benders decomposition technique applied in this research work could not converge to the optimal solution. However, most of the science and engineering mathematical problems are non-convex with a very small duality gap in most of the cases [83]. Moreover, the convexity is a solid mathematical assumption, and the convexity is not necessarily restrictive from the practical viewpoint, as many science and engineering problems in the region where the solution of the interest is located are convex; in other words, where the solution is meaningful from the viewpoint of the science and engineering [83].

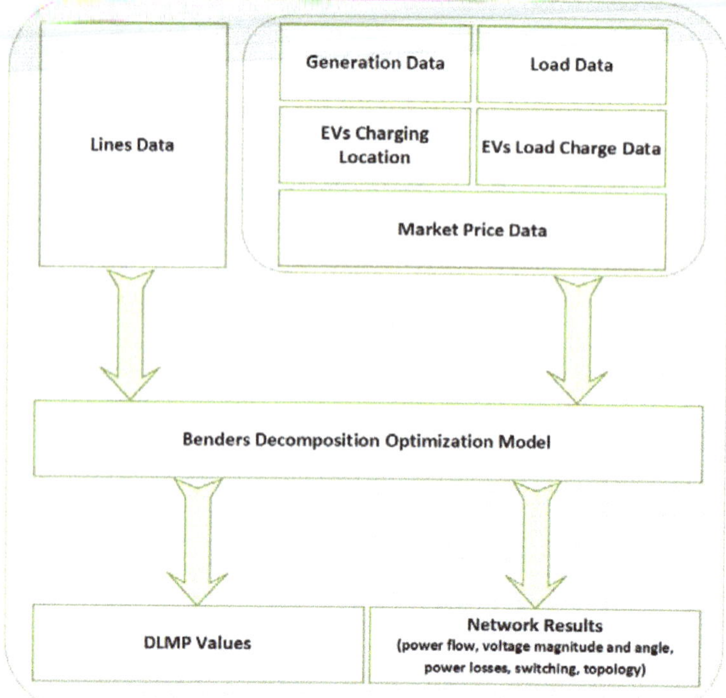

Figure 4. Diagram of the distribution operation optimization model.

2.2.1. Master Problem

The master subproblem goal consists of finding the network topology configuration for each considered period by opening/closing tie-switches (using binary variables {0,1}) to:

- Minimize the power losses' cost;
- Minimize the power not supplied cost;
- Minimize the lines' congestion cost;
- Minimize the power generation curtailment cost;
- Minimize the power from external suppliers' cost.

At this level, every binary variables must be included in the optimization problem. The master subproblem objective function minimizes the operation cost (MOC) and can be formulated as (3):

$$
MOC = \begin{bmatrix}
\displaystyle\sum_{i\in\Omega_B}\sum_{j\in\Omega_B}\left[\left(CongM^2_{(i,j)} + CongM_{(i,j)}\right)\cdot Cost^{Cong}\right] + \\[1em]
\displaystyle\sum_{i\in\Omega^b_{BS}}\left(ExtSup_{(i)}\cdot price^{Mk}\right) + \\[1em]
\displaystyle\sum_{i\in\Omega^b_L}\left(PNSM_{(i)}\cdot Cost^{PNS}\right) + \\[1em]
\displaystyle\sum_{i\in\Omega_B}\sum_{j\in\Omega_B}\left(r_{(i,j)}\cdot FlowM^2_{(i,j)}\cdot Cost^{Loss}\right) + \\[1em]
\displaystyle\sum_{i\in\Omega^{nd}_{DG}}\left(P_{PGCM(i)}\cdot P^{Cost}_{PGC}\right) + \omega^*
\end{bmatrix}
\tag{3}
$$

In the case of infeasibility, one variable is added to the master problem (ω^*), which is called linear Benders' cuts. In ideal circumstances, the value for this variable is zero, which means that the network topology along with its components fulfills every technical constraint. Otherwise, the value presented in this variable represents the minimal value cost change of the master solution.

The master subproblem (3) is subjected to constraints (4)–(25).

Network constraints

Power balance: First Kirchhoff law

Constraint (4) guarantees the power balance in each distribution network bus.

$$
\begin{aligned}
&\sum_{i\in\Omega^{nd}_{DG}}\left(p_{DG(i)} - p_{PGCM(i)}\right) + \sum_{i\in\Omega^b_{BS}}ExtSup_{(i)} - \\
&\sum_{i\in\Omega^b_L}\left(p_{Load(i)} - PNSM_{(i)}\right) - \sum_{i\in\Omega^b_V}EVP_{(i)} + \\
&\sum_{i\in\Omega^b_E}\left(STdchM_{(i)} - STchM_{(i)}\right) + \\
&\sum_{i\in\Omega_B}\left(FlowM_{(i,j)} - FlowM_{(j,i)}\right) = 0 \quad \forall j \in \Omega_B
\end{aligned}
\tag{4}
$$

Maximum admissible line flow

The maximum power flowing in each line of the network is guaranteed by (5).

$$
0 \leq FlowM_{(i,j)} \leq Flow^{max}_{(i,j)}\cdot X^{stat}_{(i,j)} \quad \forall X^{stat}\in\{0,1\}, \forall(i,j)\in\Omega_l
\tag{5}
$$

Unidirectionality of power flow

Constraint (6) guarantees unidirectionality between buses i and j.

$$
X^{stat}_{(i,j)} + X^{stat}_{(j,i)} \leq 1 \quad \forall X^{stat}\in\{0,1\}, \forall(i,j)\in\Omega_l
\tag{6}
$$

Radial topology

To ensure the radial topology, Constraint (7) is applied. This constraint imposes that only one line can enter in each bus.

$$
\sum_{j\in\Omega^b_j} X^{stat}_{(i,j)} = 1 \quad \forall X^{stat}\in\{0,1\}, \forall i\in\Omega_B
\tag{7}
$$

Avoid island creation

To avoid DG isolation from the substation, the constraints (8)–(11) are used. A fictitious flow $(d_{(i,j)})$ is created with a fictitious load of each DG $(D_{(g)})$ to be fed to the substation. If the island is permitted, the operator can omit these constraints.

$$\sum_{i \in \Omega_B} \sum_{j \in \Omega_B} d_{(i,j)} - \sum_{i \in \Omega_B} \sum_{j \in \Omega_B} d_{(j,i)} - D_{(g)} = 0 \quad \forall g \in \Omega_{DG} \tag{8}$$

$$D_{(g)} = 1 \quad \forall g \in \Omega_{DG} \tag{9}$$

$$D_{(g)} = 0 \quad \forall g \notin \Omega_{DG} \cup \Omega_{BS} \tag{10}$$

$$\left| d_{(i,j)} \right| \leq nDG \cdot X_{(i,j)}^{stat} \quad \forall (i,j) \in \Omega_l \tag{11}$$

Supplier constraint

Maximum and minimum limits for power supplier

The power is constrained by the maximum and minimum capacity that can be supplied (12).

$$ExtSup_{MinLimit(bs)} \leq ExtSup_{(bs)} \leq ExtSup_{MaxLimit(bs)} \quad \forall bs \in \Omega_{BS} \tag{12}$$

Curtailment constraints

Power generation curtailment

The power generation curtailment is verified when the excess generation of the generator g occurs. This variable is lower or equal to the power generation of the g generator (13).

$$0 \leq p_{PGCM(g)} \leq p_{DG(g)} \quad \forall g \in \Omega_{DG}^{nd} \tag{13}$$

Power not supplied

Constraint (14) guarantees that the power not supplied variable must be lower than or equal to the load demand.

$$0 \leq PNSM_{(lo)} \leq p_{Load(lo)} \quad \forall lo \in \Omega_L^b \tag{14}$$

Lines' congestion

Lines' power congestion

The power congestion in each line is constrained by Equation (15). In this work, we assume that the congestion occurs when the power flow $Flow_{(i,j)}$ is greater than or equal to a factor value ($CongMin$) multiplied by the maximum power line capacity ($Flow_{(i,j)}^{max}$). The factor value is a constant value between zero and one. In fact, this value represents the percentage of the line capacity that is being used. Equation $Cong_{(i,j)} \geq 0$ is used to ensure a positive or a zero value for $Cong_{(i,j)}$.

$$CongM_{(i,j)} \geq FlowM_{(i,j)} - CongMin \cdot Flow_{(i,j)}^{max} \quad \forall (i,j) \in \Omega_l$$
$$Cong_{(i,j)} \geq 0 \tag{15}$$

Energy storage systems (ESS) constraints

Discharge limit for the energy storage systems

The maximum discharge limit of each ESS is represented by the constraint (16).

$$STdchM_{(e)} \leq STdchR_{(e)} \cdot STdM_{(e)}^{stat} \quad \forall e \in \Omega_E^b, STdM^{stat} \in \{0,1\}, STdchM \geq 0 \tag{16}$$

Charge limit for the energy storage systems

The maximum charge limit for each ESS is represented by the constraint (17).

$$STchM_{(e)} \leq STchR_{(e)} \cdot STcM_{(e)}^{stat} \quad \forall e \in \Omega_E^b, STcM^{stat} \in \{0,1\}, STchM \geq 0 \tag{17}$$

Discharge level limit considering the state of the energy storage system

The maximum discharge limit considering energy storage systems' capacity constraint for each ESS is given by (18). The Δt is represented in units of hours.

$$STdchM_{(e)} \cdot \frac{1}{def_{(e)}} \leq STdM_{(e)}^{stat} \cdot STstoM_{(e)} \cdot \frac{1}{\Delta t} \quad \forall e \in \Omega_E^b, STdM^{stat} \in \{0,1\}, STdchM \geq 0 \tag{18}$$

Charge level limit considering energy storage systems' capacity

The maximum charge limit considering energy storage systems' capacity constraint for each ESS is given by (19). The Δt is represented in units of hours.

$$STstoM_{(e)} + STchM_{(e)} \cdot cef_{(e)} \cdot \Delta t \leq STcM_{(e)}^{stat} \cdot STcap_{(e)}$$
$$\forall e \in \Omega_E^b, STcM^{stat} \in \{0,1\}, STchM \geq 0 \tag{19}$$

State of charge of the energy storage systems

The state of charge of each ESS is given by (20). The Δt is represented in units of hours.

$$STstoM_{(e)} - \left(\Delta t \cdot STchM_{(e)} \cdot cef_{(e)}\right) + \left(\Delta t \cdot STdchM_{(e)} \cdot \frac{1}{def_{(e)}}\right) = STstoM_{(e)}^{t-1}$$
$$\forall e \in \Omega_E^b, STchM \geq 0, STdchM \geq 0 \tag{20}$$

Maximum ESS capacity limit

The maximum capacity limit of each ESS is represented by (21).

$$STstoM_{(e)} \leq STcap_{(e)} \tag{21}$$

Minimum ESS capacity limit

The minimum capacity limit of each ESS is represented by (22).

$$STstoM_{(e)} \geq STstoM_{(e)}^{min} \tag{22}$$

Charging and discharging status

The charging and discharging status of the ESSs are represented by $STcM_{(e)}^{stat}$ and $STdM_{(e)}^{stat}$, respectively. Charging and discharging cannot occur simultaneously (23).

$$STcM_{(e)}^{stat} + STdM_{(e)}^{stat} \leq 1 \tag{23}$$

where $STcM_{(e)}^{stat}$ is a binary variable. ESS are able to charge at any moment.

$STdM_{(e)}^{stat}$ is a variable that assumes zero or one according to the study period market price value and is given by (24).

$$STdM_{(e)}^{stat} = 1 \iff price^{Mk} \geq price_{min}^{Mk}$$
$$STdM_{(e)}^{stat} = 0 \iff price^{Mk} \leq price_{min}^{Mk} \tag{24}$$

Linear Benders' cut

To support the decomposition technique, a linear cuts constraint (25) is used. This constraint represents feasibility cuts in the problem. These cuts are updated in each iteration applying new constraints to the problem. The linear cuts establish the link between the master subproblem and the slave subproblem. To better understand, let us imagine the existence of a cut. Thus, the master subproblem receives and considers the infeasibility data costs of the previous iteration ω^* and the sensitivities $\lambda_{(i,j)}^{m-1}$ and $\mu_{(i)}^{m-1}$. Those sensitivities are linked to the subproblem master decision in the previous iteration $\left(X_{(i,j)}^{stat}\right)^{m-1}$ and $\left(STcM_{(i)}^{stat}\right)^{m-1}$ already known. To make a new decision, the master subproblem is fed these new data.

$$
\begin{aligned}
\omega^* \geq Z_{up}^{m-1} + \\
\sum_{i\in\Omega_B}\sum_{\substack{j\in\Omega_B \\ j\neq i}} \lambda_{(i,j)}^{m-1} \cdot \left[\left(X_{(i,j)}^{stat}\right)^m - \left(X_{(i,j)}^{stat}\right)^{m-1}\right] + \\
\sum_{i\in\Omega_{BS}^b} \mu_{(i)}^{m-1} \cdot \left[\left(STcM_{(i)}^{stat}\right)^m - \left(STcM_{(i)}^{stat}\right)^{m-1}\right]
\end{aligned}
\tag{25}
$$

where index m represents the current iteration and $m-1$ represents the previous iteration.

2.2.2. Slave problem

One of the goals of the slave subproblem is to verify the feasibility of the master problem. Moreover, through AC optimal power flow, the slave subproblem provides the optimal value for the operation variables. The slave subproblem objective function is represented by (26), where the operation costs and the slack variables ZA, ZQ, and ZF are minimized. Slack variables ZA and ZQ (for active and reactive power balance) and ZF (for thermal lines capacity) can take any positive value to make the optimization problem feasible. The value of these variables represents how much some constraints are being violated. The slave sub-problem cannot change the binary variables, but is free to explore the continuous variables in order to satisfy the several constraints, while minimizing the objective function and the value of the slack variables.

$$
SOC = \begin{bmatrix}
\sum_{i\in\Omega_B}\sum_{j\in\Omega_B} \left[\left(CongS_{(i,j)}^2 + CongS_{(i,j)}\right) \cdot Cost^{Cong}\right] + \\[2ex]
\sum_{i\in\Omega_{BS}^b} \left(P_{Supplier(i)} \cdot price^{Mk}\right) + \\[2ex]
\sum_{i\in\Omega_L^b} \left(PNSs_{(i)} \cdot Cost^{PNS}\right) + \\[2ex]
\sum_{i\in\Omega_{DG}^{nd}} \left(P_{PGCs(i)} \cdot P_{PGC}^{Cost}\right) + \\[2ex]
\sum_{i\in\Omega_B}\sum_{j\in\Omega_B} \left(SLoss_{(i,j)} \cdot Cost^{Loss}\right) + \\[2ex]
\sum_{i\in\Omega_B} \left(ZA_{(i)} + ZQ_{(i)}\right) \cdot Cost^{Inf} + \\[2ex]
\sum_{i\in\Omega_B}\sum_{\substack{j\in\Omega_B \\ \neq i}} \left(ZF_{(i,j)} \cdot Cost^{Inf}\right)
\end{bmatrix}
\tag{26}
$$

The slave subproblem (26) is subjected to Constraints (27)–(52).

Network constraints

Voltage magnitude

The voltage magnitude of each bus is constrained by a maximum and minimum deviation (27).

$$V_{(i)}^{min} \leq V_{(i)} \leq V_{(i)}^{max} \qquad \forall i \in \Omega_B \tag{27}$$

Voltage angle

The maximum and minimum angle deviation is constrained by (28).

$$\theta_{(i)}^{min} \leq \theta_{(i)} \leq \theta_{(i)}^{max} \qquad \forall i \in \Omega_B \tag{28}$$

Active power balance

Constraint (29) guarantees the active power balance in each distribution network bus.

$$\begin{aligned}
&\sum_{i \in \Omega_{DG}^{nd}} \left(P_{DG(i)} - P_{PGCs(i)} \right) + \sum_{i \in \Omega_{BS}^b} P_{Supplier(i)} - \\
&\sum_{i \in \Omega_L^b} \left(P_{Load(i)} - PNSs_{(i)} \right) - \sum_{i \in \Omega_V^b} EVP_{(i)} + \\
&\sum_{i \in \Omega_{BS}^b} \left(STdchS_{(i)} - STchS_{(i)} \right) - \\
&\sum_{i \in \Omega_B} P_{Inj(i)} + ZA_{(i,j)} = 0
\end{aligned} \tag{29}$$

Reactive power balance

Constraint (30) guarantees the reactive power balance in each distribution network bus.

$$\begin{aligned}
&\sum_{i \in \Omega_{BS}^b} Q_{Supplier(i)} + \sum_{i \in \Omega_{CB}^b} Q_{Cbanks(i)} - \sum_{i \in \Omega_L^b} Q_{Load(i)} - \\
&\sum_{i \in \Omega_B} Q_{Inj(i)} + ZQ_{(i,j)} = 0
\end{aligned} \tag{30}$$

Injected active power

This Equation (31) represents the injected active power in each bus of the network.

$$P_{Inj(i)} = V_{(i)} \sum_{j \in \Omega_B} V_{(j)} \left(G_{(i,j)} \cdot \cos \theta_{(i,j)} + B_{(i,j)} \cdot \sin \theta_{(i,j)} \right) \qquad \forall i \in \Omega_B, \forall (i,j) \in \Omega_l \tag{31}$$

Injected reactive power

The injected reactive power in each bus is represented by the Equation (32).

$$Q_{Inj(i)} = V_{(i)} \sum_{j \in \Omega_B} V_{(j)} \left(G_{(i,j)} \cdot \sin \theta_{(i,j)} - B_{(i,j)} \cdot \cos \theta_{(i,j)} \right) \qquad \forall i \in \Omega_B, \forall (i,j) \in \Omega_l \tag{32}$$

Active power flow

The active power flow for each network line is given by the Equation (33).

$$\begin{aligned}
&P_{(i,j)} = (V_{(i)}^2 - V_{(i)} \cdot V_{(j)} \cdot \cos \theta_{(i,j)}) \cdot G_{(i,j)} - (V_{(i)} \cdot V_{(j)} \cdot sen \theta_{(i,j)}) \cdot B_{(i,j)} \\
&\forall i \in \Omega_B, \forall j \in \Omega_B, \forall (i,j) \in \Omega_l
\end{aligned} \tag{33}$$

Reactive power flow

Equation (34) gives the reactive power flow for each line.

$$Q_{(i,j)} = -(V_{(i)}^2 - V_{(i)} \cdot V_{(j)} \cdot \cos\theta_{(i,j)}) \cdot B_{(i,j)} - (V_{(i)} \cdot V_{(j)} \cdot \text{sen}\,\theta_{(i,j)}) \cdot G_{(i,j)}$$
$$\forall i \in \Omega_B, \forall j \in \Omega_B, \forall (i,j) \in \Omega_l \tag{34}$$

Apparent power flow

The apparent power flow equation, as can be seen in Equation (35), is given by the square root of the active power flow and reactive power flow squares.

$$S_{(i,j)} = \sqrt{P_{(i,j)}^2 + Q_{(i,j)}^2} \qquad \forall (i,j) \in \Omega_l \tag{35}$$

Active power losses

The active power loss of each line is represented by Equation (36).

$$PLoss_{(i,j)} = \frac{P_{(i,j)}^2 + Q_{(i,j)}^2}{V_{(i)}^2} \cdot r_{(i,j)} \qquad \forall i \in \Omega_B, \ \forall (i,j) \in \Omega_l \tag{36}$$

Reactive power losses

To represent the reactive power loss, the following Equation (37) is used.

$$QLoss_{(i,j)} = \frac{P_{(i,j)}^2 + Q_{(i,j)}^2}{V_{(i)}^2} \cdot x_{(i,j)} \qquad \forall i \in \Omega_B, \ \forall (i,j) \in \Omega_l \tag{37}$$

Apparent power loss

To obtain the apparent power loss in each line, the following equation is used (38).

$$SLoss_{(i,j)} = \sqrt{PLoss_{(i,j)}^2 + QLoss_{(i,j)}^2} \qquad \forall (i,j) \in \Omega_l \tag{38}$$

Maximum admissible line flow

The maximum power flow in each line is constrained by (39).

$$0 \leq FlowS_{(i,j)} \leq Flow_{(i,j)}^{max} + ZF_{(i,j)} \qquad \forall (i,j) \in \Omega_l \tag{39}$$

Supplier constraints

Maximum and minimum limits for active power supplier

The active power is constrained by the maximum and minimum capacity that can be supplied (40).

$$P_{SMinLimit(bs)} \leq P_{Supplier(bs)} \leq P_{SMaxLimit(bs)} \qquad \forall bs \in \Omega_{BS} \tag{40}$$

Maximum and minimum limits for the reactive power supplier

The reactive power is constrained by the maximum and minimum capacity that can be supplied (41).

$$Q_{SMinLimit(bs)} \leq Q_{Supplier(bs)} \leq Q_{SMaxLimit(bs)} \qquad \forall bs \in \Omega_{BS} \tag{41}$$

Maximum and minimum limits for capacitor banks

The reactive power of a capacitor bank is considered a continuous variable in this model and is constrained by the maximum and minimum (zero) capacity that can be supplied (42).

$$0 \leq Q_{Cbanks(cb)} \leq Q^{max}_{Cbanks(cb)} \qquad \forall cb \in \Omega_{CB} \tag{42}$$

Curtailment constraints

Power generation curtailment

Power generation curtailment occurs when the generator generates an excess of power g. This variable cannot be higher than the generation of the g generator (43).

$$0 \leq P_{PGCs(g)} \leq P_{DG(g)} \qquad \forall g \in \Omega^{nd}_{DG} \tag{43}$$

Power not supplied

Constraint (44) guarantees that the power not supplied variable must be lower or equal to the load demand.

$$0 \leq PNSs_{(lo)} \leq P_{Load(lo)} \qquad \forall lo \in \Omega^{b}_{L} \tag{44}$$

Lines' congestion

Lines' power congestion

The power congestion in each line is constrained by the Equation (45). The same considerations are taken into account in (15) and in (45).

$$Cong_{(i,j)} \geq FlowS_{(i,j)} - CongMin \cdot Flow^{max}_{(i,j)} \qquad \forall (i,j) \in \Omega_{l}$$
$$Cong_{(i,j)} \geq 0 \tag{45}$$

Energy storage system constraints

Discharge limit of the energy storage systems

The maximum discharge limit determined by the constraint of each ESS (46).

$$STdchS_{(e)} \leq STdchR_{(e)} \qquad \forall e \in \Omega^{b}_{E}, STdchS \geq 0 \tag{46}$$

Charge level limit for the energy storage systems

The maximum charge level limit determined by the constraint of each ESS (47).

$$STchS_{(e)} \leq STchR_{(e)} \qquad \forall e \in \Omega^{b}_{E}, STchS \geq 0 \tag{47}$$

Discharge limit considering energy storage systems' state

The maximum discharge limit considering the capacity constraint of each energy storage system (48). Δt is represented in units of hours.

$$STdchS_{(e)} \cdot \frac{1}{def_{(e)}} \leq STstoS_{(e)} \cdot \frac{1}{\Delta t} \qquad \forall e \in \Omega^{b}_{E}, STdchS \geq 0 \tag{48}$$

Charge limit considering energy storage systems' capacity

The maximum charge level limit is determined considering the capacity constraint of each energy storage system (49). The Δt is represented in units of hours.

$$STstoS_{(e)} + STchS_{(e)} \cdot cef_{(e)} \cdot \tfrac{1}{\Delta t} \leq STcap_{(e)} \quad \forall e \in \Omega_E^b, STchS \geq 0 \tag{49}$$

State of charge of the energy storage systems

The state of charge of each ESS is given by (50). Δt is represented in units of hours.

$$STstoS_{(e)} - \left(\Delta t \cdot STchS_{(e)} \cdot cef_{(e)}\right) + \left(\Delta t \cdot STdchS_{(e)} \cdot \tfrac{1}{def_{(e)}}\right) = STstoS_{(e)}^{t-1}$$
$$\forall e \in \Omega_E^b, STchS \geq 0, STdchS \geq 0 \tag{50}$$

Maximum energy storage systems' capacity limit

The maximum capacity limit for each ESS is represented by (51).

$$STstoS_e \leq STcap_{(e)} \quad \forall e \in \Omega_E^b \tag{51}$$

Minimum energy storage systems' capacity limit

The minimum capacity limit for each ESS is represented by (52).

$$STstoS_{(e)} \geq STstoS_{(e)}^{min} \quad \forall e \in \Omega_E^b \tag{52}$$

3. Case Study

To show how the proposed methodology is applied, a medium voltage (MV) distribution network of an SC (mock-up) located at BISITE laboratory has been developed for this study (the schematic of the SC is presented in Figure 5, and the coordinates of each building can be seen in Table 3). In this case study, a high DER penetration is considered to represent a realistic scenario in the near future. The single-line diagram of the 13-bus 30-kV distribution network is presented in Figure 6.

Table 3. Building coordinates on the xy plane.

Building		L1	L2	L3	L4 to L18	L19	L20	L21	L22	L23	L24	L25	PL1 to PL2	PL3 to PL4	PL5 to PL6	PL7
Coordinates (km)	X Axis	10.50	0.50	9.00	3.75 to 8.25	0.50	0.50	2.50	3.00	4.50	6.00	8.00	1.00	7.00	6.00	11.00
	Y Axis	3.50	2.00	5.00	1.00 to 3.00	3.50	5.50	2.00	4.50	3.50	5.00	4.00	3.50	5.00	0.50	4.00

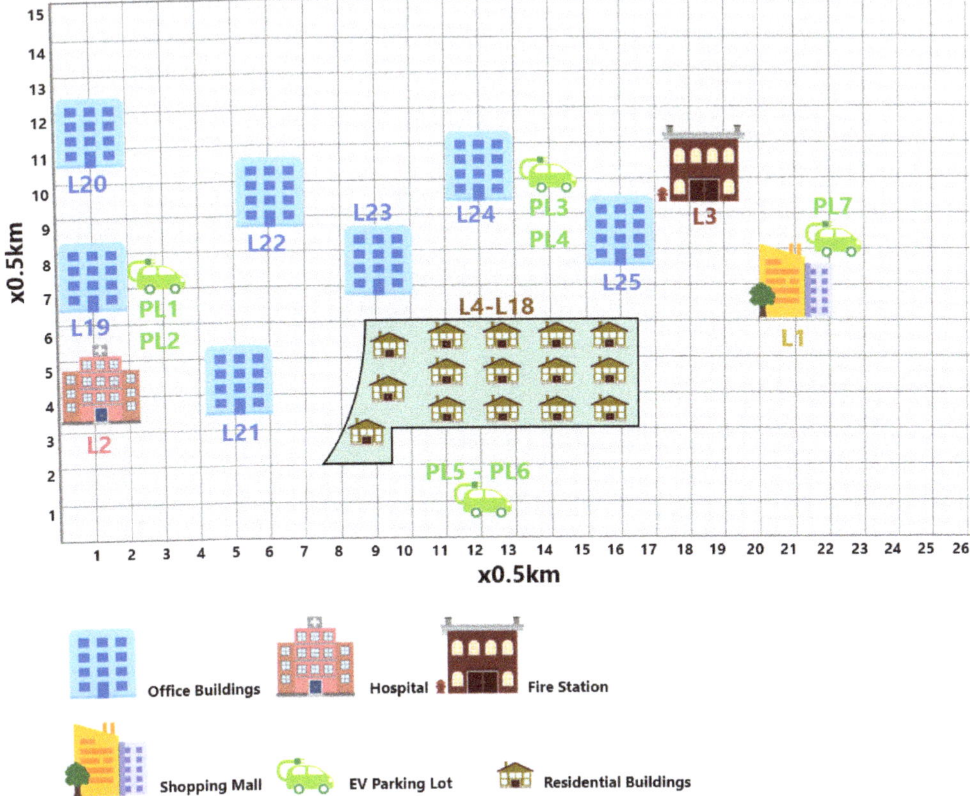

Figure 5. Smart city schematic.

This DN has one 30 MVA substation, 25 load points, and 3×35.88 km of underground cables. For the connections between the substation and the network (bus 1 to bus 2; bus 1 to bus 7), a cable of type LBHIOV 3×150 mm^2 (svrweb.cabelte.pt) has been used, while for the remaining connections, the cable type LBHIOV 3×70 mm^2 (svrweb.cabelte.pt) has been used. A total of 15 DG units (i.e., two wind farms and 13 PV parks) and four capacitor banks of 1 Mvar are included in the network, as can be seen in Figure 6.

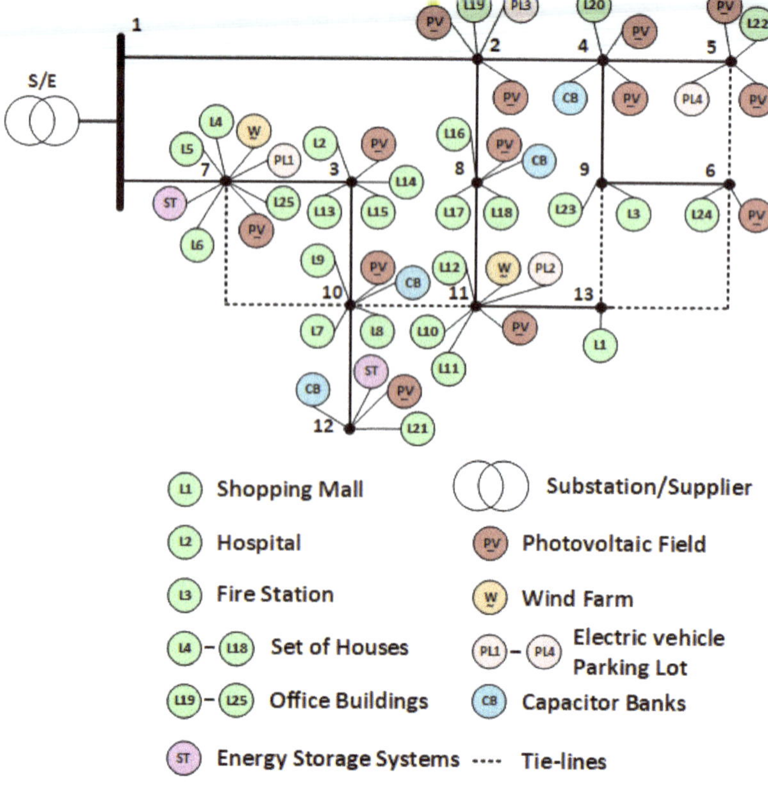

Figure 6. Single-line diagram of the 13-bus distribution network.

The DG penetration corresponds to 27% (10.925 MW) of the total installed power (24% corresponds to wind generation and 3% to PV). Each wind farm has six E48 800 kW ENERCON wind turbines (www.enercon.de). The characteristics of PV parking lots are presented in Table 4. The line congestion cost was 0.02 €/kW when power flow was above 50% of the thermal line rating capacity (*CongMin*).

The considered smart city presents five types of loads, namely:

- Residential buildings (1375 homes);
- Office buildings (seven buildings);
- Hospital;
- Fire Station;
- Shopping Mall.

Table 4. PV characteristics.

Nominal Power (W)	85.00
Short Circuit Current (A)	1.62
Nominal Operating Temperature of the Cell (°C)	45.00
Open Circuit Voltage (V)	56.70
Current at the Maximum Power Point (A)	1.41
Voltage at the Maximum Power Point (V)	45.50
Voltage High Temperature Coefficient (>25 °C) (V/°C)	−0.1531
Voltage Low Temperature Coefficient (−40 °C to 25 °C) (V/°C)	−0.1134
Current Temperature Coefficient (A/°C)	6.4800×10^{-4}
PV park at bus 12	
Number of Modules	104
Number of Panels	120
Total Number of Modules	12,480
PV parks at Buses 2–8, 10, and 11	
Number of Modules	104
Number of Panels	30
Total Number of Modules	3120

This study considered one week of input data for every 15-min period with the aim of showing the effectiveness of the proposed methodology (i.e., 672 periods were considered in the simulation process). The chosen week was 19 March 2017–25 March 2017. The total renewable generated power for each period is presented in Figure 7.

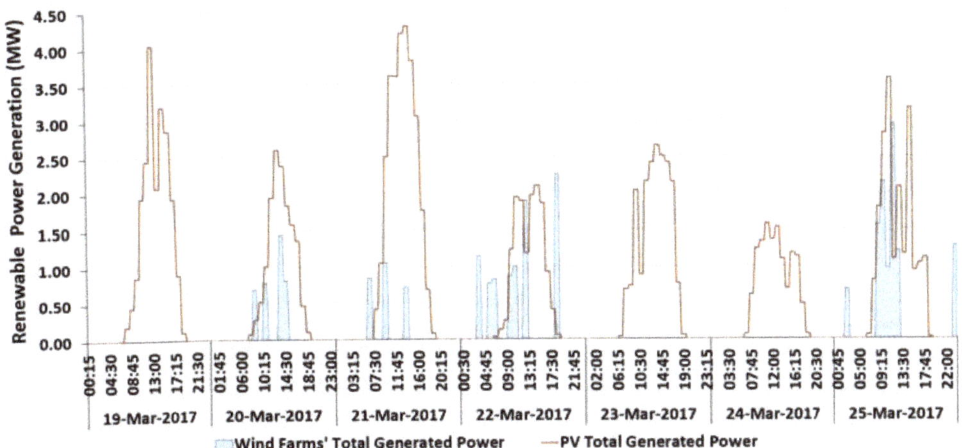

Figure 7. Renewable power generation.

Figures 8 and 9 present the power demand and the roof generation, respectively, of the office buildings, residential buildings, a shopping mall, a hospital, and a fire station. It is important to note that the power demand presented in these two figures corresponds to the subtraction of the initial demand for PV power generation, i.e., all the power generated by the PVs is consumed by the building. The generated power is therefore not sent to the grid in the present study.

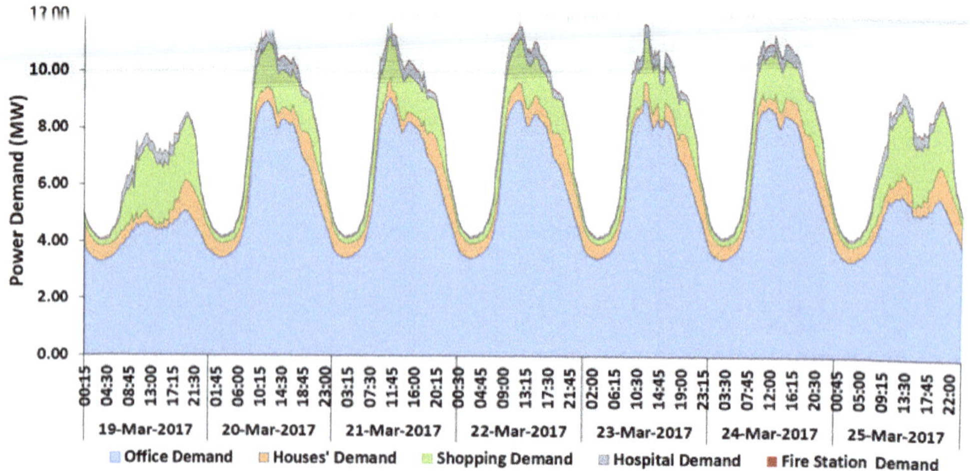

Figure 8. Power demand from office, residential, hospital, fire station, and shopping mall buildings.

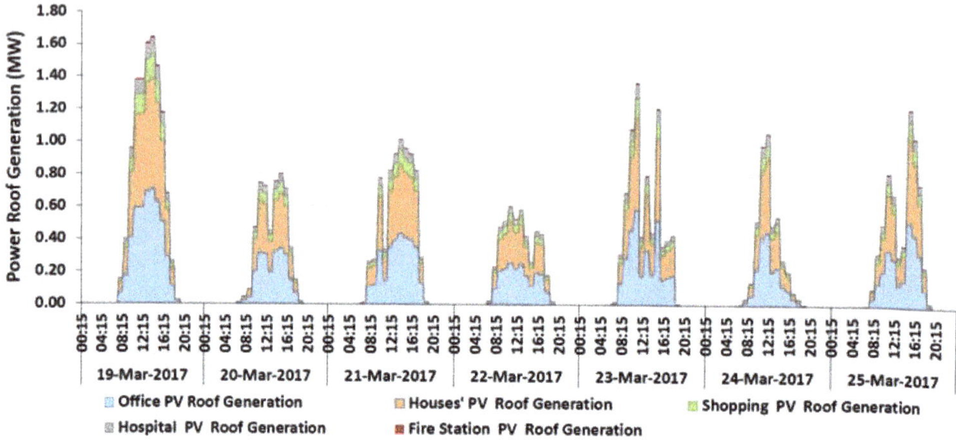

Figure 9. Roof PV generation from office, residential, hospital, fire station, and shopping mall buildings.

The market price for the chosen week was obtained from the Iberian electricity market operator (OMIE) (www.datosdelmercado.omie.es/pt-pt/datos-mercado) and can be seen in Figure 10.

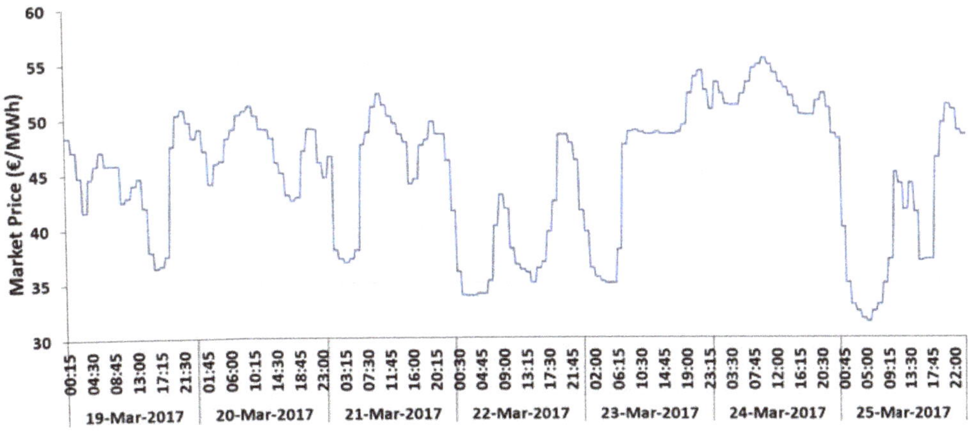

Figure 10. Market price.

Moreover, the SC has seven parking lot buildings for EV charging, four (two in bus 7 and two in bus 11) slow charging lots (7.2 kW for each connection point) and three (two in bus 2 and one in bus 5) fast charging lots (50 kW for each connection point). Each slow charging parking lot has 250 spaces for EVs, while each fast charging parking lot has 80 spaces. In this case study, we assumed that each parking lot had a 30% occupation rate (OPR). Thus, in the following equation (2), the additional cost related to the fixed term of network price rate to be charged to the customer (ACNR) for a slow charging parking space is 0.0132 €/kWh, while for a fast charging parking space, it is 0.0919 €/kWh. Furthermore, the parking owner charges an additional 5% fee and 23% of value-added tax (VAT). Moreover, consider that 50% of the EV users can charge their EV at home (3.7 kW charge point) with a fixed cost of 0.2094 €/kWh. A total of 5000 EVs were considered in this study, and the initial battery level was randomly generated between 40% and 65% of the battery capacity. The considered EV models and their characteristics are listed in Table 5. The weights ($w1$ and $w2$) attributed to the distance and price preference are presented in Tables 6 and 7, respectively. Two possible scenarios are considered: in one, the user's priority is to charge his/her EV at a charging station located as close as possible to them (Table 6); in the second scenario, the users prefer to find charging stations where they can charge their EV at a low price (Table 7).

Table 5. EV types.

Model	Battery (kWh)	Slow Charge Power (kW)	Fast Charge Power (kW)	Consumption (kWh/km)
Nissan Leaf	40.00	6.60	50.00	0.1553
Tesla Model S 70D	75.00	7.40	50.00	0.2100
BMW i3	33.20	7.40	50.00	0.1584
Renault Zoe	41.00	7.40	-	0.1460
Renault Kangoo	33.00	7.40	-	0.1926
VW e-Golf	24.20	7.20	40.00	0.1584
Ford Focus	33.50	6.60	50.00	0.1926
Hyundai IONIQ	30.50	6.60	50.00	0.1429

Table 6. Weights for the user distance preference scenario.

Preference	Weight (%)		Probability (%)
	w1	**w2**	
Price	40	60	30
Distance	85	15	70

Table 7. Weights for the user price preference scenario.

Preference	Weight (%)		Probability (%)
	w1	**w2**	
Price	15	85	70
Distance	60	40	30

Furthermore, two energy storage systems managed by the DSO were considered in the present case study, each one with 1 MWh of capacity and 0.5 MW of charge/discharge rate. Moreover, in this case, the ESS are able to charge at any moment and discharge when the energy market price is greater than or equal to 45 €/MWh. Is assumed that the ESS had a minimum of 5% of power stored, i.e., the power stored in the ESS cannot be less than 5%. The input data used in the case study can be found by the readers in [84].

In this research work, thirty different case studies were performed. Table 8 summarizes the characteristics of those studies. They have been divided into two types of EV user preference scenarios, namely the price preference scenario and distance preference scenario. For each of those scenarios, we considered DG, EV, ESS, dynamic EV charging price, and fixed prices (with three different price levels) and combined them in the case study. The purpose of these case studies was to determine in which situations dynamic charging prices were advantageous for DSO and EV users.

Table 8. Case study sets.

	User Price Preference Scenario							User Distance Preference Scenario						
	DG	EV	ESS	Dynamic EV Charging Price	Fixed Price (€/kWh)			DG	EV	ESS	Dynamic EV Charging Price	Fixed Price (€/kWh)		
					SCh= 0.15 FCh= 0.25	SCh = 0.2 FCh = 0.3	SCh = 0.3 FCh = 0.4					SCh = 0.15 FCh = 0.25	SCh = 0.2 FCh = 0.3	SCh = 0.3 FCh = 0.4
Case A	No	No	No	Yes	No	No	No	No	No	No	Yes	No	No	No
Case B	Yes	No	No	Yes	No	No	No	Yes	No	No	Yes	No	No	No
Case C	Yes	No	Yes	Yes	No	No	No	Yes	No	Yes	Yes	No	No	No
Case D	No	Yes	No	Yes	No	No	No	No	Yes	No	Yes	No	No	No
Case E	No	Yes	No	No	Yes	No	No	No	Yes	No	No	Yes	No	No
Case F	No	Yes	No	No	No	Yes	No	No	Yes	No	No	No	Yes	No
Case G	No	Yes	No	No	No	No	Yes	No	Yes	No	No	No	No	Yes
Case H	Yes	Yes	No	Yes	No	No	No	Yes	Yes	No	Yes	No	No	No
Case I	Yes	Yes	No	No	Yes	No	No	Yes	Yes	No	No	Yes	No	No
Case J	Yes	Yes	No	No	No	Yes	No	Yes	Yes	No	No	No	Yes	No
Case K	Yes	Yes	No	No	No	No	Yes	Yes	Yes	No	No	No	No	Yes
Case L	Yes	Yes	Yes	Yes	No	No	No	Yes	Yes	Yes	Yes	No	No	No
Case M	Yes	Yes	Yes	No	Yes	No	No	Yes	Yes	Yes	No	Yes	No	No
Case N	Yes	Yes	Yes	No	No	Yes	No	Yes	Yes	Yes	No	No	Yes	No
Case O	Yes	Yes	Yes	No	No	No	Yes	Yes	Yes	Yes	No	No	No	Yes

4. Results and Discussion

The proposed methodology has been applied to the case study presented in Section 3 to show its applicability. The proposed research work has been developed on a computer with one Intel Xeon E5-2620 v2 processor and 16 GB of RAM running Windows 10 Pro using the MATLAB R2016a and TOMLAB 8.1 64 bits with CPLEX and SNOPT solvers. As can be seen in Table 9, in each period, the optimization model dealt with the master problem, which had 566 constraints and 744 variables, where 171 were integer variables, and with the slave problem, which had 199 constraints (116 non-linear constraints) and 286 variables.

Table 9. Computational execution results.

Problem Level	Constraints			Number of Variables Per Period			Average Execution Time Per Period (s)	Peak Memory (kB)
	Linear	Non-Linear	Total	Continuous	Integer	Total		
Master problem	566	-	566	573	171	744	1.2	4656
Slave problem	83	116	199	286	-	286		

The average execution time was compatible with operation/reconfiguration time-frame, presenting an average value of 1.2 s (considering all case studies). The analysis of computer system resource impact was also evaluated with a memory test for which the MATLAB memory profiler tool was used. This tool shows the peak memory for each function in the code. The highest computer resource value is 4656 kB, which is perfectly compatible with today's computers.

This section looks at the results of the analysis from two perspectives: that of the operator (Section 4.1) and that of the EV user (Section 4.2).

4.1. The Operator's Perspective

In this subsection, the results are discussed from the perspective of the operator. Figure 11 presents the total operation and congestion cost (672 periods, one week) for all case studies. This figure makes evident the advantages in terms of cost when the DG and ESS systems are used in the network. (a) gives the total operation cost and the total congestion cost for the reference case, i.e., without EVs. Operation costs and congestion costs are reduced significantly when combined with distributed resources, namely with RES and ESS.

(b) (RES and ESS are not considered) verifies that with dynamic EV charging price, operation costs were reduced by 1.20%, 1.20%, and 2.10% when compared to the E, F, and G cases, respectively, for the user price preference scenario. In the user distance preference scenario, costs were reduced by 0.28%, 0.28%, and 3.20%. Moreover, congestion costs were reduced by 8.35%, 8.35%, and 15.20% thanks to dynamic EV charging prices in the user price preference scenario and by 2.29%, 2.29%, and 4.59% in the user distance preference scenario. From the analysis of (c) in Figure 11, compared to fixed prices (Cases I, J, and K), the dynamic EV charging prices presented a cost reduction in the user preference scenario by 1.43%, 1.43%, and 2.52% and in the user distance preference scenario by 0.24%, 0.24%, and 3.43%. Congestion costs were reduced by 13.87%, 13.87%, and 22.62% in the user price preference scenario and by 1.53%, 1.53%, and 4.60% in the user distance preference scenario. In (d), operation costs with fixed EV charging prices (Cases M, N, and O) were reduced by 1.47%, 1.47%, and 2.53% with dynamic EV charging prices. In the user distance preference scenario, cost was reduced by 0.29%, 0.29%, and 3.49%. Congestion costs were reduced by 5.25%, 15.25%, and 23.64% in the user price preference scenario and by 1.41%, 1.41%, and 4.48% in the user distance preference scenario. It is noted that there was no difference in operation costs between slow charging of 0.15 €/kWh or 0.20 €/kWh and fast charging of 0.25 €/kWh or 0.30 €/kWh. Thus, the operator was indifferent to the charging price for the EV user.

The use of dynamic prices for EV charging is beneficial in terms of reduced operation and congestion costs when compared to fixed price options. The reductions are more evident when the fixed prices are higher. Thanks to dynamic EV charging, different charging prices were offered to the users in the parking lots, and this helped alleviate certain power lines, contributing in this way to the operational cost reduction.

Total power loss, power generation curtailment, and power not supplied costs in each user preference scenario are presented in Figure 12a, i.e., with no electric vehicles. It has been verified that the costs associated with those three terms reduced once the distributed energy resources were included (RES and ESS). In fact, the power not supplied (PNS) cost was reduced to zero when the RES were considered alone or together with ESS. However, with RES and ESS, power generation curtailment (PGC) was present, but the costs were lower than with the PNS.

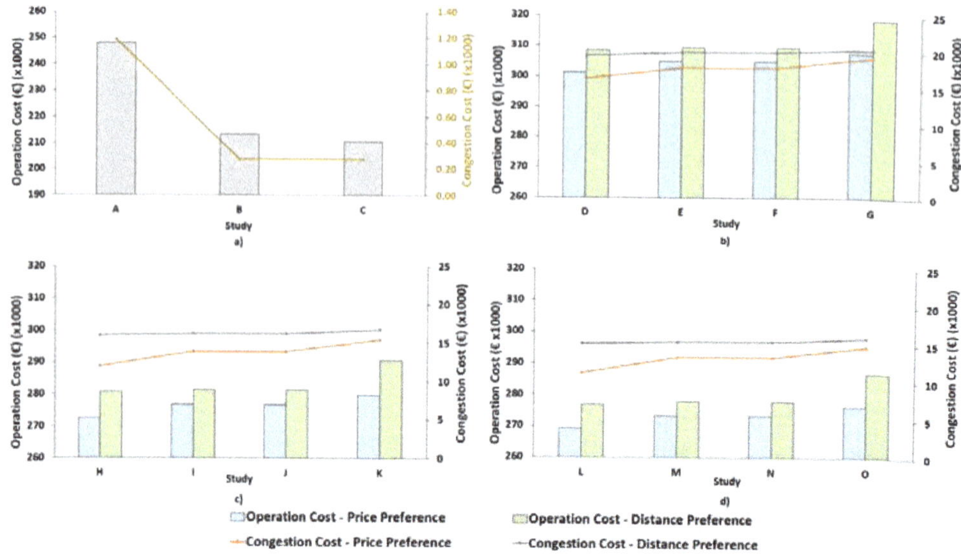

Figure 11. Total operation and congestion costs. (**a**) For cases without EVs. (**b**) For cases with EVs, but without DER. (**c**) For cases with RES, but without ESS. (**d**) For cases with DER (RES and ESS).

Through the analysis of (b) (RES and ESS were not considered) in Figure 12, it can be observed that the total power loss (PL) cost was equivalent to the three fixed-price cases with a cost of around 3662 € in the user price preference scenario. Through the use of the dynamic EV charging price method, the PL cost reduced by around 17%. Considering the user distance preference scenario, the dynamic EV charging prices presented a reduction of only 1.03% in Cases E and F and of 1.91% in Case G. The PNS occurred only in the user distance preference scenario. When the dynamic EV charging prices were included, the PNS cost was small compared to the fixed price (83.54% smaller than in Cases E and F and 98.67% smaller than in Case G). Considering the user price preference scenario in (c) of Figure 12, the observed PL cost reduction with dynamic energy pricing was of 16.75% in Cases I and J and 18.08% in Case K. Cost reductions were lower in the distance user preference scenario, reducing by 0.21% in Cases I and J and 1.52% in Case K. The PNS occurred only for the fixed price cases in the user distance preference scenario, being zero when the dynamic EV charging prices was used. The presence of RES will create the necessity of PGC in some periods. The dynamic EV charging prices can mitigate the costs associated with the PGC in the user price preference scenario, by 3.46% in Cases I and J and 4.32% in Case K. If the user distance preference scenario were considered, it would not be possible to benefit from dynamic EV charging prices. In (d), the presence of ESS was also considered, and its advantages in reducing PGC cost were evident. Through the use of dynamic EV charging prices, PL was reduced by 16.24% in Cases M and N and 18.03% in Case O in the user price preference scenario and by 2.19% in Cases M and N and 2.29% in Case O in the user distance preference scenario. With dynamic EV charging prices in the user price preference scenario, the PGC costs were reduced by 6.86% in Cases M and N and by 8.48% in Case O. In the user distance preference scenario, the cost of PGC did not reduce with dynamic EV charging prices. As can be seen, the use of dynamic EV charging prices is of great advantage in the PGC, leading to a zero value.

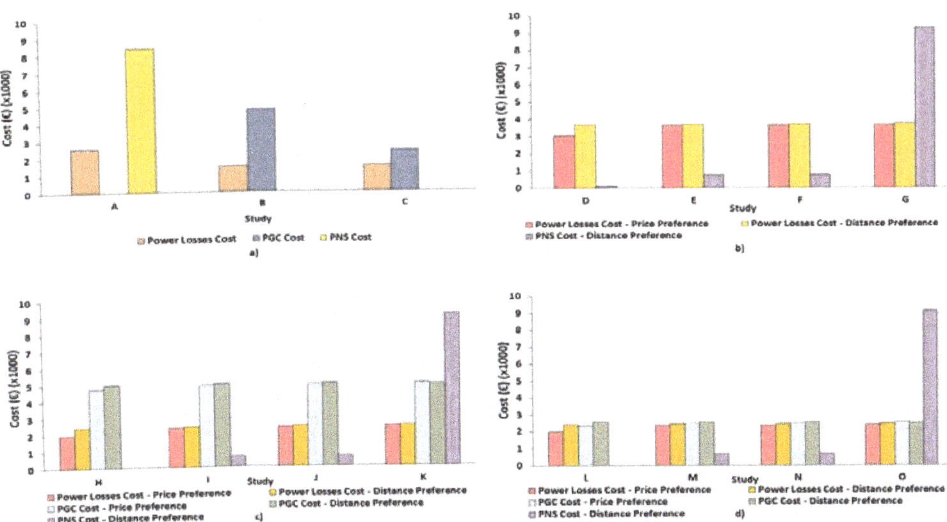

Figure 12. Total power loss, power generation curtailment, and power not supplied costs in each user preference scenario. (**a**) For cases with no EVs. (**b**) For cases with EVs, but without DER. (**c**) For cases with RES, but without ESS. (**d**) For cases with DER (RES and ESS).

Once again, the conclusion drawn from the above analysis is that using dynamic energy pricing for EVs' charging contributed greatly to a reduction in costs associated with power loss, power generation curtailment, and power not supplied. The reductions were more evident for PNS, where they reached 100% in Cases H and L.

Table 10 presents the maximum and the minimum voltage reached in each study. It also presents the buses where those values are verified. As can be seen in this table, the worst voltage values for the user price preference scenario and for the user distance preference scenario were verified in Case G at bus 6 and bus 5, respectively, mainly because these cases did not consider DG and ESS. When adopting dynamic pricing combined with the use of EVs (Cases D, H, and L in the user price preference scenario), it is possible to verify that this leads to better voltage levels (i.e., min. voltage), demonstrating the advantage of dynamic charging prices when EVs react to charging price.

Case L (which is a dynamic EV charging price case) and the case with 0.20 €/kWh for slow charge and 0.30 €/kWh for fast charge (fixed energy charging price) were chosen as an example to present the total energy charge consumption, the average charge power, and the preference percentages of the EV users for each bus that had parking lots. Figure 13 illustrates Case L, and Figure 14 presents the fixed price case.

The preference for a bus with an EV parking lot was counted from the moment the EV began to charge until the time it left the parking lot (one charging session).

In (a), it is possible to see that when the user price preference scenario was considered, the total energy consumed when charging an EV in bus 7 (slow charging parking lot) was 88,037 kWh, which in comparison to the other three buses was 69%, 88%, and 91% more, meaning that the energy price to charge at this bus was better than at the others. Thus, the average charging power followed the same trend as energy consumption. In the user distance preference scenario, energy consumption during charging was spread more evenly over the other parking lot buses. In this case, the highest consumption was the one in bus 2 (fast charging parking lot) with around 45,500 kWh. This bus consumed 19%, 35%, and 14% more energy than the remaining parking lot buses. Once again, the average charge power followed the energy charge consumption trend.

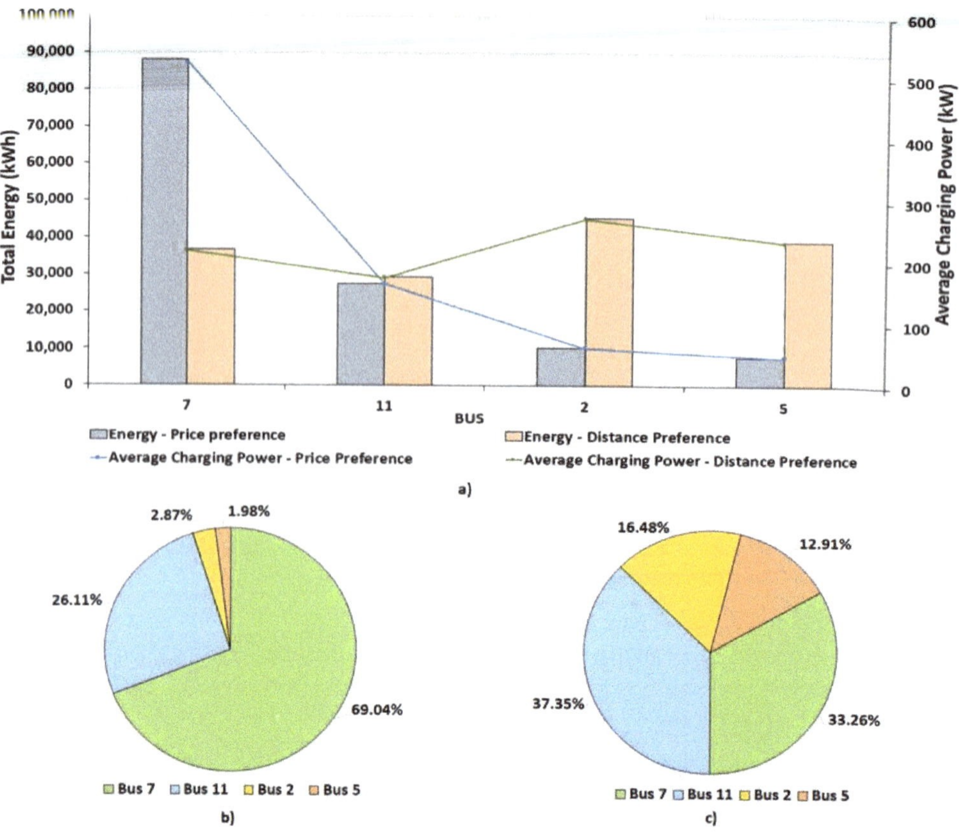

Figure 13. Energy and preference results in each bus that had parking lots for EV charging, considering the dynamic EV charging price in Case L. (**a**) Energy and average charge power at each EV parking lot bus. (**b**) Preference percentage for each parking lot bus considering the user price preference scenario. (**c**) Preference percentage for each parking lot bus considering the user distance preference scenario.

Figure 13b,c shows the preference percentages of the EV users for each bus with a parking lot, considering the user price and distance preferences scenarios, respectively. Figure 13b shows that the parking lot located at bus 7 was the one preferred by EV users, with 69.04% of charged EVs. The parking lot located in bus 11 was the second most chosen, while the fast charging parking lots were the ones least used, with a total of 4.85%. The slow charging parking lots were those that had the lowest energy charging price when compared to the fast charging parking lots. Then, since the user price preference scenario is being considered here, the choice of the less expensive parking lot was logical.

In Figure 13c, the user distance preference scenario is considered. In this scenario, the user preference was to find a parking lot that was as close as possible to the total route that the user would have to travel, i.e., the lowest summation distance between the current EV location and the parking lot and the distance between the parking lot and the next destination. In this user preference scenario, the fast charging parking lots obtained a higher preference when compared to the case where the user preference was defined by the price. This indicates that when the price was not the most important factor, fast charging parking lots could attract users who were located close to them. Nevertheless, we arrived at the conclusion that the location of those parking lots was not optimal, because even when considering the user distance preference scenario, the majority of the users chose the slow charging parking lots: the lease expensive ones.

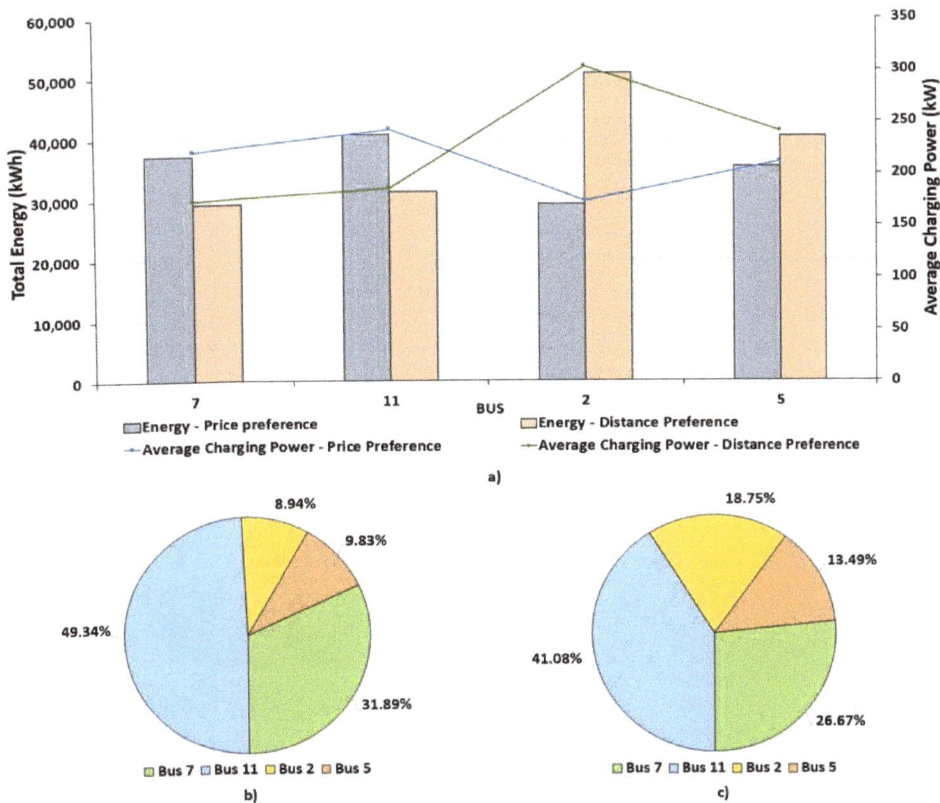

Figure 14. Energy and preference results at each bus that had parking lots at a fixed price for EV charging, with 0.20 €/kWh for slow charging and 0.30 €/kWh for fast charging. (**a**) Energy and average charging power in each EV parking lot bus. (**b**) Preference percentage for each parking lot bus considering the user price preference scenario. (**c**) Preference percentage for each parking lot bus considering the user distance preference scenario.

It is also important to note in the user distance preference scenario that even the parking lots located at bus 11 presented higher user charging preference when compared to the parking lots located at bus 7; the energy consumption and the average charging power at the parking lots of bus 11 were not higher than those at bus 7. This means that the energy price in bus 11 presented higher variations, and it was worse in general when compared to the energy price in bus 7 (it is possible to observe this in Section 4.2, second box plot figure), which contributed to higher energy charge consumption in bus 7 and a considerable charging preference (37.35%) even though there was only a 30% probability in the user price preference scenario (see Table 6). Moreover, due to the higher charge preference at bus 11, it is possible to conclude that the location of the parking lots at this bus was better (advantageous because EV users were at a shorter distance from them) when compared to the parking lots at bus 7.

The total energy charge consumption, the average charge power, and the preference percentages of the EV users for the fixed energy prices (0.20 €/kWh for slow charge and 0.30 €/kWh for fast charge) are presented in Figure 14.

In Figure 11a, it can be observed that in the user price preference scenario, the bus with the highest total energy consumption for EV charging was bus 11 with 40,733 kWh, that is 9%, 28%, and 13% more than buses 7, 2, and 5, respectively. Thus, the average charging power followed the same trend of the energy consumed during charging. In comparison, (a) in Figure 13 shows that the energy consumed by the charging EVs was spread more evenly among all the parking lot buses, while in the dynamic EV charging price case, energy consumption due to charging was more concentrated in bus 7 than the others. This means that bus 7 with respect to dynamic EV charging price presented a better charging price. In the user distance preference scenario, the energy charge consumption followed the same trend as in the dynamic EV charging price case, indicating energy consumption of 50,815 kWh at bus 2; the consumption was higher by 42%, 38%, and 21% in relation to the remaining buses. Regarding the average charging power, the trend was the same as for energy consumption.

Analyzing Figure 14b, which presents the preference percentages of the EV users for each bus with parking lots, considering the user price preference scenario, it can be seen that the parking lot in 11 was preferred among users with 49.34% of EV users choosing this lot. The parking lots located at bus 7 were in second place, while the fast charging parking lots had a total of around 19% preference among users, quite higher when compared with (b) of Figure 13. This means that in the dynamic EV charging price case, the most attractive prices were on the buses that had slow charging parking lots, leading to a great number of users choosing them over the fast charging parking lots. The majority of the users preferred slow charging due to the lower charging price (0.20 €/kWh).

Table 10. Maximum and minimum voltage magnitude for each case study.

| Case | User Price Preference Scenario | | | | User Distance Preference Scenario | | | |
| | Max Voltage | | Min Voltage | | Max Voltage | | Min Voltage | |
	bus	Value (p.u.)	bus	Value (p.u.)	bus	Value (p.u.)	bus	Value (p.u.)
A	2	0.9996	9	0.9819	2	0.9996	9	0.9819
B	7	0.9998	9	0.9844	7	0.9998	9	0.9844
C	7	0.9998	9	0.9844	7	0.9998	9	0.9844
D	2	0.9996	9	0.9814	2	0.9996	6	0.9690
E	2	0.9996	13	0.9761	2	0.9996	6	0.9688
F	2	0.9996	13	0.9761	2	0.9996	6	0.9688
G	2	0.9996	6	0.9685	2	0.9996	5	0.9623
H	7	0.9999	13	0.9826	7	0.9998	6	0.9692
I	7	0.9998	13	0.9763	7	0.9999	6	0.9690
J	7	0.9999	13	0.9763	7	0.9999	6	0.9690
K	7	0.9999	6	0.9687	7	0.9999	5	0.9624
L	7	0.9999	12	0.9832	7	0.9999	6	0.9710
M	7	0.9998	13	0.9763	7	0.9999	6	0.9690
N	7	0.9999	13	0.9763	7	0.9999	6	0.9690
O	7	0.9999	6	0.9687	7	0.9999	5	0.9624

Once again, in the user distance preference scenario, the fast charging parking lots were a more popular choice among users than in the user price preference scenario ((c) of Figure 14)). This also indicates that those parking lots could attract users who find themselves closer to the fast charging parking lots, if the price is not the most important factor. However, we arrived at the conclusion that those parking lost cannot be located optimally because the slow charging parking lots were highly preferred among users due to more attractive EV charging prices (even in the user distance preference scenario).

4.2. User Perspective

This subsection looks at the results of the case studies from the perspective of the EV users. Figures 15 and 16 present the box plots for the dynamic EV charging price cases considering the user price and distance preference scenarios, respectively. By comparing these two figures, it is possible to see that the differences between the same cases in each figure were small. The verified variations were mainly in Quartile 3 (Q3) and were higher in the user price preference scenario, in which the users gave priority to price.

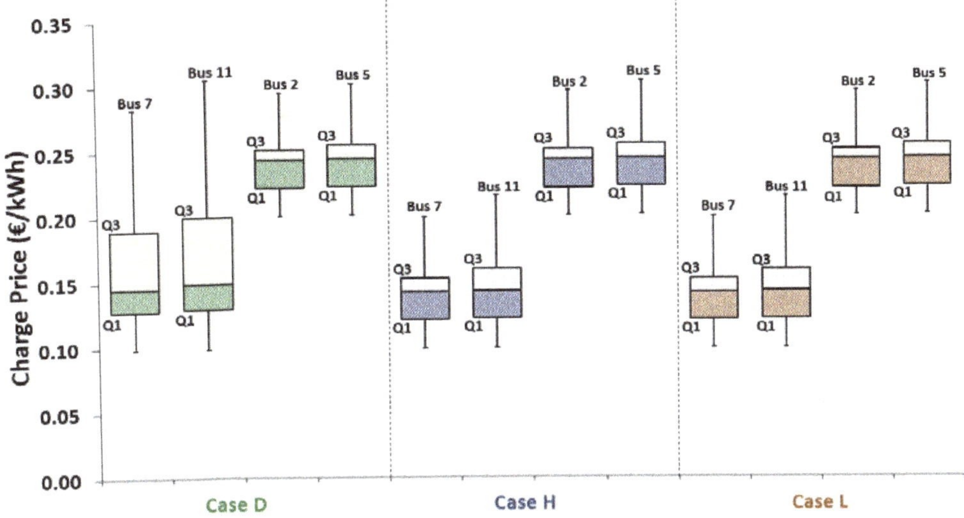

Figure 15. Electric vehicle charge price variation for the user price preference scenario.

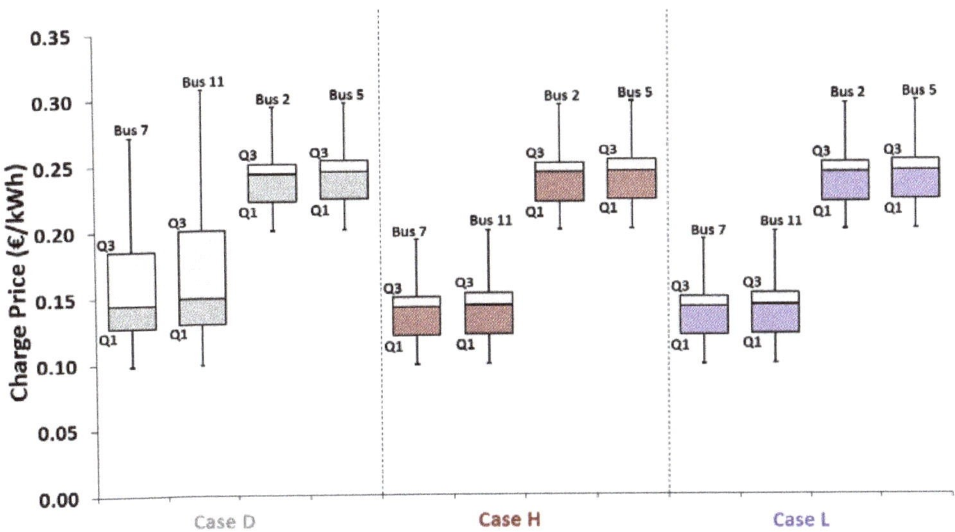

Figure 16. Electric vehicle charging price variation in the user distance preference scenario.

Let us take bus 11 in Case L as an example: it can be seen that the charge price variation in the user price preference scenario was between 0.0990 €/kWh and 0.2150 €/kWh, while in the user distance preference scenario, it was between 0.0990 €/kWh and 0.2000 €/kWh, corresponding to a 0.0150 €/kWh of difference. Fifty percent of the charge price values (interquartile range) were located between 0.1210 €/kWh and 0.1600 €/kWh for the user price preference scenario and between 0.1210 €/kWh and 0.1510 €/kWh for the user distance preference scenario (0.0090 €/kWh of difference). Twenty-five percent of the values varying between 0.0990 €/kWh and 0.1210 €/kWh for both user preference scenarios were located in the first quartile (Q1). Seventy five percent of the EV charging price values were represented by the third quartile (Q3) and varied between 0.0990 €/kWh and 0.1600 €/kWh in the user price preference scenario and between 0.0990 €/kWh and 0.1510 €/kWh in the user distance preference scenario. These two figures show that the highest variation in the charge prices among the dynamic EV charging price cases occurred specifically in slow charge buses in Case D. This is mainly due to the wind farms (one of them at bus 7 and the other one at bus 11, corresponding to 24% of the total installed power), which were not considered in Case D (it did not consider RES nor ESS).

Table 11 presents the results collected over a one-week period during which the case study was conducted; the average prices paid by EV users in the case of both dynamic EV charging prices and fixed charging prices. In these average prices values, the home charging price is included (0.2094 €/kWh). All dynamic EV charging price cases in the user price preference scenario show that the prices paid by EV users for EV charging were on average lower than what EV users normally would pay if the charging prices were fixed. However, this was not the case in the user distance preference scenario. To better understand the values presented in this table, let us analyze Tables 12–14, which stress the dynamic EV charging price cases with their homologous fixed price cases, presenting the gains in terms of the percentage of the EV users.

Table 11. Spent average charge price of the EV users for dynamic and fixed prices.

User Preference Scenario	Average Price (€/kWh)					
	Cases					
	D	H	L	SCh = 0.15 €/kWh FCh = 0.25 €/kWh	SCh = 0.20 €/kWh FCh = 0.30 €/kWh	SCh = 0.30 €/kWh FCh = 0.40 €/kWh
Price	0.1925	0.1877	0.1867	0.2005	0.2281	0.2907
Distance	0.2414	0.2180	0.2178	0.2087	0.2370	0.2955

In Table 12, it is possible to see that the dynamic EV charging price Case D for the user price preference scenario presented gains of 4.03%, 16.63%, and 33.79% over all the homologous fixed price cases (E, F, and G), respectively. Even comparing a dynamic EV charging price case that did not consider distributed resources with the lowest fixed prices case (0.15 €/kWh for slow charge and 0.25 €/kWh for fast charge) verified the charge prices' advantages. Regarding the user distance preference scenario, the dynamic EV charging price case did not present advantages in terms of charge price for the EV users when compared with fixed Cases E and F, which had 0.15 €/kWh for slow charge and 0.25 €/kWh for fast charge and 0.20 €/kWh for slow charge and 0.30 €/kWh for fast charge, respectively. Comparing with these two fixed prices cases, if the dynamic EV charging price were applied, the EV users would have had a loss of 15.66% and 1.88%, respectively, but obtained a gain of 18.30% when compared with the fixed charge price Case G.

Case H also presented charge price gains when compared with the homologous fixed charge prices, as can be seen in Table 13. In this case, the gains were 6.42%, 17.73%, and 35.45%, respectively, and when compared with Case D, it is possible to see a growth in those gains. For the user distance preference scenario, it can be seen that the dynamic EV charging prices were not also advantageous for the EV users when compared with the lowest considered fixed energy charge prices, but with a strong reduction when compared with the case that did not consider RES. Furthermore, a gain of 8%

can be seen with Case H (a growth of 9.88%) over the charge fixed energy price considered for Case J (0.20 €/kWh for slow charge and 0.30 €/kWh for fast charge), as well as a growth of 7.92% over the 0.30 €/kWh (slow charge) and 0.40 €/kWh (fast charge) fixed charge prices.

Table 12. Average charge price differences between dynamic Case D and the homologous fixed cases. The average prices paid by the EV users when Case D was used were 0.1925 €/kWh and 0.2414 €/kWh for the user price preference scenario and the user distance preference scenario, respectively. Blue color means that Case D is advantageous for the EV user, whereas the red color means that Case D is not advantageous for the EV user.

Dynamic Price	Fixed Prices		
	Case E	Case F	Case G
	Price preference		
Case D	4.03%	15.63%	33.79%
	Distance preference		
	−15.66%	−1.88%	18.30%

With the RES and ESS presented in the distribution network, the results' tendency was similar to Case H. Comparing the differences, it is possible to see through Table 14 that the gains of Case L were higher than the gains of Case H, namely due to the ESS consideration.

Table 13. Average charge price differences between dynamic Case H and the homologous fixed cases. The average prices paid by the EV users when Case H was used were 0.1877 €/kWh and 0.2180 €/kWh for the user price preference scenario and the user distance preference scenario, respectively. Blue color means that Case H was advantageous for the EV user, whereas the red color means that Case H was not advantageous for the EV user.

Dynamic Price	Fixed Prices		
	Case I	Case J	Case K
	Price preference		
Case H	6.42%	17.73%	35.45%
	Distance preference		
	−4.45%	8.00%	26.22%

Table 14. Average charge price differences between dynamic Case L and the homologous fixed cases. The average prices paid by the EV users when Case L was used were 0.1867 €/kWh and 0.2178 €/kWh for the user price preference scenario and the user distance preference scenario, respectively. Blue color means that Case L was advantageous for the EV user, whereas the red color means that Case L is not advantageous for the EV user.

Dynamic Price	Fixed Prices		
	Case M	Case N	Case O
	Price preference		
Case L	6.92%	18.17%	35.79%
	Distance preference		
	−4.33%	8.10%	26.30%

5. Conclusions

In this research work, the authors investigated if the dynamic EV charging prices have a positive impact on the smart distribution network operation and on the EV user behavior. To this end, the authors combined an EV behavior simulator with a proposed innovative smart DLMP-based distribution network operation/reconfiguration. The main contributions of the conducted study can be summarized as follows: (a) an EV user behavior simulator has been adopted to generate a realistic population, considering the network size and parking lots; (b) a distribution network operation/reconfiguration optimization model has been created in an SG context with high DER penetration concerning the behavior of the EV users and the dynamic EV charging price considering DLMPs using the Benders decomposition method; (c) the positive impact of the dynamic EV charging prices on the smart distribution network operation and on the electric vehicles users has been assessed.

The proposed methodology was tested in a case study, which has been conducted on a mock-up model of an SC located at the BISITE laboratory with a 13-bus distribution network. Furthermore, the distribution network operation/reconfiguration optimization model considering two user preference scenarios (price and distance preference) and using the dynamic EV charging prices were compared with the model using the EV fixed charging prices to demonstrate the advantage of the former.

It was verified that the use of dynamic pricing for EV charging is advantageous for the network operator in all of the considered cases due to reduced cost of operation and the user preference scenarios. These benefits are even more evident when considering high fixed charging prices (0.30 €/kWh for slow charging and 0.40 €/kWh for fast charging, −35.79% in the user price preference scenario, Case L). The lowest cost reduction was 0.24% in Case H of the distance preference scenario. Moreover, when the distance preference scenario and dynamic price were considered, it was verified that the PNS was zero, with exception of Case D, which presented an insignificant value (123.35 €).

For the EV users, the dynamic pricing also presented considerable cost advantages, namely when the price preference was considered. In this scenario, the lowest advantage (4.03% better) was verified in Case D compared with the lowest considered fixed charging prices (0.15 €/kWh for slow charge and 0.25 €/kWh for fast charge). Furthermore, for this scenario, the advantages can reach 35.75% (Case L), i.e., around 0.10 €/kWh of savings if the fixed charging prices are 0.30 €/kWh for slow charge and 0.40 €/kWh for fast charge. If the distance preference was considered, the dynamic EV charging price cases did not present savings in comparison with the lowest fixed charging price cases, namely when the fixed charging prices were 0.15 €/kWh for slow charge and 0.25 €/kWh for fast charge. Here, the user lost up to 15.66% for the dynamic EV charging price Case D. Nevertheless, the dynamic price still presented considerable savings when fixed prices were higher, reaching up to 26.30%.

The results suggest that the dynamic energy pricing for EVs' charge can be used as an efficient approach in smart cities that allows important monetary savings for both the distribution system operator and EV users.

The main drawbacks of the proposed work are: (a) the EV users' profiles were not adapted to the different weekdays; (b) the decision charge method was only based on the battery charge level; (c) vehicle-to-grid was not considered; (d) the ESS charge/discharge decision was limited and based on rules.

As future work, the authors suggest this research work include more EV user profiles, an additional charging decision method that depends on the energy price, an optimized ESS charge/discharge decision, an optimization model for EV users' costs minimization, solar-powered charging infrastructures in the parking lots, and also the possibility of vehicle-to-grid.

Author Contributions: Conceptualization, B.C. and J.S.; Data curation, B.C.; Formal analysis, B.C. and J.S.; Investigation, B.C.; Methodology, B.C. and J.S.; Project administration, Z.V.; Resources, Z.V. and J.M.C.; Software, B.C.; Supervision, Z.V. and J.M.C.; Validation, B.C.; Visualization, B.C. and J.S.; Writing—original draft, B.C.; Writing—review and editing, B.C., J.S., Z.V. and J.M.C.

Funding: This research has received funding from FEDER funds through the Operational Programme for Competitiveness and Internationalization (COMPETE 2020), under Project POCI-01-0145-FEDER-028983; and by National Funds through the FCT—Portuguese Foundation for Science and Technology, under Projects PTDC/EEI-EEE/28983/2017 (CENERGETIC), UID/EEA/00760/2019, and SFRH/BD/110678/2015 PhD scholarship.

Acknowledgments: This work has received funding from FEDER Funds through the COMPETE program, from National Funds through FCT under the project UID/EEA/00760/2019 and from the project PTDC/EEI-EEE/28983/2017-CENERGETIC. Bruno Canizes is supported by FCT Funds through the SFRH/BD/110678/2015 PhD scholarship.

Conflicts of Interest: The authors declare no conflict of interest.

Abbreviations

The following abbreviations are used in this manuscript:

DER	Distributed energy resources
DG	Distributed generators
DLMP	Locational marginal pricing
DN	Distribution network
DNR	Distribution network reconfiguration
DSO	Distribution system operator
ESS	Energy storage systems
EU	European Union
EV	Electric vehicle
FCh	Fast charge
LMP	Locational marginal pricing
MINLP	Mixed-integer nonlinear programming
MOC	Master subproblem objective function
MV	Medium voltage
OMIE	Iberian electricity market operator
PGC	Power generation curtailment
PL	Power losses
PNS	Power not supplied
PV	Photovoltaic
RES	Renewable energy sources
SC	Smart city
SCh	Slow charge
SG	Smart grid
VAT	Value-added tax

Indices

c	Line options
i	Electrical buses
j	Electrical buses
lo	Loads
bs	External supplier
cb	Capacitor bank
g	Distributed generator unit
e	Energy storage systems
v	Electric vehicles parking lot
m	Bender's cuts iteration

Parameters

P_{PGC}^{Cost}	Power generation curtailment cost [€/MW]
$price^{MK}$	Market price [€/MWh]
$Cost^{PNS}$	Power not supplied cost [€/MW]
$Cost^{Loss}$	Power losses cost [€/MW]
cef	Charge efficiency of energy storage systems
def	Discharge efficiency of energy storage systems
$Cost^{Cong}$	Lines congestion cost [€/MW]
$price^{Mk}$	Market price [€/MW]
$Cost^{PNS}$	Power not supplied cost [€/MW]
$r_{(i,j)}$	Resistance for i,j line [Ω]
$Cost^{Loss}$	Power losses cost [€/MW]
P_{PGC}^{Cost}	Power generation curtailment [€/MW]
$P_{DG(g)}$	Power generation for g DG unit [MW]
$EVP_{(i)}$	Power charge for EV parking lot in the bus i [MW]
$Flow^{max}{}_{(i,j)}$	Maximum admissible line flow between bus i and bus j [MW]
nDG	Number of DG units
$ExtSup_{MinLimit(bs)}$	Minimum limit of power supplied by substation/supplier bs [MW]
$ExtSup_{MaxLimit(bs)}$	Maximum limit of power supplied by substation/supplier bs [MW]
$P_{DG(g)}$	Generated power of distributed generation g [MW]
$P_{Load(lo)}$	Active power demand for load lo [MW]
$CongMin$	Power congestion factor
$STdchR_{(e)}$	ESS discharge rate [MW]
$STchR_{(e)}$	ESS charge rate [MW]
$STdM_{(e)}^{stat}$	Decision for ESS e discharge {0,1}
$STcap_{(e)}$	ESS e capacity [MWh]
$STdM_{(e)}^{stat}$	Decision for ESS e discharge {0,1}
Δt	Duration of the period [hours]
$STstoM_{(e)}^{t-1}$	Energy stored in e ESS in previous period for master subproblem [MWh]
$STstoS_{(e)}^{t-1}$	Energy stored in e ESS in previous period for slave subproblem [MWh]
$STsto_{(e)}^{min}$	Minimum capacity limit of the ESS e
$price_{min}^{Mk}$	Minimum market price value that will permit the ESS discharge
$\lambda_{(i,j)}^{m-1}$	Sensitivities associated to the radiality decision taken by the master problem in the previous iteration
$\mu_{(i)}^{m-1}$	Sensitivities associated to the ESS charge decision taken by the master problem in the previous iteration
$Cost^{Inf}$	Slave problem infeasibilities cost [€]
$V_{(i)}^{min}$	Minimum voltage magnitude limit in the bus i [V]
$V_{(i)}^{max}$	Maximum voltage magnitude limit in the bus i [V]
$\theta_{(i)}^{min}$	Minimum voltage angle limit in the bus i [rad]
$\theta_{(i)}^{max}$	Maximum voltage angle limit in the bus i [rad]
$Q_{Load(lo)}$	Reactive power demand for load lo [Mvar]
$Q_{Cbanks(cb)}^{max}$	Maximum limit of the capacitor bank cb [Mvar]
$G_{(i,j)}$	Real term of the element i,j in the bus admittance matrix
$B_{(i,j)}$	Imaginary term of the element i,j in the bus admittance matrix
$x_{(i,j)}$	Reactance for i,j line [Ω]
$Ps_{MinLimit(bs)}$	Minimum limit of active power supplied by substation/supplier bs [MW]
$Ps_{MaxLimit(bs)}$	Maximum limit of active power supplied by substation/supplier bs [MW]
$Qs_{MinLimit(bs)}$	Minimum limit of reactive power supplied by substation/supplier bs [Mvar]
$Qs_{MaxLimit(bs)}$	Maximum limit of reactive power supplied by substation/supplier bs [Mvar]

Variables

$D_{(g)}$	Fictitious load for each distributed generator g
$ExtSup_{(bs)}$	Power supplied by substation bs [MW]
$PNSM_{(lo)}$	Power not supplied for load lo in the master subproblem [MW]
$FlowM_{(i,j)}$	Power flow in the line i,j for the master subproblem [MW]
$P_{PGCM(g)}$	Power generation curtailment for master subproblem in the g DG unit [MW]
ω^*	Linear Benders' cut variable
$STdchM_{(e)}$	Power discharge of ESS e for master subproblem [MW]
$STchM_{(e)}$	Power charge of ESS e for master subproblem[MW]
$X^{stat}_{(i,j)}$	Binary decision variable $\{0,1\}$ for the line usage between bus i and bus j
$d_{(i,j)}$	Fictitious flow associated with branch i,j
$CongM_{(i,j)}$	Power congestion for line i,j in the master subproblem [MW]
$STcM^{stat}_{(e)}$	Binary decision variable $\{0,1\}$ for ESS e charge
$STdM^{stat}_{(e)}$	Binary decision variable $\{0,1\}$ for ESS e discharge
$STstoM_{(e)}$	Energy stored in e ESS for master subproblem [MWh]
Z^{m-1}_{up}	Sum of the infeasibilities of the slave problem
ZA	Slack variable for active power balance
ZQ	Slack variable for reactive power balance
ZF	Slack variable for thermal lines capacity
$CongS_{(i,j)}$	Power congestion for line i,j in the salve subproblem [MW]
$P_{Supplier(bs)}$	Active power supplied by substation bs[MW]
$Q_{Supplier(bs)}$	Reactive power supplied by substation bs[Mvar]
$P_{PGCs(g)}$	Power generation curtailment for slave subproblem in the g DG unit [MW]
$PNSs_{(lo)}$	Power not supplied for slave subproblem in the load lo [MW]
$SLoss_{(i,j)}$	Apparent power loss in the line i,j [MVA]
$V_{(i)}$	Voltage magnitude in the bus i [V]
$\theta_{(i)}$	Voltage angle in the bus i [rad]
$P_{Inj(i)}$	Active injected power in the bus i [MW]
$Q_{Inj(i)}$	Reactive injected power in the bus i [Mvar]
$Q_{Cbanks(cb)}$	Reactive power from capacitor bank cb [Mvar]
$P_{(i,j)}$	Active power flow in the i,j line [MW]
$Q_{(i,j)}$	Reactive power flow in the i,j line [Mvar]
$S_{(i,j)}$	Apparent power flow in the i,j line [MVA]
$PLoss_{(i,j)}$	Active power loss in the i,j line [MW]
$QLoss_{(i,j)}$	Reactive power loss in the i,j line [Mvar]
$FlowS_{(i,j)}$	Power flow in the i,j line for slave subproblem [MW]
$STdchS_{(e)}$	Power discharge of ESS e for slave subproblem [MW]
$STchS_{(e)}$	Power charge of ESS e for slave subproblem[MW]
$STstoS_{(e)}$	Energy stored in e ESS for slave subproblem [MWh]
DEP	Dynamic EV charging price for each period [€/kWh]
$TariffMV$	Energy tariff price for each period [€/kWh]
PLG	Additional profit margin of the parking owner
$ACNR$	Additional cost related to the fixed term of network price rate to be charged to the customer [€/kWh]

Sets

Ω_B	Set of buses
$\Omega^b{}_{BS}$	Set of substation buses
$\Omega^b{}_{CB}$	Set of capacitor banks buses
$\Omega^b{}_L$	Set of load buses
$\Omega^b{}_E$	Set of ESS buses
$\Omega^b{}_V$	Set of EV parking lot buses
$\Omega_B S$	Set of substations
$\Omega_C B$	Set of capacitor banks
Ω_l	Set of lines
$\Omega^{nd}{}_{DG}$	Set of non-dispatchable DG buses

References

1. Mokryani, G.; Hu, Y.F.; Papadopoulos, P.; Niknam, T.; Aghaei, J. Deterministic approach for active distribution networks planning with high penetration of wind and solar power. *Renew. Energy* **2017**. doi:10.1016/j.renene.2017.06.074. [CrossRef]
2. The European Union Leading in Renewables. Available online: https://ec.europa.eu/energy/sites/ener/files/documents/cop21-brochure-web.pdf (accessed on 8 February 2019).
3. Paaso, A.; Kushner, D.; Bahramirad, S.; Khodaei, A. Grid Modernization Is Paving the Way for Building Smarter Cities [Technology Leaders]. *IEEE Electr. Mag.* **2018**, *6*, 6–108. doi:10.1109/MELE.2018.2816848. [CrossRef]
4. Curiale, M. From smart grids to smart city. In Proceedings of the IEEE 2014 Saudi Arabia Smart Grid Conference (SASG), Jeddah, Saudi Arabia, 14–17 December 2014; pp. 1–9. doi:10.1109/SASG.2014.7274280. [CrossRef]
5. Tuballa, M.L.; Abundo, M.L. A review of the development of Smart Grid technologies. *Renew. Sustain. Energy Rev.* **2016**, *59*, 710–725. doi:10.1016/j.rser.2016.01.011. [CrossRef]
6. Kazmi, S.A.A.; Shahzad, M.K.; Khan, A.Z.; Shin, D.R. Smart Distribution Networks: A Review of Modern Distribution Concepts from a Planning Perspective. *Energies* **2017**, *10*, 501. doi:10.3390/en10040501. [CrossRef]
7. Blesl, M.; Das, A.; Fahl, U.; Remme, U. Role of energy efficiency standards in reducing CO2 emissions in Germany: An assessment with TIMES. *Energy Policy* **2007**. doi:10.1016/j.enpol.2006.05.013. [CrossRef]
8. Erickson, L.E.; Robinson, J.; Brase, G.; Cutsor, J. *Solar Powered Charging Infrastructure for Electric Vehicles: A Sustainable Development*; CRC Press: Boca Raton, FL, USA, 2016.
9. Alam, M.M.; Mekhilef, S.; Seyedmahmoudian, M.; Horan, B. Dynamic Charging of Electric Vehicle with Negligible Power Transfer Fluctuation. *Energies* **2017**, *10*, 1–20, doi:10.3390/en10050701. [CrossRef]
10. Foley, A.M.; Winning, I.; O Gallachoir, B.; Ieee, M. State-of-the-art in electric vehicle charging infrastructure. In Proceedings of the IEEE Vehicle Power and Propulsion Conference, Lille, France, 1–3 September 2010. doi:10.1109/VPPC.2010.5729014.
11. Heydt, G.T. The Impact of Electric Vehicle Deployment on Load Management Strategies. *IEEE Power Eng. Rev.* **1983**. doi:10.1109/MPER.1983.5519161. [CrossRef]
12. Rahman, S.; Shrestha, G.B. An investigation into the impact of electric vehicle load on the electric utility distribution system. *IEEE Trans. Power Deliv.* **1993**. doi:10.1109/61.216865. [CrossRef]
13. Salihi, J.T. Energy Requirements for Electric Cars and Their Impact on Electric Power Generation and Distribution Systems. *IEEE Trans. Ind. Appl.* **1973**, *IA-9*, 516–532. doi:10.1109/TIA.1973.349925. [CrossRef]
14. Geske, M.; Komarnicki, P.; Stötzer, M.; Styczynski, Z.A. Modeling and simulation of electric car penetration in the distribution power system—Case study. In Proceedings of the 2010 Modern Electric Power Systems, Wroclaw, Poland, 20–22 September 2010; pp. 1–6.
15. Juanuwattanakul, P.; Masoum, M.A. Identification of the weakest buses in unbalanced multiphase smart grids with plug-in electric vehicle charging stations. In Proceedings of the 2011 IEEE PES Innovative Smart Grid Technologies, ISGT Asia 2011 Conference: Smarter Grid for Sustainable and Affordable Energy Future, Perth, WA, Australia, 13–16 November 2011. doi:10.1109/ISGT-Asia.2011.6167155.
16. Zhang, C.; Chen, C.; Sun, J.; Zheng, P.; Lin, X.; Bo, Z. Impacts of Electric Vehicles on the Transient Voltage Stability of Distribution Network and the Study of Improvement Measures. In Proceedings of the IEEE PES Asia-Pacific Power and Energy Engineering Conference, Hong Kong, China, 7–10 December 2014. doi:10.1109/APPEEC.2014.7066085.
17. Dharmakeerthi, C.H.; Mithulananthan, N.; Saha, T.K. Impact of electric vehicle fast charging on power system voltage stability. *Int. J. Electr. Power Energy Syst.* **2014**. doi:10.1016/j.ijepes.2013.12.005. [CrossRef]
18. Ul-Haq, A.; Cecati, C.; Strunz, K.; Abbasi, E. Impact of Electric Vehicle Charging on Voltage Unbalance in an Urban Distribution Network. *Intell. Ind. Syst.* **2015**, *1*, 51–60. doi:10.1007/s40903-015-0005-x. [CrossRef]
19. de Hoog, J.; Muenzel, V.; Jayasuriya, D.C.; Alpcan, T.; Brazil, M.; Thomas, D.A.; Mareels, I.; Dahlenburg, G.; Jegatheesan, R. The importance of spatial distribution when analysing the impact of electric vehicles on voltage stability in distribution networks. *Energy Syst.* **2014**. doi:10.1007/s12667-014-0122-8. [CrossRef]

20. Zhang, Y.; Song, X.; Gao, F.; Li, J. Research of voltage stability analysis method in distribution power system with plug-in electric vehicle. In Proceedings of the 2016 IEEE PES Asia-Pacific Power and Energy Engineering Conference (APPEEC), Xi'an, China, 25–28 October 2016; pp. 1501–1507. doi:10.1109/APPEEC.2016.7779740. [CrossRef]

21. Staats, P.T.; Grady, W.M.; Arapostathis, A.; Thallam, R.S. A statistical analysis of the effect of electric vehicle battery charging on distribution system harmonic voltages. *IEEE Trans. Power Deliv.* **1998**. doi:10.1109/61.660951. [CrossRef]

22. Gómez, J.C.; Morcos, M.M. Impact of EV battery chargers on the power quality of distribution systems. *IEEE Trans. Power Deliv.* **2003**. doi:10.1109/TPWRD.2003.813873. [CrossRef]

23. Basu, M.; Gaughan, K.; Coyle, E. Harmonic distortion caused by EV battery chargers in the distribution systems network and its remedy. In Proceedings of the 39th International Universities Power Engineering Conferences, Bristol, UK, 6–8 September 2004.

24. Jiang, C.; Torquato, R.; Salles, D.; Xu, W. Method to assess the power-quality impact of plug-in electric vehicles. *IEEE Trans. Power Deliv.* **2014**. doi:10.1109/TPWRD.2013.2283598. [CrossRef]

25. McCarthy, D.; Wolfs, P. The HV system impacts of large scale electric vehicle deployments in a metropolitan area. In Proceedings of the 2010 20th Australasian Universities Power Engineering Conference, Christchurch, New Zealand, 5–8 December 2010.

26. Putrus, G.; Suwanapingkarl, P.; Johnston, D.; Bentley, E.; Narayana, M. Impact of electric vehicles on power distribution networks. In Proceedings of the 2009 IEEE Vehicle Power and Propulsion Conference, Dearborn, MI, USA, 7–10 September 2009. doi:10.1109/VPPC.2009.5289760.

27. Fan, Y.; Guo, C.; Hou, P.; Tang, Z. Impact of Electric Vehicle Charging on Power Load Based on TOU Price. *Energy Power Eng.* **2013**, *05*, 1347–1351. doi:10.4236/epe.2013.54B255. [CrossRef]

28. Di Silvestre, M.L.; Riva Sanseverino, E.; Zizzo, G.; Graditi, G. An optimization approach for efficient management of EV parking lots with batteries recharging facilities. *J. Ambient Intell. Hum. Comput.* **2013**. doi:10.1007/s12652-013-0174-y. [CrossRef]

29. Sehar, F.; Pipattanasomporn, M.; Rahman, S. Demand management to mitigate impacts of plug-in electric vehicle fast charge in buildings with renewables. *Energy* **2017**. doi:10.1016/j.energy.2016.11.118. [CrossRef]

30. Hüls, J.; Remke, A. Coordinated charging strategies for plug-in electric vehicles to ensure a robust charging process. In Proceedings of the ACM 10th EAI International Conference on Performance Evaluation Methodologies and Tools, Taormina, Italy, 25–28 October 2017. doi:10.4108/eai.25-10-2016.2266997.

31. Biegel, B.; Andersen, P.; Stoustrup, J.; Bendtsen, J. Congestion Management in a Smart Grid via Shadow Prices. *IFAC Proc. Volumes* **2012**. doi:10.3182/20120902-4-FR-2032.00091. [CrossRef]

32. Bohn, R.E.; Caramanis, M.C.; Schweppe, F.C. Optimal Pricing in Electrical Networks over Space and Time. *RAND J. Econ.* **1984**. doi:10.2307/2555444. [CrossRef]

33. Sotkiewicz, P.M.; Vignolo, J.M. Nodal pricing for distribution networks: Efficient pricing for efficiency enhancing DG. *IEEE Trans. Power Syst.* **2006**. doi:10.1109/TPWRS.2006.873006. [CrossRef]

34. Singh, R.K.; Goswami, S.K. Optimum allocation of distributed generations based on nodal pricing for profit, loss reduction, and voltage improvement including voltage rise issue. *Int. J. Electr. Power Energy Syst.* **2010**. doi:10.1016/j.ijepes.2009.11.021. [CrossRef]

35. Meng, F.; Chowdhury, B.H. Distribution LMP-based economic operation for future Smart Grid. In Proceedings of the 2011 IEEE Power and Energy Conference at Illinois, PECI 2011, Champaign, IL, USA, 25–26 February 2011. doi:10.1109/PECI.2011.5740485.

36. Heydt, G.T.; Chowdhury, B.H.; Crow, M.L.; Haughton, D.; Kiefer, B.D.; Meng, F.; Sathyanarayana, B.R. Pricing and control in the next generation power distribution system. *IEEE Trans. Smart Grid* **2012**. doi:10.1109/TSG.2012.2192298. [CrossRef]

37. Shaloudegi, K.; Madinehi, N.; Hosseinian, S.H.; Abyaneh, H.A. A novel policy for locational marginal price calculation in distribution systems based on loss reduction allocation using game theory. *IEEE Trans. Power Syst.* **2012**. doi:10.1109/TPWRS.2011.2175254. [CrossRef]

38. O'Connell, N.; Wu, Q.; Østergaard, J.; Nielsen, A.H.; Cha, S.T.; Ding, Y. Day-ahead tariffs for the alleviation of distribution grid congestion from electric vehicles. *Electric Power Syst. Res.* **2012**. doi:10.1016/j.epsr.2012.05.018. [CrossRef]

39. Li, R.; Wu, Q.; Oren, S.S. Distribution Locational Marginal Pricing for Optimal Electric Vehicle Charging Management. *IEEE Trans. Power Syst.* **2014**, *29*, 203–211. doi:10.1109/TPWRS.2013.2278952. [CrossRef]

40. Liu, W.; Wu, Q.; Wen, F.; Ostergaard, J. Day-ahead congestion management in distribution systems through household demand response and distribution congestion prices. *IEEE Trans. Smart Grid* **2014**. doi:10.1109/TSG.2014.2336093. [CrossRef]

41. Huang, S.; Wu, Q.; Oren, S.S.; Li, R.; Liu, Z. Distribution Locational Marginal Pricing Through Quadratic Programming for Congestion Management in Distribution Networks. *IEEE Trans. Power Syst.* **2015**. doi:10.1109/TPWRS.2014.2359977. [CrossRef]

42. Clement-Nyns, K.; Haesen, E.; Driesen, J. The impact of Charging plug-in hybrid electric vehicles on a residential distribution grid. *IEEE Trans. Power Syst.* **2010**. doi:10.1109/TPWRS.2009.2036481. [CrossRef]

43. Leemput, N.; Geth, F.; Van Roy, J.; Delnooz, A.; Buscher, J.; Driesen, J. Impact of electric vehicle on-board single-phase charging strategies on a flemish residential grid. *IEEE Trans. Smart Grid* **2014**. doi:10.1109/TSG.2014.2307897. [CrossRef]

44. Rezaee, S.; Farjah, E.; Khorramdel, B. Probabilistic analysis of plug-in electric vehicles impact on electrical grid through homes and parking lots. *IEEE Trans. Sustain. Energy* **2013**. doi:10.1109/TSTE.2013.2264498. [CrossRef]

45. Bin Humayd, A.S.; Bhattacharya, K. Assessment of distribution system margins to accommodate the penetration of plug-in electric vehicles. In Proceedings of the 2015 IEEE Transportation Electrification Conference and Expo, ITEC 2015, Dearborn, MI, USA, 14–17 June 2015. doi:10.1109/ITEC.2015.7165764.

46. Möller, F.; Meyer, J.; Schegner, P. Load model of electric vehicles chargers for load flow and unbalance studies. In Proceedings of the 9th International: 2014 Electric Power Quality and Supply Reliability Conference, PQ 2014—Proceedings, Rakvere, Estonia, 11–13 June 2014. doi:10.1109/PQ.2014.6866774.

47. Min, Z.; Qiuyu, C.; Jiajia, X.; Weiwei, Y.; Shu, N. Study on influence of large-scale electric vehicle charging and discharging load on distribution system. In Proceedings of the IEEE 2016 China International Conference on Electricity Distribution (CICED), Xi'an, China, 10–13 August 2016; pp. 1–4. doi:10.1109/CICED.2016.7576208. [CrossRef]

48. Sarker, M.R.; Ortega-Vazquez, M.A.; Kirschen, D.S. Optimal Coordination and Scheduling of Demand Response via Monetary Incentives. *IEEE Trans. Smart Grid* **2015**. doi:10.1109/TSG.2014.2375067. [CrossRef]

49. Mou, Y.; Xing, H.; Lin, Z.; Fu, M. Decentralized optimal demand-side management for PHEV charging in a smart grid. *IEEE Trans. Smart Grid* **2015**. doi:10.1109/TSG.2014.2363096. [CrossRef]

50. Zou, S.; Ma, Z.; Liu, X.; Hiskens, I. An Efficient Game for Coordinating Electric Vehicle Charging. *IEEE Trans. Autom. Control* **2017**. doi:10.1109/TAC.2016.2614106. [CrossRef]

51. Veldman, E.; Verzijlbergh, R.A. Distribution grid impacts of smart electric vehicle charging from different perspectives. *IEEE Trans. Smart Grid* **2015**. doi:10.1109/TSG.2014.2355494. [CrossRef]

52. Li, C.; Schaltz, E.; Vasquez, J.C.; Guerrero, J.M. Distributed coordination of electric vehicle charging in a community microgrid considering real-time price. In Proceedings of the 2016 18th European Conference on Power Electronics and Applications, EPE 2016 ECCE Europe, Karlsruhe, Germany, 5–9 September 2016. doi:10.1109/EPE.2016.7695693.

53. Akbari, H.; Fernando, X. Futuristic model of Electric Vehicle charging queues. In Proceedings of the IEEE 2016 3rd International Conference on Signal Processing and Integrated Networks (SPIN), Noida, India, 11–12 February 2016; pp. 789–794. doi:10.1109/SPIN.2016.7566721. [CrossRef]

54. Wang, R.; Wang, P.; Xiao, G. Two-Stage Mechanism for Massive Electric Vehicle Charging Involving Renewable Energy. *IEEE Trans. Veh. Technol.* **2016**. doi:10.1109/TVT.2016.2523256. [CrossRef]

55. Nguyen, H.N.T.; Zhang, C.; Zhang, J. Dynamic Demand Control of Electric Vehicles to Support Power Grid With High Penetration Level of Renewable Energy. *IEEE Trans. Transp. Electr.* **2016**, *2*, 66–75. doi:10.1109/TTE.2016.2519821. [CrossRef]

56. Jin, C.; Sheng, X.; Ghosh, P. Optimized Electric Vehicle Charging With Intermittent Renewable Energy Sources. *IEEE J. Sel. Top. Signal Process.* **2014**, *8*, 1063–1072. doi:10.1109/JSTSP.2014.2336624. [CrossRef]

57. Schuelke, A.; Erickson, K. The potential for compensating wind fluctuations with residential load shifting of electric vehicles. In Proceedings of the 2011 IEEE International Conference on Smart Grid Communications, SmartGridComm 2011, Brussels, Belgium, 17–20 October 2011. doi:10.1109/SmartGridComm.2011.6102342.

58. Sourkounis, C.; Einwachter, F. Smart charge management of electric vehicles in decentralized power supply systems. In Proceedings of the IEEE 11th International Conference on Electrical Power Quality and Utilisation, Lisbon, Portugal, 17–19 October 2011; pp. 1–6. doi:10.1109/EPQU.2011.6128863. [CrossRef]

59. Xiong, Y.; Chu, C.C.; Gadh, R.; Wang, B. Distributed optimal vehicle grid integration strategy with user behavior prediction. In Proceedings of the 2017 IEEE Power & Energy Society General Meeting, Chicago, IL, USA, 16–20 July 2017; pp. 1–5. doi:10.1109/PESGM.2017.8274327. [CrossRef]

60. Gan, L.; Topcu, U.; Low, S.H. Optimal decentralized protocol for electric vehicle charging. *IEEE Trans. Power Syst.* **2013**. doi:10.1109/TPWRS.2012.2210288. [CrossRef]

61. Wang, B.; Wang, Y.; Qiu, C.; Chu, C.C.; Gadh, R. Event-based electric vehicle scheduling considering random user behaviors. In Proceedings of the 2015 IEEE International Conference on Smart Grid Communications, SmartGridComm 2015, Miami, FL, USA, 2–5 November 2016. doi:10.1109/SmartGridComm.2015.7436319.

62. Cao, Y.; Tang, S.; Li, C.; Zhang, P.; Tan, Y.; Zhang, Z.; Li, J. An optimized EV charging model considering TOU price and SOC curve. *IEEE Trans. Smart Grid* **2012**. doi:10.1109/TSG.2011.2159630. [CrossRef]

63. Franke, T.; Krems, J.F. Interacting with limited mobility resources: Psychological range levels in electric vehicle use. *Transp. Res. Part A Policy Pract.* **2013**, *48*, 109–122. doi:10.1016/j.tra.2012.10.010. [CrossRef]

64. Franke, T.; Bühler, F.; Cocron, P.; Neumann, I.; Krems, J.F. Enhancing Sustainability of Electric Vehicles: A Field Study Approach to Understanding User Acceptance and Behavior. Available online: https://www.tu-chemnitz.de/hsw/psychologie/professuren/allpsy1/pdf/Franke_et_al._2012_SustainEV.pdf (accessed on 8 February 2019).

65. Franke, T.; Neumann, I.; Bühler, F.; Cocron, P.; Krems, J.F. Experiencing Range in an Electric Vehicle: Understanding Psychological Barriers. *Appl. Psychol.* **2012**, *61*, 368–391. doi:10.1111/j.1464-0597.2011.00474.x. [CrossRef]

66. Franke, T.; Krems, J.F. Understanding charging behavior of electric vehicle users. *Transp. Res. Part F Traffic Psychol. Behav.* **2013**, *21*, 75–89. doi:10.1016/j.trf.2013.09.002. [CrossRef]

67. Lu, X.; Chen, B.; Chen, C.; Wang, J. Coupled Cyber and Physical Systems: Embracing Smart Cities with Multistream Data Flow. *IEEE Electr. Mag.* **2018**, *6*, 73–83. doi:10.1109/MELE.2018.2816845. [CrossRef]

68. Hung, D.Q.; Mithulananthan, N. Loss reduction and loadability enhancement with DG: A dual-index analytical approach. *Appl. Energy* **2014**, *115*, 233–241. doi:10.1016/j.apenergy.2013.11.010. [CrossRef]

69. Santos, S.F.; Fitiwi, D.Z.; Cruz, M.R.; Cabrita, C.M.; Catalão, J.P. Impacts of optimal energy storage deployment and network reconfiguration on renewable integration level in distribution systems. *Appl. Energy* **2017**, *185*, 44–55. doi:10.1016/j.apenergy.2016.10.053. [CrossRef]

70. Song, I.K.; Jung, W.W.; Kim, J.Y.; Yun, S.Y.; Choi, J.H.; Ahn, S.J. Operation Schemes of Smart Distribution Networks With Distributed Energy Resources for Loss Reduction and Service Restoration. *IEEE Trans. Smart Grid* **2013**, *4*, 367–374. doi:10.1109/TSG.2012.2233770. [CrossRef]

71. Hung, D.Q.; Mithulananthan, N.; Bansal, R. A combined practical approach for distribution system loss reduction. *Int. J. Ambient Energy* **2015**, *36*, 123–131. doi:10.1080/01430750.2013.829784. [CrossRef]

72. Dorostkar-Ghamsari, M.R.; Fotuhi-Firuzabad, M.; Lehtonen, M.; Safdarian, A. Value of Distribution Network Reconfiguration in Presence of Renewable Energy Resources. *IEEE Trans. Power Syst.* **2016**, *31*, 1879–1888. doi:10.1109/TPWRS.2015.2457954. [CrossRef]

73. Hung, D.Q.; Mithulananthan, N.; Bansal, R. An optimal investment planning framework for multiple distributed generation units in industrial distribution systems. *Appl. Energy* **2014**, *124*, 62–72. doi:10.1016/j.apenergy.2014.03.005. [CrossRef]

74. Munoz-Delgado, G.; Contreras, J.; Arroyo, J.M. Joint Expansion Planning of Distributed Generation and Distribution Networks. *IEEE Trans. Power Syst.* **2015**, *30*, 2579–2590. doi:10.1109/TPWRS.2014.2364960. [CrossRef]

75. Capitanescu, F.; Ochoa, L.F.; Margossian, H.; Hatziargyriou, N.D. Assessing the Potential of Network Reconfiguration to Improve Distributed Generation Hosting Capacity in Active Distribution Systems. *IEEE Trans. Power Syst.* **2015**, *30*, 346–356. doi:10.1109/TPWRS.2014.2320895. [CrossRef]

76. Wu, Y.K.; Lee, C.Y.; Liu, L.C.; Tsai, S.H. Study of Reconfiguration for the Distribution System With Distributed Generators. *IEEE Trans. Power Deliv.* **2010**, *25*, 1678–1685. doi:10.1109/TPWRD.2010.2046339. [CrossRef]

77. Abu-Mouti, F.S.; El-Hawary, M.E. Optimal Distributed Generation Allocation and Sizing in Distribution Systems via Artificial Bee Colony Algorithm. *IEEE Trans. Power Deliv.* **2011**, *26*, 2090–2101. doi:10.1109/TPWRD.2011.2158246. [CrossRef]

78. Lueken, C.; Carvalho, P.M.; Apt, J. Distribution grid reconfiguration reduces power losses and helps integrate renewables. *Energy Policy* **2012**, *48*, 260–273. doi:10.1016/j.enpol.2012.05.023. [CrossRef]

79. Esmaeilian, H.R.; Fadaeinedjad, R. Energy Loss Minimization in Distribution Systems Utilizing an Enhanced Reconfiguration Method Integrating Distributed Generation. *IEEE Syst. J.* **2015**, *9*, 1430–1439. doi:10.1109/JSYST.2014.2341579. [CrossRef]

80. Conejo, A.J.; Carrión, M.; Morales, J.M. *Decision Making Under Uncertainty in Electricity Markets*; International Series in Operations Research & Management Science; Springer US: Boston, MA, USA, 2010; Volume 153. doi:10.1007/978-1-4419-7421-1.

81. Soares, J.; Canizes, B.; Ghazvini, M.A.F.; Vale, Z.; Venayagamoorthy, G.K. Two-Stage Stochastic Model Using Benders' Decomposition for Large-Scale Energy Resource Management in Smart Grids. *IEEE Trans. Ind. Appl.* **2017**, *53*, 5905–5914. doi:10.1109/TIA.2017.2723339. [CrossRef]

82. Benders, J.F. Partitioning procedures for solving mixed-variables programming problems. *Numerische Mathematik* **1962**, *4*, 238–252. doi:10.1007/BF01386316. [CrossRef]

83. Conejo, A.; Castillo, E.; Minguez, R.; Garcia-Bertrand, R. *Decomposition Techniques in Mathematical Programming*; Springer: Berlin/Heidelberg, Germany, 2006. doi:10.1007/3-540-27686-6.

84. Canizes, B.; Soares, J. Data 13 bus Distribution Network. 2019. Available online: https://doi.org/10.5281/zenodo.2556871 (accessed on 8 February 2019).

Article

Optimal Charging Navigation Strategy Design for Rapid Charging Electric Vehicles

Wangyi Mo, Chao Yang *, Xin Chen, Kangjie Lin and Shuaiqi Duan

School of Automation, Guangdong University of Technology, Guangzhou 510006, China;
2111704310@mail2.gdut.edu.cn (W.M.); xinchen@gdut.edu.cn (X.C.); KongJetLin@163.com (K.L.);
Duanshuaiqi96@163.com (S.D.)
*   Correspondence: yangchaoscut@aliyun.com; Tel.: +20-39322946

Received: 12 January 2019; Accepted: 7 March 2019; Published: 13 March 2019

Abstract: Electric vehicles (EVs) have become an efficient solution to making a transportation system environmentally friendly. However, as the number of EVs grows, the power demand from charging vehicles increases greatly. An unordered charging strategy for huge EVs affects the stability of a local power grid, especially during peak times. It becomes serious under the rapid charging mode, in which the EVs will be charged fully within a shorter time. In contrast to regular charging, the power quality (e.g.,voltages deviation, harmonic distortion) is affected when multiple EVs perform rapid charging at the same station simultaneously. To reduce the impacts on a power grid system caused by rapid charging, we propose an optimal EV rapid charging navigation strategy based on the internet of things network. The rapid charging price is designed based on the charging power regulation scheme. Both power grid operation and real-time traffic information are considered. The formulated objective of the navigation strategy is proposed to minimize the synthetic costs of EVs, including the traveling time and the charging costs. Simulation results demonstrate the effectiveness of the proposed strategy.

Keywords: electric vehicle; rapid charging; charging navigation; internet of things

1. Introduction

With the increasing concern about environmental protection and the energy supply problems, more attention has been focused on the development of electric vehicles (EVs) [1]. In contrast to the normal gasoline/diesel powered vehicles, EVs are considered as a kind of zero-emission transportation. Meanwhile, EVs can function as moving electric storage equipment to help the power balance between the supply and demand sides [2–4]. The vehicle performances are improved via realizing the advantages of environmental protection and energy conservation. However, as the number of EVs grows, the power charging demand increases greatly. The larger-scale and disorganized charging strategies cause several serious problems for the local power grid. The charging terminals will be overloaded and the performances of power grids, including the efficiency, stability and reliability will be directly affected [5,6]. Along with the development of internet of things (IoT) technologies, how to design an optimal and efficient charging strategy for EVs becomes a critical problem.

Normally, the existing EV charging modes can be classified into: regular (slow) charging, rapid charging and battery switching [7,8]. In the regular charging mode, the charging power is low. It takes a longer time for EVs to be fully charged, which can extend EVs' battery life and reduce the impact of charging behaviors on the local power grid. Since the charging process is slow, this charging mode requires a long parking time and the charging stations are often located in the large public/commercial parking lots or residential underground garages.

Given the limited capacity of EV battery, the drivers who have a long-distance trip should select a fast charging mode to refuel their vehicles. Compared with regular charging, the charging time durations of EVs under the rapid charging and the battery switching modes are reduced. As for the battery switching mode, the battery parameters and interface standards are different for the various EV types and manufacturers. Under the battery switching mode, several professional and particular EV battery switching stations are needed to support the EV traffic [8,9]. It is suitable for public transport buses, and it has a limited influence on the power grid. Besides battery switching, rapid charging is another faster charging mode. Under this charging mode, the charging process will be completed in the charging stations with high power level. It is more convenient and flexible, so the normal drivers prefer this mode to recharge their EVs to continue driving more quickly.

However, rapid charging may pose threats to the power grid system. A large number of disordered and random rapid charging behaviors will cause negative effects on the local power grid, especially during the peak periods. Compared with regular charging, excessive rapid charging loads on the electric power distribution network can cause more serious problems, such as voltages deviation, overload of network components (e.g., cables or transformers), and the increase of the harmonic distortion level of the local power grid [10–12].

To reduce the EV charging impacts on the power grid, many studies have been carried out, focusing on the charging scheduling design to flatten the peak loads [13,14], via the help of the real time IoT technologies, such as vehicle-to-grid (V2G) communication, crowd sensing. Ref. [15] proposes a coordinated architecture to shift the charging loads by dynamic price regulating. Furthermore, in [16], a fair energy scheduling scheme is proposed to control the charging loads of EVs. By coordinating both the charging and discharging behaviors, an optimal charging approach is presented in [17] to achieve peak shaving and minimize the charging power losses. Most of these studies mentioned are based on the regular charging mode, the main objectives of these scheduling schemes are to change the charging behaviors of EV drivers. However, when urgent charging demands arrive for the moving EVs on roads, adjusting the start time of charging reduces the drivers' satisfaction sharply. When the EVs are driving on the road outside, the rapid/faster charging modes are needed. It is necessary to develop a charging navigation strategy for rapid charging EVs.

In the literature about charging navigation, Ref. [18] analyzes the EV navigation problem while the traveling cost is minimized. Ref. [19] formulates the charging-scheduling problem and proposes an optimal method to reduce the total charging time of EVs. In terms of navigation in transport sectors, some necessary navigation applications (e.g., GPS) are installed to guide the EVs to the charging stations. However, the mentioned navigation strategies [18,19] do not consider the power system information. Then, the power system performance will be reduced when the mentioned strategies are utilized for the rapid charging EVs directly. In detail, as the power demand for rapid charging increases, it causes charging congestion when a large number of EVs arrive at one rapid charging station at the same time. In China, since the rapid charging stations are generally built in bustling downtown areas, such behaviors may lead to traffic congestion around rapid charging stations as well as the power overload of these areas. Likewise, heavy traffic situations also greatly affect the charging of EVs, including the traveling time on the roads and the waiting time at the charging stations. An integrated rapid-charging navigation strategy is proposed in [20], that considers both the traffic condition and power grid status. It develops a strategy considering the total time for charging, however, the charging expenses are neglected. Ref. [15] formulates an optimized charging model to minimize the charging cost in response to time-of-use (TOU) price in a regulated market. Even in the same time period, the charging stations in different areas can adjust their rapid charging prices according to their operation status. Adjusting the charging prices in real time becomes an efficient method to control the number of charging EVs. Therefore, besides considering the power grid and traffic system constraints, it is necessary to satisfy the EVs' demands with the minimal synthetic cost (i.e., traveling time and charging expenses) under the rapid charging mode, via the deployed IoT network in both the traffic and power grid systems.

In this paper, we develop an optimal EV navigation strategy under the rapid charging mode, while the synthetic costs, including the traveling time and charging expenses are minimized.

Meanwhile, the traffic congestion and the charging station overloading are mainly considered. In our work, we only focus on private electric vehicles and specifically aim at the Chinese situation. The real time traffic data and the power grid operation information are the basis of our proposed navigation strategy. We propose an intelligent transport system framework for the rapid charging EV system, the data collection and transmission are processed in the distributed computing manner. The rapid charging stations regulate the charging power of each charging pole dynamically according to the power grid operation information, and adjust their charging prices base on TOU tariffs and the queuing number of EVs. During the driving process of EVs, the charging navigation path is updated in real time, according to the power grid status and the traffic conditions. Therefore, the charging navigation paths are always optimal in the driving process. In summary, the contributions of this paper are listed as follows.

1. We design an optimal charging navigation strategy for rapid charging EVs to reduce EV users' traveling time and charging expenses flexibly.
2. We propose an intelligent transport system framework for our optimal charging navigation strategy, while the traffic and power grid operation information are considered.
3. To implement our optimal charging navigation strategy, we use the charging power regulation scheme to reduce the influence for power grid, and charging price adjustment scheme to balance the number of EVs at each rapid charging station.

The remainder of this paper is organized as follows. The system model is presented in Section 2. The optimal charging navigation scheme is proposed in Section 3. Simulation results are proposed in Section 4. In Section 5, we present the conclusions of this paper.

2. System Model

In this section, we propose a comprehensive system model as illustrated in Figure 1. It contains three parts: Intelligent Transport System center (ITS center), Rapid-Charging Station (RCS), and EV terminals. Real-time traffic data and power grid operation information are collected and calculated in a distributed model via the IoT networks (e.g., crowd sensing). The EV navigation system operates in the slot-by-slot fashion. The structure of each slot is shown in Figure 2. Each slot has three frames, including calculation and broadcast frame t_1, decision frame t_2, and monitoring frame t_3. Normally, $t_3 >> t_1$ and $t_3 >> t_2$. In order to obtain optimal rapid-charging navigation strategies for EVs, the navigation paths should be updated in each slot. In the monitoring frame t_3, the RCS collects the power grid operation data, meanwhile, the ITS center collects real-time traffic data. In frame t_1, the RCS and ITS center calculate the collected information respectively and then broadcast the charging station information and road conditions to the EVs. After receiving the broadcasting data, the EVs determine the navigation paths in frame t_2. The integrated system is described in detail as follows.

2.1. Rapid-Charging Station

As shown in Figure 1, there are multiple RCSs deployed around the road. A distribution management system (DMS) unit is included in each RCS, which is used to connect the local power grid. In frame t_3 of each slot, the RCS collects the power grid operation data via the DMS. Then, in frame t_1, in order to ensure safe and stable operation of the local power grid, after collecting the power grid information, the DMS calculates the maximum permissible charging load based on the power loading level of the distribution power system. According to the maximum accessible charging capacity released by DMS, each RCS calculates and adjusts their charging plans. Then, it distributes the charging powers to each charging pole dynamically. In addition, according to the TOU electricity price issued by power grid company in advance, combined with their profitability and the number of queuing vehicles, the RCSs adjusts their charging prices in each slot. After that, these planned charging power data and the station information (i.e., charging prices, arrival rate and service rate at RCSs) will be broadcast to the moving EVs.

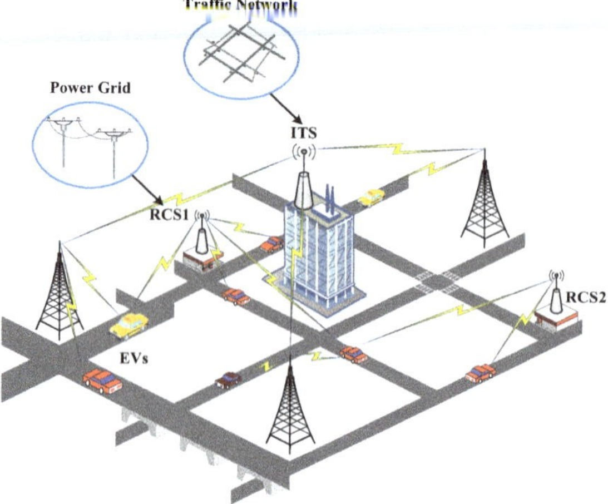

Figure 1. System model overview.

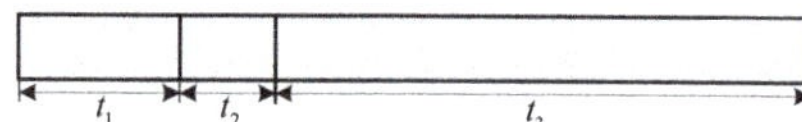

Figure 2. Structure of a slot.

2.2. Intelligent Transport System Center

With the emerging communication and information techniques, the ITS center becomes a transport management system, which has the remarkable benefits in reducing traffic congestion. In our work, the ITS center is connected with the traffic communication network. In frame t_3, the ITS center collects real-time traffic data via crowd sensing or other IoT data collecting technologies. Then, in frame t_1 of the next slot, the ITS center calculates the road conditions (e.g., average driving velocities in each road segment in the slot) and broadcasts them to the moving EVs.

2.3. EV Terminal

We set that the EVs can obtain the electricity from both the regular and rapid charging modes. In order to balance the gap between the traveling time and cost, we consider that the EVs obtain the charging energy via the rapid charging only on the trip. When they arrive the destination, their batteries will be charged fully via using the regular charging mode. On the basis of the received road condition data and charging station information, in the frame t_2 of each slot, the EV terminals calculate the synthetic costs for all reachable charging stations and develop optimal navigation strategies for EVs. Thus, with the dynamic information of charging station and traffic conditions, the navigation strategies can be proposed dynamically and display the result to EVs in each slot.

3. Optimal Charging Navigation Scheme

Assume that the EV starts traveling from the origin and passes through the complex traffic network to reach the destination, as shown in Figure 3. During the traveling, at the beginning of each slot in frame t_2, the EV terminal determines whether it needs to be charged. When the remaining amount of state of charge (SoC) is not enough to finish the trip, the EV terminal should design an optimal charging navigation strategy with the minimum synthetic cost. During the driving process of charging navigation, the navigation path will be updated at the beginning of each slot, based on the dynamic information EV terminal receives. The EV arrives at the selected RCS and obtains the suitable charging energy.

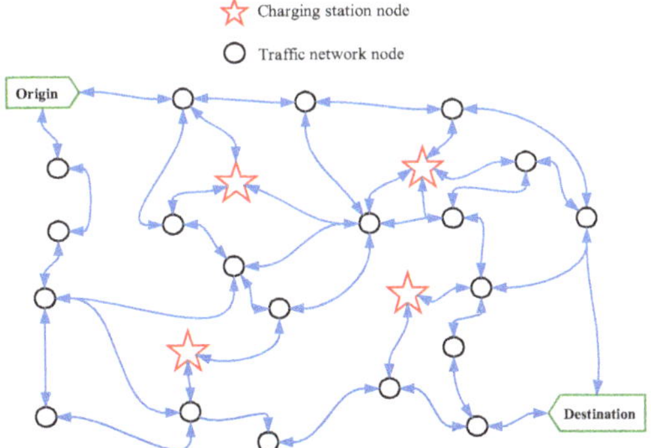

Figure 3. Topology structure of road network model.

In the proposed optimal navigation strategy, both the time cost and the charging cost are considered. In detail, the time cost includes the EV driving time, the waiting time and charging time at the RCSs. Moreover, the charging cost includes the rapid charging cost in the RCSs and the regular charging cost at the destination. The details are shown below, and the mathematical symbols in this paper are listed in Table 1.

Table 1. Mathematical symbols used in this paper.

Symbol	Description
$T_{drive} / T_{wait} / T_{charge}$	Driving/waiting/charging time
$C / C_{rapid} / C_{regular}$	Total cost/rapid charging cost/regular charging cost
φ	Weight coefficient
T_{ij}	Traveling time between locations i and j
d_{ij}	Distance between locations i and j
$\bar{v}_{ij,k}$	Average driving velocity between locations i and j during time slot k
$(*)_{j,k}$	At RCS j during time slot k
$\delta_{j,k}$	EVs' arrival rate (EV number per minute)
$\eta_{j,k}$	EVs' service rate (EV number per minute)
$p_{j,k}$	Occupation rate per charger
$P_{j,k}$	Rapid charging power of each charging pole
$Q_{j,k}$	Charging capacity of RCS j during time slot k
$n_{j,k}$	Number of charging EVs
$w_{j,k}$	Queuing number of EVs
$\bar{\rho}_{j,k}$	Rapid charging price
c_j	Total number of EV chargers at RCS j
ρ_t	TOU charging price in electricity market(Regular charging price)
ρ_*	Additional charging charges
E_{ca} / E_{ca}^{min}	Rated battery capacity/minimum storage of battery capacity
ϑ_{rapid}	Rapid charging efficiency
P_{max}	Maximum charging power of charging pole
e_c	Battery energy consumption
$E_{ch}^{rapid} / E_{ch}^{regular}$	Rapid charging amount/regular charging amount
SOC_0	State of charge at origin
D_{path}^1 / D_{path}^2	Driving distance from origin to selected RCS/driving distance from selected RCS to destination

3.1. Time Cost

3.1.1. Driving Time

Firstly, let T_{drive} represent the total driving time from the origin to the destination, which can be calculated as

$$T_{drive} = \sum_{i,j \in \{\Omega \cup \Phi\}} T_{ij},$$ (1)

where T_{ij} is the traveling time between locations i and j. Ω and Φ are the sets for all visited locations and RCSs.

The driving time between locations i and j is expressed as

$$T_{ij} = \frac{d_{ij}}{\bar{v}_{ij,k}},$$ (2)

where d_{ij} is the distance between locations i and j. $\bar{v}_{ij,k}$ is the average driving velocity in time slot k between locations i and j. Here, $k \in \{1, 2, \cdots, K\}$, K is the total number of time slots. We set that the ITS center broadcasts the average driving velocity $\bar{v}_{ij,k}$ to all EVs at the beginning of each slot in frame t_1.

We set that the EV can complete its journey with only one charging at the RCS. Let $T^1_{j,drive}$ denote the driving time from origin to RCS j, $T^2_{j,drive}$ denote the driving time from RCS j to destination. The total traveling time for RCS j is obtained via the below formula

$$
\begin{aligned}
T_{j,drive} &= T^1_{j,drive} + T^2_{j,drive} \\
&= \sum_{i,j \in \{\Omega \cup \Phi\}} T_{ij} \\
&= \sum_{i,j \in \{\Omega \cup \Phi\}} \frac{d_{ij}}{\bar{v}_{ij,k}}.
\end{aligned}
$$ (3)

3.1.2. Waiting Time

We estimate the waiting time T_{wait} at RCS j based on Queue Theory. Let T_k denote the time duration of slot k. The number of arriving EVs during T_k is denoted by $n^{arrive}_{j,k}$ and the number of EVs being charged during T_k is denoted by $n^{service}_{j,k}$. These data can be easily obtained via multiple deployed sensors or devices in the charging station [21]. Then, at RCS j, the EV arrival rate $\delta_{j,k}$ (EV number per minute) and service rate $\eta_{j,k}$ (EV number per minute) at time slot k are given as

$$\delta_{j,k} = \frac{n^{arrive}_{j,k}}{T_k}, \qquad \eta_{j,k} = \frac{n^{service}_{j,k}}{T_k}.$$ (4)

According to the queuing model $M/M/s$ of queue theory mentioned in [21–24], the arrival rate is subject to the poisson distribution with parameter $\delta_{j,k}$, the service rate is subject to the negative exponential distribution with parameter $\eta_{j,k}$. The total idle probability of the RCS j at time slot k is shown as

$$P_{0,j,k} = \left[\sum_{n=0}^{c_j-1} \frac{1}{n!} \left(\frac{\delta_{j,k}}{\eta_{j,k}} \right)^n + \frac{1}{c_j} \left(\frac{\delta_{j,k}}{\eta_{j,k}} \right)^{c_j} \left(\frac{1}{1-p_{j,k}} \right) \right]^{-1},$$ (5)

where n is the number of charging EVs, c_j is the total number of EV chargers at RCS j. $p_{j,k}$ is the occupation rate per charger of RCS j and is expressed as

$$p_{j,k} = \frac{\delta_{j,k}}{\eta_{j,k} c_j}.$$ (6)

The length of queue $L_{q,j}$ is calculated as

$$L_{q,j} = \frac{p_{j,k}}{c_j!(1 - p_{j,k})^2} \left(\frac{\delta_{j,k}}{\eta_{j,k}}\right)^{c_j} P_{0,j,k},$$ (7)

thus, when a EV arrives at RCS j, the waiting time is

$$T_{j,wait} = \frac{L_{q,j}}{\delta_{j,k}}.$$ (8)

3.1.3. Charging Time

The charging time is related to the charging power of EVs at RCSs. For the safety consideration of local power grid, we give a rapid charging power regulation scheme similar as Ref. [20]. We define the maximum rapid charging load of each RCS as the available charging capacity $Q_{j,k}$. The power system allocates available charging capacity on the basis of the load condition of power nodes at each slot. At the RCS, based on the rapid-charging characteristics, the total rapid charging power of all charging pole should not exceed the available charging capacity $Q_{j,k}$. Thus, to ensure the safety of a distribution power grid system, the rapid charging power $P_{j,k}$ of each charging pole can be obtained from

$$P_{j,k} = \min\left\{\frac{Q_{j,k}}{n_{j,k}}, P_{max}\right\},$$ (9)

where $n_{j,k}$ denotes the number of charging EVs at RCS j at time slot k, and P_{max} is the maximum charging power that charging pole can provide.

Then, the charging time $T_{j,ch}$ at RCS j is

$$T_{j,ch} = \frac{E_{ch}^{rapid}}{P_{j,k}\vartheta_{rapid}},$$ (10)

where ϑ_{rapid} is the rapid charging efficiency and E_{ch}^{rapid} is the amount of rapid charging electricity.

3.2. Charging Cost

3.2.1. Rapid Charging Cost

The rapid charging cost is associated with the charging price. From the formulation Equation (9) of $P_{j,k}$, when more EVs perform rapid charging in a RCS simultaneously, the rapid charging power will be affected. To balance the rapid charging demand of each RCS and avoid overcrowding at the RCS, we propose a rapid charging price adjustment scheme. We set the rapid charging price base on TOU tariffs. A parameter ρ_* is introduced to express the additional charging charges. The charging price $\bar{p}_{j,k}$ at RCS is designed as

$$\bar{p}_{j,k} = \begin{cases} \lambda_j \rho_t & (w_{j,k} = 0) \\ \lambda_j \rho_t + w_{j,k}\rho_* & (w_{j,k} > 0) \end{cases},$$ (11)

where ρ_t denotes the TOU price in electricity market, λ_j denotes the profit coefficient for RCS j. The queuing number of EVs is denoted as $w_{j,k}$ (i.e., the number of EVs that waiting to be charged). If $n_{j,k} \leq c_j$, we have $w_{j,k} = 0$, otherwise $w_{j,k} > 0$. The rapid charging price adjustment scheme is shown in Figure 4. The charging price will be updated and broadcast to the moving EVs at every time slot. Once the EV starts charging, the charging price of this EV will remain unchanged during the whole charging process.

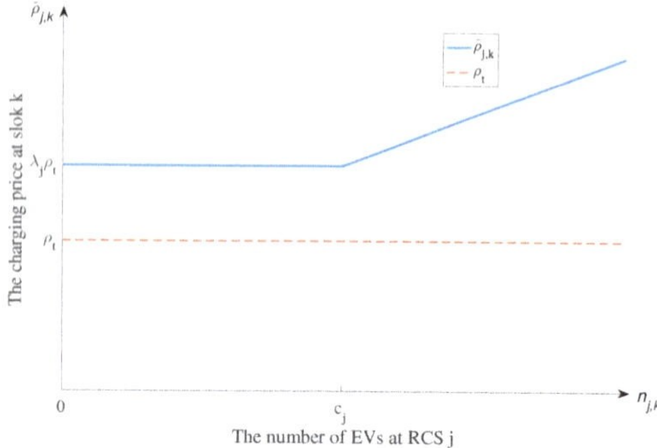

Figure 4. The rapid charging price adjustment scheme.

The rapid charging cost is calculated as

$$C_{rapid} = E_{ch}^{rapid} \bar{\rho}_{j,k}. \tag{12}$$

3.2.2. Regular Charging Cost

When arriving at the destination, the EV is parked in the residential area or parking lot with charging poles. The EV will be charged to the rated capacity of battery using the regular charging mode. The regular charging price is the TOU price ρ_t in the electricity market. We denote $E_{ch}^{regular}$ as the regular charging amount, we have

$$C_{regular} = E_{ch}^{regular} \rho_t. \tag{13}$$

3.3. Objective Function and Constraints

In summary, the goal of charging navigation scheduling is to minimize the synthetic cost C. The objective function is formulated as

$$\min C = \varphi \left(T_{drive} + T_{wait} + T_{ch} \right) + C_{rapid} + C_{regular}, \tag{14}$$

where φ is the weight coefficient, which is adjusted according to users' requests. For example, a higher φ means that the driver is more concerned about the time consumption.

The constraints for the objective function are

$$E_{ca} SOC_0 \geq e_c D_{path}^1 + E_{ca}^{min}, \tag{15}$$

$$E_{ca} SOC_0 - e_c D_{path}^1 + E_{ch}^{rapid} \geq e_c D_{path}^2 + E_{ca}^{min}, \tag{16}$$

$$E_{ca} SOC_0 + E_{ch}^{rapid} + E_{ch}^{regular} - e_c(D_{path}^1 + D_{path}^2) = E_{ca}, \tag{17}$$

$$0 < E_{ch}^{regular} < E_{ca}, \tag{18}$$

$$0 < E_{ch}^{rapid} < E_{ca}, \tag{19}$$

where E_{ca} is the rated battery capacity, E_{ca}^{min} is the minimum storage of battery capacity, SOC_0 is the state of charge at origin, e_c is the energy consumption per kilometer. D_{path}^1 is the driving distance from origin to the selected RCS and D_{path}^2 is the driving distance from the selected RCS to destination. In detail, constraint (15) means that the remaining battery energy of EVs at origin should be greater than the energy consumption from origin to the selected RCS. Constraint (16) means that the battery energy after rapid charging at RCS should be greater than the energy requirement from the selected RCS to destination. We assume that the EV will be charged to the rated capacity E_{ca} using the regular charging mode after arriving at the destination, the battery capacity should satisfy the Equation (17). The regular charging amount and rapid charging amount should satisfy constraint (18), (19).

3.4. Solution

The optimization problem (14) is a typical mixed integer nonlinear programming (MINP) problem, it is hard to solve directly. The Dijkstra algorithm is a common tool to solve the shortest path problem [25]. According to Dijkstra algorithm, the traffic network can be modeled with a weighted directed graph. The weighted lines of the graph represent the road segment and the nodes represent the intersection of road network. Normally, the attribute of weight value can be either the distance of road segment or the average time to drive through the road segment. In this paper, we use the synthetic cost of traveling the road segment as the weight value. We can rewrite the objective function (14) as the following formulation:

$$Weight = \varphi Time + Expense. \tag{20}$$

Time represent the time traveling the road segment, *Expense* represent the expense of electricity consumption through road segment. φ is the weight coefficient, which is also used in the objective function (14). According to the expression of objective function, the *Weight* is expressed as

$$W_{ij} = \varphi T_{ij} + \rho_t E_{ij}, \tag{21}$$

where W_{ij} is the weight value of road segment between location i and j, E_{ij} is the electricity consumption through road segment between location i and j. According to the weight value of each road segment, we can use the Dijkstra algorithm to plan the optimal driving path. By using the *Optimal Path Planning Algorithm* illustrated in Algorithm 1, we can obtain the optimal driving path and the selected RCS for EV in each slot. According to the proposed charging navigation strategy, we can obtain the solution procedure of optimal rapid-charging navigation. The solution procedure is illustrated in Figure 5. After the real-time data is updated, the Optimal Path Planning Algorithm is employed to search for optimal driving path and the RCS for EV. During the driving process of charging navigation, the optimization of path planning will be repeated at the beginning of each time slot. The charging navigation path is also updated through optimization until destination has been reached.

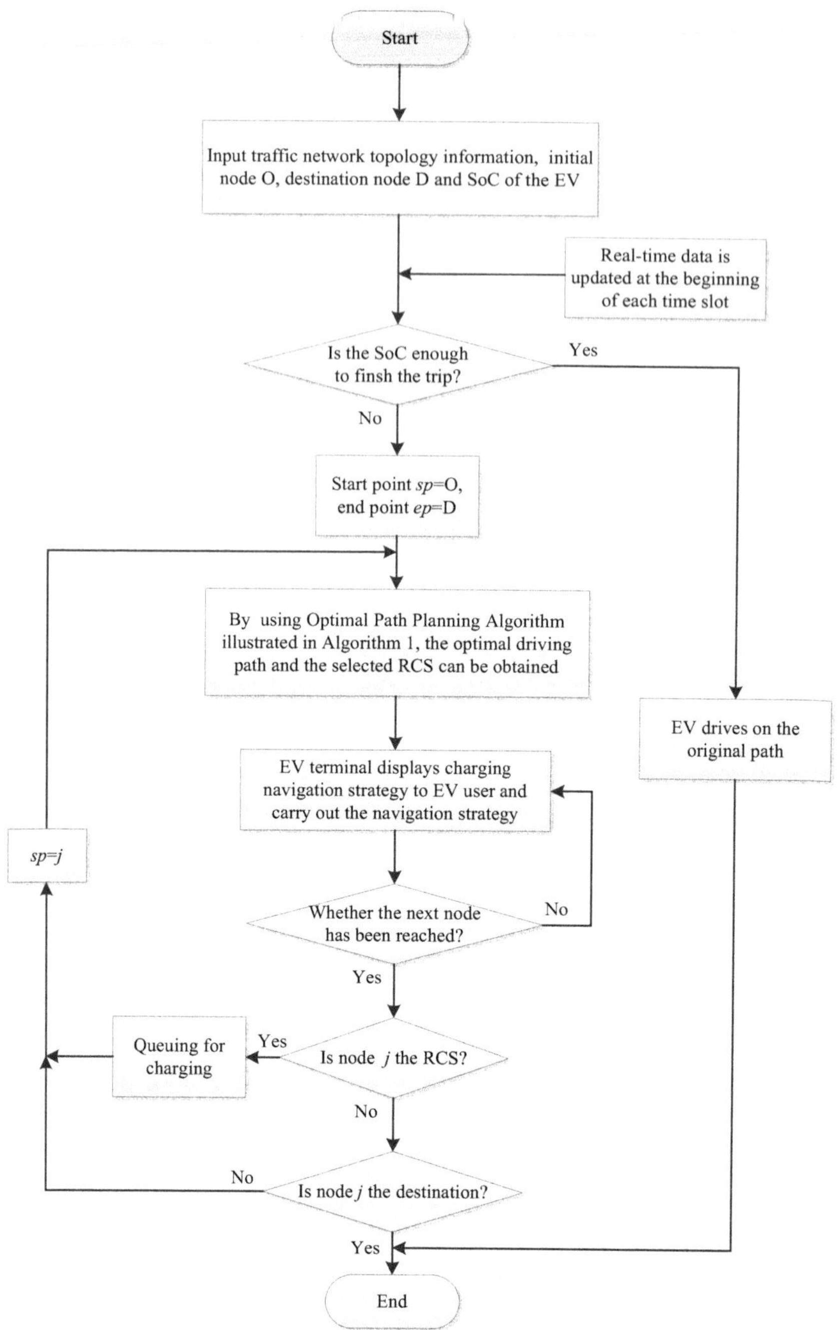

Figure 5. Solution procedure of optimal rapid-charging navigation.

Algorithm 1 *Optimal Path Planning Algorithm* at slot k.

Require:

    The traffic network topology information.

    The parameters listed in Table 2.

    The real-time data $\delta_{j,k}, \eta_{j,k}, P_{j,k}, \bar{p}_{j,k}$

    The start point sp and the end point ep.

Ensure:

 1: **for** each $j \in \Phi$ **do**

 2:    Simulate driving path from sp to RCS j and the path from RCS j to ep based on Dijkstra algorithm.

 3:    **if** The driving path satisfy the constraints (15) and (16). **then**

 4:        Calculate the total cost

$$C_j = \varphi \left(T_{j,drive} + T_{j,wait} + T_{j,ch} \right) + C_{j,rapid} + C_{regular},$$

        by solving the optimization model proposed in Section 2.

 5:    **else**

 6:        $C_j = inf.$

 7:    **end if**

 8: **end for**

 9: Select the driving path and RCS with the minimum total cost $C = \min\{C_j \mid j \in \Phi\}$.

---

4. Simulation Results

We consider a 25×25 km region of a city-center road network, which is similar to Guangzhou Higher Education Mega Center in Guangzhou, Guangdong Province, China. The topological graph of the transportation network is shown in Figure 6. The number of RCSs is 4. The RCSs (labeled with red circle) are located at transportation nodes 17, 21, 37 and 11. Each RCS contains 8 charging poles with the maximum charging power of 120 kW. The capacity of each EV is 54.75 kWh, and the energy consumption is 0.147 kWh/km(these data are based on the GACNE Trumpchi GE530 [26]). We assume the initial SoC of each EV is 20%. The parameters of studied EVs are listed in Table 2. The TOU charging prices in electricity market are listed in Table 3 [27]. We set the profit coefficient λ_j and additional charging charge parameter ρ_* of each RCS as listed in Table 4. The conventional load curve over one day is shown in Figure 7. We set that the permissible maximum load of this area is 25 MW, which means, if the actual load exceeds the maximum load, it will cause a harmful effect on the power grid. As the conventional load increases, the maximum available charging capacity of all RCSs decreases as is shown in Figure 8. The charging capacity of each RCS is also dynamically adjusted as the total available charging capacity changes.

Table 2. Parameters in simulation.

Parameter	Values	Parameter	Values
E_{ca}(kWh)	54.75	E_{ca}^{min}(kWh)	2.5
e_c(kWh/km)	0.147	ϑ_{rapid}	0.9

Table 3. Time-of-use (TOU) Price of charging in China's electricity market.

Periods	Bottom	Flat	Peak
	(00:00–06:59)	(07:00–09:59)	(10:00–14:59)
	(23:00–23:59)	(15:00–17:59), (21:00–22:59)	(18:00–20:59)
ρ_t(Yuan/kWh)	0.3818	0.8395	1.3222

Table 4. Profit coefficient and additional charging charge.

Stations	RCS1	RCS2	RCS3	RCS4
λ_j	1.4	1.3	1.25	1.35
ρ_*(Yuan/kWh)	0.15	0.15	0.15	0.15

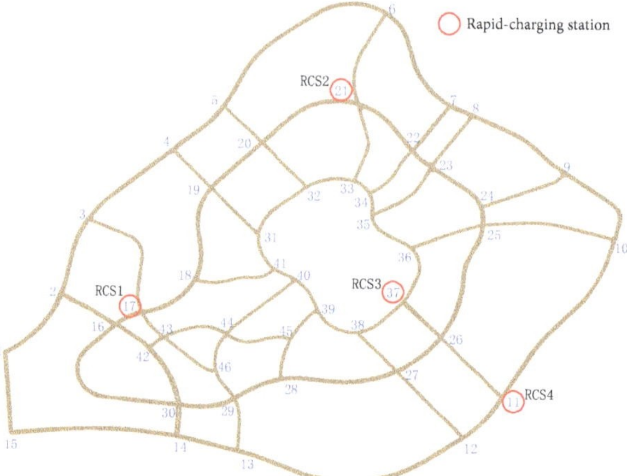

Figure 6. Transportation network and its topological graph based on Guangzhou Higher Education Mega Center.

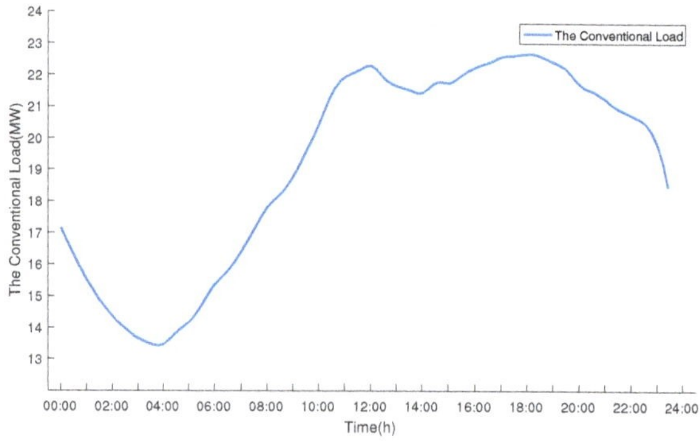

Figure 7. The conventional load curve.

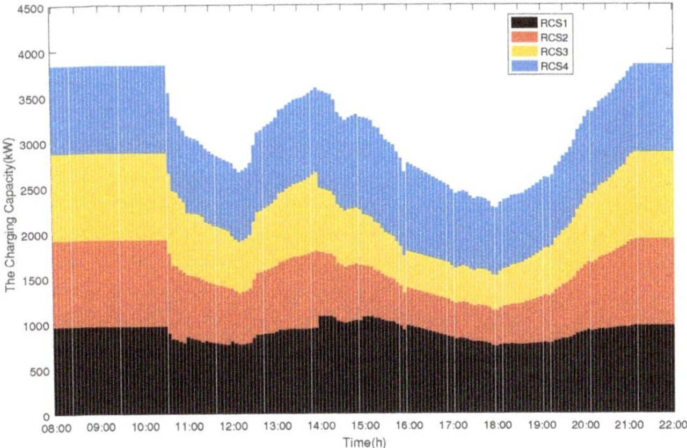

Figure 8. The charging capacity of Rapid-Charging Stations (RCSs) at different time.

4.1. Impact Analysis of the Charging Power Regulation Scheme

In our simulation, we test the impacts of two charging approaches: the uncoordinated EV charging approach and the charging approach using our proposed charging navigation strategy based on the charging power regulation scheme, as shown in Figure 9. We can find that the uncoordinated EV charging approach can cause an exceeded load peak over the maximum load. However, with the proposed charging approach, the charging load can be controlled according to the state of the power grid, which can effectively avoid overload. Thus, in order to reduce the influence of rapid charging on the power grid, it is necessary to design an efficient and optimal charging strategy to limit the rapid charging power, especially at the peak hours.

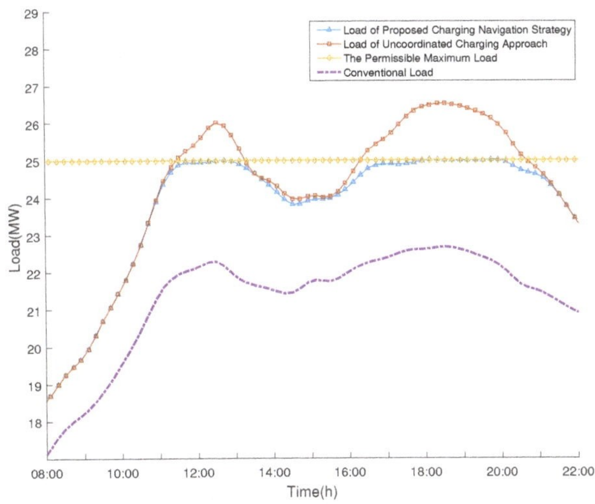

Figure 9. Comparison of load curves to demonstrate the impacts of Electric vehicle (EV) charging approaches.

4.2. Impact Analysis of the Proposed Rapid Charging Price Adjustment Scheme

In Figure 10, we simulate 7 EVs that start from node 1 to node 9. We set their weight coefficients $\varphi = 0.2$ (Yuan/min). They start every 5 min one by one from 18:00 to 18:30 under two cases as follows.

Case 1 : Using the proposed rapid charging price adjustment scheme. The rapid charging price is obtained by Equation (11).

Case 2 : Using the conventional rapid charging price scheme. The rapid charging price is based on TOU price, i.e., $\rho_{j,k} = \lambda_j \rho_t$.

From the Figure 10, we find that, in Case 1, the proposed rapid charging price adjustment scheme adjusts the charging price according to TOU price and the number of queuing vehicles. The studied EVs select different RCSs based on minimum synthetic cost. In Case 2, the EVs choose the same RCS based on the conventional rapid charging price. Obviously, the proposed rapid charging price adjustment scheme can balance the rapid charging demand of each RCS effectively. When the charging demand is large, the EVs can be more balanced distributed in each RCS, reducing their waiting time and the operation pressure of stations.

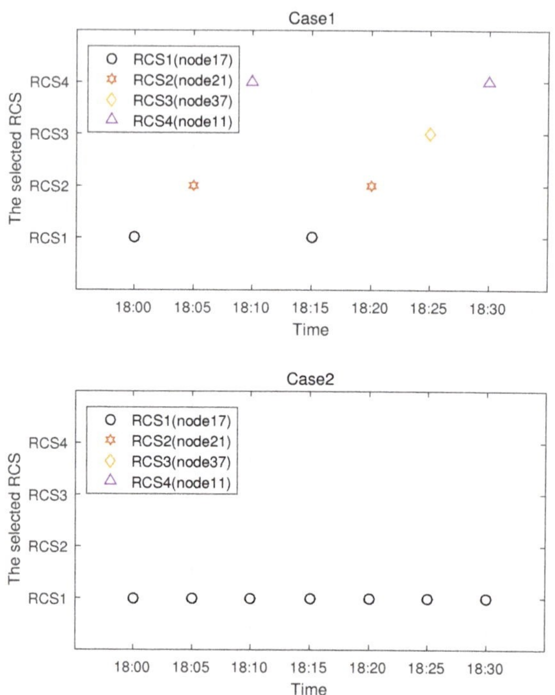

Figure 10. Comparison of two rapid charging price scheme.

4.3. Impact Analysis of the Proposed Optimal Navigation Strategy with Different Weight Coefficient

In Figures 11 and 12, the performance of the proposed optimal navigation strategy is compared with the simple shortest path strategy with different coefficients. The shortest path strategy is that the EVs choose the path with shortest distance to travel. We select one EV that starts from node 1 to node 9 during the hours between 08:00 and 22:00. Under these two different charging strategies, the total cost comparisons are shown in Figure 11. The comparisons of average traveling time and charging cost is shown in Figure 12.

In Figure 11, with different weight coefficient φ, we can find that the total cost of the proposed charging navigation strategy is lower than the simple shortest path strategy. When $\varphi = 0.5$ (Yuan/min), at 18:00, the total cost is reduced about 25% using the proposed charging navigation strategy. This is because at peak times, when the congestion occurs in some road segments or charging stations, the proposed charging strategy can update the driving path timely and select other RCS to minimize the overall cost with a weight coefficient.

In Figure 12, we compare the time costs and charging costs separately under different weight coefficients. When $\varphi = 0$ (Yuan/min), the objective is to minimize the charging cost. The average charging cost of the proposed charging strategy is lower than the shortest path strategy, while their average time costs are very close. This is because when $\varphi = 0$ (Yuan/min), the proposed navigation strategy only considers the charging cost. It selects the path with minimum charging cost even though the time cost is high. As the weight coefficient increases, the objective is more concerned with the time consumption and the average traveling time of proposed charging strategy is reduced a lot. When $\varphi = 0.2$ (Yuan/min), the average traveling time is reduced about 11 (min). When $\varphi = 0.5$ or $\varphi = 1$ (Yuan/min), the average traveling time is reduced about 15 (min). This is because when $\varphi > 0.5$, it is difficult to further reduce the time cost under the same road conditions. As the results show, the proposed strategy can optimize charging navigation path and reduce EV users' traveling time and charging expenses flexibly depending on different weight coefficients.

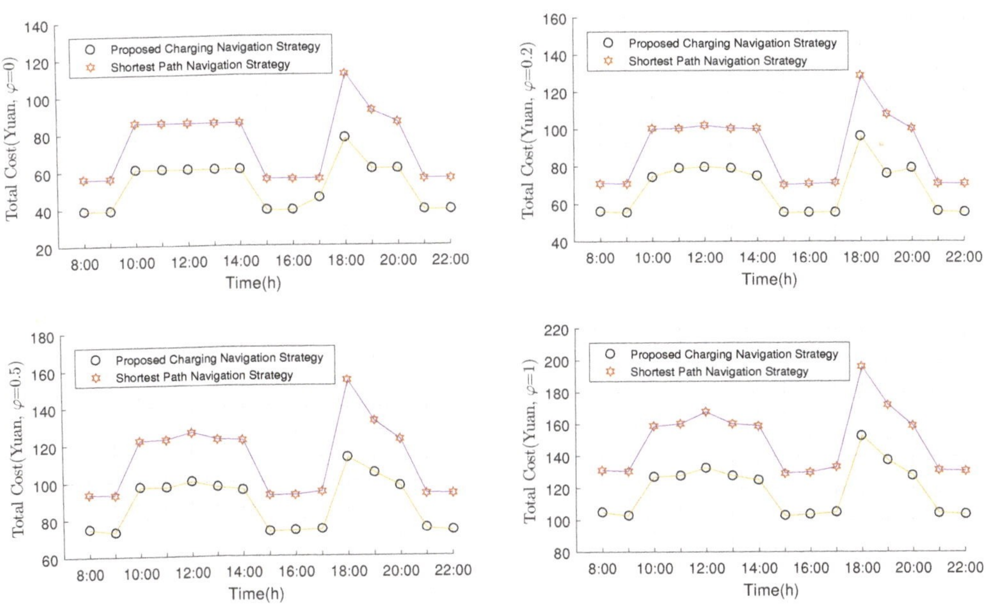

Figure 11. Total cost comparison of two different navigation strategies.

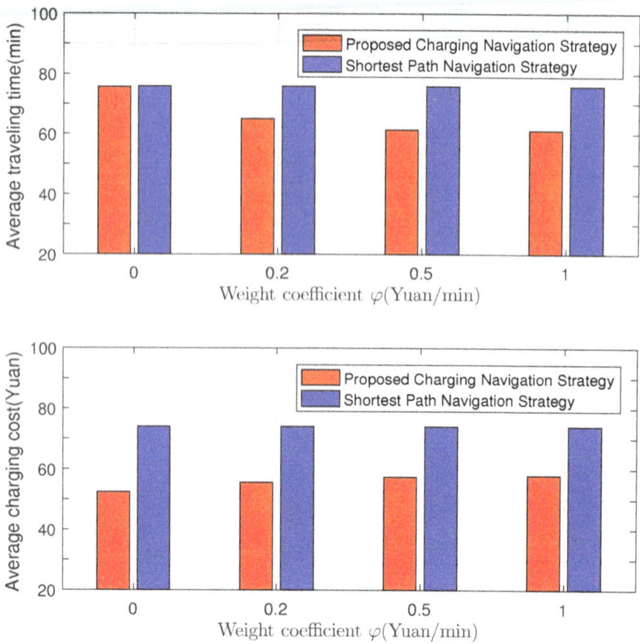

Figure 12. Comparison of traveling time and charging cost with different weight coefficients by two navigation strategies.

5. Conclusions

In this paper, we propose an optimal charging navigation strategy for rapid charging EVs, taking into consideration power grid operation and the real-time traffic information. With the proposed charging strategy based on the charging power regulation scheme, the proposed strategy can effectively avoid overload and mitigate the peak load in the distribution network. The proposed rapid charging price adjustment scheme can balance the rapid charging demand of each RCS, thus reducing waiting time of EV users and the operation pressure of stations. Moreover, we compare the total cost with shortest path strategy under different weight coefficients at different times. Both the time cost and the charging cost with different coefficient are analyzed. The modified Dijkstra algorithm is used to find the optimal solution. The simulation shows that the proposed strategy can effectively reduce EV users' charging navigation cost depending on different optimization objectives.

Author Contributions: Data curation, W.M., K.L. and S.D.; Formal analysis, W.M.; Investigation, W.M. and C.Y.; Methodology, C.Y. and X.C.; Project administration, X.C.; Supervision, C.Y.

Funding: This research was funded by National Natural Science Foundation of China under Grant number 61603099, 61773126, 61727810, and 61701125, Natural Science Foundation of Guangdong under Grant number 2018A0303130080; Pearl River S&T Nova Program of Guangzhou under Grant number 201806010176.

Conflicts of Interest: The authors declare no conflict of interest.

Abbreviations

The following abbreviations are used in this manuscript:

EV   Electric Vehicle
IoT   Internet of Things
V2G   vehicle-to-grid
TOU   time-of-use
ITS   Intelligent Transport System
RCS   Rapid Charging Station
DMS   Distribution Management System
SoC   State of Charge

References

1. Hannan, M.; Hoque, M.; Mohamed, A.; Ayob, A. Review of energy storage systems for electric vehicle applications: Issues and challenges. *Renew. Sustain. Energy Rev.* **2017**, *69*, 771–789. [CrossRef]
2. Zhang, K.; Mao, Y.; Leng, S.; Zeng, M.; Xu, L.; Jiang, L.; Vinel, A. Optimal energy exchange schemes in smart grid networks: A contract theoretic approach. In Proceedings of the IEEE/CIC International Conference on Communications in China (ICCC), Chengdu, China, 27–29 July 2016; pp. 1–6.
3. Debnath, U.K.; Ahmad, I.; Habibi, D.; Saber, A.Y. Improving Battery Lifetime of Gridable Vehicles and System Reliability in the Smart Grid. *IEEE Syst. J.* **2015**, *9*, 989–999. [CrossRef]
4. Liu, Y.; Yang, C.; Jiang, L.; Xie, S. Intelligent Edge Computing for IoT-Based Energy Management in Smart Cities. *IEEE Netw.* **2019**, accepted.
5. De Quevedo, P.M.; Muñoz-Delgado, G.; Contreras, J. Joint expansion planning of distribution networks, EV charging stations and wind power generation under uncertainty. In Proceedings of the IEEE Power Energy Society General Meeting, Chicago, IL, USA, 16–20 July 2017; pp. 1–5.
6. Marcincin, O.; Medvec, Z.; Moldrik, P. The impact of electric vehicles on distribution network. In Proceedings of the 18th International Scientific Conference on Electric Power Engineering (EPE), Ostrava, Czech Republic, 17–19 May 2017; pp. 1–5. [CrossRef]
7. Yang, H.; Pan, H.; Luo, F.; Qiu, J.; Deng, Y.; Lai, M.; Dong, Z.Y. Operational Planning of Electric Vehicles for Balancing Wind Power and Load Fluctuations in a Microgrid. *IEEE Trans. Sustain. Energy* **2017**, *8*, 592–604. [CrossRef]
8. You, P.; Yang, Z.; Zhang, Y.; Low, S.H.; Sun, Y. Optimal Charging Schedule for a Battery Switching Station Serving Electric Buses. *IEEE Trans. Power Syst.* **2016**, *31*, 3473–3483. [CrossRef]
9. Wang, Y.; Huang, Y.; Xu, J.; Barclay, N. Optimal recharging scheduling for urban electric buses: A case study in Davis. *Transport. Res. Part E Logist. Transport. Rev.* **2017**, *100*, 115–132. [CrossRef]
10. Akhavan-Rezai, E.; Shaaban, M.F.; El-Saadany, E.F.; Zidan, A. Uncoordinated charging impacts of electric vehicles on electric distribution grids: Normal and fast charging comparison. In Proceedings of the IEEE Power and Energy Society General Meeting, San Diego, CA, USA, 22–26 July 2012; pp. 1–7.
11. Alshareef, S.M.; Morsi, W.G. Impact of fast charging stations on the voltage flicker in the electric power distribution systems. In Proceedings of the IEEE Electrical Power and Energy Conference (EPEC), Saskatoon, SK, Canada, 22–25 October 2017; pp. 1–6.
12. Febriwijaya, Y.H.; Purwadi, A.; Rizqiawan, A.; Heryana, N. A study on the impacts of DC Fast Charging Stations on power distribution system. In Proceedings of the International Conference on Electrical Engineering and Computer Science (ICEECS), Sanur-Bali, Indonesia, 24–25 November 2014; pp. 136–140.
13. Wang, K.; Gu, L.; He, X.; Guo, S.; Sun, Y.; Vinel, A.; Shen, J. Distributed Energy Management for Vehicle-to-Grid Networks. *IEEE Netw.* **2017**, *31*, 22–28. [CrossRef]
14. He, Y.; Venkatesh, B.; Guan, L. Optimal Scheduling for Charging and Discharging of Electric Vehicles. *IEEE Trans. Smart Grid* **2012**, *3*, 1095–1105. [CrossRef]
15. Cao, Y.; Tang, S.; Li, C.; Zhang, P.; Tan, Y.; Zhang, Z.; Li, J. An Optimized EV Charging Model Considering TOU Price and SOC Curve. *IEEE Trans. Smart Grid* **2012**, *3*, 388–393. [CrossRef]

16. Yin, S.; Zhong, W.; Xie, K.; Yu, R.; Zhang, Y. Fair Energy Scheduling for Vehicle-to-Grid Networks Using Adaptive Dynamic Programming. *IEEE Trans. Neural Netw. Learn. Syst.* **2016**, *27*, 1697–1707. [CrossRef] [PubMed]

17. Khatiri-Doost, S.; Amirahmadi, M. Peak shaving and power losses minimization by coordination of plug-in electric vehicles charging and discharging in smart grids. In Proceedings of the IEEE International Conference on Environment and Electrical Engineering and 2017 IEEE Industrial and Commercial Power Systems Europe (EEEIC/I CPS Europe), Milan, Italy, 6–9 June 2017; pp. 1–5.

18. Liu, C.; Zhou, M.; Wu, J.; Long, C.; Wang, Y. Electric Vehicles En-Route Charging Navigation Systems: Joint Charging and Routing Optimization. *IEEE Trans Control Syst. Technol.* **2017**, *27*, 906–914. [CrossRef]

19. Zhu, M.; Liu, X.Y.; Kong, L.; Shen, R.; Shu, W.; Wu, M.Y. The charging-scheduling problem for electric vehicle networks. In Proceedings of the IEEE Wireless Communications and Networking Conference (WCNC), Istanbul, Turkey, 6–9 April 2014; pp. 3178–3183.

20. Guo, Q.; Xin, S.; Sun, H.; Li, Z.; Zhang, B. Rapid-Charging Navigation of Electric Vehicles Based on Real-Time Power Systems and Traffic Data. *IEEE Trans. Smart Grid* **2014**, *5*, 1969–1979. [CrossRef]

21. Yang, H.; Deng, Y.; Qiu, J.; Li, M.; Lai, M.; Dong, Z.Y. Electric Vehicle Route Selection and Charging Navigation Strategy Based on Crowd Sensing. *IEEE Trans. Ind. Inf.* **2017**, *13*, 2214–2226. [CrossRef]

22. Hafez, O.; Bhattacharya, K. Integrating EV Charging Stations as Smart Loads for Demand Response Provisions in Distribution Systems. *IEEE Trans. Smart Grid* **2018**, *9*, 1096–1106. [CrossRef]

23. Bae, S.; Kwasinski, A. Spatial and Temporal Model of Electric Vehicle Charging Demand. *IEEE Trans. Smart Grid* **2012**, *3*, 394–403. [CrossRef]

24. Gong, Q.; Midlam-Mohler, S.; Serra, E.; Marano, V.; Rizzoni, G. PEV charging control for a parking lot based on queuing theory. In Proceedings of the American Control Conference, Washington, DC, USA, 17–19 June 2013; pp. 1124–1129.

25. Zhang, X.; Chen, Y.; Li, T. Optimization of logistics route based on Dijkstra. In Proceedings of the 6th IEEE International Conference on Software Engineering and Service Science (ICSESS), Beijing, China, 23–25 September 2015; pp. 313–316.

26. GACNE Trumpchi GE3 530. 2018. Available online: http://www.gacne.com.cn/vehicles/ge3_530 (accessed on 21 December 2018).

27. TOU Tariff Standard for Commercial Electricity. 2018. Available online: http://www.docin.com/p-804976213.html (accessed on 21 December 2018).

Article

Electric Vehicles' User Charging Behaviour Simulator for a Smart City

Bruno Canizes [1], **João Soares** [1], **Angelo Costa** [2], **Tiago Pinto** [1,*], **Fernando Lezama** [1], **Paulo Novais** [2] and **Zita Vale** [3]

1   GECAD—Knowledge Engineering and Decision Support Research Center—Polytechnic of Porto (IPP), R. Dr. António Bernardino de Almeida, 431, 4200-072 Porto, Portugal; brmrc@isep.ipp.pt (B.C.); jan@isep.ipp.pt (J.S.); flzcl@isep.ipp.pt (F.L.)

2   ALGORITMI Centre, University of Minho, 4710-057 Braga, Portugal; acosta@di.uminho.pt (A.C.); pjon@di.uminho.pt (P.N.)

3   Polytechnic of Porto (IPP), R. Dr. António Bernardino de Almeida, 431, 4200-072 Porto, Portugal; zav@isep.ipp.pt

*   Correspondence: tcp@isep.ipp.pt

Received: 18 February 2019; Accepted: 12 April 2019; Published: 18 April 2019

Abstract: The increase of variable renewable energy generation has brought several new challenges to power and energy systems. Solutions based on storage systems and consumption flexibility are being proposed to balance the variability from generation sources that depend directly on environmental conditions. The widespread use of electric vehicles is seen as a resource that includes both distributed storage capabilities and the potential for consumption (charging) flexibility. However, to take advantage of the full potential of electric vehicles' flexibility, it is essential that proper incentives are provided and that the management is performed with the variation of generation. This paper presents a research study on the impact of the variation of the electricity prices on the behavior of electric vehicle's users. This study compared the benefits when using the variable and fixed charging prices. The variable prices are determined based on the calculation of distribution locational marginal pricing, which are recalculated and adapted continuously accordingly to the users' trips and behavior. A travel simulation tool was developed for simulating real environments taking into account the behavior of real users. Results show that variable-rate of electricity prices demonstrate to be more advantageous to the users, enabling them to reduce charging costs while contributing to the required flexibility for the system.

Keywords: electric charging behaviour; electric mobility; energy prices; EVs; travel simulator

---

1. Introduction

The need to reduce greenhouse gas emissions is ever increasing, and several nations have agreed on ambitious targets in the Paris Agreement Treaty [1]. This treaty has the aim to limit global temperature 2 °C above the pre-industrial levels. The transportation and its infrastructure represents 23% of greenhouse gas emissions and is only surpassed by fossil fuel emissions (e.g., energy production) [2]. This shows that the electrification of transport plays a significant role in making the planet a greener place, reducing dependence on fossil fuels.

The use of electric vehicles (EVs) not only has the potential to change individual mobility but also to reduce pollutant emissions, which is considered a major cause of air pollution and causes serious health problems in the global population. However, as an increasing number of charges will ideally be covered by renewable production to achieve decarbonization of the transportation sector, the introduction of dynamic electricity prices could increase the risk of substation overloads [3].

In Europe, growth in the use of EVs will result in extra energy demand, with consumption increasing from approximately 0.03% in 2014 to 9.5% in 2050 [4].

Generally, the population is accustomed to deal with fossil energies and with the convenience of easy to find service stations and fast refueling times. Thus, there are no concerns regarding waiting times or about the fuel needed to reach the intended destination. When using an electric vehicle it is essential to consider these factors as the current range of the vehicles is limited and changing stations are few. Also, there are other challenges such as increasing peak power demand if the charging events occur at the same instant as residential or industrial peak consumption [5]. The electrical network reacts according to the level of loads connected to it, and with growing usage of this mean of transportation in the future, it is necessary to study how the impact of the extra energy required can be mitigated. Understanding the behavior of electric vehicle users while at the same time recognize the changes in the network will be a crucial part.

Recent studies suggest that dynamic electricity prices can spur demand and help electric companies avoid costly investments in infrastructures [6]. However, the lack of variability in electricity prices does not allow the studies to be completely realistic or in line with the actual variability of renewable energy generation. In this context, it is crucial to address the following research question: can electric vehicle users change their charging patterns as a result of varying electricity prices?

Providing incentives to EV users in a way that behavior and charging patterns are changed and adapted accordingly to the variation of electricity prices is essential to ensure the EV's flexibility balances the variation of renewable energy sources. It is in this scope that this paper brings its main contributions, by presenting a study on the impact of electricity prices variation on EV users' charging habits.

This study compares the benefits when using a variable and fixed charging prices. The variable prices are determined based on the calculation of distribution locational marginal pricing (DLMP) using distribution network operation and reconfiguration optimization model, which enables achieving prices that are not only continuously recalculated and adaptable to the ongoing changes in the power network (variation of consumption and generation at each time) but also reflect the situation and needs in each different location of the network. These prices are used to incentive EV users to change their charging habits according to the variation of renewable generation in different places of the power network. A travel simulation tool specifically developed for this study is also presented. The simulator takes into account the behavior of real users to simulate their trips from the origin place (e.g., house or workplace) to multiple destinations, and back. The tool also considers different types of users and vehicles, thus allowing to create personalized profiles, destinations, and schedules. Moreover, the simulator enables defining the position of the vehicles in a power network continuously throughout time. In this way, the proposed tool simulates a real environment, with trips and charging stations (CS). Considering the defined scenarios, users make decisions regarding their charging process, i.e., if they charge their vehicles or not at each time, according to the behaviors previously analyzed. For this, intelligent charging is simulated considering variables such as distance and the price of electricity. In this way, it is possible to test the impact of different types of incentives on EV users' behavior. A physical laboratory model of a smart city (SC) located at BISITE laboratory with a 13 buses distribution network with high distributed energy resources (DER) penetration is used to demonstrate the application of the proposed methodology. Results show that variable charging prices prove to be more advantageous to the EV users, enabling them to reduce charging costs, while contributing to the required flexibility for the system. This allows mitigating the problems introduced with the large-scale penetration of distributed, variable renewable energy sources.

The rest of the paper is organized as follows: Section 2 presents a brief review of state of the art. The proposed simulator tool is described in Section 3, along with the methodology for the calculation of the variable electricity prices. The case studies are discussed in Section 4. Finally, the conclusions are shown in Section 6.

2. State of the Art

2.1. Electric Mobility

In 2017, the number of EVs on the road was about 3.1 million, an increase of 57% compared to 2016 (according to Figure 1). This increase was similar to that registered between 2015 and 2016, of 60% [7]. It is also possible to verify that purely electric vehicles (battery electric vehicle - BEV), had a more significant growth than the hybrid vehicles (plug-in hybrid electric vehicle - PHEV), representing two-thirds of the total. China is the country with the largest share, accounting for 40% of the total [7].

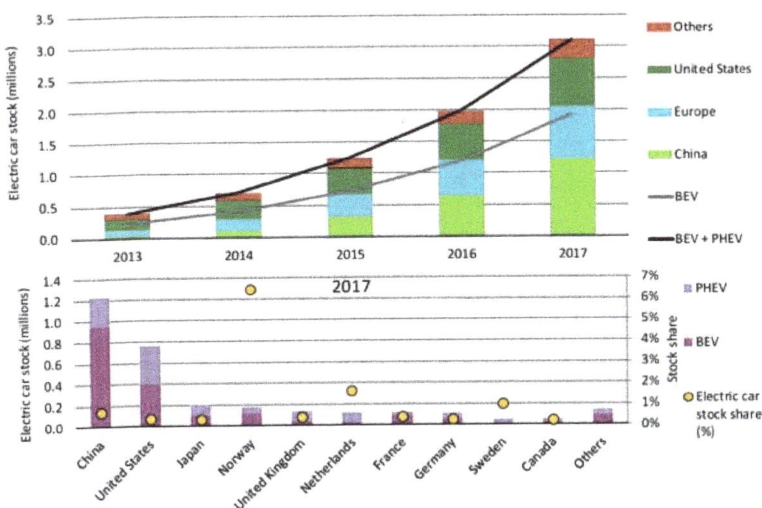

Figure 1. Number of EVs globally [7].

With the increasing popularity of EVs, there is the need to improve charging infrastructures and offer more affordable models. While governments offer incentives for adopting EVs and continue to invest in infrastructure, the motive that drives people to opt for this type of transportation is socially driven: it is the cleanest solution that will help sustain a habitable planet. This is reflected in the satisfaction of EVs' users, where 51% say that the most significant incentive to buy one is to contribute to a more sustainable future [8].

Overall, the results show that this adoption does not only depend on incentives but also fewer obstacles to a more comfortable driving. In this sense, it is essential that charging is accessible, both monetarily and geographically. It is important to have homes, shopping malls, workplaces, and parking lot buildings with charging stations.

Another aspect to consider is the type of charging since time spent stopped is perhaps the variable that the consumer values most. The high power capacity of fast chargers (with a power greater than 40 kW) makes them difficult to implement in residential homes due to possible hazards, which even being addressed, are still mostly underdeveloped. To this, the implementation of fast CS will facilitate the user by reducing waiting times.

2.2. Charging Behaviour

Between 2011 and 2013 data about driving and EV demand patterns were collected on a study conducted in Europe [9]. More than 230,000 charges have been registered. The average state of charge (SoC) of the battery was of 60% when users recharged it, which shows that users do not let the battery discharge completely and charge it whenever they have the opportunity and not when the battery is low. The average percentage of users who started a journey or a charge with a SoC level of less than 20% is less than 5%. Regarding the moment of the charging, it is verified that the majority are performed between 18:00 and 22:00, which corresponds to the peak hours of energy demand.

Franke and Krems [10] analyzed the charging behavior of users in Germany. They concluded that the true vehicle range affects charging decisions. They have also developed a conceptual model based on principles of self-regulation and control theory where it is possible to understand the use of energy resources. This model is based on the premise that whenever users interact with limited power sources, they continuously monitor and manage their mobility needs and their mobility resources. For instance, the mobility needs relate to the distance of a trip and the mobility resources relate to the remaining battery. Users often feel "range anxiety" that can be described as the experience of never having enough battery to reach a location and getting stranded. Thus, as the anxiety increases so does the likelihood of resorting to strategies to handle this situation, e.g., driving economically or charging the car more often.

Marmaras et al. [11] developed two behavioral profiles to be used in a simulation environment: unaware and aware. The unaware profile tries to find the best possible solution with limited access to information and minimal interaction with the environment and other EVs' users. In this case, the level of range anxiety is strong, and the user is always seeking to charge the vehicle even when it is not needed. The aware profile has more access to information and interacts with the environment and other EVs. This profile has a low anxiety level, charging the vehicle only when needed. The results of this research show that the unaware profile starts charging the vehicle as soon as it is parked, typically at home between 17:30 and 18:00, whereas the aware profile waits for the off-peak hours between 22:00 and 06:00 h.

Neubauer and Wood [12] applied a battery life analysis and simulation tool for vehicles (BLAST-V) of the National Renewable Energy Laboratory to study the sensitivity of EVs concerning range anxiety and different scenarios of different charging infrastructures. The results showed that the effects of range anxiety might be significant but reduced with access to additional charging infrastructures.

Nicholas et al. [13] studied the charging behavior when simulating trips and charging in public stations. The results show that more than 5% of trips would require recharging in a public charger for different charging autonomy and assumptions.

Xu et al. [14] used a mixed logic model to study which factors influence BEV users in the decision-making of the type of charging (normal or fast) and local. The results suggest that the battery capacity, the initial state of the battery and the number of fast charges carried out are the predictive factors for the choice of type and place of charging of the users. Also, the day range between the current and next trip positively affects normal charging at home/business.

2.3. Distribution Locational Marginal Pricing

The distribution network congestion may occur, with the high penetration level of EVs. However, the congestion problems can be handled by the distribution system operator (DSO) with the employment of market-based congestion control methods [15]. The way how locational marginal pricing (LMP) in transmission systems are obtained can be extended to the distribution systems [16], usually named as distribution locational marginal pricing (DLMP). It is known that the resistance of the distribution network lines is higher than that of transmission lines. Thus, the distribution system losses can be considered one of the main factors that affect the DLMP [17]. To deal with the EV demand congestion in distribution networks, Reference [18] proposes step-wise congestion management developed whereby the DSO predicts congestion for the next day and publishes day-ahead tariff before the clearing of the day-ahead market. Reference [19] solves the social welfare optimization of the distribution system considering EV aggregators as price takers in the local DSO market and demand price elasticity. Reference [20] presents a market-based mechanism using the DLMP concept to alleviate possible distribution system congestion due to EVs and heat pump integration. Additional,

Reference [21] propose a DLMP-based algorithm with quadratic programming to deal with the congestion in distribution networks with high penetration of EVs and heat pumps.

2.4. Simulation Tools

SUMO (Simulation of Urban MObility) [22] is perhaps the best-known traffic simulator. It is an intermodal and multimodal traffic flow simulation platform, which includes vehicles, public transportation, and pedestrians. SUMO has several tools that allow it to perform tasks such as locating routes, importing networks and calculating emissions. It can be enhanced with custom templates and provides multiple application programming interfaces (API) to control the simulation remotely.

MatSim [23] is a framework for large-scale, agent-based simulations. Each agent has a transport demand represented by a chain of activities that must be done in a day at different times and locations. Decisions on how to travel between places are planned before the simulation. [24] presents a method for the synthesis and animation of realistic traffic flows in large-scale road networks. It uses a technique based on a model of continuous traffic flow. Other multi-agent models are often used to create drivers behavior models.

When incorporating EVs into the simulation aspects such as power consumption, charging stations available and the charging duration must be considered [25]. The problem of the shortest path and travel planning is studied in [26], where the authors designed an approximation scheme to calculate the most energy-efficient path. In [27] it is possible to do traffic simulations using only electric vehicles, where EVs are simulated on roads with online charging. A similar case is that of [28], in which a spatial and temporal model was constructed to charge EVs in highway public chargers. Soares et al. [29] presents a probabilistic simulator that generates driving and charging profiles of EVs that can be customized to adapt to different distribution networks. It simulates how vehicles move to estimate the impacts that charging may have on each configuration of the system, the energy consumed or emissions.

There are also other simulation tools related to EVs. FASTSim [30] is a simulation tool that compares vehicles powertrain and estimates the impact of technological improvements to vehicle efficiency, performance, cost, and battery life. V2G-Sim uses individual driving and charging models of EVs to generate spatial and temporal impact/opportunity provisions in the electric grid [31]. Alegre et al. [32] proposes a pure and hybrid EV model, using a Matlab/Simulink environment, focusing on different aspects of the vehicle such as engine power, battery, and observing how the distance traveled and performance can be affected by the changes of the vehicles' features.

Table 1 presents a summary of the main characteristics of the reviewed tools. It shows that the reviewed simulation tools share some limitations, such as the lack of charging decisions using learned to charge behaviors, and missing variable prices. Our proposed model overcomes both these limitations, by incorporating dynamic adaptation of charging behaviors from EV users, and the application of variable charging prices in the simulation model. Moreover, the proposed model includes several components also considered by other simulators, such as the simulation and analysis of trips, and the modeling and analysis of charging stations. The proposed model only partially considers the electrical network distribution impact of the EV user decisions. The effect of changes in demand and generation throughout the time of the electricity prices is considered using the DLMP-based distribution network operation and reconfiguration optimization model. The aim is to overcome several limitations in the current state-of-the-art developments.

Table 1. Analysed tools.

Tool	Charging Decisions Using Learned Charging Behaviours	Variable Prices	Simulation/ Trip Analysis	Model/Charging Stations Analysis	Electrical Network Impact
[24]	No	No	Yes	No	No
MATSim [23]	No	No	No	No	No
SUMO [22]	No	No	Yes	No	No
[26]	No	No	Yes	Yes	No
[25]	No	No	No	No	Yes
[28]	No	No	No	Yes	Yes
EVeSSi [29]	No	No	Yes	No	Yes
V2G-Sim [31]	No	No	No	No	Yes
Proposed tool	**Yes**	**Yes**	**Yes**	**Yes**	**Partially**

3. Proposed Simulation Tool

In this section, the simulation tool parameters and algorithm are described. The tool allows the simulation of electric vehicle trips in a simple way, and it was developed using *R* language using *RStudio* integrated development environment [33].

3.1. Parameters

The global parameters of the simulator are described in Table 2. These parameters mean that they are applied to all the generated profiles, i.e., for any moment of the simulation they are the same. These are default values but can be changed according to user preferences.

Table 2. Global tool parameters.

Parameter	Description	Example Value
ncars	Number of EVs	5000
cdist	Compensatory distance between two points	20%
sf	Map scale	1
hcpower	Home charging power	3.7 kW
chargingeff	Charge efficiency	85%

3.2. Simulator Algorithm

The simulator consists of two main parts: data generation and simulation of trips. Data is generated concerning the profile of each user, such as vehicle features (battery, consumption, etc.), trips to be performed (locations and departure times) and behavioral parameters.

3.3. Data Generation

Population generation is an iterative process in which each of the variables is generated randomly from a sample of values with individual probabilities. Initially, each profile is assigned an initial location, depending on the available positions in the city map. This location will be a residence or a point of exit/entry into the city, considering users that live the city. Values are generated for the initial SoC, the preferred charging level, and the travel profile. It also generated the value of the battery capacity that will determine the rest of the characteristics of the vehicle. In the same way, a weight is assigned for the distance in terms of the charging station choosing, being the remaining weights attributed according to this value. The last data sets to be generated are the trips and times as well as their importance. Algorithm 1 has the following structure:

Algorithm 1 Data generation algorithm.

```
 1: for each of the cars do
 2:     Add an x coordinate to variable x
 3:     if x equal to some of the correspondent existent x available on the map then
 4:         Add y coordinate to y variable
 5:     end if
 6:     Generate initial SoC, available range preference, battery capacity and trip importance
 7:     Random generate w1
 8:     if w1 equals to a specific value then
 9:         w2 = 1-w1-w3
10:         w3 = 1-w1-w2
11:     end if
12:     if cars battery = value then
13:         Attribute all data to this car model in the cars data frame
14:     end if
15:     for i:=0 to 5 do
16:         Number of trips = 2, 3, 4 or 10-15
17:         Generate trips importance
18:         Generate locations for the number of trips
19:         Generate work day times, night times and/or leisure times
20:     end for
21: end for
```

3.4. Trip Simulation

The trips simulation runs in periods of 15 min, totalling to 96 ($j = 96$) for a full day. Its entire structure and mode of operation are described through a flow chart in Figure 2. Each vehicle has an initial location and a series of trips to be performed during the day. Each trip has a departure time, the period j in which the user will make that trip. When this happens, the Euclidean distance is calculated between the start location and the end location, with a margin of 20%, since the calculated distance is straight, and then it is multiplied by the scaling factor sf. Knowing the distance, the travel time is determined according to the average speed of the vehicle. For instance, if the calculated distance is 9000 m, and the average speed is 35 km/h, the travel time will be 15 min and 26 s, which is longer than one interval, and thus it will consume two periods. However, if the average speed is 40 km/h, the travel time will be 13 min and 30 s, which is equivalent to one period. The following equation determines the travel time:

$$T = \frac{\frac{d}{Vm \times \frac{1000}{3600}}}{60} \tag{1}$$

where:

T—Travel time (minutes)

d—Distance between destinations (meters)

Vm—Average vehicle speed (km/h)

3.5. Charging Stations

To simulate charging, four public charging stations (parking lot buildings) were created along with domestic chargers. Of the public stations, two are a slow charge (of 7.2 kW), and two are a fast charge (of 50 kW). The domestic chargers have a power of 3.7 kW.

The location of the stations was not chosen following a specific methodology. Their distribution covered all points of the city, with some randomness. In this sense, the objective is to understand what and how the various factors can influence the choice of charging sites and how energy prices influence EV users behavior.

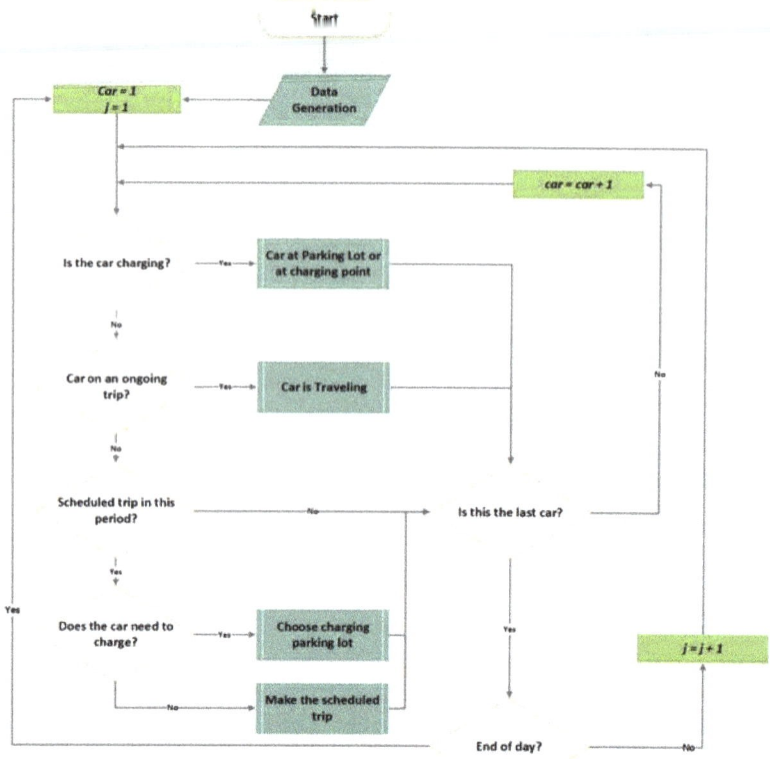

Figure 2. Flowchart of the travel simulation algorithm.

3.6. Charging Decisions

When the user decides to charge the EV, a location must be chosen (a charging station or house). For this simulation, three variables were considered: distance, energy price and charging time (slow or fast). After determining the scores of each of those variables and considering the preferences of each user, it is obtained the final score (Equation (2)). The charging station with the highest score is the one chosen to charge the vehicle.

$$FinalScore = Ds \times w_1 + Ps \times w_2 + Cts \times w_3 \tag{2}$$

where:

Ds—Distance score from 0 to 100

Ps—Price score from 0 to 100

Cts—Charge time score from 0 to 100

w—Weight for each of the variables (w_1 for distance preference; w_2 for price preference; w_3 for time preference)

The process of selecting the preferred place to charge follows the structure described in Figure 3. The distances to slow charging stations are calculated. These values, together with the energy price (€/kWh) and the charging time that the user has, allows a final score between 0 and 100 for each station. If the vehicle allows fast charging, this process is repeated for these types of charging stations. Finally, the scores of the charging stations are compared, and the highest is chosen.

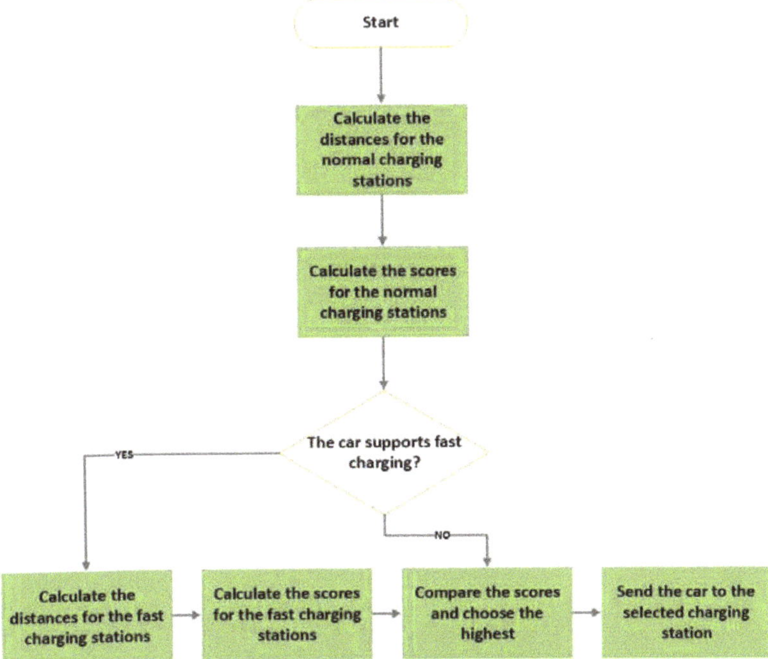

Figure 3. Flowchart for the charging station selection.

To determine how long each user can delay or not a trip to charge his vehicle a variable was introduced that defines its importance. Thus, three different levels were assigned:

1. *Low importance*: this trip is discarded, and the car is charged until the next trip;
2. *Medium importance*: the user delays the trip, and all subsequent ones, until a time limit that varies according to the user profile;
3. *High importance*: the user must carry out this trip, not being able to charge, unless the level of battery charge reaches critical values.

To ensure that each user always has sufficient charge to make the trips, it was considered a critical battery level. Following the results of [9] this value is set to 20%. Whenever a user reaches a level lower than this, regardless of the trip, the car must be charged. In this case, there are two options: either the user finds a place near the workplace (1st destination) and leaves the car there until the next trip, or looks for the nearest parking place and leave it there overnight until the next scheduled trip. It is assumed that the user leaves the car at this location and, hypothetically, does the rest of its travel using another mean of transportation.

3.7. Energy Prices

One of the variables that the users consider when deciding where to charge their vehicle is the energy price. This price differs between the type of station (slow or fast) and if they are public or domestic. Also, there are two domains where prices differ: fixed prices or variable prices.

In terms of simulation, in the case of fixed prices the user always pays the same regardless of the charging time. The only difference is whether it is a fast charging station or a domestic one, as a fast charging station is more costly. The variable prices vary by 15-min intervals. This is accomplished using the model described in Section 2.3. Firstly, an additional cost which is related to the fixed term of network price rate to be charged to the customer (Equation (3)) is calculated:

$$ACNR = \frac{\left(\frac{0.397 \cdot CP}{720}\right)}{OPR} \tag{3}$$

where:

$ACNR$—additional cost related to the fixed term of network price rate [€/kWh]. The contracted power cost is 0.397 €/kW/month paid to the DSO monthly (www.erse.pt)

CP—charging power of the charging station. 720 h per month

OPR—the park occupation rate.

Then the final energy price for the consumer is calculated (Equation (4)). This value is the sum of the DLMP received, with the energy tariff and the additional price previously calculated (Equation (3)). To Equation (4) a fee of 5% is added by the owner of the charging station and the VAT value.

$$Final\ Price = (DLMP + Tariff\,MV + ACNR) \times PLG \times VAT \tag{4}$$

where:

$DLMP$—Distribution locational marginal pricing [€/kWh]

$Tariff\,MV$—Energy tariff price for each period [€/kWh]

PLG—Additional profit margin of the parking owner

VAT—Value added tax

3.8. DLMP Optimisation Model Description

In this research work, the DLMPs (which will permit to determine the variable charging price (see Equation (4))) are defined through Lagrangian multipliers of the corresponding constraints (power balance) of the optimisation problem which has the goal to minimise the DSO expenditures [34]. Thus, the DSO seeks to:

- Minimize the power losses cost;
- Minimize the power not supplied cost;
- Minimize the power lines congestion cost;
- Minimize the power generation curtailment cost;
- Minimize the power from external suppliers cost.

The DLMPs optimization problem is classified as mixed-integer nonlinear programming (MINLP) due to the non-linearity features. To solve complex problems like this, Benders decomposition is an adequate technique [35,36]. The following constraints are considered:

- Network constraints:

  - Voltage;
  - Power balance;
  - Power flow equations;
  - Maximum admissible line flow.

- Supplier constraints:

  - Maximum and minimum limits for the power supplier;
  - Maximum and minimum limits for capacitor banks.

- Curtailment constraints:

  - Power generation curtailment;
  - Power not supplied.

- Lines congestion;

- Energy storage systems constraints:

    - Charge and discharge limit;
    - Charge and discharge limit considering energy storage systems state;
    - State of charge;
    - Maximum and minimum energy storage systems capacity limit.

4. Case Studies

To carry out the case studies a physical model of SC by GECAD-BISITE [34] was used. The considered SC presents five types of loads, namely:

- Residential buildings (1375 homes);
- Office buildings (7 buildings);
- Hospital;
- Fire Station;
- Shopping Mall.

The schematic of the SC is presented in Figure 4 and the coordinates of each building can be seen in Table 3). The distribution network that feeds the entire city has one 30MVA substation and 25 load points. A total of 15 DG units (i.e., 2 wind farms and 13 PV parks), four capacitor banks of 1 Mvar, and are included in the network, as can be seen in Figure 5. Moreover, the SC has seven parking lot buildings (commonly referred as charging stations in this research work) for EV charging, four (two in bus 7 and two in bus 11) slow charging lots (7.2 kW for each connection point) and three (two in bus 2 and one in bus 5) fast charging lots (50 kW for each connection point). Each slow charging parking lot has 250 spaces for EVs and 80 spaces for each fast charging parking lot building. The considered value for OPR is 30% leading to an ACNR value of 0.0132 €/kWh for a slow charging parking space and a value of 0.0919 €/kWh for a fast charging parking space. Additionally, the parking owner charges an additional 5% fee and 23% of value-added tax (VAT). Furthermore, it is considered that 50% of the EV users can charge their EVs at home (3.7 kW charge point) with a fixed cost of 0.2094 €/kWh. The initial EV battery level is randomly generated between 40% and 65% of the battery capacity and the considered EV models, and their characteristics can be found in Table 4.

Table 3. Building coordinates on the xy plane.

Building		L1	L2	L3	L4 to L18	L19	L20	L21	L22	L23	L24	L25	PL1 to PL2	PL3 to PL4	PL5 to PL6	PL7
Coordinates (km)	X Axis	10.50	0.50	9.00	3.75 to 8.25	0.50	0.50	2.50	3.00	4.50	6.00	8.00	1.00	7.00	6.00	11.00
	Y Axis	3.50	2.00	5.00	1.00 to 3.00	3.50	5.50	2.00	4.50	3.50	5.00	4.00	3.50	5.00	0.50	4.00

Table 4. EVs types.

Model	Battery (kWh)	Slow Charge Power (kW)	Fast Charge Power (kW)	Consumption (kWh/km)
Nissan Leaf	40.00	6.60	50.00	0.1553
Tesla Model S 70D	75.00	7.40	50.00	0.2100
BMW i3	33.20	7.40	50.00	0.1584
Renault Zoe	41.00	7.40	-	0.1460
Renault Kangoo	33.00	7.40	-	0.1926
VW e-Golf	24.20	7.20	40.00	0.1584
Ford Focus	33.50	6.60	50.00	0.1926
Hyundai IONIQ	30.50	6.60	50.00	0.1429

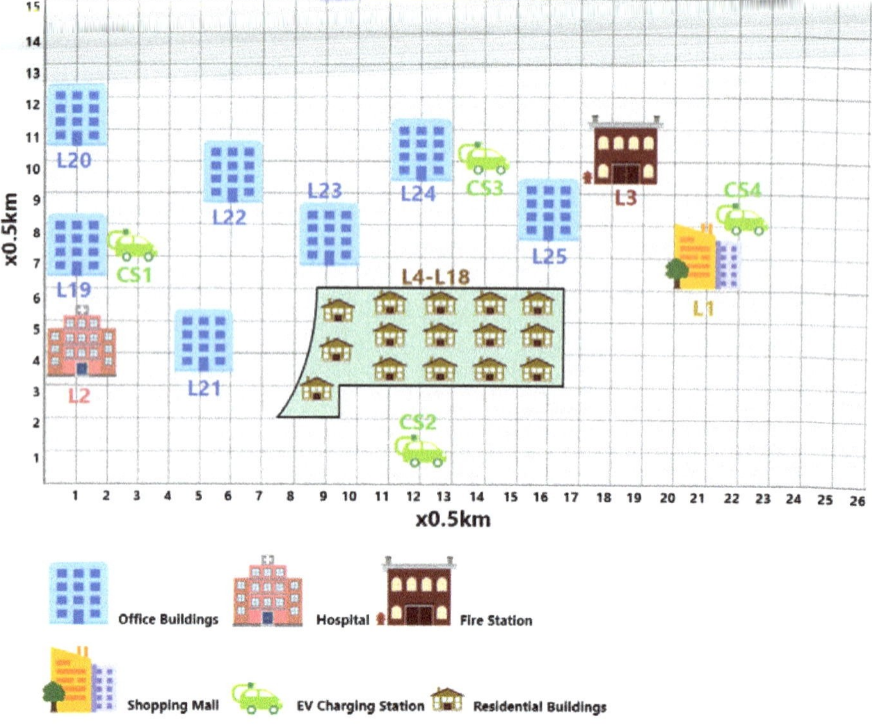

Figure 4. Smart city diagram [34].

Three user preference scenarios are considered, in one, the user's priority is to charge their EV at a charging station located as close as possible to them. In the second scenario, the users prefer to find charging stations where they can charge their EV at a low price. In the third scenario, the EV user gives its preference to the charging time.

The line congestion cost is 0.02 €/kW when power flow is above 50% of the thermal line rating capacity.

The study presented in this research paper considers one week of input data for every 15 min with the aim of showing the effectiveness of the proposed model (i.e., 672 periods are considered in the simulation process). The chosen week is the 19 March 2017 to 25 March 2017.

This work has been developed on a computer with one Intel Xeon E5-2620 v2 processor and 16 GB of RAM running Windows 10 Pro. In addition to *R* language (for EV user behavior simulator), the MATLAB R2016a and TOMLAB 8.1 64 bits with CPLEX and SNOPT solvers were used for the optimization problems.

Simulations were performed using fixed and variable energy charging prices for two different populations scenarios, i.e., considering 2500 EVs and 5000 EVs. For each simulation, the following user preferences were changed: distance, price and charging time. The following features are fixed for all simulations in each population scenario:

- The amount of vehicles and their models;
- The initial battery charge;
- The amount of trips;
- The trips schedule;
- The starting locations.

The fixed charging prices are equal for all periods of the day and are 0.15 €/kWh for slow charge and 0.25 €/kWh for fast charge.

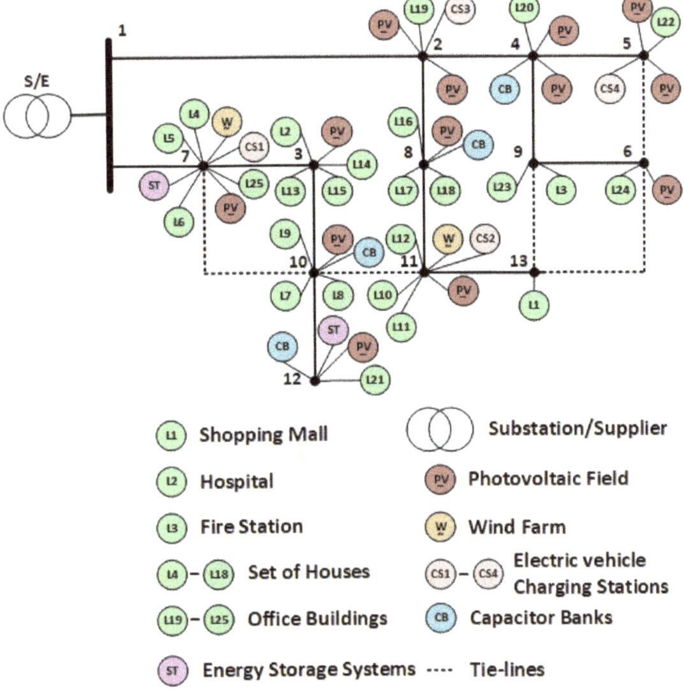

Figure 5. 13-bus distribution network diagram [34].

5. Results and Discussion

This section presents the results for the carried out simulations. Section 5.1 presents the results for a scenario with 2500 EVs, while Section 5.2 provides the results for a 5000 EVs population scenario.

5.1. Population Scenario with 2500 EVs

Considering a 2500 EVs scenario and using the fixed charging price, it is possible to see in Figure 6 the correspondent charging sessions percentages for each user scenario preference (distance Figure 6a, price Figure 6b, and time Figure 6c). It is worthy to note, that one charging session is counted from the moment the EV beings to charge until the time of it leaves the charging station. Analysing Figure 6a (where the preference is the charging stations proximity to the total path that the user will have to do, i.e., the lowest sum value of the distance between the current location and the CS and the distance between the CS and the next EV user destination), it can be seen that the charging station 2 was preferred by users, with 37% of charging sessions. Since for this user preference scenario, the only differentiation between normal charging stations is the distance, and it can be concluded that CS 2 will be nearest to the users' destinations when compared to the remaining CSs. Figure 6b) (EV users' gives priority to the price over the distance and charging time at the moment to chose a CS) shows that the CS 2 presents the higher charging sessions around 47% while the CS 1 was the second chosen one. The CS 3 and CS 4 (fast charging stations) presents together only a percentage around 21% of the total charging sessions. This is as expected result once the slow charging stations present lower charging prices compared to the fast charging stations. When the user time preference scenario is considered the majority of the users prefer the fast charging stations. As can be seen in Figure 6c fast charging stations present 55.2% of the total charging sessions. However, this value also shows that the influence of the distance (CS 2 lower distance) and charging price (slow charging stations—lower price) have a strong influence on the users'.

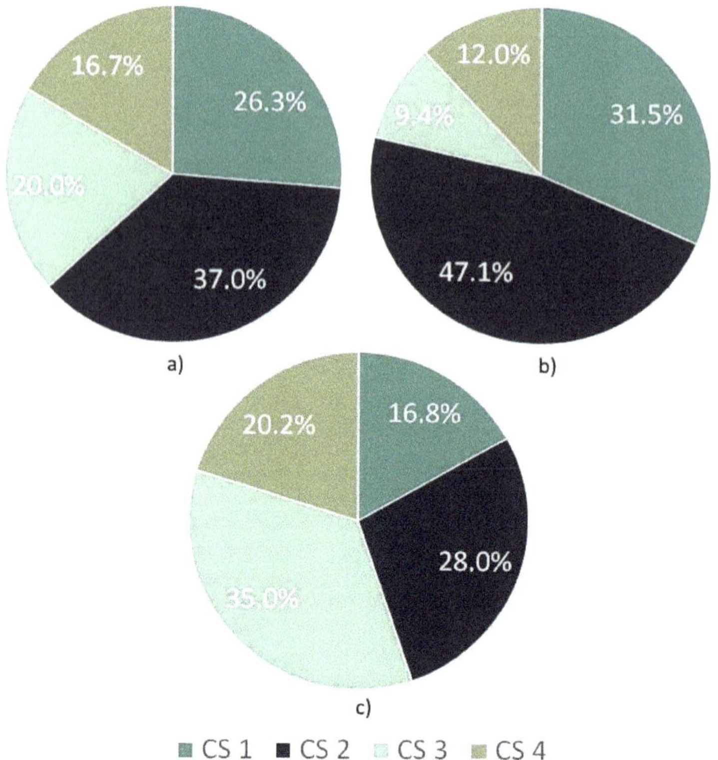

Figure 6. Charging sessions using fixed charging prices considering the 2500 EVs scenario. (**a**) User distance preference scenario charging sessions. (**b**) User price preference scenario charging sessions. (**c**) User time preference scenario charging sessions.

Figure 7 depicts the charging sessions when the variable charging prices model is used. Like the fixed prices method when user distance preference is considered the variable model charging price also gives more sessions to CS 2 (Figure 7a), leading to the same conclusion—this is the CS nearest to the users' destinations. Checking Figure 7b it is possible to conclude that CS 1 is the cheapest one due to the high percentage of charging sessions (71.7%). Regarding to the user time preference scenario (Figure 7c), the results are very similar to the case where the fixed prices are considered, i.e., the majority of the users preferring the fast charging stations.

For the user distance and time preference scenarios, the fast charging station obtained a higher preference when compared to the case where the user preference is defined by the price. This indicates that when the price is not the most important factor, fast charging stations can attract users who are located close to them. Nevertheless, we concluded that the location of those charging stations is not optimal, because even when considering the user distance preference scenario, the majority of the users choose the slow charging stations—the cheapest ones. Moreover, to highlight this conclusion, when charging time is the most important factor the slow charging stations also present high preference, being the CS 2 the second most preferred.

A comparison between the fixed charging prices and the proposed variable charging prices model are shown in Figure 8. As can be seen, the proposed model presents advantages in all scenarios for the EVs users in terms of charging prices. When the user distance preference scenario is considered, the proposed model presents 4% of gains for users' (0.0083 €/kWh), for price and time user preference scenario the benefits are 10% (0.0210 €/kWh), and 2% (0.0046 €/kWh), respectively.

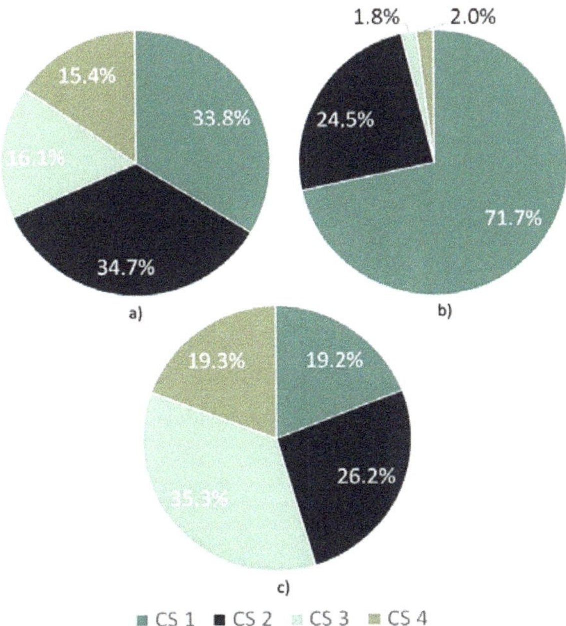

Figure 7. Charging sessions using variable charging prices considering the 2500 EVs scenario. (**a**) User distance preference scenario charging sessions. (**b**) User price preference scenario charging sessions. (**c**) User time preference scenario charging sessions.

5.2. Population Scenario with 5000 EVs

In this subsection, it is presented the simulation results for a scenario with 5000 EVs. Figure 9 presents the charging session results for each Charging station in the three EV user scenarios preference considering fixed charging prices. The achieved conclusions are the same when the scenario with 2500 EVs is considered. Seeing Figure 9a it is also checked through the presented values that CS 2 is near one to the users' destinations. Also, when the price preference is considered, the slow charging stations are proffered (Figure 9b), since they present the lower charging prices. When the charging time is crucial for the EVs users, the set of charging stations presents a percentage of charging sessions around 51% (Figure 9c). Nevertheless, it is also verified that slow charging stations have a strong influence on the users' choice.

The charging sessions result considering the variable charging price model is depicted in Figure 10. Once again, the results are very similar to the scenario with 2500 EVs, with more charging sessions in CS 2 when it is considered the user distance preference scenario (Figure 10a)—CS 2 is the near one to the users' destinations. For the user price preference scenario, the CS 1 presents the higher percentage of charging sessions (Figure 10b). Thus, it can be concluded that CS presents more competitive charging prices compared to the remaining CSs. Considering the user time preference scenario, the higher percentage of EVs users' (Figure 10c) prefer the fast charging stations (51%).

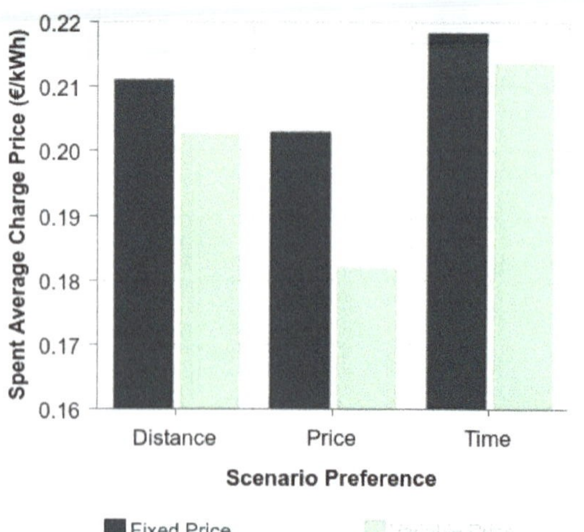

Figure 8. Average charging price comparison considering the 2500 EVs scenario.

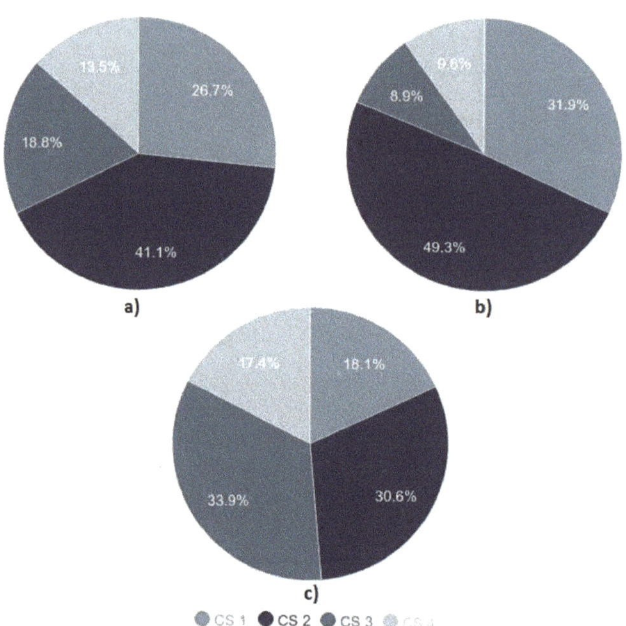

Figure 9. Charging sessions using fixed charging prices considering the 5000 EVs scenario. (**a**) User distance preference scenario charging sessions. (**b**) User price preference scenario charging sessions. (**c**) User time preference scenario charging sessions.

The conclusions are the same when the scenario with 2500 EVs is considered, i.e., if time is not the most important factor, the users can be attracted by the fast charging station which can be closer to them. However, once again, it can be seen that the location of fast charging stations is not optimal (the majority of the users choose the slow charging stations even when the user distance preference scenario is considered). Also, as in the 2500 EVs scenario, when the most important factor for the users is the charging time, the slow charging stations also present a considerable preference, with the CS 2 as the second most preferred.

Figure 11 presents a comparison between the fixed charging prices and the proposed variable charging prices model. The proposed variable charging price model presents considerable advantages for the EVs users' when distance and price preference scenarios are considered, with gains of 5% (0.0120 €) and 18% (0.0418 €), respectively. Regarding user charging time preference scenario the variable charging price model does not present advantages in terms of charge price for the EVs users when compared with fixed charging price (3% higher—0.0073 €).

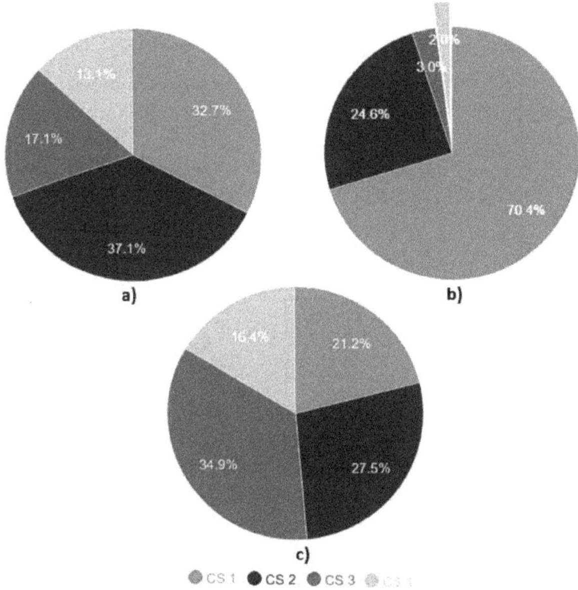

Figure 10. Charging sessions using variable charging prices considering the 5000 EVs scenario. (**a**) User distance preference scenario charging sessions. (**b**) User price preference scenario charging sessions. (**c**) User time preference scenario charging sessions.

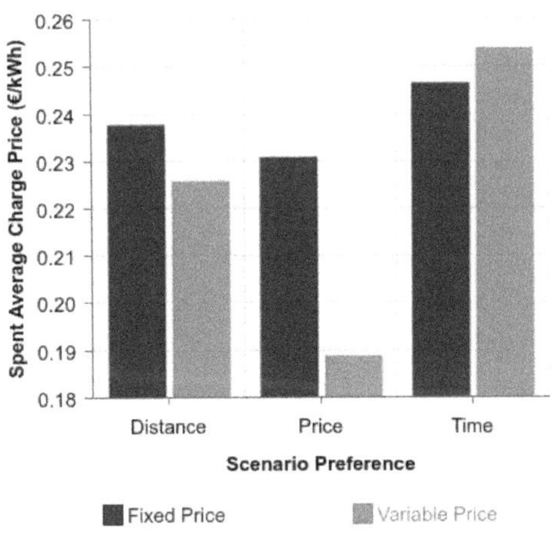

Figure 11. Average charging price comparison considering the 5000 EVs scenario.

6. Conclusions

This research paper presents a study of the impact of the variation of the energy charging prices on the behavior of electrical vehicle users. It also compared its benefits when using the variable and fixed charging prices. To this end, the authors developed an EV behavior simulator and combined it with an DLMP-based distribution network operation and reconfiguration optimization model. The main contributions of the conducted study can be summarized as follows: (1) EV user behavior simulator has been developed to generate a realistic population, considering the city size, and charging stations; (2) the positive impact of the variable EV charging prices on the electric vehicles users has been assessed.

The proposed methodology was tested in a case study which has been conducted on a mock-up model of a SC located at the BISITE laboratory with a 13 buses distribution network. Moreover, three users scenarios preferences (distance, price and time) were considered and were used to compare the results of the variable EV charging prices and EV fixed charging prices to demonstrate the advantage of the former.

It was verified that the use of variable pricing for EV charging is advantageous for the EV users in all scenarios when it is considered 2500 EVs. The gains are 4%, 10%, and 2%, respectively for distance, price, and time preferences. With 5000 EVs, the variable pricing does not present savings in comparison with the fixed charging prices when time scenario preference is considered. However, the proposed variable charging price model still presents considerable savings when the distance and price preference scenarios are considered. These two scenarios present for EV users 5% and 18% of gains, respectively.

The results suggest that the use of variable prices is promising, and can be used as an efficient approach in smart cities by offering to EVs' users more options (in terms of price) when deciding where to charge their EVs.

The main disadvantages of the proposed model are: (a) the EV users profiles are not adapted to the different weekdays; (b) the decision charge method is only based on the battery charge level; (c) vehicle-to-grid is not considered.

In terms of future work, the authors will address more user profiles and additional charging decisions that depend on the energy price (increasing the flexibility), and also the possibility of vehicle-to-grid.

Author Contributions: The authors contributions were: Conceptualization, B.C., J.S., A.C., T.P., F.L., P.N., Z.V.; Methodology, B.C., J.S., A.C., T.P., F.L., P.N., Z.V.; Software, B.C., J.S., A.C., T.P., F.L., P.N., Z.V.; Validation, B.C., J.S., A.C., T.P., F.L., P.N., Z.V.; Formal analysis, B.C., J.S., A.C., T.P., F.L., P.N., Z.V.; Investigation, B.C., J.S., A.C., T.P., F.L., P.N., Z.V.; Writing—original draft preparation, B.C., J.S., A.C., T.P., F.L., P.N., Z.V.; Writing—review and editing, B.C., J.S., A.C., T.P., F.L., P.N., Z.V.; Visualization, B.C., J.S., A.C., T.P., F.L., P.N., Z.V.

Funding: This work has received funding from FEDER Funds through COMPETE program, from National Funds through FCT projects UID/EEA/00760/2019 and UID/CEC/00319/2019 and PTDC/EEI-EEE/28983/2017-CENERGETIC. Bruno Canizes is supported by FCT Funds through SFRH/BD/110678/2015 PhD scholarship. Angelo Costa is supported by the FCT Post-Doc Grant SFRH/BPD/102696/2014.

Conflicts of Interest: The authors declare no conflict of interest.

References

1. United Nations. Chapter XXVII—Environment—7. d Paris Agreement, 2015. Available online: https://trea ties.un.org/doc/Publication/MTDSG/Volume%20II/Chapter%20XXVII/XXVII-7-d.en.pdf (accessed on 17 April 2019).

2. European Commission. *Greenhouse Gas Emission Statistics—Emission Inventories*; Technical Report; European Commission: Brussels, Belgium, 2018. Available online: https://ec.europa.eu/eurostat/statistics-explained /pdfscache/1180.pdf (accessed on 17 April 2019).

3. Salah, F.; Ilg, J.P.; Flath, C.M.; Basse, H.; van Dinther, C. Impact of electric vehicles on distribution substations: A Swiss case study. *Appl. Energy* **2015**, *137*, 88–96. [CrossRef]

4. European Environment Agency. *Electric Vehicles and the Energy Sector—Impacts on Europe's Future Emissions*; Technical Report; European Union: Copenhagen , Denmark, 2017. Available online: https://www.eea.europa.eu/themes/transport/electric-vehicles/electric-vehicles-and-energy (accessed on 17 April 2019).

5. Daina, N.; Sivakumar, A.; Polak, J.W. Electric vehicle charging choices: Modelling and implications for smart charging services. *Transp. Res. Part C Emerg. Technol.* **2017**, *81*, 36–56. [CrossRef]

6. Latinopoulos, C.; Sivakumar, A.; Polak, J. Response of electric vehicle drivers to dynamic pricing of parking and charging services: Risky choice in early reservations. *Transp. Res. Part C Emerg. Technol.* **2017**, *80*, 175–189. [CrossRef]

7. International Energy Agency. *Global EV Outlook 2018*; Technical Report; International Energy Agency: Paris, France, 2018. Available online: https://webstore.iea.org/global-ev-outlook-2018 (accessed on 17 April 2019).

8. EVBox. *Manifesto of Electric Mobility*; Technical Report; EVBox: Amsterdam, The Netherlands, 2017. Available online: https://info.evbox.com/manifesto-electric-mobility (accessed on 17 April 2019).

9. Corchero, C.; Gonzalez-Villafranca, S.; Sanmarti, M. European electric vehicle fleet: Driving and charging data analysis. In Proceedings of the 2014 IEEE International Electric Vehicle Conference (IEVC), Florence, Italy, 17–19 December 2014.

10. Franke, T.; Krems, J.F. Understanding charging behaviour of electric vehicle users. *Transp. Res. Part F Traffic Psychol. Behav.* **2013**, *21*, 75–89. [CrossRef]

11. Marmaras, C.; Xydas, E.; Cipcigan, L. Simulation of electric vehicle driver behaviour in road transport and electric power networks. *Transp. Res. Part C Emerg. Technol.* **2017**, *80*, 239–256. [CrossRef]

12. Neubauer, J.; Wood, E. The impact of range anxiety and home, workplace, and public charging infrastructure on simulated battery electric vehicle lifetime utility. *J. Power Sources* **2014**, *257*, 12–20. [CrossRef]

13. Nicholas, M.A.; Tal, G.; Turrentine, T.S. *Advanced Plug-in Electric Vehicle Travel and Charging Behavior Interim Report*; Research Report—UCD-ITS-RR-16-10; Institute of Transportation Studies—University of California: Davis, CA, USA, 2017.

14. Xu, M.; Meng, Q.; Liu, K.; Yamamoto, T. Joint charging mode and location choice model for battery electric vehicle users. *Transp. Res. Part B Methodol.* **2017**, *103*, 68–86. [CrossRef]

15. Biegel, B.; Andersen, P.; Stoustrup, J.; Bendtsen, J. Congestion Management in a Smart Grid via Shadow Prices. *IFAC Proc. Vol.* **2012**, *45*, 518–523. [CrossRef]

16. Bohn, R.E.; Caramanis, M.C.; Schweppe, F.C. Optimal Pricing in Electrical Networks over Space and Time. *RAND J. Econ.* **1984**, 360–376. [CrossRef]

17. Sotkiewicz, P.M.; Vignolo, J.M. Nodal pricing for distribution networks: Efficient pricing for efficiency enhancing DG. *IEEE Trans. Power Syst.* **2006**, *21*, 1013–1014. [CrossRef]

18. O'Connell, N.; Wu, Q.; Østergaard, J.; Nielsen, A.H.; Cha, S.T.; Ding, Y. Day-ahead tariffs for the alleviation of distribution grid congestion from electric vehicles. *Electr. Power Syst. Res.* **2012**, *92*, 106–114. [CrossRef]

19. Li, R.; Wu, Q.; Oren, S.S. Distribution Locational Marginal Pricing for Optimal Electric Vehicle Charging Management. *IEEE Trans. Power Syst.* **2014**, *29*, 203–211. [CrossRef]

20. Liu, W.; Wu, Q.; Wen, F.; Ostergaard, J. Day-ahead congestion management in distribution systems through household demand response and distribution congestion prices. *IEEE Trans. Smart Grid* **2014**, *5*, 2739–2747. [CrossRef]

21. Huang, S.; Wu, Q.; Oren, S.S.; Li, R.; Liu, Z. Distribution Locational Marginal Pricing Through Quadratic Programming for Congestion Management in Distribution Networks. *IEEE Trans. Power Syst.* **2015**, *30*, 2170–2178. [CrossRef]

22. Krajzewicz, D.; Erdmann, J.; Behrisch, M.; Bieker-Walz, L. Recent Development and Applications of SUMO—Simulation of Urban MObility. *Int. J. Adv. Syst. Meas.* **2012**, *5*, 128–138.

23. Axhausen, K.W. *The Multi-Agent Transport Simulation MATSim*; Ubiquity Press: London, UK, 2016.

24. Sewall, J.; Wilkie, D.; Merrell, P.; Lin, M.C. Continuum Traffic Simulation. *Comput. Graph. Forum* **2010**, *29*, 439–448. [CrossRef]

25. de Arcaya, A.D.; Lázaro, G.A.; González-González, A.; Sanchez, V. Simulation Platform for Coordinated Charging of Electric Vehicles. In Proceedings of the 6th IC-EpsMsO: International Conference on "Experiments/Process/System Modelling/Simulation/Optimization", Athens, Greece, 8–11 July 2015; pp. 8–11.

26. Strehler, M.; Merting, S.; Schwan, C. Energy-efficient shortest routes for electric and hybrid vehicles. *Transp. Res. Part B Methodol.* **2017**, *103*, 111–135. [CrossRef]

27. Mou, Y.; Xing, H.; Lin, Z.; Fu, M. Decentralized Optimal Demand-Side Management for PHEV Charging in a Smart Grid. *IEEE Trans. Smart Grid* **2015**, *6*, 726–736. [CrossRef]
28. Bae, S.; Kwasinski, A. Spatial and Temporal Model of Electric Vehicle Charging Demand. *IEEE Trans. Smart Grid* **2012**, *3*, 394–403. [CrossRef]
29. Soares, J.; Canizes, B.; Lobo, C.; Vale, Z.; Morais, H. Electric Vehicle Scenario Simulator Tool for Smart Grid Operators. *Energies* **2012**, *5*, 1881–1899. [CrossRef]
30. Brooker, A.; Gonder, J.; Wang, L.; Wood, E.; Lopp, S.; Ramroth, L. *FASTSim: A Model to Estimate Vehicle Efficiency, Cost, and Performance*; SAE Technical Paper 2015-01-0973; SAE International: Warrendale, PA, USA, 2015. [CrossRef]
31. V2G-Sim. Available online: http://v2gsim.lbl.gov/ (accessed on 17 April 2019).
32. Alegre, S.; Míguez, J.V.; Carpio, J. Modelling of electric and parallel-hybrid electric vehicle using Matlab/Simulink environment and planning of charging stations through a geographic information system and genetic algorithms. *Renew. Sustain. Energy Rev.* **2017**, *74*, 1020–1027. [CrossRef]
33. RStudio. Available online: https://www.rstudio.com/ (accessed on 17 April 2019).
34. Canizes, B.; Soares, J.; Vale, Z.; Corchado, J.M. Optimal Distribution Grid Operation Using DLMP-Based Pricing for Electric Vehicle Charging Infrastructure in a Smart City. *Energies* **2019**, *12*, 686. [CrossRef]
35. Conejo, A.J.; Carrión, M.; Morales, J.M. *Decision Making Under Uncertainty in Electricity Markets*; Volume 153, International Series in Operations Research & Management Science, Springer: Boston, MA, USA, 2010. [CrossRef]
36. Soares, J.; Canizes, B.; Ghazvini, M.A.F.; Vale, Z.; Venayagamoorthy, G.K. Two-Stage Stochastic Model Using Benders' Decomposition for Large-Scale Energy Resource Management in Smart Grids. *IEEE Trans. Ind. Appl.* **2017**, *53*, 5905–5914. [CrossRef]

Article

Electric Vehicle Charging Process and Parking Guidance App

Gonçalo Alface [1,*] [iD], **João C. Ferreira** [1,2] [iD] **and Rúben Pereira** [1,*] [iD]

1   Instituto Universitário de Lisboa (ISCTE-IUL), ISTAR-IUL, 1649-026 Lisboa, Portugal; jcafa@iscte.pt
2   INOV INESC Inovação—Instituto de Novas Tecnologias, 1000-029 Lisboa, Portugal
*   Correspondence: gpaea@iscte-iul.pt (G.A.); rfspa@iscte-iul.pt (R.P.); Tel.: +351-210-464-277 (G.A.)

Received: 16 May 2019; Accepted: 30 May 2019 ; Published: 3 June 2019

Abstract: This research work presents an information system to handle the problem of real-time guidance towards free charging slot in a city using past date and prediction and collaborative algorithms since there is no real-time system available to provide information if a charging spot is free or occupied. We explore the prediction approach using past data correlated with weather conditions. This approach will help the driver in the daily use of his electric vehicle, minimizing the problem of range anxiety, provide guidance towards charging spots with a probability value of being available for charging in a context for the app and smart cities. This work handles the uncertainty of the drivers to get a suitable and vacant place at a charging station because missing real-time information from the system and also during the driving process towards the free charging spot can be taken. We introduce a framework to allow collaboration and prediction process using past related data.

Keywords: electric vehicle; charging station; prediction; probability; mobile App

---

1. Introduction

Electric Vehicles (EV) provide an environment-friendly solution to combustion engine vehicles and modern cities' transportation [1], but there still exists some disadvantages of driving EVs, namely the autonomy of the batteries still being to low when compared to combustion engine vehicles [2], the time it takes to charge an EV, usually six to eight hours at home and thirty minutes at a high voltage Charging Station (CS) [1] and also because recharge rate occurs at different rates depending on the initial state of the charge [3], and the number of stations providing full recharge of batteries is quite rare [4]. A diversity of challenges and opportunities are identified in [5] and one of the first work to show the importance of EV information in mobile Apps [6].

Another solution this study tries to tackle is the availability of the CS to recharge the EVs. As we know, charging an EV could be very time-consuming [7], so reaching a CS and all of the connectors are being used could be frustrating and make us wait in line. This type of problem could well be compared to the prediction problem for vehicles open parking spaces, as generally people with EVs leave their car parked in the CS and go on with their lives, as well as the meteorological effects on the willingness of users to leave their car at those places [8].

Taking into considerations we try to provide one of the first solutions to this problem, creating a probabilistic recommended system in a mobile app.

In this research work, we handle the problem of the need for real-time information of free charging spots for EV without a connection to the proprietary charging system. Since this connection is not available we apply prediction algorithms taking into account past charging sessions data correlated with hour, day, month and atmospheric conditions, city location and also a collaborative approach. We propose a kind of parking recommendation system that aims to present guidance towards this charging spots during the driving process, while considering the maximum distance the EV can reach and the braking regeneration that the EVs currently have.

This is helpful information for EV drivers daily journey and the is one of the most important problems for the EV drivers when the availability for charging points is not enough taking into account the increasing number of EV.

Our approach combines information, mobile devices, prediction algorithms and communication technologies, Web services, geo-positioning techniques for deliverance and exchange of information and linear programming for optimal scheduling, guidance and prediction of electrical charging processes.

2. Literature Review

In this section we created a section related to EV (Section 2.1) and consider similar existing solution systems to guidance for EV towards a CS, where is needed a prediction process because most of the cases we do not have sensor information to check if a place is occupied or not (Section 2.2).

2.1. Optimal EV Path and Important Constraints

Energy efficient and optimal routing, when applied to EVs, are not the same for a gas powered vehicles, since for the last one the only worry is simply understanding edge costs as energy values and applying standard algorithms. This new paradigm of energy-optimal routing for EVs must take into account multiple different heuristics, namely the recovery of some energy due to their kinetic and/or potential energy during deceleration phases, able to increase the cruising range of EVs up to 20% when in typical urban setting and hilly areas [4], meaning that edge costs could be negative [9], the costs of the edges may depend on parameters only known at query time [10], and at last, battery capacity limitations that implies that the cost of path is no longer just the sum of its edge costs, since additional energy losses or gains can occur from taking a path [7].

So, we can conclude that rather than finding the fast or short routes, there is a bigger necessity of finding energy-efficient routes and the goal of a guidance route is to start at a certain point and with a battery charge level and find a route to the destination point that respects the EVs constraints and reaching the endpoint with the maximum battery charge possible. If we ignore the battery constraints, the problem of finding the most energy-optimal path from a point to other can be described as just finding the shortest path between the two points, which can be typically solved using a regular shortest path algorithm, like Dijkstra's algorithm [4]. Dijkstra's algorithm cannot be directly used in this type of problems, as edge costs for an EV can be negative and the battery constraints which seem to require special algorithmic treatment [9], also has the disadvantage of expanding more vertices than the actual need, as it can expand all vertices before the shortest path is found, showing a big running time [10].

An algorithm is developed in [4] that extends the generic shortest path algorithm by taking the initial constraints into account, namely, the route segments can only be feasible if the required energy to go through does not exceed the charging level of the battery. Other heuristics to take into account is there are road sections where the energy consumption is negative where the EVs battery can be recharged, due to elevation differences or changes in the cruising speed. Just like in [4], for [9] the battery constraints are never running out of energy, there can not be overcharging, meaning that the battery current charge status can never exceed the maximum charge level and that an EV cannot take a trip and end up with a higher charge status than the one it started with. So, to perform the decision, a route only is feasible if there is no point where the required energy charge exceeds the current charge level from the EV, and a route is less preferred if there is a point where the maximum capacity is already reached and energy can be recuperated. This algorithm computes the shortest path tree using one of the four expansion strategies Dijkstra, Expand, Expand-Distance and FIFO. The shortest path tree represents the solution of all shortest path problems with the source vertex and the strategy determines the order in which vertices are processed.

In [9] the battery constraints are modeled by cost function on the edges obeying the FIFO property. Results show that the current implementation of Dijkstra and the results from Contraction Hierarchies outperform those found by [4], where CH shows the lowest running time for a significant amount.

Results also show that battery capacity constraints can be modelled as cost functions on the edges, and a transformation of the edge cost functions permits the application of Dijkstra's algorithm.

The model created in [10] takes into consideration some default parameters to calculate the amount of energy consumed or gained by the EV when passing by the edge of the network like the cars speed is always the speed limit of each road section. Other further parameters that influence vehicle energy consumption, such as its mass, wind resistance, friction coefficients, and more, are also taken into account. With these constraints in mind, the authors created a definition for the energy-optimal routing problem to find a path P in the graph from one source point to a destination point with a minimal path cost. Those paths correspond to energy-optimal routes which are feasible to use and where the remaining battery charge at the destination is maximal. The second contribution of this study is applying the A* algorithm, that determines the shortest path between two points, pursuing paths that appear to be the best routes to the destination based on the available information, with a non-trivial and consistent heuristic rather than the common Dijkstra's algorithm. A* is also know to be optimally efficient, as it expands the fewest numbers of vertices among all the search algorithms having access to the same heuristics. The weight function of the A* algorithm is then modified to take into account the battery constraints. The Energy-A* algorithm shows promising results, as it is faster than others algorithms used as a baseline, especially for smaller distances where the improvement is significant, and also shows fewer node expansions and consequently fewer evaluations of the energy cost function created. The approach described in [10] avoids the use of preprocessing techniques so that edge costs can be calculated dynamically, and it achieves an order of magnitude reduction in the time complexity of the algorithm from [4].

None of the methods used in [4,9,10] consider recharging decisions at the nodes, and this is where [3] has an input. So, the model generated in this study is modelled as a dynamic program based on a network with recharging capabilities at every node of the graph, except at the destination, recurring to two algorithms, Backward Recursion, that can be applied when the state space is discrete, and Approximate Dynamic Programming, that is useful for more general instances of the model. The underline goal is to create a minimum-cost path from the start point to the end, where the charge level of the vehicle always remains between a given minimum threshold and the maximum capacity of the EV battery, removing every unnecessary node where recharge does not happen, when there exists an optimal path where only recharging nodes exists.

A new method called "EV Assist Route" is proposed in [2], to show a route for EVs that will take into account stops with over CS to have extra battery charge when current remaining charge is not enough to reach the destination. This method starts by identifying the potential CS and then searches the most cost effective route passing through some selected CS, where the cost refers to the travelling distance or travel time from one point to another. This method takes into account various parameters, namely the remaining battery level, the battery level when fully charged, electric mileage, location information from the CS, charging efficiency from the CS, departure and destination points. Recurring to those parameters and applying the "EV Assist Route" method results on the optimal route for the EV, with information like the travel distance, travel time, the remaining battery level after reaching the destination point and the potential stopovers in the CS. Various cases are taken into account when developing this method, namely if the vehicle has enough battery to reach the destination, if it needs to stop for at least one CS to recharge, stopping in two or more CS and the case there is no route leading to the destination, where one of the options of travel distance and travel time can be prioritized, as quick CS can be farther away than a normal CS, which has more travel distance, but less time is needed to charge the battery levels.

For [7] EV-reachable node sets are composed by the feasible paths from the source node and strongly EV-connected node sets is a subset of the EV-reachable node sets, containing only the nodes that allow the returning to the source without running out of energy. Strong EV-connectivity is not an equivalence relation as the reflexivity could be compromised due to the existence of negative cost edges, as EVs are able to recuperate energy when going downhill and/or braking, requiring to model those as edge cost functions to obtain efficient running times. Picking up the techniques developed in [4,9], this study tries to answer the fundamental questions of route planning for EVs, namely the computing of a set of EV-reachable node and, also determine the minimal battery necessary to reaches a certain point. To give the shortest path to the destination node recurring to the most optimal route, Battery Switch Stations (BSS) are taken into account, those enable batteries exchanged in order to make the recharging not more time-consuming than a normal gas refuelling. Graph preprocessing techniques that allow efficient computation of EV-reachable and EV-connected sets at query time even in the presence of BSSs are also proposed. The generated paths to the destination are energy-optimized, hence avoiding going uphill as far as possible, only showing the necessary detours to visit the BSS when really needed, resulting on the most useful as possible routes.

2.2. Comparison to Forecasting in Open Car Parks

The small and sparse network of CS also prove to be a disadvantage, when ideally is to have a dense network of CS, that can allow drivers to travel from one arbitrary source to an arbitrary target destination and have always an energetically reasonable route [11], but right now, CS is relatively scarce [3] and until a dense network of CS is established there is a requirement of planning the route of the trip taking into account the battery-powered EVs energy constraints, or the possibility of getting stranded with an empty battery is real [7], as well as doing a good management of the charging sessions on the CS. As we can see, for the process of electrification of the transport sector, the infrastructure is going to play a key role [12], so CS must have a good availability to encourage the adoption of EVs [13] and their locations must be well thought out as well, since an infrastructure dedicated to the charging processes, whether fast or not, requires a large investment [2] and can impact the queuing time in CS and the traffic conditions in the road network [1], important aspects that have revealed to significantly impacts the adoption of EVs.

In the past, EV infrastructures were limited to single and slow charging points in urban streets or parking lots, with little or none attention to their grid operation. With the continuous increase of EVs, the attention to those infrastructures has increased resulting in the implementation of fast-charging options for EVs and the option to parallel charge at different charging power levels, so a plan of CS in low-voltage networks for EV parallel charging is proposed [12], resulting in smaller charging sessions, thus increasing the availability of CS. Another aspect to take into account when building such infrastructures is the strategic location that needs to be suitable for EV users. Urban grids have a big potential to integrate CS, thanks to the proximity of houses and parking lots, however, the CS can have different impacts on the grid voltage, so the grid constraints should be taken into account when planning the installation of new CS.

There is a need to identify reasonable locations for the CS with few instances as possible so the EVs can achieve reachability and connectivity in the network [11]. The authors of [14] say that EVs will only prevail if a road trip with an EV can be undertaken without taking no significant detour from the current paths. So, this study proposed a way to solve the problem of placement of the CS in a way that for every shortest path there are enough CS, meaning that the EV will not get stranded when starting with a fully loaded battery, called EV Shortest Path Cover. For [15] CS need to be extensively installed, especially on residential areas, in order to satisfy the need for the number of EVs. Two objectives for optimal placement of fast CS in residential areas are taken into account, namely the total-costs of installation for fast CS, the initial investment cost, the annual variance cost of operation and the travelling cost, and the real power loss in the transmission line of the distribution grid.

Behavioral models are used to predict when and where the vehicles are likely to be parked in the future in [13], using parking demand as a variable, as a way to help satisfy the demand for public charging of EVs, as well as land use attributes and trip characteristics. This study took a three-step approach, and in the first step, the parking locations and durations for all trips and for all stop that were 15 min or longer in duration, as those represent feasible candidates for public charging. The second step consists of using parking information for regression models that relate zone-level parking demands with land use attributes, as well as trip-level parking demand to individual trip characteristics. In the last step, the parking information is used as an input to identify the optimal location for the CS to satisfy as much demand as it is possible. With the help of Ordinary Least Squares, regression predictions were made to get the parking demand, also revealing that the more important factors are the number of population and job density in the area. Taking this into account, and drivers are unlikely to walk long distances for parking, the current locations with no existing CS, the parking demand and travel costs and applying a General Algebraic Modeling System, a location for the CS is revealed. Later, the authors apply for a mixed-integer program as a way to reduce total access cost.

A variant of placing CS in the network is proposed in [16], while taking into account the reasonability of routes and sparsity of the CS set, with the goal of placing CS in certain locations where the EV drivers do not need to leave the shortest path trips and make large detours for recharging. So a new and more practical model for placing CS for EVs is proposed, but with a larger complexity. For the development of this model, the authors looked to this problem as a Hitting Set problem, and used a modified greedy algorithm to compute the placement of loading stations, as the standard one showed many memory problems but maintaining the same approximation guarantee for small networks, showing up to a reduction of 40% of required loading stations by allowing detours.

Just like a normal parking space in an open parking lot or on the street-park, the parking spaces with electric charging properties can be influenced by weather factors. Temperature, rain and wind intensity can affect the parking occupancy [8], as bad weather conditions could lead to lower traffic flow than expected [17], meaning that people may be less susceptible to driving. The period of the day and time of year are also important [18], like holidays, weekdays and hour of the day could have a direct impact on park occupancy, in this case in the CS occupancy. The location of the CS can influence the usability of the station, [19], since if a parking spot is in the proximity of some type of shopping mall or close to an important public highway, or even if events happen regularly around the parking lot, like football games and concerts, those can cause a significant increase in the amount of traffic, consequently increasing the demand for free parking spaces [20]. For [21] the parking cost and estimated queuing time outside the parking lot are important factors to be taken into consideration, which can be compared to the charging costs and the queue time to charge the EV.

The prediction of park availability on CS can be very useful and helpful as could improve the quality of life of the users, not wasting time on queues to charge their vehicle. Historical data is really important as it could give a notion of seasonal variations over time of charging sessions, as well as give a more insightful idea of the loading routines, allowing us to build a more robust and precise predictive model.

To build the predictive model [17] used methods like Gradient Boosting Decisions Trees, as it is very effective in training and on scoring. Algorithms used on model training for the [20] were Decision Trees, Support Vector Machine, Multilayer Perceptrons and Gradient Boosted Trees, concluding that Extreme Gradient Boosting has had the best accuracy result of all of them. For modelling the occupancy rates after applying different features and for prediction, methods like Regression Trees, Support Vector Regression and Neural Networks were used, and to prevent overfitting a 5-fold cross validation was made for training in [18]. Predictions are made for periods of 15 min ahead and all three algorithms were used, showing better results on Regression Trees when comparing with the NN and SVR.

3. Conceptual Model

Some of the current problems faced by EVs previously stated like the autonomy of the batteries [2] and the charging times [1] are significant to cause some resistance to the global adoption of EVs. Another problem that occurs is the time that a user can spend waiting for a CS to be freed. While a combustion engine vehicle can go to a gas station and fully fill the gas tank in a matter of minutes, the amount of time it takes an EV to fully charge generally takes around 30 min at a high voltage CS [1]. Taking this to account, it is unpleasant to reach a CS after driving and spending battery to it and all charging points are being used, leaving the user to, for at least 30 min, wait for a free charging connector. So, in this work, we present a based android application solution for this kind of problem, as we can see in Figure 1.

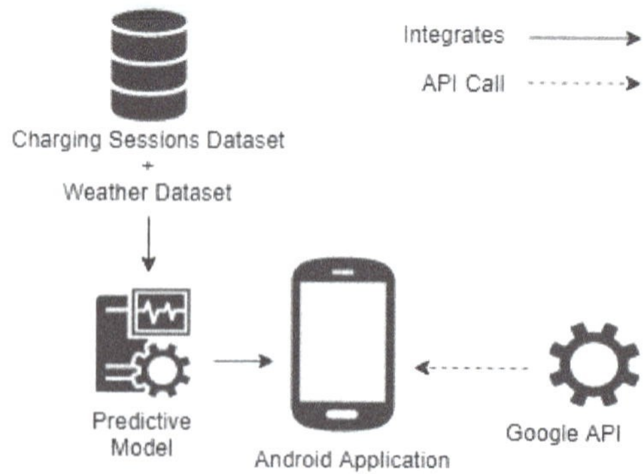

Figure 1. Conceptual Model Diagram.

At the center of the solution is the "Charge Time" android application that incorporates two main components, services from Google, namely the Maps API, Directions API, Elevation API and Firebase, and a predictive model developed in this study with the help of a charging sessions dataset and weather data. All of these components are explained in the following sections of the paper.

3.1. Android Application

The Android application, being the center of the solution, gives the user a way to interact and receive information about the CS. The first step when entering the application is the login, where the user needs to put his credentials to log in the application, and in case it is the first time logging in, the user can register a new account by providing the email, username and a password. This login is important so the user can keep points which are explained later on the study. After filling the login credentials and entering the app, the user is greeted with a three option menu, shown by the Figure 2, being explained in the next paragraphs.

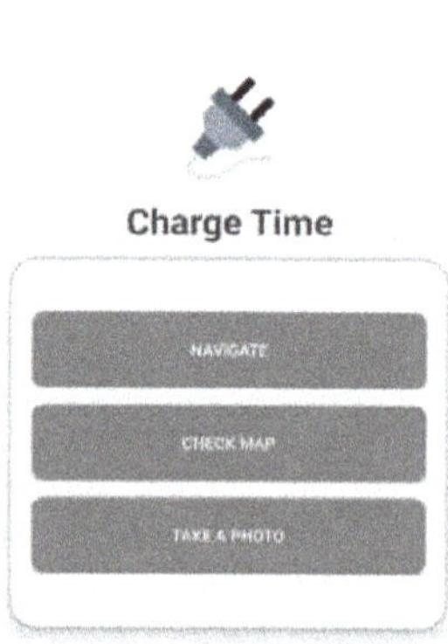

Figure 2. Charge Time application menu.

The first option of the application "Navigate", opening a new window with two input boxes that the user must fill. The first input being the destination where the user wants to travel to and the second the current battery level the EV has (in an ideal scenario, this information could be accessed trough the vehicle API). After filling those inputs, the user can confirm and a route to the closest and more likely to have an available place to charge is selected in the surroundings of the searched destination, by a decision algorithm created in this research. After that, a route is created using the Google Directions API, starting at the current location of the user to the respective CS. The CS is chosen by the decision algorithm, that takes into account various conditions: the first and most important is the maximum possible distance the EV can travel, meaning that all CS further than the maximum possible distance can not be taken into account, so they are automatically discarded and a pop-up window is sent to the user informing that the desired destination cannot be reached with the current vehicle battery. The second most important factor to take into account is the probability to have a free charging spot on each CS taking into consideration the time it takes the user to reach the CS. This can be achieved by the predictive model also being developed in this study. The route generated, just like in [2], provides information like the battery level the EV has at the end of the route, considering regenerative braking, if possible, the occupancy of the CS at arrival and the time it would take to reach the CS. The time it would take to reach the CS makes it possible to feed the predictive model with the time of arrival and is also used as the third condition, already taking into account the intensity of traffic, where the CS with less travel time has a higher possibility to be chosen. The last heuristic used is the distance to the CS, where the closest station to the destination the user provided has a bigger probability to be chosen.

The second option shown in the menu is "Check Map" and when clicked a map focused in the current user location appears. This map shows markers that represent the location of all of the CS, shown in the Figure 3, and when a marker is clicked a popup information board appears with information of the CS, as we can see in Figure 4. The information goes from the probability of having a free charging spot to the name, charging types, number of charging stations and connectors, working period, charging fees and a button that creates a route to the respective CS. In this case, the current state of the battery is not taken into account, since the initial battery level is not supplied, but information about the percentage of occupation in the CS at the end of the route is shown, as well as the duration it would take to reach the CS.

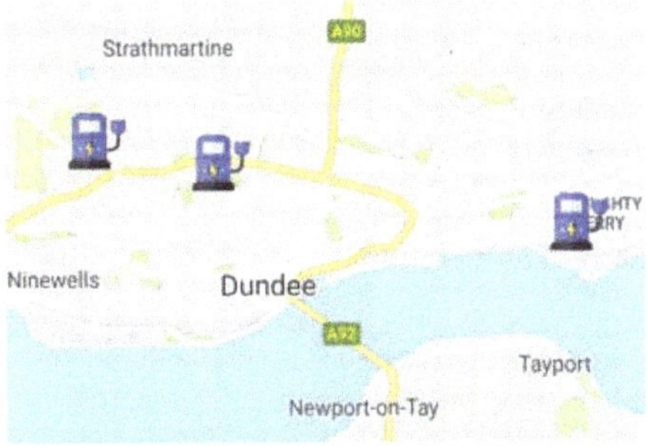

Figure 3. Location of CS on the map.

The last option "Take a Photo" gives the user the option to collaborate with the system by taking a photo to a free charging spot on a CS, this way the system will know that a CS is free, making a re-validation on the probability of having a free charging spot. The photo taken is then sent to the server, being validated if it really represents an empty spot in the CS by a trained model, that it is out of scope on this study, and by the current location of the users, if it is in the surroundings of the CS. In case the photo veracity is checked, then the users will be rewarded with points that could result in discounts on the charging fees.

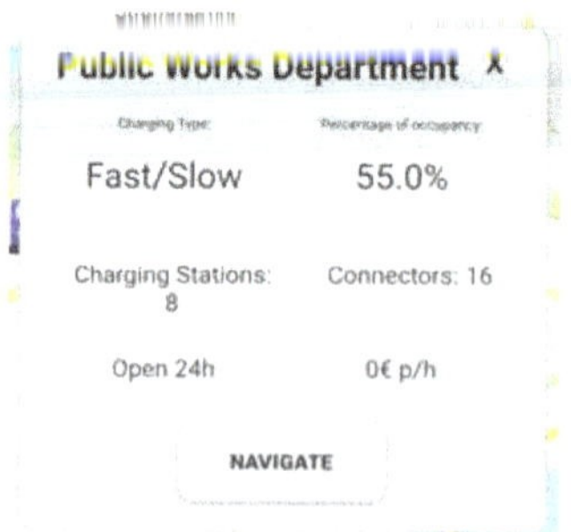

Figure 4. Pop-up information board on a CS.

3.2. Google API Services

The Google API services are of great importance in this application, as the entire application works around the maps service provided by Google. The Maps API allows the presence of a map based on Google Maps data, that the user can explore by clicking the "Check Map" or showing the route to the CS suggested by the "Navigate" option. Another service used in the application is the Directions API, that allows the application to obtain direction information and draw a route between two points, taking into consideration traffic stats. This service provides us with information for different transport modes, waypoints and travel times, as well as one or more travelling routes to the destination while checking the distance and time it takes from one point to another. In this application, the only considered transport mode is a car, since all of the studies focus on the EV. One more service used is the Elevation API, that provides a simple interface to query locations on the earth for elevation data and with this information we can calculate if any energy was gained and the exact amount of energy gained by the EV from the source to the destination by the braking, also called by regenerative braking, calculating the difference in elevation at the start and end of the path, just like in [9] where the height-dependent part of the cost of a path only depends on the point of start and the destination point, because all others height-dependent parts cancel each other out, unless overcharging takes place. In order to calculate the energy gains through the regenerative braking we use the Equation (1), resulting on the conversion of the vehicle kinetic energy into chemical energy stored in the battery, where it can be used later to drive the vehicle.

$$KE = \frac{1}{2} * m * v^2 \tag{1}$$

The kinetic energy (*KE*) stored in a moving vehicle is related to the mass in kilograms (*m*) and speed of the vehicle in meters per second (*v*). In order to know the regeneration gained by braking along the route of the vehicle we would need to have access to information on how the route was made, namely when the brakes were applied and in which situations. Since we do not have access to this information at the time of executing the "Navigate" option of the application, we only calculate the potential energy gains when the difference between the heights of the starting point and the end point is greater than 0 by a simplified Equation (2).

$$PE = m * a * h * 0.7 \tag{2}$$

To get the potential energy (*PE*) gains we use the Equation (2), achieved by multiplying the mass (*m*) of the EV by the acceleration (*a*) the vehicle has by the difference in height (*h*) from the starting point to the destination point. There is always some loss of energy, either by the rolling resistance, mechanical friction and aerodynamics in the EV, resulting on the dissipation of energy into heating the road, the surrounding air and various spinning parts in the EV. In order to interpret these losses in a very simplistic way, we decided to multiply the PE by an efficiency factor of 0.7.

$$a = g * sin(slopeangle) \tag{3}$$

The acceleration is calculated by multiplying the earth surface gravitational value (g), with the value of 9.8, by the slope angle between the starting and ending points of the route.

By applying the (1) and (2) equations we can calculate the potential gains from regenerative braking, which can results in an increase of up to 25% in the distance travelled by the EV [22].

Other Google service used is Firebase that supplies the means to build an authentication system for the login and register option, as well as the database to keep information about the users and the CS.

3.3. Predictive Model

The predictive model is built on to the android application, but in a commercial use, it would be advisable to have this model in a server to reduce the total size of the application and the battery drainage. It would also increase performance and help build a more robust model. This predictive model is going to be built with the help of the open datasets of EV charging sessions in the city of Dundee, Scotland [23]. The city of Dundee was awarded in 2018 as the most visionary city in Europe for EVs at a ceremony in Japan, due to its pioneering initiatives to encourage the use of EVs. The city council was able to develop the largest fleet of council electric cars in the UK, with almost 40% of council cars and vans being electric. On top of that, 15% of the city's taxis are electric [24]. That said, we have access to four datasets, that combined make a total of a year of charging sessions around the Dundee city, as well as a dataset that gives us more information of each charging point. To help with the predictive power of the model, information about the historical weather in the city of Dundee was obtained with the development of a Web Crawler with the help of the Python Library Selenium. The historical data was obtained from 1 of September of 2017 up to 6 of September of 2018, matching the first and last record of the charging sessions in the Dundee combined dataset, taking 1 h and 20 min to gather all data. Further details from the predictive model and the data used in the creation of the model are explained in the following section.

4. Development of the Predictive Model

As was said in the previous chapter, the predictive model was built using two combined datasets and applying predictive algorithms, one of the datasets is from the Dundee City representing the EV charging sessions and the other is a weather dataset gathered by a web crawler. The algorithm chosen is evaluated in this chapter, as various algorithms are tested and examined to decide which one is the most desirable, taking into consideration the accuracy levels and the performance, since the execution time is important. The developed approach can be applied to any case with the previous datasets of the charging process and weather dataset.

4.1. Charging Sessions Dataset

Firstly, there are five datasets in public access from the Dundee City Council in the matters of the EV Charging Sessions divided by charging points. Four of the datasets correspond to the charging sessions, representing each a time set of three months, a year in total data, and the last dataset represents information about each CS.

The charging dataset was combined using the four datasets, with data from 1 of September of 2017 to 6 of September of 2018, resulting in a total of 67,112 records with the following columns: CP ID, that uniquely identifies each charging point; Connector ID, the identifier of the connector since each charging point can have more than one connector; starting time and end time of the charging session; and the address of the charging point. In this study, we focused on three areas of Dundee, namely

Queen Street Park, Dundee Ice Arena and Public Works Department, all of those charging points are open 24 h per day with zero cost associated and for public access. Removing rows with missing values resulted in a total of 66,783 records. In short, the CS under study is characterized by the information presented in Table 1.

Table 1. Charging Points Information.

Charging Point ID	Location	Number of Connectors	Type of Charging
50,692	Dundee Ice Arena	2	Fast
50,911	Queen Street Park	2	Fast
51,087		2	Fast
50,339	Public Works Department	2	Fast
50,235		2	Slow
50,236		2	Slow
50,238		2	Slow
50,270		2	Slow
50,271		2	Slow
50,272		2	Slow
50,232		2	Slow

Having said that, we then removed every row with missing values, as well as every charging sessions with a duration of less than one minute or of more than two days, as we found those to most probably be wrong measures or misunderstandings by the users. After that, we have chosen three locations to continue the study and filtered the complete dataset to have a dataset per location, namely in the Queen Street Park (QSP) location, for Dundee Ice Arena (DIA) location and for Public Works Department location (PWD). Those locations were chosen because these are the locations with the highest number of recharging sessions, more specifically the dataset from QSP has a total amount of 8147 rows, the DIA dataset has 5209 rows and the PWD dataset has a total of 6264 rows.

We then divided the data from the charging session into intervals of fifteen minutes, as a way to predict the occupancy in the charging sessions every fifteen minutes, making it more dynamic and immediate, in case any connector of the station turns unoccupied. The average charging time is 3 h and 26 min for all CS and, more specifically, 51 min for QSP, 29 min for DIA station and 2 h and 35 min for PWD, and so, it is interesting to keep a low time interval for prediction to realize when the stations have a place available. To all of those records, a column called 'occupied' was added, and for the respective datetime, meaning that for every fifteen minutes we would count how many connectors from the CS were occupied, as a way to know the total number of connectors being used for that location.

4.2. Weather Dataset

As mentioned earlier, weather conditions can be important in forecasting parking spaces, which can be equated with the problem we are trying to address. Having said that, we are concentrating on obtaining meteorological data to help predict the occupation of CS, and this data was obtained via Web Crawler, developed using the Selenium Python Library from [25], within range of day 1 of September of 2017 to day 6 of September of 2018, matching the earliest and latest date in the charging sessions dataset. This historical data was gathered with an interval of 3 h between each measurement, collecting information like: datetime, showing the time and date of the measurement; temperature, in Celsius; wind speed, in kilometers per hour; clouds, cloud cover amount in percentage; humidity, in percentage; precipitation, in millimeters of water per hour; and atmospheric pressure in millibars.

4.3. Data Analysis

Before starting to apply algorithms, we combined the three datasets, from each location with the weather dataset, through the datetime column. All of the three datasets where composed by nine columns, namely occupied and datetime from the CS, as well as the temperature, humidity, wind speed, clouds, pressure and precipitation from the weather dataset.

After having combined the datasets, we started to check if the weather features had an influence in the occupation of the CS by analysing the correlation between every weather features and the target, 'occupied'. Correlation can give us a relationship between two values indicating that as one variable changes in value, the other variable tends to change its value in a specific direction. In this case, we used the Pearson's Correlation Coefficients, where the correlation coefficient value can range between -1 and 1, measuring both the strength and direction of the linear relationship between two continuous variables. Strength reveals that the greater the absolute the correlation coefficient is, the stronger is the relationship, meaning that as one value changes, the other will also change, where a coefficient of zero represents no linear relationship. As for direction, the sign of the correlation coefficient reveals the direction of the relationship, where positive coefficients indicate that when a value increases, the value of the other variable also tends to increase, and when the coefficient is negative represents that as a one variable increases the other tends do decreases. Having said that the correlation values between the target 'occupied' and the weather features can be seen in Table 2.

Table 2. Correlation of weather features with the target 'occupied'.

Dataset Feature	QSP	DIA	PWD
Temperature	−0.07	−0.19	−0.18
Humidity	−0.14	−0.12	0.07
Cloud	−0.02	0.02	0.07
Precipitation	0.01	0.01	0.01
Wind	0.07	0.12	0.09
Pressure	−0.05	−0.11	−0.10

By analyzing Table 2 we can conclude that the weather features do not show a high correlation with the loading sessions. The charging of the EV is necessary in case of lack of battery in the EV, where weather conditions do not condition it, as the drivers need to have the vehicle operational. Having said that, we decided to remove the weather data for the creation of the predictive model.

4.4. Testing Algorithms

In this section and taking into consideration the above findings, we started the testing of algorithms to predict the occupancy of the CS, but firstly we transformed the occupied columns to represent an occupation tax, instead the true value, as it results in better predictions, meaning that for the case of the DIA location, if there are 2 cars charging in it at the same time, that means the CS is full, having a 100% occupation tax. The algorithms chosen to test this were Distributed Random Forest (DRF), Neural Networks (NN) and Gradient Boosting Machine (GBM), from the python library H2O [26]. Those algorithms were applied to the three different locations datasets and evaluated using a five cross-validation, resulting in the outcomes for the accuracy metric present in Table 3.

Table 3. Accuracy and average run time results.

Algorithm Location	DRF	Run Time (Seconds)	NN	Run Time (Seconds)	GBM	Run Time (Seconds)
QSP	64.64%	40	45.61%	117	54.38%	13
DIA	73.91%	28	69.44%	127	71.37%	9
PWD	69.93%	66	33.62%	114	48.78%	23

Considering the results obtained, we can verify that the algorithm that presented the best results was the DRF, reaching around 70% accuracy values and having the second lowest average run time of all the three algorithms tested. The best value as we can see is using the DRF algorithm with the DIA location, reaching a robust and efficient accuracy value of 74%, the other locations, QSP and PWD also had good values, almost 65% and 70%, respectively. The NN algorithm showed the worst values in accuracy, while also presenting high run times, and so this option was discarded. The GBM algorithm presents the lowest run time values, and although this metric is important for the efficiency of the application and the predictive models, the accuracy difference is high enough to choose the DRF as the adopted algorithm.

To conclude, the algorithm chosen for the creation of the predictive models is the Distributed Random Forest. These models are built using historical information with an interval of 15 min about the charging sessions, namely the occupancy rate in each charging station for a given month, day, hour and minutes the session occur.

5. Results Evaluation

For this assessment we considered two examples of EV, the first being the Renault Zoe and the second being the Nissan Leaf. These vehicles were chosen because they are two of the most common EV in Europe [27]. The way in which the EV is driven has a lot of influence on its energy consumption, from the velocity, braking force, number of times it has been braked, etc, and as we do not have information on consumption along the route, we then use an average consumption for each of the EVs taking into account their characteristics.

5.1. Test Example: Renault Zoe

The Renault Zoe R90 model has a total weight of 1455 kg and a reach of 260 km with a battery of 41 kWh and an average consumption of 15.8 kWh per 100 km [28].

So, in the following example we started at Braemar, Ballater, UK, with an elevation of 323 m with a Renault Zoe at 55% of battery level. We then used the "Navigate" option with Broughty Ferry, UK as the final destination which is 44 m above the water level.

As we can see in the Figure 5, the user inserted Broughty Ferry on the first input box and the current battery percentage has 55%. After this, the user would click on navigate and the decision algorithm would run to calculate the weight of each option. This algorithm takes into consideration the three heuristics previously announced for each CS, firstly the occupancy of the CS at the time of arrival, secondly the duration it takes from the current location of the user to the CS, and lastly the distance from the CS to the destination provided by the user, in this case Broughty Ferry.

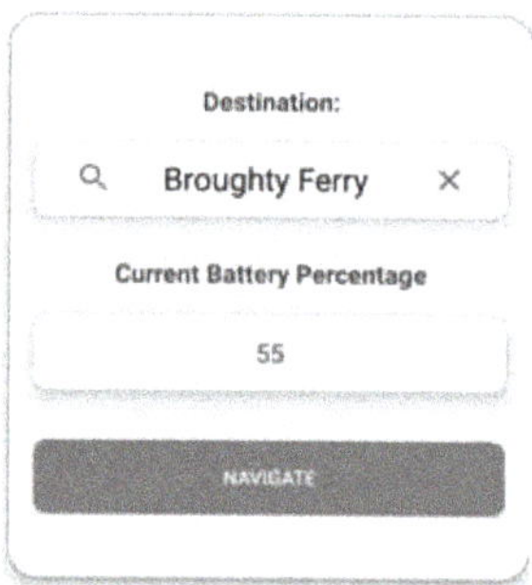

Figure 5. Navigate option with Broughty Ferry as the destination.

The algorithm features and weight results for the Renault Zoe can be seen on Table 4.

Table 4. Navigate option with Broughty Ferry as the destination.

CS	Route Duration (Seconds)	Distance to CS (Meters)	Distance from Destination to CS (Meters)	CS Occupancy (Percentage)	Weight
QSP	6819	68,855	841	25	202.80
PWD	6202	64,819	7364	6	201.95
DIA	6038	63,508	9786	0	200.80

The algorithm takes the percentage of occupancy as a higher priority, and the CS with the lowest occupancy was the DIA location with 0% occupancy at the time of arrival, while PWD location had a value of 6% and the QSP location an occupancy of 25%. In the case of the duration, the route with less time was for the DIA location at a distance of 6038 s, about 1 h and 38 min. For the rest of the CS, we had a total duration of 6202 s (1 h and 43 min) for the PWD location and 6819 s (1 h and 53 min) for the QSP location. For the distance from destination to the CS, the shortest is the QSP CS with only 841 m away, while the rest of the CS are more than 7 kilometers away. So, taking in those values, we generated the weight value, to decide which CS suits best, where CS with the highest weight value is chosen as the optimal CS and the route to that CS is made. The CS with the highest value was the QSP, with a total of 202.80 weight, so the route to this CS is made while showing the duration to the CS in bold, as well as the battery level at the destination and the occupancy percentage of the CS while arriving the CS. In this case the route to the CS takes around 1 h and 38 min, resulting on an occupancy of 25% at the estimated time of arrival, with a total of 13% battery level, seen in Figure 6.

As the elevation is higher at the start (323.13 m) then in the end (9.73 m) there is a reduced energy gain, more specifically 4 watts. It is not a very large gain, but it tends to have a greater contribution in cases of routes with a greater difference in height and where the route is not so long. In this case, the EV ended the route on the QSP CS with 13 percent battery charge.

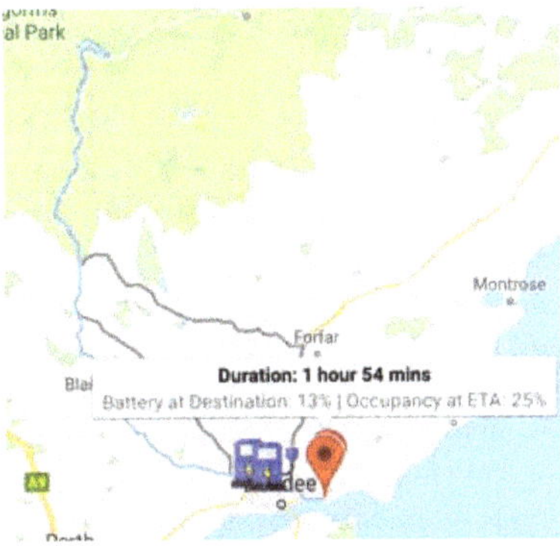

Figure 6. The route to the CS chosen from the decision algorithm.

5.2. Test Example: Nissan Leaf

The Nissan Leaf weights 1580 kg with a 38 kWh battery capacity as a total of range of 230 km, resulting on an average of 16.5 kWh consumption per 100 km [29].

Taking into consideration the Nissan Leaf characteristics and starting at Leuchars Station, UK with a total of 30% battery level. In this example we executed the "Navigate" option with the Dundee Ice Arena as the destination, as a way to evaluate once more the decision algorithm. In Table 5 we can check the features and weight results from the decision algorithm for the Nissan Leaf example.

Table 5. Decision algorithms heuristics results for the Nissan Leaf example.

CS	Route Duration (s)	Distance to CS (m)	Distance from Destination to CS (m)	CS Occupancy (Percentage)	Weight
QSP	1902	9744	9443	25	132.38
PWD	1884	13,651	2429	18	193.97
DIA	2037	15,549	54	50	194.55

In this case the difference between the two points is 89.65 m, where the final destination has a height of 100.60 m and a total of 10.94 m at the starting point, so there are no energy gains taken into consideration.

As previously mentioned, the heuristic with the highest priority is the occupancy rate of the stations, and the station with the lowest occupancy rate in this case was PWD with an occupancy rate of 18%. The QSP station had an occupancy rate of 25% and the DIA station an occupancy rate of 50%. In terms of duration of the route, the route to PWD is the shortest with 1884 s, a total of 31 min. For QSP and DIA, we have 1902 s (approximately 32 min) and 2037 s (34 min), respectively. In terms of distance from the final destination to the CS, the shortest distance is at the DIA station with 54 m distance from the provided destination, to the PWD station we have a distance of 2429 m and to the QSP station we have a total of 9443 m distance. After finding all these values, we were able to calculate the weight of each station to choose the best option for charging, and in this case we verified that the station with the highest weight is the DIA station with a value of 194.55, being this the selected one. The Leaf reaches the DIA CS with a total of 19% of battery and a occupancy at the estimated time of arrival of 50%.

For the Check Map feature, the user decides which CS wants to move to. For example, the user chooses the location to navigate to manually and the application takes care of creating the route to the selected CS. As the battery information is not provided in this case, the application does not take this information into account, only showing the occupancy information upon arrival.

6. Conclusions

This work introduces a novel approach to handle the problem of real-time guidance to free CS near the desired destination location for EV drivers. As the number of electric cars gets bigger this problem will be increasingly more important, where the allocation of new CS will be a necessity. If this increase does not happen in a controlled and intelligent way, the current CS will become too crowded which can present a huge challenge for EV drivers as well as for the traffic surrounding the CS, as most of the times they do travels that can be performed with available charge because there is a big uncertainty about getting a free charging spot. A prediction approach can handle and mitigate this problem when no real-time connection is available to these proprietary systems. This information is vital when the CS implementation is lower than the EV market penetration. This predictions model can reduce this problem proving probabilistic information about the availability and, as we were able to conclude with this study. A predictive model was built for the three studied locations showing good accuracy results for the CS, by combining a total of one year of charging sessions data proving to be a reliable solution. The model developed could also be implemented in other cities, requiring only the input of datetime features, namely the day, month, hour and minutes of future charging, thus allowing the model to return a value of the occupancy rate. Current solutions also provide information about the usage of CS and with this, it is possible to identify zones where there is a need for more CS because several EV can not charge in the desirable location. This could be useful information for future studies of the location of CS.

Author Contributions: G.A. conducted the investigation, developed the application and implemented the algorithms. J.C.F. was responsible for the technical support. R.P. was responsible for the methodological support.

Funding: This work has been partially supported by Portuguese National funds through FITEC programa Interface, with reference CIT "INOV - INESC Inovação - Financiamento Base" and "FCT - Fundação para a Ciência e a Tecnologia" through the project UID/Multi/04466/2019.

Acknowledgments: FCT - Fundação para a Ciência e a Tecnologia.

Conflicts of Interest: The authors declare no conflict of interest.

Abbreviations

The following abbreviations are used in this manuscript:

EV	Electric Vehicle
CS	Charging Station
BSS	Battery Switch Stations
QSP	Queen Street Park
DIA	Dundee Ice Arena
PWD	Public Works Department
DRF	Distributed Random Forest
NN	Neural Networks
GBM	Gradient Boosting Machine
API	Application Programming Interface
FIFO	First In First Out

References

1. Xiong, Y.; Gan, J.; An, B.; Miao, C.; Bazzan, A.L. Optimal electric vehicle charging station placement. *IJCAI Int. Joint Conf. Artif. Intell.* **2015**, *2015*, 2662–2668. [CrossRef]
2. Kobayashi, Y.; Kiyama, N.; Aoshima, H.; Kashiyama, M. A route search method for electric vehicles in consideration of range and locations of charging stations. In Proceedings of the 2011 IEEE Intelligent Vehicles Symposium (IV), Baden-Baden, Germany, 5–9 June 2011; pp. 920–925. [CrossRef]
3. Sweda, T.M.; Klabjan, D. Finding minimum-cost paths for electric vehicles. In Proceedings of the 2012 IEEE International Electric Vehicle Conference, IEVC 2012, Greenville, SC, USA, 4–8 March 2012; pp. 1–4. [CrossRef]
4. Artmeier, A.; Haselmayr, J.; Leucker, M.; Sachenbacher, M. The shortest path problem revisited: Optimal routing for electric vehicles. In *Annual Conference on Artificial Intelligence*; Springer: Berlin, Germany, 2010; pp. 309–316. [CrossRef]
5. Monteiro, V.; Afonso, J.A.; Ferreira, J.C.; Afonso, J.L. Vehicle electrification: New challenges and opportunities for smart grids. *Energies* **2019**, *12*. [CrossRef]
6. Ferreira, J.C.; Monteiro, V.; Afonso, J.L. Vehicle-to-anything application (V2Anything App) for electric vehicles. *IEEE Trans. Ind. Inform.* **2014**, *10*, 1927–1937. [CrossRef]
7. Storandt, S.; Funke, S. Cruising with a Battery-Powered Vehicle and Not Getting Stranded. *Assoc. Adv. Artif. Intell.* **2012**, *3*, 46.
8. Lijbers, J. Predicting Parking Lot Occupancy Using Prediction Instrument Development for Complex Domains. Master's Thesis, University of Twente, Enschede, The Netherlands, 2016.
9. Eisner, J.; Funke, S.; Storandt, S. Optimal Route Planning for Electric Vehicles in Large Networks. *IEEE Trans. Appl. Supercondu.* **2011**, 108–1113. [CrossRef]
10. Sachenbacher, M.; Leucker, M.; Artmeier, A.; Haselmayr, J. Efficient Energy-Optimal Routing for Electric Vehicles. In Proceedings of the Association for the Advancement of Artificial Intelligence (AAAI), San Francisco, CA, USA, 7–11 August 2011; The AAAI Press: San Francisco, CA, USA, 2011; pp. 1402–1407. [CrossRef]
11. Funke, S.; Storandt, S. Enabling E-Mobility: Facility Location for Battery Loading Stations. In Proceedings of the Association for the Advancement of Artificial Intelligence (AAAI), Washington, DC, USA, 14–18 July 2013; The AAAI Press: Bellevue, WA, USA, 2013; pp. 1341–1347.
12. Marra, F.; Traeholt, C.; Larsen, E. Planning Future Electric Vehicle Central Charging Stations Connected to Low-Voltage Distribution Networks. In Proceedings of the 3rd International Symposium on Power Electronics for Distributed Generation System, Aalborg, Denmark, 25–28 June 2012.

13. Chen, T.; Kockelman, K.; Khan, M. Locating Electric Vehicle Charging Stations. *Transp. Res. Rec. J. Transp. Res. Board* **2013**, *2385*, 28–36. [CrossRef]

14. Funke, S.; Nusser, A.; Storandt, S. Placement of loading stations for electric vehicles: No detours necessary! *J. Artif. Intell. Res.* **2015**, *53*, 633–658. [CrossRef]

15. Phonrattanasak, P.; Leeprechanon, N.; Member, S. Optimal placement of EV fast charging stations considering the impact on electrical distribution and traffic condition. In Proceedings of the International Conference and Utility Exhibition on Green Energy for Sustainable Development (ICUE), Pattaya, Thailand, 19–21 March 2014; pp. 19–21. [CrossRef]

16. Funke, S.; Nusser, A.; Storandt, S. Placement of Loading Stations for Electric Vehicles: Allowing Small Detours. In Proceedings of the International Conference on Automated Planning and Scheduling (ICAPS), London, UK, 12–17 June 2016; The AAAI Press: London, UK, 2016; pp. 131–139.

17. Rong, Y.; Xu, Z.; Yan, R.; Ma, X. Du-parking: Spatio-temporal big data tells you realtime parking availability. In Proceedings of the ACM SIGKDD International Conference on Knowledge Discovery and Data Mining, London, UK, 19–23 August 2018; pp. 646–654. [CrossRef]

18. Zheng, Y.; Rajasegarar, S.; Leckie, C. Parking Availability Prediction for Sensor-Enabled Car Parks in Smart Cities. In Proceedings of the 2015 IEEE Tenth International Conference on Intelligent Sensors, Sensor Networks and Information Processing (ISSNIP), Singapore, 7–9 April 2015; pp. 7–9.

19. Predicting the Availability of Parking Spaces with Publicly Available Data. Available online: https://pdfs.semanticscholar.org/f70d/6ebf231994090c58baeecf1e64d2b199f942.pdf (accessed on 5 May 2019).

20. Ionita, A.; Pomp, A.; Cochez, M.; Meisen, T.; Decker, S. Where to Park?: Predicting Free Parking Spots in Unmonitored City Areas. In Proceedings of the 8th International Conference on Web Intelligence, Mining and Semantics, Novi Sad, Serbia, 25–27 June 2018; pp. 22:1–22:12. [CrossRef]

21. Shin, J.H.; Jun, H.B. A study on smart parking guidance algorithm. *Transp. Res. Part C Emerg. Technol.* **2014**, *44*, 299–317. [CrossRef]

22. Energy Efficient Electric Vehicle Using Regenerative Braking. 7. Available online: http://large.stanford.edu/courses/2017/ph240/leis-pretto1/docs/lakshmi.pdf (accessed on 5 May 2019).

23. Electric Vehicle Charging Sessions Dundee. Available online: www.data.dundeecity.gov.uk/dataset/ev-charging-data (accessed on 1 April 2019).

24. Dundee is Europe's Most Visionary EV City. Available online: www.dundeecity.gov.uk/news/article?article_ref=3284 (acessed on 5 May 2019).

25. Dundee Historical Weather. Available online: www.worldweatheronline.com/dundee-weather-history/dundee-city/gb.aspx (accessed on 1 April 2019).

26. Welcome to H2O 3. Available online: http://docs.h2o.ai/h2o/latest-stable/h2o-docs/data-science.html (accessed on 8 April 2019).

27. Electric Vehicle Sales Jump 67% In Europe - CleanTechnica EV Sales Report. Available online: https://cleantechnica.com/2019/03/04/electric-vehicle-sales-jump-67-in-europe-cleantechnicas-europe-ev-sales-report/ (accessed on 29 May 2019).

28. Renault Zoe R90—Battery Electric Vehicle. Available online: https://ev-database.org/car/1150/Renault-Zoe-R90 (accessed on 29 May 2019).

29. Nissan Leaf—Battery Electric Vehicle. Available online: https://ev-database.org/car/1106/Nissan-Leaf (accessed on 29 May 2019).

Article

IoT and Blockchain Paradigms for EV Charging System

Jose P. Martins [1], Joao C. Ferreira [1,2,*], Vitor Monteiro [3], Jose A. Afonso [4] and Joao L. Afonso [3]

1   DCTI, Instituto Universitário de Lisboa (ISCTE-IUL), ISTAR-IUL, 1649-026 Lisboa, Portugal
2   INOV INESC Inovação—Instituto de Novas Tecnologias, 1000-029 Lisboa, Portugal
3   ALGORITMI Research Centre, University of Minho, 4800-058 Guimarães, Portugal
4   CMEMS-UMinho Center, University of Minho, 4800-058 Guimarães, Portugal
*   Correspondence: jcafa@iscte-iul.pt; Tel.: +351-210-464-27

Received: 13 June 2019; Accepted: 31 July 2019; Published: 2 August 2019

Abstract: In this research work, we apply the Internet of Things (IoT) paradigm with a decentralized blockchain approach to handle the electric vehicle (EV) charging process in shared spaces, such as condominiums. A mobile app handles the user authentication mechanism to initiate the EV charging process, where a set of sensors are used for measuring energy consumption, and based on a microcontroller, establish data communication with the mobile app. A blockchain handles financial transitions, and this approach can be replicated to other EV charging scenarios, such as public charging systems in a city, where the mobile device provides an authentication mechanism. A user interface was developed to visualize transactions, gather users' preferences, and handle power charging limitations due to the usage of a shared infrastructure. The developed approach was tested in a shared space with three EVs using a charging infrastructure for a period of 3.5 months.

Keywords: electric vehicle; EV charging process; blockchain; IoT; mobile app

1. Introduction

One of the big challenges related with electric vehicle (EV) market penetration is the charging process, where the main problems are related to the lack of proper infrastructure in residential buildings (condominiums) since they are not prepared for this new reality. Condominiums have the problem of shared electricity, which does not meet the EV owner's requirements. Based on new advances in the Internet of Things (IoT) [1], and the associated sensing devices and communication platforms, blockchain and information systems have the potential to create new solutions for these problems. Another facet of this challenge is the problem associated with rental houses and the eventual need for supporting EV charging in these cases.

In condominiums, unfortunately, there is a general reluctance regarding the installation of EV charging stations that will only be used by a few homeowners [2]. In addition, there is also an issue regarding the safety of the electrical installations, since they are not built proactively to support EV charging stations, and, adapting the condominium electrical infrastructure will require not only that a consensus between the majority of the owners is reached, which may be hard to achieve, but also authorizations issued by the government building safety entities.

Taking into consideration that most residential buildings have shared spaces with common electrical installations and are not prepared for the installation of new EV charging systems, this is a barrier to EV uptake [3]. A study by Lopez-Behar et al. [4] identified four main problem domains in the context of sharing EV charging solutions in buildings: unavailable charging infrastructure, building limitations, regulation issues and parking availability.

In this work, we propose a new IoT-based approach for handling the EV charging process, which can be used in the context of a shared energy infrastructure without requiring a supervision entity to control the process.

The proposed solution is supported by a decentralized blockchain approach, running on a mobile device app. Figure 1 shows an overview of a condominium with the proposed EV charging platform. This work allows the following features: (1) A pre-registration with a local EV charging provider is not required, avoiding the problem of different cards in different charging infrastructures (every charging infrastructure has its own cards, and this is a problem for EV owners because they need several charging cards when different providers are available); (2) it can work with digital currency using a peer-to-peer (P2P) framework on the same homogeneous blockchain infrastructure and technology; and (3) reduced cost (almost zero fees), because there is no requirement for a third party management entity, apart from the condominium, which would create additional costs.

As illustrated in Figure 1, the major features of the proposed system are: (1) User authentication with a mobile device using Bluetooth Low Energy (BLE) communication and, based on this, release of energy for the EV charging process; and (2) energy consumption is monitored by Internet of Things (IoT) sensors and a microcontroller board transmits the data to a web server (Raspberry Pi with Raspbian operating system), which acts as the management unit, storing the data, handling the transactions in a blockchain implementation and managing the charging according to the power limitations.

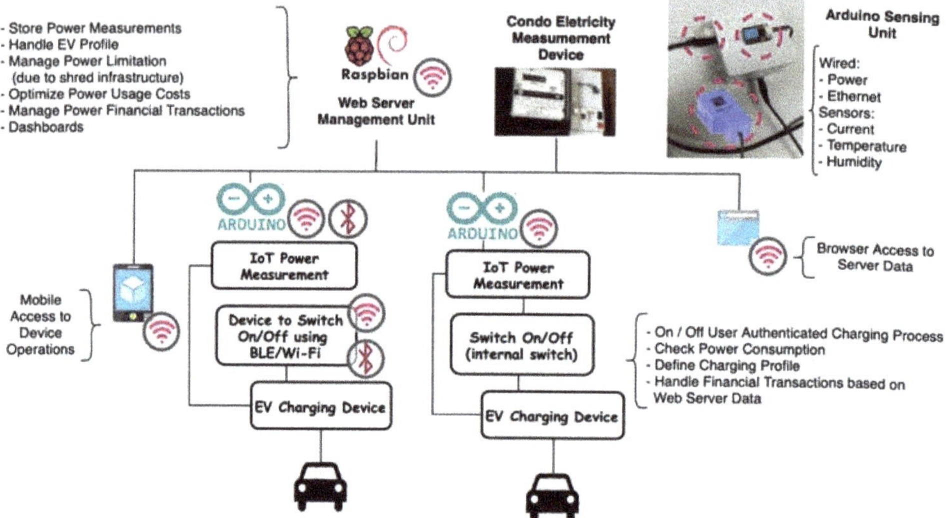

Figure 1. Overview of the proposed electric vehicle (EV) charging platform in shared spaces.

Complementary to the setup presented in Figure 1, which is suitable for deploying the solution at the local level, in the context of a single condominium, an equivalent model can be applied to scale the solution to a wider geographical area with an increased number of charging locations. In this sense, Figure 2 expands the proposed model to an IoT architecture that is suitable to explore cloud paradigms, such as Infrastructure as a Service (IaaS) or Software as a Service (SaaS), where the local management unit is replaced by a shared cloud computing platform. Without loss of generality and instantiating the model with existing platforms, the mobile app can be deployed on the Google Play store or Apple's App Store, the Management Unit can be packaged in a Docker container [5], and deployed on the AWS (Amazon Web Services) cloud computing platform, and the Ethereum open blockchain network can be used to support the financial transactions originated by the EV charging operation.

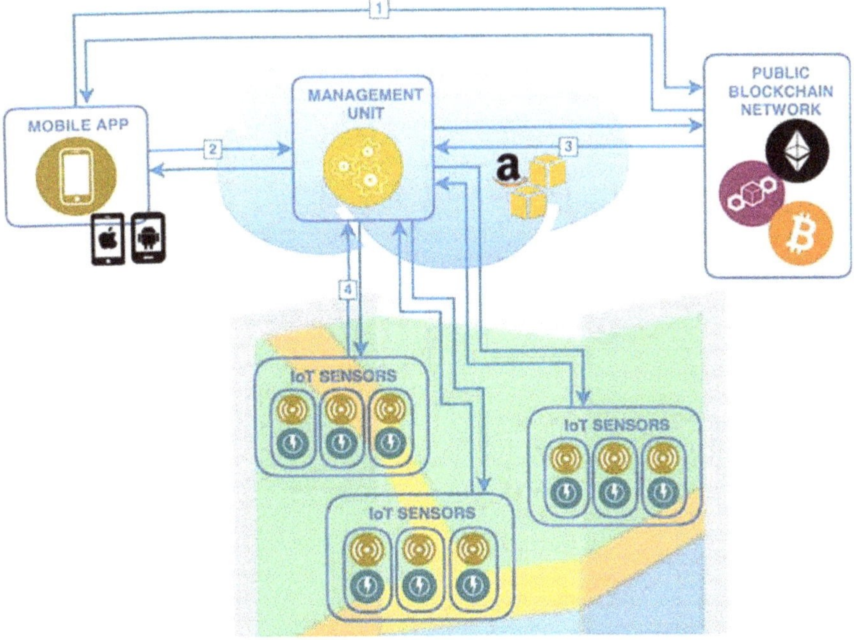

Figure 2. Overview of an IoT/cloud model solution to handle the EV.

Figure 2 also enumerates the sequence steps to initiate a charging process: (1) Using the internet connection, the payment is sent from the mobile device to the open blockchain network (Ethereum); (2) the information related to the operation is exchanged between the mobile device and the Management Unit hosted on the AWS; (3) payment is received from the blockchain network, triggering the charging process on the Management Unit; and (4) the EV charging process is enabled on the IoT device (installed on the parking facilities), and the information related to the energy being delivered is sent back to the Management Unit on the AWS.

This paper is organized as follows. Section 2 presents the state of the art in related work. An overview of the proposed approach is presented in Section 3, and Section 4 describes the system implementation. Section 5 presents a case study at a condominium, and Section 6 discusses future implications of the presented work. Finally, Section 7 presents the conclusions.

2. State of the Art

The proposed approach explores a set of works in several domain areas to create a new approach to handle the EV charging process in shared spaces, including the use of IoT sensing information to measure electricity taken on the EV charging process. Concerning driver profiles and EV charging with power limitations, several studies have been performed, and we apply an approach based on our previous work described in [6]. In our implementation, it was also considered an implicit authentication mechanism [7], applied on user's mobile devices, which confirms the user authentication based on actions that he had performed on a daily basis. This implicit authentication mechanism can be used to prevent fraudulent credit transactions on a mobile device, verifying that the user is who he claims to be during the transaction. After researching systems that meet our criteria, we found some promising work [8–12]. We apply a solution with user privacy (no identification is performed) in an approach based on the system proposed by Frank et al., called Touchalytics [13]. We also apply the blockchain approach to handle distributed transactions without central supervision. The primary goal of the blockchain is to allow decentralized transactions with a digital currency, such as Bitcoin [14] or Etherum [15], without the need of a public authority to control the process. From the technical perspective, a blockchain is a sequence of blocks associated with transactional data using encryption based on a private and public key [16]. User *A* performs a transaction, and this process is associated with a block encrypted with his private key, in a hash process. User *B* checks the transaction using the public key of user *A*, allowing the following properties:

- Decentralization, since we need confirmation from some party of each block transaction without central control;
- Anonymity, since it allows for the authentication of transactions without giving up any personal information;
- Auditability, which is performed based on the fact that each of the transactions is recorded and validated with a timestamp, where users can trace the previous transactions by accessing any node in the distributed network.

The application of blockchain in the domain of smart grids has great potential, providing a decentralized approach to implement management systems [17] and handle power transactions. Due to the large space occupied by the meter sampling information on a blockchain block, [17] presents a design to balance the amount of information kept onchain/offchain while keeping the properties of a block chain implementation. The authors of [18] note the use of an open public cryptocurrency network, such as Bitcoin or Ethereum, can introduce a high transactional cost, due to the fees associated with cryptocurrency transaction processing (eventually similar to the cost of the energy supplied), and propose the development of a private Bitcoin-based blockchain network for EV charging purposes. Other relevant application cases include micro-generation [19,20], as well as the contribution to handle the EV charging payment process without the use of propriety company payment systems.

The EV charging payment process is more frequent than fossil fuel refuelling and more complex due to the immaturity of the service. Specifically, the following issues are fairly common: (1) Transparency and clarity of rates and charges before they are incurred; (2) ability to pick-and-choose best rates and location of available charging points on the go; (3) ability to request priority charging and pay for it, when other EVs do not need priority; (4) ability to select a supplier or source of electricity, which would also enable greater competition and increase trust of customers; and (5) preferences for various types of payment, such as post-paid, pre-paid, or one-off payment.

We complement this work with our previous work on an EV charging system [21,22] and IoT energy measurements using local sensors [23], as well as new challenges of energy markets [19]. Some issues identified are also addressed in [24], which proposes a blockchain-based model with recourse to a bid to identify charging stations (and eventually schedule the charging), complementary to the approach suggested in [21]. Another issue originated by the increase of the EV charging needs is the impact on the energy demands and the power limitation of the existing infrastructure [25], which may not only increase the operational costs to fulfil the required demand, but also affects the voltage stability of the network. In [25], the authors introduced the AdBEV, which is an algorithm to optimize the EV charging schedule, maximizing the voltage stability at the power grid side, and minimizing the charging costs. In [26] the application of a blockchain-based process is suggested to support the EV charging queue management.

Together with mobile device authentication and a payment system, we developed a new approach to be used in shared EV charging spaces. Another interesting output is to use mobile devices to provide authentication and payment services in the context of the public EV charging systems, exploring recent advances in mobile device payment systems for public transportation [27] and other application areas [28]. As a new topic of research, new publications are appearing in the literature concerning the use of a blockchain approach to handling the EV charging process, such as: testing pilots to use digital currency for the EV charging process [29,30]; proposal of a P2P energy transaction model to handle the EV vehicle-to-grid (V2G) operation in smart grids [31]; handling the EV authentication issues based on a blockchain approach [32]; proposal of a cross-domain authentication scheme with blockchain [33];

and handling of security and privacy issues for energy transactions based on blockchain. Moreover, in this context, the EV is identified as part of the energy market [34], and as a contribution to the contextualization of the local energy market [35], where the blockchain plays an important role in the decentralization process, as well as for optimization purposes [36].

3. Proposed Approach—Conceptual Model

The EV charging platform is composed of the elements presented in Figure 3, whose roles are briefly described below, and the implementation details for each component is detailed in the next section:

- IoT Units. Sensor and power management units that support the interaction with the EV charger, being used to enable or disable it (on/off switch), to measure the amount of power consumed, gather environment temperature and humidity (complementary measures), and to upload all the information to the management unit. Implemented with COTS (commercial off-the-shelf) components, Arduino microcontrollers, actuators and sensors. Depending on the installation requirements, different components can be combined to set up the IoT Unit.
- Mobile App. The element that establishes the interaction between the EV owner and the platform, authenticates the user, starts/stops the charging process, and provides some common operations, such as configuration management, usage dashboards, transactions lists, etc.
- Management Unit. This element is the heart of the platform, providing not only all the backend services to support the required operations, but also the management console for the platform. In the prototype presented in this paper, the management unit was implemented using a Raspberry Pi, which also acts as a Wi-Fi access point, providing network access to the sensor units and to the mobile app, but it could also be implemented using a cloud computing platform.

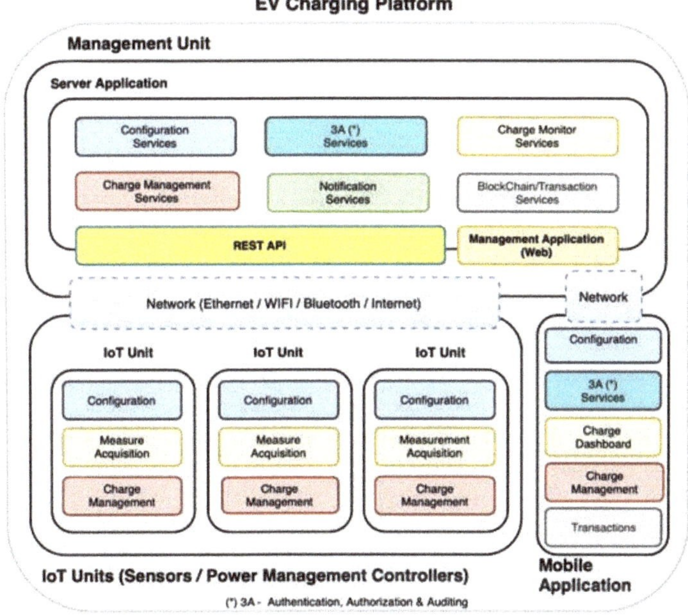

Figure 3. Main architecture of the proposed EV charging platform: Server Application, IoT Units and Mobile Application.

4. System Implementation

As previously described, the proposed EV charging platform is composed of three major elements: IoT units (sensors/actuators devices), a mobile app and a management unit. This section explores the implementation details of each element.

4.1. IoT Unit

The IoT unit was developed considering the approach described in our previous work [19], with improvements to the hardware and transmission process, as well as the creation of a prototype towards a possible commercial system. The first steps were the assessment of the surrounding environment and context, aiming to review the system design approach. The goal of reaching a potential commercial system's architecture led to the consideration of a flexible design, where different network transmission requirements/devices, current sensor devices and power switching devices should be available to use, tailoring their combination to match a specific installation requirement. After an initial period of checking and testing hardware, we implemented a solution based on an Arduino Uno (microcontroller) combined with the devices listed in Table 1, where only one component for each type was used to assemble the IoT unit.

Table 1. List of IoT hardware add-ons.

Component Type	Device
Network (Shields)	Sparkfun ESP8266 (Wi-Fi) WIZnet's W5100 (Ethernet)
Current Sensors	SCT-013-000 (non-intrusive) ACS712 20A (intrusive)
Power Switching (*)	SRD-05VDC-SL-C (generic network switch)
Temperature and Humidity	DHT11
NFC RFID (**) Wireless Module	PN532

(*) A generic network-controlled switch can be controlled by the management unit. Approaches such as BLE-controlled switches can eventually also be used, providing that a BLE add-on is added to the IoT unit. (**) Near Field Communication e Radio-Frequency IDentification.

4.1.1. Configuration of Variations

Wired (Ethernet) vs. Wireless (Wi-Fi) Network: Taking into account that most existing condominiums do not have a wired network infrastructure, the use of a Wi-Fi network simplifies the deployment of the platform, as no other infrastructure components are required, particularly when using the Wi-Fi network provided by the management unit. For new installations or for larger installations, a cable-based approach may be more suitable and less error-prone.

Intrusive vs. Non-Intrusive Power Sensing: The non-intrusive approach offers the capability to measure the energy that passed through a specific IoT unit, allowing the measures to be gathered without any major changes to the existing infrastructure, as the sensor only needs to be "hooked" around the power cable that powers the EV charger device socket. However, since no physical devices are installed between the power plug and the EV charging device, the capability to enable/disable the charging process needs to be implemented by the EV or by the charging device and exposed as a service to the charging platform; eventually, the platform network or other communication technologies, like BLE, could allow the management unit to start/stop the process based on user commands. On the other hand, although the intrusive approach forces the platform owner to introduce the IoT device between the power grid and the power socket, which requires some intervention in the existing infrastructure, it is able to provide a sound solution to the platform owner, as it provides a "one-in-a-box" unit that is able to measure and control the energy delivery (enabling/disabling) simultaneously, while providing energy only to authenticated users or inside of a blockchain transactional context.

Built-In vs. COTS (Commercial Off-The-Shelf) Power Switching: To enable/disable the EV charging devices, we have considered using the SRD-05VDC-SL-C (Ningbo Song Relay Co., Ningbo, China) device (see Figure 4f), which, when connected to the Arduino device, can be used as a switch. A different approach to support this requirement is to use a standard TCP/IP-based (Transmission Control Protocol—Internet Protocol) switch commonly available as COTS on the market.

4.1.2. Hardware Components

The most relevant characteristics of the hardware components used for the prototype implementation are:

- Arduino R3 Uno Microcontroller (Figure 4a): based on the microcontroller ATmega328P, it has the following characteristics (from the Arduino R3 Uno dataset):

    - 14 digital input/output pins (the first 2 are commonly used for serial RX/TX (Receive and Transmit), 6 can be used as pulse-width modulation (PWM) outputs that mimic an analogue output) and 6 analogue input pins (A0–A6).
    - 16 MHZ clock speed (memory: flash, 32 K; SRAM, 2 K; EEPROM, 1 K).
    - USB type B connection, ICSP Header.
    - Power input 9 V (operating voltage 5 V), built-in LED, reset button.

- Sparkfun Wi-Fi Arduino Shield (based on ESP8266) (Figure 4b): manufactured by Sparkfun this Arduino shield is commonly used to connect the Arduino microcontroller to a Wi-Fi network and use the "standard" internet protocols (TCP or UDP).
- Arduino Ethernet Shield (based on Wiznet W5100) (Figure 4c): designed for embedded applications where ease of integration, stability, performance and cost are required, as well as ease of internet connection without the need for an operating system to implement. This chip complies with IEEE 802.3 10Base-T and 802.3u 1000Base-TX standards and includes a TCP/IP hardwired stack, supports up to four simultaneous socket connections, integrated MAC and PHY Ethernet, and 16 kilobytes of internal buffer for data transmission. The standard RJ45 connection allows speeds from 10 to 100 megabytes.
- Non-Intrusive Current Sensor SCT-013-000 (non-intrusive) (Figure 4d): a non-intrusive sensor used to measure the current passing through a conductor without the need to cut or modify the conductor itself. The measurements are collected from the electromagnetic induction, which is proportional to the intensity of the current passing through the conductor. This sensor collects measurements up to 100 A, outputting at 50 mA. In terms of accuracy, it may deviate from 1% to 2% of the actual value.
- Intrusive Current Sensor 20 A (based on ACS712) (Figure 4e): based on ACS712 this intrusive Hall effect current sensor can be used to measure currents between −20 A and +20 A, with an output ratio of 100 mV/A.
- Power Switch 10 A (based on SRD-05VDC-SL-C) (Figure 4f): a mechanical relay which operates a switch. Powered by the standard Arduino 5 Vcc, it has a control line (+5 V) that when powered, establishes a connection between the terminals common (C) and normally open (NO). The used part also includes a small LED which is enabled when the circuit between the terminals C and NO is established.
- Temperature and Humidity Sensor (DHT11 based) (Figure 4g): from DFRobot, can work from 0 to 50 °C and humidity from 20% to 90%, and has low power consumption, with a precision of 2 °C.
- RFID/NFC Reader/Writer (PN532) (Figure 4h): has several wireless capabilities, it can be used to read and write RFID and to exchange data with Near Field Communication (NFC) enabled devices.

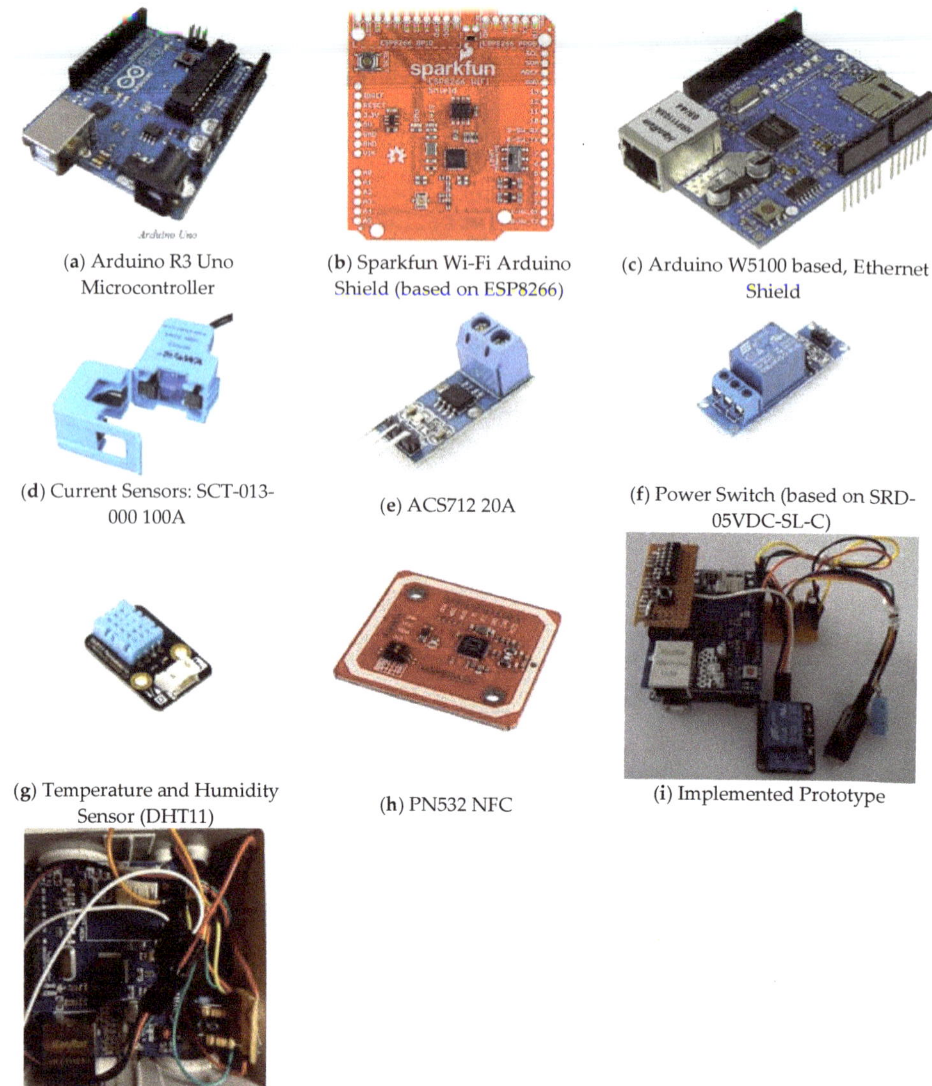

(**a**) Arduino R3 Uno Microcontroller

(**b**) Sparkfun Wi-Fi Arduino Shield (based on ESP8266)

(**c**) Arduino W5100 based, Ethernet Shield

(**d**) Current Sensors: SCT-013-000 100A

(**e**) ACS712 20A

(**f**) Power Switch (based on SRD-05VDC-SL-C)

(**g**) Temperature and Humidity Sensor (DHT11)

(**h**) PN532 NFC

(**i**) Implemented Prototype

(**j**) Implemented Prototype

Figure 4. Hardware components used in the proposed EV charging platform.

4.1.3. IoT Unit Software

The software implemented in the Arduino Uno microcontroller was developed in C++ through the Arduino IDE, where the main methods are used to read the sensor data, enable/disable the EV charging device, authenticate the user, initialize the network shield and obtain an IP (Internet Protocol) address via DHCP (Dynamic Host Configuration Protocol) or configure a static IP address, buffer the collected sensor data, and send the data to the management unit. Once the IoT Unit is powered on, it performs the following steps:

1. When the device starts, it checks if it contains configuration information stored in the EEPROM. In this case, it will automatically go to step 3 (if the sensor is started with the reset button pressed the EEPROM configuration is deleted, Figure 4i).
2. In the absence of a stored configuration, the device contacts the server to obtain the configuration data, receiving the following parameters in response: (a) Data transmission frequency; (b) sampling frequency; (c) target server; (d) IP configuration (static or dynamic); and (e) time server. To identify the sensor together with the central application, the sensor Id is read from the dip-switches shown in Figure 4i, allowing a total of 64 (2^6) sensors configured to obtain configuration.
3. After reading the configuration data, the device is ready for operation.

The communication with the server to obtain the configuration and sending of readings is done using the TCP and Hypertext Transfer Protocol (HTTP) protocols, using the GET and POST methods, respectively.

4.2. Mobile App

As an integral part of the project and to allow the EV owner to interact with the platform, a mobile app was developed in C# using the framework Xamarin.Forms, which allows multiplatform development for Android, iOS and UWP (Universal Windows Platform). Figure 5 presents the use case diagram that enumerates the most relevant features implemented.

Figure 5. Use case diagram of the developed mobile app.

4.2.1. EV Charging Process (Automatic vs. Manual Starting Process)

Initiating the EV charging process is the key function of the mobile app, being simultaneously the most frequent operation, as charging the EV is the purpose of the entire system. Requiring the EV owner to connect to a network where the management unit is reachable, as is illustrated in Figure 3 (assuming that the system operates in a closed network), to be able to start the charging process adds a non-practical, time-consuming process. Aiming to improve the user experience, while performing the operation, we have implemented two approaches: one automatic approach supported by the NFC capabilities of the user's mobile device and one manual approach relying exclusively on the implemented mobile app. Using a more automatic approach, the user initiates the charging process using the NFC capabilities of his mobile device to authenticate the operation, starting the process by placing his mobile device near to the NFC reader attached to the IoT unit. In this case, the process will use the statistical information collected from the previous operations to confirm the user authentication, estimate the power needs and the amount of time that the EV will be connected to the charging plug (to forecast the price/time usage). Complementary to this process, a more controlled approach can be used, when the NFC device is not available or not attached to the sensor unit, or if the vehicle owner needs to configure the charging process (setting parameters such as the amount of the battery energy according to the state of charge (SoC), the amount of time connected to the platform, time-window for charging, etc.). In this case, the charging process can be initiated by connecting to the network where the management unit is reachable, eventually to the Wi-Fi network provided by the management unit, and manually starting the process, providing the required information. Figure 6 shows the application interfaces to initiate a charging process and to stop the charging process.

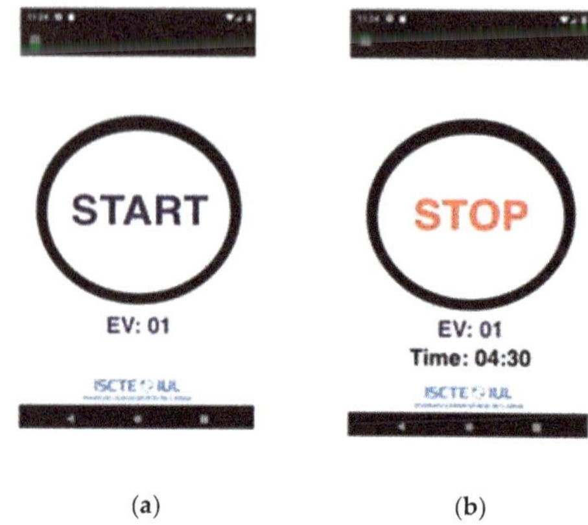

Figure 6. Mobile app interfaces for starting the EV charging (**a**) process and for stopping the EV charging process (**b**).

4.2.2. General App Features

Apart from the EV charging process, which can be considered the crux of the system, the mobile app also implements several features that, although not as relevant, are required to achieve a production-grade design stage. Figures 7 and 8 shows screenshots for some of the implemented features:

1. Application splash screen, Figure 7a.
2. Current usage pattern, Figure 7b.
3. Application settings, Figure 7c.
4. Energy costs calculated on the basis of kWh and sensor statistical measures, Figure 8a.
5. List of sensor readings received, Figure 8b.
6. Sensor configuration details, Figure 8c.
7. About screen, Figure 8d.

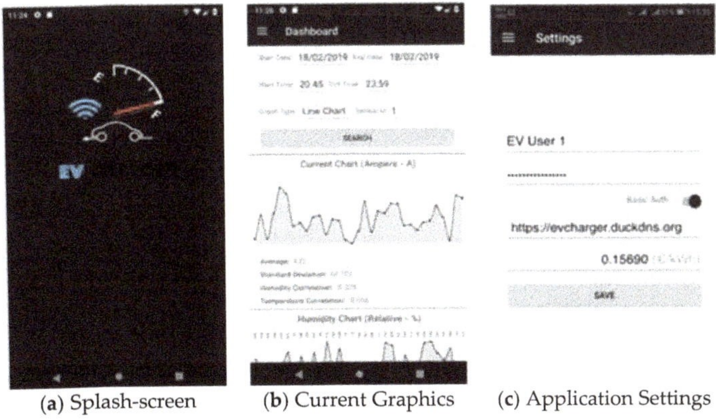

(**a**) Splash-screen   (**b**) Current Graphics   (**c**) Application Settings

Figure 7. Mobile app screenshots.

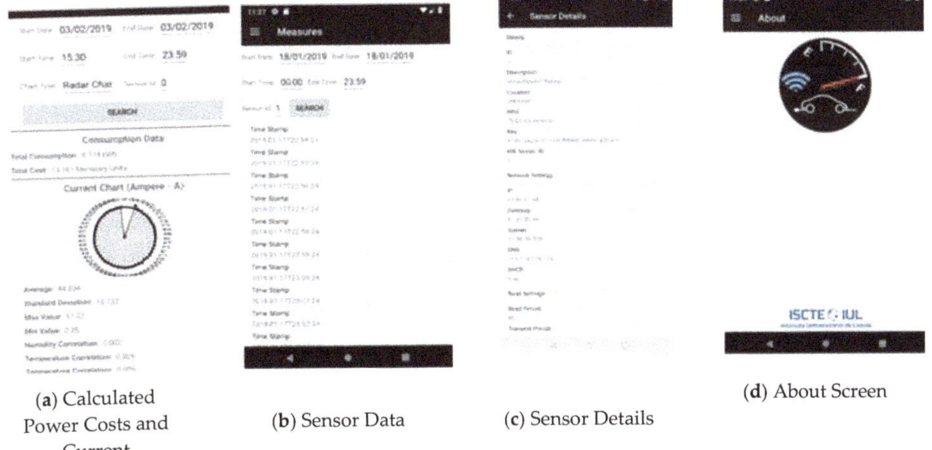

(**a**) Calculated Power Costs and Current   (**b**) Sensor Data   (**c**) Sensor Details   (**d**) About Screen

Figure 8. Other mobile app major functionalities.

4.2.3. Software Architecture

Figure 9 displays the mobile app software organization. From a functional perspective, the mobile app is split into several modules enforcing the separation of concerns between each functional unit. From the software architecture perspective, the mobile app is implemented following the Model-View-ViewModel (MVVM) pattern, which can be considered one extension of the Presentation Model (PM) pattern [37], frequently used in Xamarin.Forms applications (as well as other mobile apps) where each logical layer has a clear separation of concerns, as described:

- View, implemented with XAML (eXtensible Application Markup Language), a declarative language used to design and structure the user interface.
- View-Model is the binding element that intermediates the relationship between the View and the Model, mapping the information and actions between both elements.
- Model is the representation of the data.
- Rest Adapter since the mobile app entirely relies on services provided by the management unit, this component acts as a proxy between the mobile application and the services exposed. Due to the financial nature of transactions, the information exchanged between the mobile app and the central management unit uses a secure Hypertext Transfer Protocol Secure [38] (HTTPS)

transmission (secured by a server certificate) and Hypertext Transfer Protocol [39] (HTTP) standard authentication mechanisms. Stronger authentication schemes can be supported by use of client certificates to authenticate the mobile app requests on the server; however, this was not considered for the current implementation to avoid the complexity of introducing a Public Key Infrastructure (PKI) in the platform.

- Local Storage consists of a small information repository to store local configuration data in the mobile device.

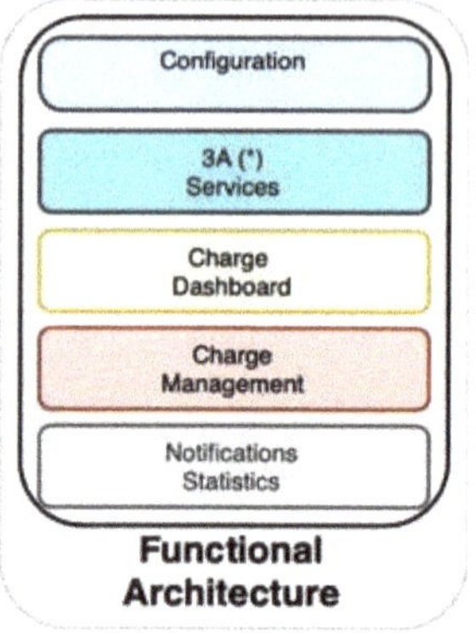

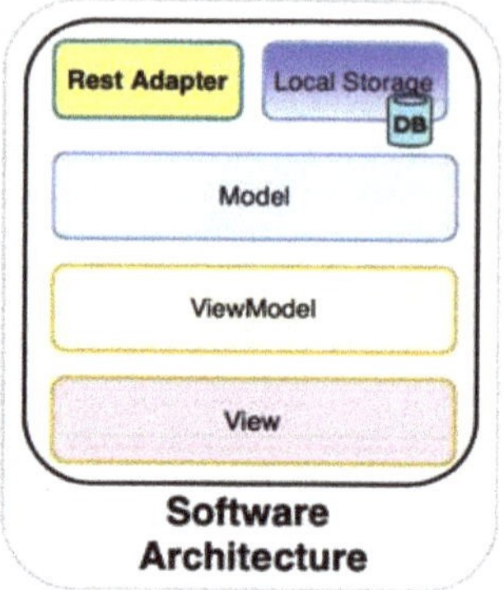

(*) 3A—Authentication, Authorization and Auditing

Figure 9. Mobile app functional and software architecture views.

4.3. Management Unit

The management unit is the heart of the platform. This section is divided into the following subsections: Hardware and Network Infrastructure; Software and Services Infrastructure; Management Services; Management Web Application; and the Blockchain.

4.3.1. Hardware and Network Infrastructure

The management unit was built using a Raspberry Pi 3 Model B+ hardware, and the Raspbian operating system. The unit was configured as a Wi-Fi access point, setting up the network to allow Wi-Fi communications between all the platform components (management unit, sensor units and mobile app). This configuration allows the deployment of a completely self-contained, pluggable, low-cost solution, without requiring any other infrastructure components (apart from the energy power grid), while increasing the security of the overall solution by reducing its exposure to external network threats. Complementarily, if deployed in a location with existing network support, the management unit can be connected to the network using the RJ45 Ethernet connector of the Raspberry Pi, allowing the platform to benefit from the existing infrastructure and to eventually be deployed in setups where the use of a Wi-Fi network may not be available or the most suitable option, for instance, a multi-level condominium parking lot, or a parking lot spread over several areas and sharing only one management unit.

4.3.2. Software and Services Infrastructure

Figure 10 displays the software infrastructure that supports the services exposed to the platform elements (IoT unit, mobile app).

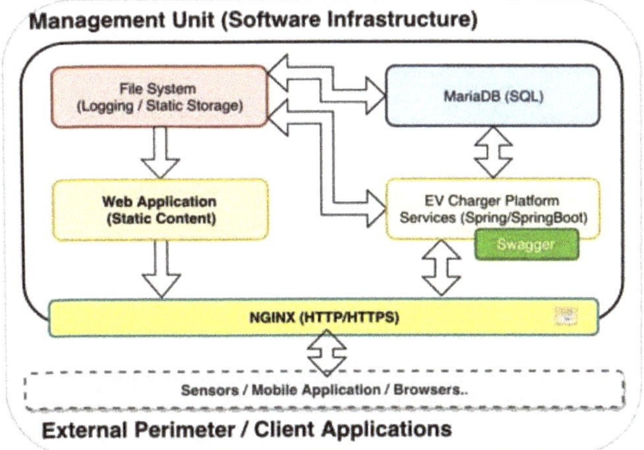

Figure 10. Software infrastructure and flows of information.

The platform services, built using the Spring Framework and SpringBoot, are exposed as a set of Representational State Transfer (REST) endpoints, self-documented through the use of the Swagger Framework, as presented in Figure 11. This approach exposes an API (Application Programming Interface) that can be easily used by third-party applications, using standard interoperability tools, allowing the development of custom-made integrations (for instance, to integrate the platform with a condominium management system). The platform data is stored in a local MariaDB database server. Aiming to guarantee the security of the communications between the management console, the mobile app and the central unit use a standard HTTPS protocol that has been archived by the installation, and configuration of HTTPS certificates (freely provided by Let's Encrypt), deployed in an NGINX (Engine X) web server, which acts as a proxy between the "external world" and the services layer.

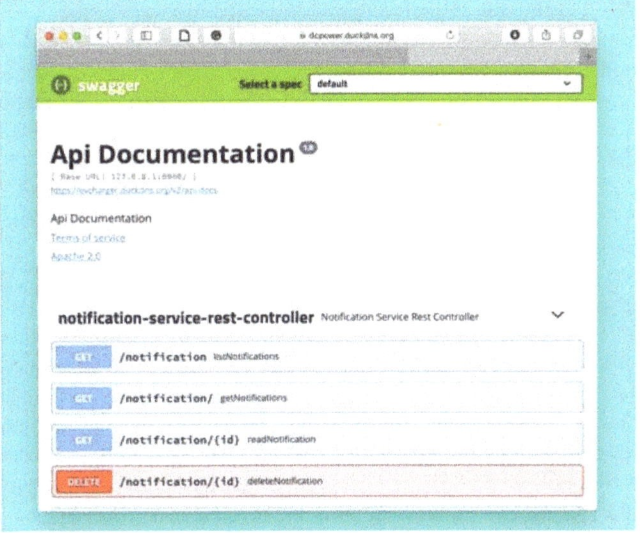

Figure 11. Self-generated API (to support interfacing with third-party applications).

4.3.3. Management Services

Figure 12 presents the application level services that constitute the EV charging platform.

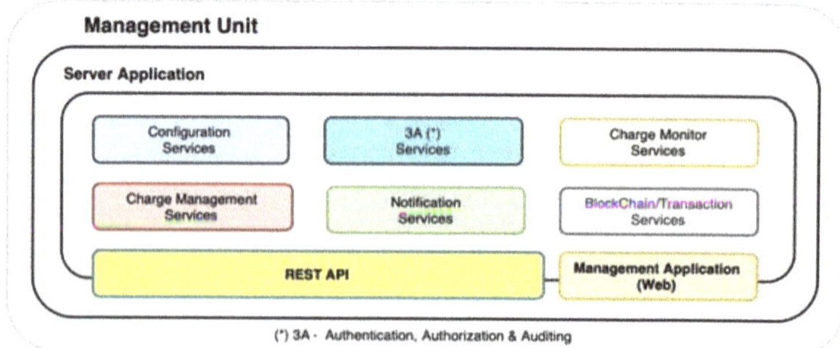

Figure 12. Management Unit Services.

A brief description of the implemented services and their contribution to the overall platform is presented below:

- Configuration: Provides a set of services required to configure the EV platform, allowing the user to define several platform parameters, such as the existing sensors and their configuration (e.g., network configuration, maximum current, accounting frequency, measure period, etc.), as well as the groups of sensors (sensors inside the same group, maximum load per group, etc.).
- 3A (Authentication, Authorization, Auditing): This module has a central role in the entire platform. It is responsible for centralizing all the operations related to user/system authentication ("who is who"), authorization ("what can do") and auditing ("what was done"). Apart from implementing the set of operations to manage the user access to the platform, it also implements the implicit authentication [6] to validate the charging request automatically, based on the current user's usage pattern.
- Charge Monitor: This module collects and processes all information generated from the installed sensors to update the EV charging records and detect the end of the charging events, as well as any anomalies on the charging process (e.g., exceeding the nominal current, temperature, charging time), and triggering eventual notifications when required. This module also collects the user's usage pattern to estimate the power needs for the current charging process, as well as estimate the leave time of the EV from the charging plug, if that information is not provided explicitly by the user.
- Charge Management: If the installation has the capability to enable or disable the EV charging process, by the use of network-controlled charging devices or by the use of charging switches attached to the sensor unit, the module enables or disables de-charging of the EV, aiming to properly distribute the available charging windows between all the EVs connected to the charging group, based on the charging requirements and the amount of time that the vehicle will be connected to the charging device and using the information provided explicitly by the user or inferred by the platform based on the users usage pattern.
- Notification Services: This module provides all the notification related services to the platform, routing the system-generated notifications to users that had subscribed to that notification (i.e., vehicle charged, abnormal charge pattern, etc.)
- Transaction/Blockchain: This module supports all the "financial" related operations, for instance, it records the changing event in the blockchain ledger; if the installation supports the charge management process (described previously), it allows the platform managers to transfer "charging tokens" to the user's wallet (if not using a public crypto-currency network); it monitors the

reception of user's transferred credit to start the charging process; and it returns the unused credit to the user's wallet. It also provides minimal reporting capabilities to allow the financial management and analysis of the platform usage.

Each service is implemented following a similar pattern to the pattern presented in Figure 13, where the responsibilities of each are defined as follows:

- Service Layer: Acts as a mapping service, translating the external representation of the information to the internal representation.
- Business Layer: All the application behaviour level is defined on this layer, and any interaction between layers made exclusively through the interface provided at this level.
- Persistence Layer: This layer maps the internal representation of the information to representation used by the database engine.

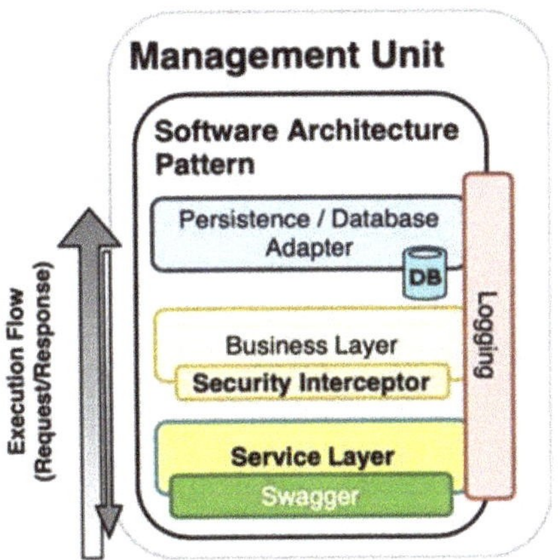

Figure 13. Software architecture pattern.

4.3.4. Web Application

The management unit exposes a web application, developed in Angular, which relies on the services exposed and allows the EV platform managers to monitor, configure and operate the platform. It also provides to the platform users a complementary user interface that, although supporting only a reduced set of the operations available on the native mobile app, allows the users to interact with the platform using browser-only technologies, available in a wider range of devices. Figure 14 displays some screenshots for the management unit web application.

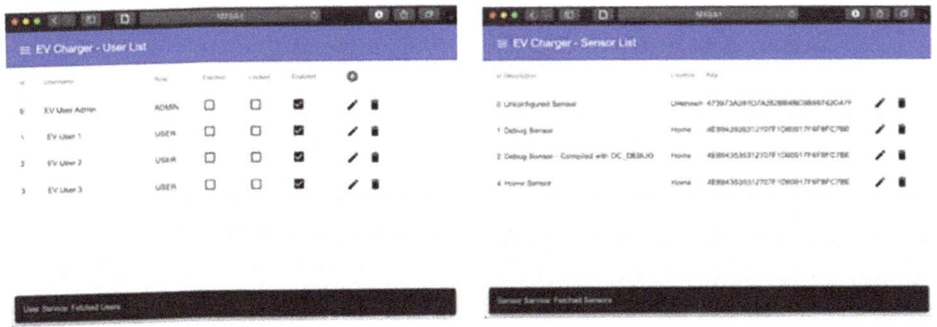

Figure 14. Web application: Users list (**left**) / Sensors list (**right**).

4.3.5. Blockchain Implementation and Integration

The transactions between the users and the platform rely on the exchange of EV charging tokens (self-generated). If using a "public" crypto-currency infrastructure like Ethereum (or Bitcoin), the trades are made using that crypto-currencies (which can be exchanged in the market). Currently, the transaction is performed with a fixed energy price or based on pre-defined rules but, in the future, the price can be negotiated dynamically in full implementation of a smart grid [19]. This implementation uses the same approach of our previous work on this topic [19]. Figure 15 presents the interactions with a blockchain network using a private/internal blockchain ledger, whereas Figure 16 shows the interactions that would be held when using an "open" cryptocurrency (e.g., Bitcoins).

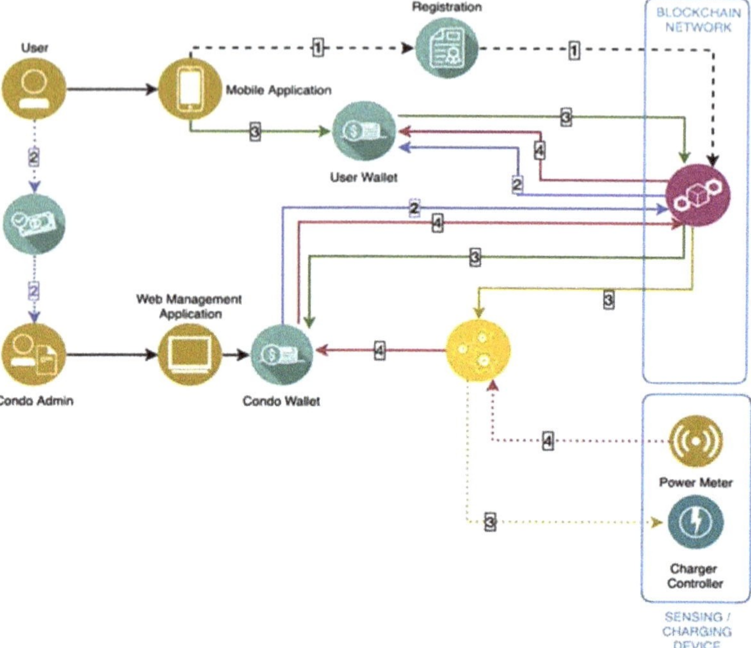

Figure 15. Blockchain interactions with an internal ledger.

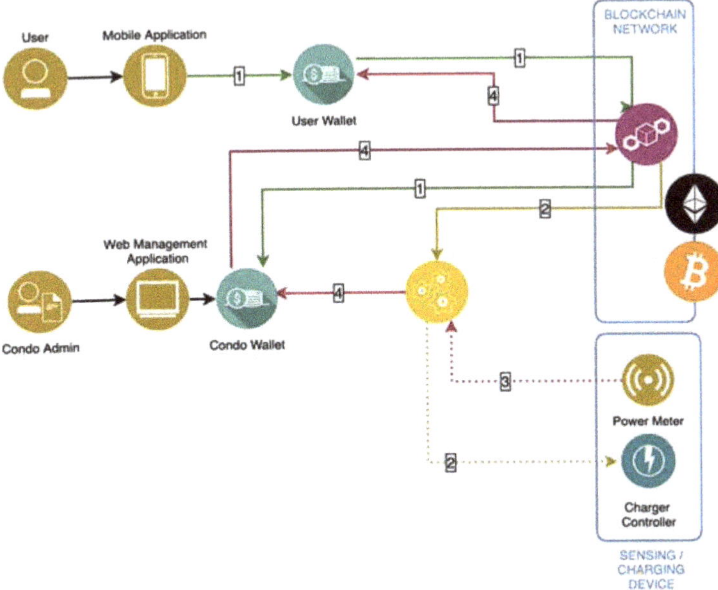

Figure 16. Blockchain interactions using an open cryptocurrency.

The sequence of operations presented in Figures 15 and 16 are explained as follows.

1. Using the mobile app, the user registers/creates his account on the blockchain network (if using public crypto-currency infrastructure, the user creates his crypto-currency wallet).
2. Using real money, the user buys EV charging tokens from the EV platform management, referring to the web interface of the Management Unit. The charging tokens are transferred from the platform wallet to the user wallet through a blockchain network (if using a public crypto-currency the user buys the currency on the market).
3. Using the mobile app, the user sends charging tokens from his wallet to the EV platform wallet, defining the maximum amount to spend and the maximum time that the vehicle will be connected to the plug (used to optimize the power distribution). The Management Unit receives the transfer from the network and triggers the power management unit to start the charging process, which may not be immediate due to the optimization of the power distribution between the used chargers.
4. The Management Server receives the power measures from the charger, stopping the charging process when the maximum amount is reached, the maximum charging time is reached, or when the vehicle is removed from the charger (detected by a reduction of the consumed power). If the charging process is interrupted, the remaining amount is returned by the Management Server to the user wallet using the blockchain network.

5. Case Study at a Condominium

We applied the current approach to a shared place in a condominium, where three EV owners shared the condominium electric installation available at parking places for a period of 3.5 months. Each sensor was configured to generate one sample each minute, allowing further study of the current load patterns during a charging event. A set of three EVs (all Leaf vehicles with 24 kWh battery capacity) and three independent sensors (Sensor 0; Sensor 1; Sensor 2) were used; Figure 17 presents the diagram of the test environment for the case study. Due to physical constraints of the installation, the charging adapter connected to Sensor 0 was directly connected to the power grid, without one intermediate switch ("always on" on the scheme).

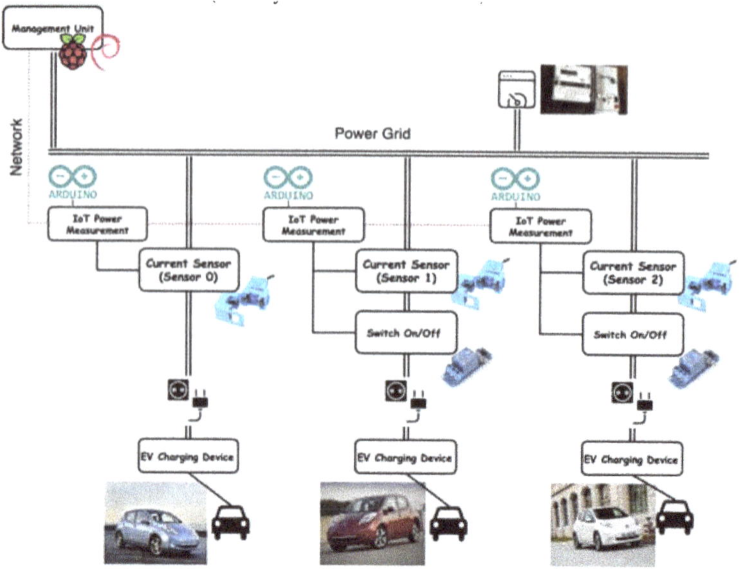

Figure 17. Setup diagram of the case study.

Figure 18 presents photos of one of the developed prototypes. Figure 18a shows a photo of one of the IoT unit prototypes installed (Label (3) in Figure 18a) of the test environment, measuring the current with the non-intrusive SCT-013-000-100A sensor (Label (1) in Figure 18a). In this case, due to the weak Wi-Fi signal at the install location and the absence of other network infrastructure, the sensor unit was connected, using the RJ45 Ethernet interface, to a Wi-Fi Range Extender (Label (2) in Figure 18a) to amplify the signal, allowing the IoT unit to reach the Management Unit accessible from the network where the Wi-Fi Range Extender was connected. Figure 18b shows the contents of the IoT unit installed in Figure 18a (Label (3)).

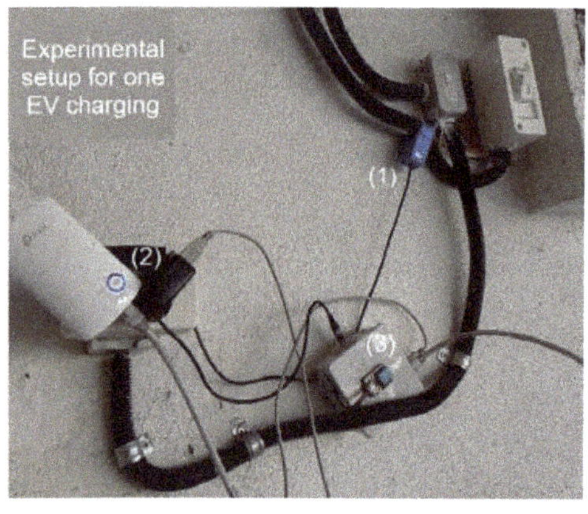

(**a**) Installed IoT unit

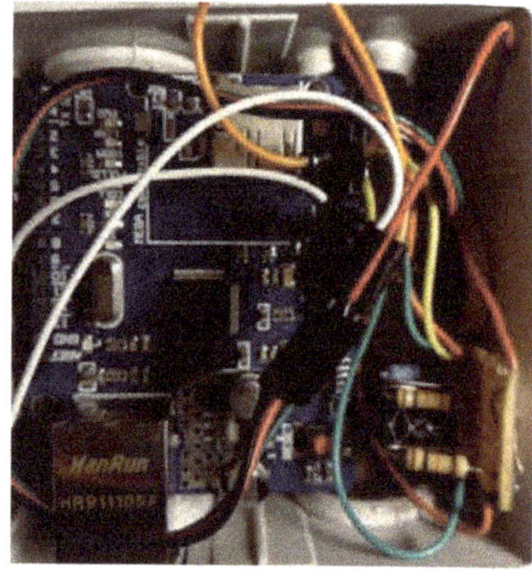

(**b**) IoT unit contents

Figure 18. Developed prototype: (**a**) IoT unit prototype deployed in one test environment to take measurements at a condominium; (**b**) contents of the IoT unit prototype.

Table 2 summarizes the data collected during the case study.

Table 2. Data collected during the case study.

Measure	Value
Data Samples	450,000 [1]
Total Time (hours)	2700
Start Date	20 January 2019
End Date	12 May 2019
Charging Data Samples	63,000
Charging Events	300
Total Charging Time (hours)	1060 h (~40%)
Unused Charging Time (hours)	1640 h (~60%)
Total Energy (kWh)	2450 kWh [2]

[1] Estimation, based on the configuration, as "empty" data samples are discarded. [2] For the current case study, it was assumed a voltage of 230 V.

Figure 19 shows the charging time and the average charging power for each charging event (for events with > 3 h of charging duration), where it is possible to identify an average value of 2.3 kW, approximately (assuming an root mean square, RMS, voltage value of 230 V). The absence of a strong correlation between the charging time and the average charging power is also observable (the correlation coefficient between the charging duration and the charging power dataset is −0.30), which suggests that the average charged power by hour load is limited by the charging device and not directly dependent of the amount of energy required to charge the EV (e.g., a charging event of 6 h has a similar average charging power as a charging event with 3 h).

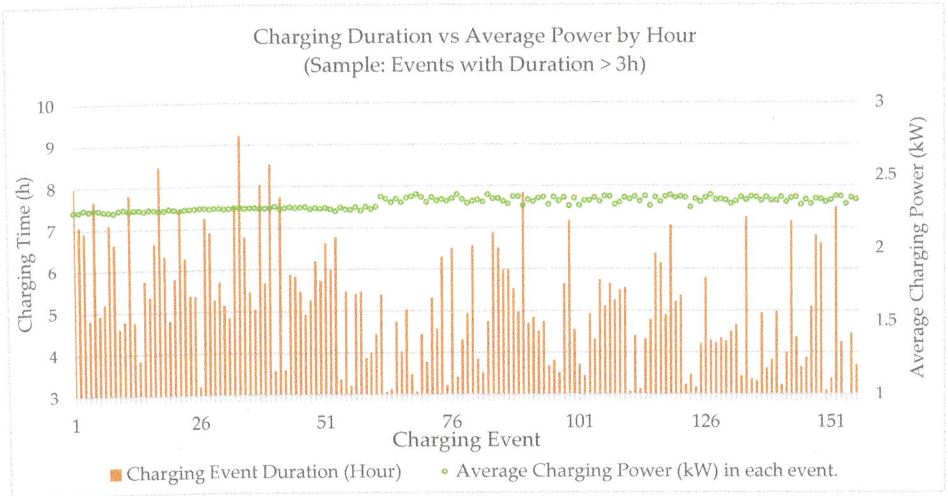

Figure 19. Average charging power and charging duration during each charging event.

Figure 20 displays simultaneous charging events for the entire period analysed (330 charging events on 20 January and 12 May). Due to the power limitations, only two EVs are allowed to be charging at the same time, using full charging power, and the power is delivered on a first-come, first-served (FCFS) basis, where the platform controls the maximum number of stations that are allowed to charge the EVs simultaneously, queueing the remaining charging requests until a charging slot is available.

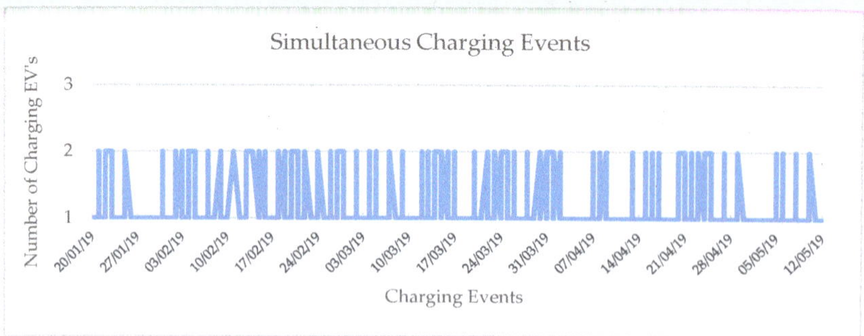

Figure 20. Simultaneous charging (during the test period).

Since the charging platform measures the supplied power continuously, it detects when the EV is fully charged. At that time, it interrupts the EV charging process, transfers the data to the blockchain network (to account for the transaction performed), and starts supplying energy to the next EV queued. Supported by the drivers' consumption profile and the statistical information about their behavior (taken from past stored data, average power required, the average number of hours before the vehicle is unplugged, etc.) a priority/utility-based resource scheduler can be applied to maximize the benefits/utility of the energy supplied.

Figure 21 shows the charging sessions of a Leaf with 24 kWh battery capacity, in a 3.5 month period, where it is possible to verify charging session periods ranging from 1 to 9 h (with an arithmetic average of 5.12 h and standard deviation of 2.03 h), and Figure 22 shows the charged energy, which varies between 2 kWh and 22 kWh (with an arithmetic average of 11.67 kWh and standard deviation of 4.58 kWh). It is possible to identify in this figure that, on average, this driver only charges 52% of the total charge and uses, on average, 5.5 h to charge the EV. From this approach, it is possible to identify driver profiles and use this for future charging processes accounting for the power limitation, as is shown in [19,22].

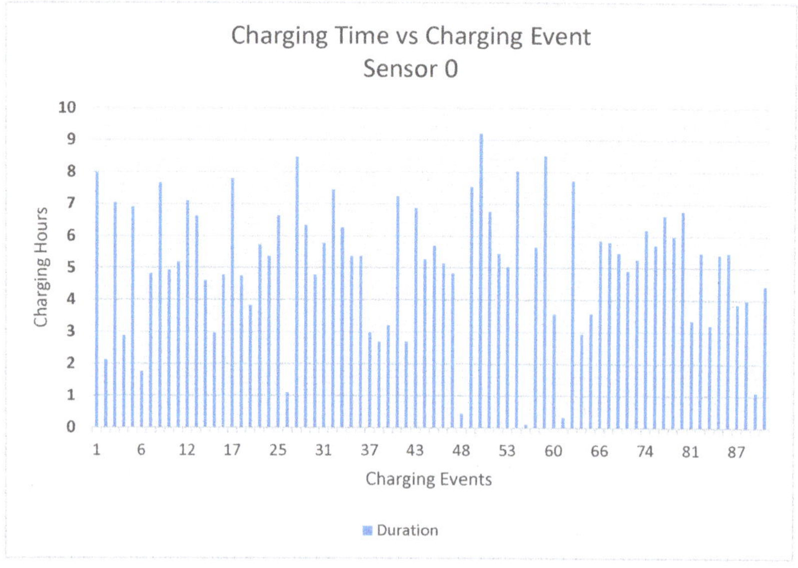

Figure 21. Charging hours (Y-axis) per charging session event in a 3.5-month period for sensor 0, used to charge a Leaf with 24 kWh battery capacity.

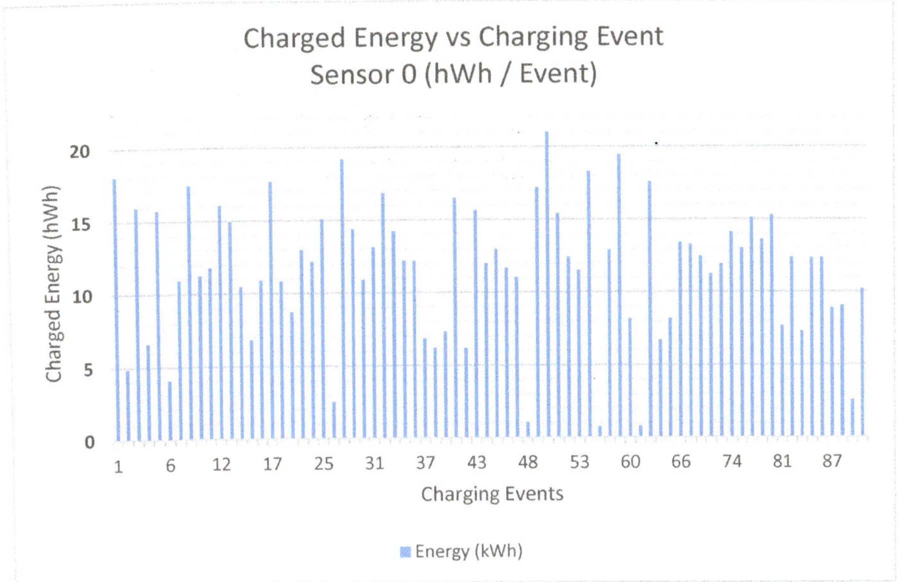

Figure 22. Energy (kWh) per each EV charging session in a Leaf with 24 kWh battery capacity.

Figure 23 shows the charging process with three EVs at the condominium, where it is possible to identify that, due to the power limitation, EV2 had to wait for an available charging window.

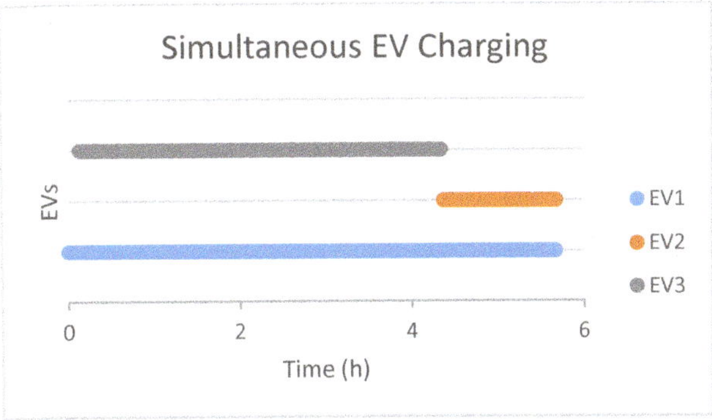

Figure 23. Charging windows (power limitation allows only two EVs to charge simultaneously).

Figure 24 presents the distributions for the charging time (left) and for the charged energy (right) for each charging event. It can be observed that for 89% of the charging events ((117 + 82 + 69)/300), the EV will be charging for 6 h or less. A similar analysis can be made for the charging energy, where for 92% ((108 + 93 + 76)/300) of the charging events, the EV will charge 15 kWh, which represents roughly 62.5% of the total battery capacity.

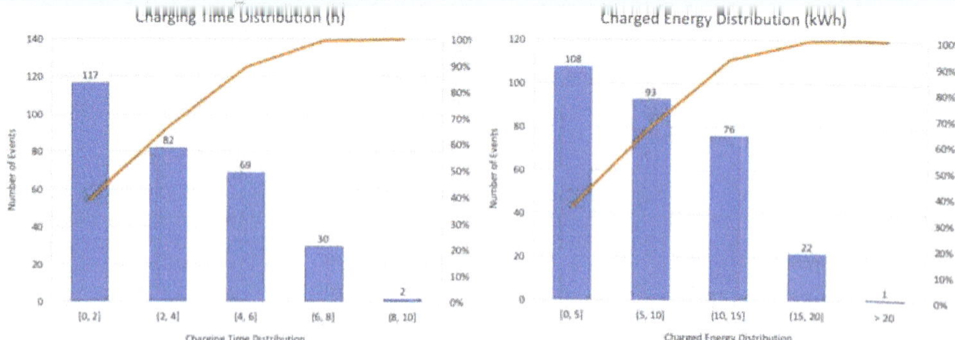

Figure 24. Charging time (**left**) and energy (**right**).

Several usages pattern also were observed. Figure 25 displays the distribution of the amount of time between each EV charging event, which shows that for 64% of the times the driver charges the EV with less than 20 h between charging events, which may be correlated with the commute journey.

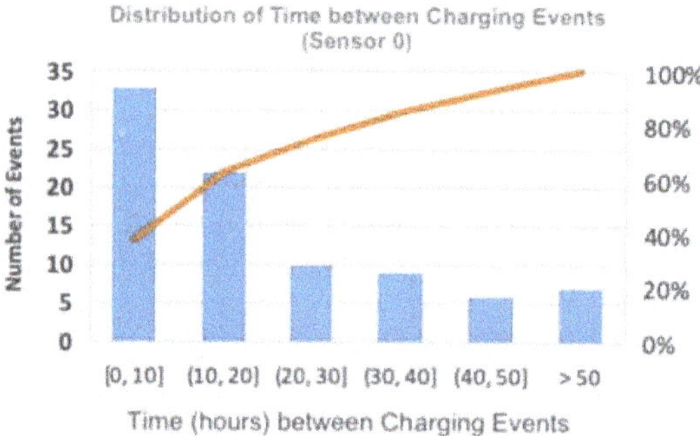

Figure 25. Distribution of time between charging events.

6. Future Implication of Mobile Devices as a Payment System for EV Charging

EVs face several problems when going abroad, due to the need for previous planning in getting charging cards from foreign operators in a process that is not easy. The developed app for the condominiums can be adapted for public charging with an identity management process and a secure environment provided by the blockchain.

This approach can be similarly applied for mobile ticketing systems, allowing users to buy public transportation tickets using mobile devices (see [27] as an example of this work), or pay motorway fees and for other services.

Some commercial approaches are currently being tested; for example, charging stations in the UK will be equipped with NFC payment technology [40]. In future work, we will perform security tests on this approach and develop an interface to a payment service.

7. Conclusions

The work presented in this paper explores different approaches based on IoT, mobile devices and blockchain to create a novel solution for the EV charging process in shared spaces with authentication and security features, accounts and a transaction system. This approach can contribute to the proliferation of EVs, because one of their current barriers is the charging process at condominiums and rented houses. Moreover, from this solution, it is possible to identify EV charging profiles, create patterns to handle power limitations and share services without the need for new individual services. This approach can also be applied to handle energy transactions in other application scenarios, such as micro-generation without a central supervision control mechanism, although the use of open public cryptocurrency platforms like Bitcoin or Ethereum, due to high transaction costs, can create some barriers to the acceptance of the model.

The proposed solution demonstrated the robustness of the developed prototype for an EV charging process in shared spaces in the context of the presented case study at a condominium. During the 3.5 month of operation, there was only one failure of an IoT sensor unit due to a general power failure, and the problem was corrected by simply delaying the start of the charging process. Although no network-related limitations were identified while using traditional wired (Ethernet) and wireless (Wi-Fi) local area network (LAN) technologies to establish communication between the IoT devices and the Management Unit for the presented case study environment, the implementation of the system in wider geographical environments or other building topologies may require the use of wireless communication technologies more suitable for that context, for instance, low-power wide-area network (LPWAN) technologies such as LoRa, Sigfox, NB-IoT or LTE-M.

Author Contributions: J.P.M. is a Master student that performed all development work. J.C.F. is a thesis supervisor and organized all work in the computer science subject, and the other authors revised the document and collaborated on energy and power electronics, as well as in IoT topics.

Funding: This work has been partially supported by Portuguese National funds through FITEC programa Interface, with reference CIT "INOV—INESC Inovação—Financiamento Base".

Conflicts of Interest: The authors declare no conflict of interest.

References

1. Al-Fuqaha, A.; Guizani, M.; Mohammadi, M.; Aledhari, M.; Ayyash, M. Internet of things: A survey on enabling technologies, protocols, and applications. *IEEE Commun. Surv. Tutor.* **2015**, *17*, 2347–2376. [CrossRef]

2. Stat of the Week: Percent of Households That Rent By Country. Available online: https://evadoption.com/stat-of-the-week-percent-of-households-that-rent-by-country/ (accessed on 21 May 2019).

3. Axsen, J.; Goldberg, S.; Bailey, J.; Kamiya, G.; Langman, B.; Cairns, J.; Wolinetz, M.; Miele, A. *Electrifying Vehicles: Insights from the Canadian Plug-in Electric Vehicle Study*; Simon Fraser University: Vancouver, BC, Canada, 2015.

4. Lopez-Behar, D.; Tran, M.; Mayaud, J.R.; Froese, T.; Herera, O.; Merida, W. Putting electric vehicles on the map: A policy agenda for residential charging infrastructure in Canada. *Energy Res. Soc. Sci.* **2019**, *50*, 29–37. [CrossRef]

5. Ismail, B.I.; Goortani, E.M.; Ab Karim, M.B.; Tat, W.M.; Setapa, S.; Luke, J.Y. Evaluation of docker as edge computing platform. In Proceedings of the 2015 IEEE Conference on Open Systems (ICOS 2015), Melaka, Malasya, 24–26 August 2015.

6. Ferreira, J.C.; Monteiro, V.; Afonsom, J.L. Vehicle-to-Anything Application (V2Anything App) for electric vehicles. *IEEE Trans. Ind. Inform.* **2014**, *10*, 1927–1937. [CrossRef]

7. Bo, C.; Zhang, L.; Li, X.-Y.; Huang, Q.; Wang, Y. Silentsense: Silent user identification via touch and movement behavioral biometrics. In Proceedings of the 19th Annual International Conference on Mobile Computing & Networking (MobiCom '13), Miami, FL, USA, 30 September–4 October 2013; ACM: New York, NY, USA, 2013; pp. 187–190. [CrossRef]

8. Patel, V.M.; Chellappa, R.; Chandra, D.; Barbello, B. Continuous User Authentication on Mobile Devices: Recent progress and remaining challenges. *IEEE Signal Process. Mag.* **2016**, *33*, 49–61. [CrossRef]

9. De Luca, A.; Hang, A.; Brudy, F.; Lindner, C.; Hussmann, H. Touch me once and iknow it's you!: Implicit authentication based on touch screenpatterns. In Proceedings of the SIGCHI Conference on Human Factors in Computing Systems (CHI '12), Austin, TX, USA, 5–10 May 2012; ACM: New York, NY, USA, 2012; pp. 987–996, ISBN 978-1-4503-1015-4. [CrossRef]

10. Feng, T.; Liu, Z.; Kwon, K.-A.; Shi, W.; Carbunar, B.; Jiang, Y.; Nguyen, N. Continuous mobile authentication using touchscreen gestures. In Proceedings of the 2012 IEEE Conference on Technologies for Homeland Security (HST 2012), Waltham, MA, USA, 13–15 November 2012; pp. 451–456, ISBN 978-1-46732709-1. [CrossRef]

11. Jakobsson, M.; Shi, E.; Golle, P.; Chow, R. Implicit authentication for mobile devices. In Proceedings of the 4th USENIX Conference on Hot Topics in Security (HotSec'09), Montreal, QC, Canada, 10–14 August 2009; USENIX Association: Berkeley, CA, USA, 2009; p. 9.

12. Sae-Bae, N.; Ahmed, K.; Isbister, K.; Memon, N. Biometric-rich gestures: A novel approach to authentication on multi-touch devices. In Proceedings of the SIGCHI Conference on Human Factors in Computing Systems (CHI '12), Austin, TX, USA, 5–10 May 2012; ACM: New York, NY, USA, 2012; pp. 977–986, ISBN 978-1-4503-1015-4. [CrossRef]

13. Frank, M.; Biedert, E.M.R.; Martinovic, D.S.I. Touchalytics: On the applicability of touchscreen input as a behavioral biometric for continuous authentication. *IEEE Trans. Inf. Forensics Secur.* **2013**, *8*, 136–148. [CrossRef]

14. Nakamoto, S. Bitcoin: A Peer-to-Peer Electronic Cash System. Available online: https://bitcoin.org/bitcoin.pdf (accessed on 8 July 2019).

15. A Next-Generation Smart Contract and Decentralized Application Platform. Available online: https://github.com/ethereum/wiki/wiki/White-Paper (accessed on 8 July 2019).

16. Panarello, A.; Tapas, N.; Merlino, G.; Longo, F.; Puliafito, A. Blockchain and IoT integration: A systematic survey. *Sensors* **2018**, *18*, 2575. [CrossRef] [PubMed]

17. Pop, C.; Antal, M.; Cioara, T.; Anghel, I.; Sera, D.; Salomie, I.; Raveduto, G.; Ziu, D.; Croce, V.; Bertoncini, M. Blockchain-based scalable and tamper-evident solution for registering energy data. *Sensors* **2019**, *19*, 3033. [CrossRef] [PubMed]

18. Erdin, E.; Cebe, M.; Akkaya, K.; Solak, S.; Bulut, E.; Uluagac, S. Building a Private Bitcoin-based Payment Network among Electric Vehicles and Charging Stations. In Proceedings of the 2018 International Conference in Blockchain, Xi'an, China, 10–12 December 2018.

19. Ferreira, J.C.; Martins, A.L. Building a community of users for open market energy. *Energies* **2018**, *11*, 2330. [CrossRef]

20. Sanseverino, E.R.; Silvestre, M.L.D.; Gallo, P.; Zizzo, G.; Ippolito, M. The blockchain in microgrids for transacting energy and attributing losses. In Proceedings of the 2017 IEEE International Conference on Internet of Things (iThings) and IEEE Green Computing and Communications (GreenCom) and IEEE Cyber, Physical and Social Computing (CPSCom) and IEEE Smart Data (SmartData), Exeter, UK, 21–23 June 2017; pp. 925–930.

21. Ferreira, J.C.; Monteiro, V.; Afonso, J.L. Smart electric vehicle charging system. In Proceedings of the 2011 IEEE Intelligent Vehicles Symposium (IV 2011), Baden, Germany, 5–9 June 2011.

22. Ferreira, J.C.; Afonso, J.L. EV-cockpit—Mobile personal travel assistance for Electric vehicles. In *Advanced Microsystems for Automotive Applications 2011*; Springer: Berlin/Heidelberg, Germany, 2011.

23. Ferreira, J.C.; Monteiro, V.; Afonso, J.; Afonso, J.L. An energy management platform for public buildings. *Electronics* **2018**, *7*, 294. [CrossRef]

24. Pustišek, M.; Kos, A.; Sedlar, U. Blockchain based autonomous selection of electric vehicle charging station. In Proceedings of the 2016 International Conference on Identification, Information and Knowledge in the Internet of Things (IIKI), Beijing, China, 20–21 October 2016; pp. 217–222. [CrossRef]

25. Liu, C.; Chai, K.K.; Zhang, X.; Lau, E.T.; Chen, Y. Adaptive blockchain-based electric vehicle participation scheme in smart grid platform. *IEEE Access* **2018**, *6*, 25657–25665. [CrossRef]

26. Thakur, S.; Breslin, J.G. Electric vehicle charging queue management with blockchain. In *Internet of Vehicles. Technologies and Services Towards Smart City. IOV 2018. Lecture Notes in Computer Science*; Skulimowski, A., Sheng, Z., Khemiri-Kallel, S., Cérin, C., Hsu, C., Eds.; Springer: Cham, Switzerland, 2018; Volume 11253, pp. 249–264.

27. Ferreira, J.C. Android as a Cloud Ticket Validator. In Proceedings of the International Conference on Cloud Ubiquitous Computing Emerging Technologies (CUBE 2013), Pune, India, 15–16 November 2013. [CrossRef]

28. Dahlberg, T.; Guo, J.; Ondrus, J. A critical review of mobile payment research. *Electron. Commer. Res. Appl.* **2015**, *14*, 265–284. [CrossRef]

29. Higgins, S. Why a German Power Company is Using Ethereum to Test Blockchain Car Charging. *CoinDesk.* 7 March 2016. Available online: http://www.coindesk.com/german-utility-company-turns-to-blockchain-amid-shifting-energy-landscape/ (accessed on 8 July 2019).

30. Allison, I. RWE and Slock.it—Electric Cars Using Ethereum Wallets Can Recharge by Induction at Traffic Lights. *International Business Times UK.* 22 Feburary 2016. Available online: http://www.ibtimes.co.uk/rwe-slock-it-electric-cars-using-ethereum-wallets-can-recharge-by-inductiontraffic-lights-1545220 (accessed on 8 July 2019).

31. Kang, J.; Yu, R.; Huang, X.; Maharjan, S.; Zhang, Y.; Hossain, E. Enabling localized peer-to-peer electricity trading among plug-in hybrid electric vehicles using consortium blockchains. *IEEE Trans. Ind. Inf.* **2017**, *13*, 3154–3164. [CrossRef]

32. Garg, S.; Kaur, K.; Kaddoum, G.; Gagnon, F.; Rodrigues, J.J. An efficient blockchain-based hierarchical authentication mechanism for energy trading in V2G environment. *arXiv* **2019**, arXiv:1904.01171.

33. Liu, D.; Li, D.; Liu, X.; Ma, L.; Yu, H.; Zhang, H. Research on a cross-domain authentication scheme based on consortium blockchain in V2G networks of smart grid. In Proceedings of the 2nd IEEE Conference on Energy Internet and Energy System Integration (EI2), Beijing, China, 20–22 October 2018.

34. Aitzhan, N.Z.; Svetinovic, D. Security and privacy in decentralized energy trading through multi-signatures, blockchain and anonymous messaging streams. *IEEE Trans. Dependable Secur. Comput.* **2018**, *15*, 840–852. [CrossRef]

35. Mengelkamp, E.; Notheisen, B.; Beer, C.; Dauer, D.; Weinhardt, C. A blockchain-based smart grid: Towards sustainable local energy markets. *Comput. Sci. Res. Develop.* **2018**, *33*, 207–214. [CrossRef]

36. Münsing, E.; Mather, J.; Moura, S. Blockchains for decentralized optimization of energy resources in microgrid networks. In Proceedings of the IEEE Conference on Control Technology and Applications (CCTA 2017), Kohala Coast, HI, USA, 27–30 August 2017; pp. 2164–2171.

37. Presentation Model. Available online: https://martinfowler.com/eaaDev/PresentationModel.html (accessed on 21 May 2019).

38. HTTP Over TLS. Available online: https://tools.ietf.org/html/rfc2818 (accessed on 21 May 2019).

39. Hypertext Transfer Protocol (HTTP/1.1): Semantics and Content. Available online: https://tools.ietf.org/html/rfc7231 (accessed on 21 May 2019).

40. Charging Solutions for Your Business. Available online: https://www.allego.eu (accessed on 21 May 2019).

Article

Developing an On-Road Object Detection System Using Monovision and Radar Fusion [†]

Ya-Wen Hsu [1], **Yi-Horng Lai [2,*]**, **Kai-Quan Zhong [1]**, **Tang-Kai Yin [3]** and **Jau-Woei Perng [1]**

1   Department of Mechanical and Electro-Mechanical Engineering, National Sun Yat-sen University, Kaohsiung 80424, Taiwan; d043020006@student.nsysu.edu.tw (Y.-W.H.); wdf4200@gmail.com (K.-Q.Z.); jwperng@faculty.nsysu.edu.tw (J.-W.P.)
2   School of Mechanical and Electrical Engineering, Xiamen University Tan Kah Kee College, Zhangzhou 363105, China
3   Department of Computer Science and Information Engineering, National University of Kaohsiung, Kaohsiung 81148, Taiwan; tkyin@nuk.edu.tw
*   Correspondence: lai81.tom@gmail.com
†   This paper is an extended version of our paper published in 2018 IEEE Image and Vision Computing New Zealand Conference, Auckland, New Zealand, 19–21 November 2018.

Received: 15 November 2019; Accepted: 23 December 2019; Published: 25 December 2019

Abstract: In this study, a millimeter-wave (MMW) radar and an onboard camera are used to develop a sensor fusion algorithm for a forward collision warning system. This study proposed integrating an MMW radar and camera to compensate for the deficiencies caused by relying on a single sensor and to improve frontal object detection rates. Density-based spatial clustering of applications with noise and particle filter algorithms are used in the radar-based object detection system to remove non-object noise and track the target object. Meanwhile, the two-stage vision recognition system can detect and recognize the objects in front of a vehicle. The detected objects include pedestrians, motorcycles, and cars. The spatial alignment uses a radial basis function neural network to learn the conversion relationship between the distance information of the MMW radar and the coordinate information in the image. Then a neural network is utilized for object matching. The sensor with a higher confidence index is selected as the system output. Finally, three kinds of scenario conditions (daytime, nighttime, and rainy-day) were designed to test the performance of the proposed method. The detection rates and the false alarm rates of proposed system were approximately 90.5% and 0.6%, respectively.

Keywords: particle filter; histogram of gradient; sensor fusion; neural network; support vector machine; object recognition

---

1. Introduction

In recent years, the development of advanced driving assist systems (ADAS) has attracted a large amount of research and funds from major car factories and universities. The key issues of ADAS include on road object detection, anti-collision technology, park assist system, etc. Three kinds of sensors (i.e., radar, Lidar, and camera) are widely adopted for object detection in front of vehicles [1–5]. Since there are limitations of single sensors, multi-sensor fusion technology can be used to compensate for the disadvantages of each single sensor [6,7].

In reference [8], by using background subtraction and a Haar wavelet translation, the foreground image was transformed into a second-order feature space. Then, based on the concept of a histogram of original gradients (HOG), horizontal and vertical high-frequency components were obtained. In a hierarchical SVM classifier architecture, the proposed system can classify pedestrians, automobiles, and two wheeled vehicles effectively. Yang et al. [9] used an optical flow method to calculate the motion vectors of the objects. Subsequently, the focus of expansion (FOE) of each object was found by using voting. By using the concept of a hierarchical decision tree, false alarms for detection (e.g., shadows on ground marking lines, etc.) can be avoided. Finally, the collision time was calculated by using the motion vectors of the objects.

Millimeter-wave (MMW) radars detect objects by transmitting electromagnetic waves onto the objects and analyzing the reflected waves that are not affected by light and weather. These radars can measure the relative distances and speeds of objects. However, millimeter-wave radars are susceptible to noise and environmental interference. To address the issues related to the microwave radar noise, Park et al. [10] proposed applying a statistical model to the radar using hybrid particle filter to track the preceding vehicle.

The laser range finder is an electronic measuring instrument that uses a laser to accurately measure the distance to the target, which exhibits the advantages of high measurement accuracy and good stability. Nashashibi et al. [4] developed a method to detect, track, and classify multiple vehicles by means of a laser range finder mounted on a vehicle. The classification was based on different criteria: sensor specifications, geometric configuration, occlusion reasoning, and tracking information. The system was tested in highways and urban centers with three different laser range finders.

In contrast with range finder sensors, camera sensors are not only cost-effective but can also provide other useful information. Many novel vision-based object detection algorithms for the front of vehicles have been proposed in the past decade. Vehicle detection and vehicle distance estimation systems were proposed in reference [11]. By using the histogram of an oriented gradient (HOG) feature and support vector machine (SVM) classifier, the authors can segment the road area and identify the shadow area under the vehicle in which to detect the vehicle position. Guo et al. [12] used a two-stage detection algorithm for pedestrian detection. First, the candidate regions were decided from foreground image, then the edge features of object were identified in the second stage. The experiment result verified the accuracy of the proposed method.

Despite the advantages exhibited by all sensors, they have limitations that affect their object detection abilities. For instance, cameras are susceptible to light and environmental factors, and the radar stability is affected by the relative speed and surrounding environment. Hence, a sensor fusion mechanism is developed to compensate for the deficiencies of relying on a single sensor.

The series type fusion architecture based on laser and vision sensors was addressed in reference [13]. The proposed system can quickly find the region of interesting objects without a huge amount of computation time. The other advantage was that after the verification and comparison of each sensor, the overall false alarm rate was reduced. Wang et al. [14] proposed a system scheme for on-road obstacle detection by fusing an MMW radar and a monocular vision sensor. An experimental method to investigate the radar-vision point alignment was proposed. In addition, a region searching method for potential target detection was proposed to reduce image processing time. Wang et al. [15] proposed a tandem sensor fusion of series connection architecture that uses MMW radar to obtain the candidate position of the detected object. The position coordinates are converted into image coordinates that considered as regions of interest to reduce the number of window searches. Then the image is used to recognize and track the vehicle in candidate areas. A Kalman filter is used to compare the tracking trajectory of the radar and camera to improve the vehicle detection rate and reduce the false positive rate.

In the aforementioned references, using a single sensor to detect objects has significantly reduced the detection system cost; however, the system stability is still a challenge when considering special weather conditions. The main purpose of using series architecture in sensor fusion is to rapidly determine the candidate area via radar or Lidar and accelerate the image search process. Another advantage of using a second layer sensor is to reduce noise interference after verification and comparison. However, the entire tandem architecture system will fail when one of the sensors fails.

This paper extends our earlier vision based research work [16] and proposes a set of MMW radar and camera fusion strategies based on a parallel architecture that can compensate for the failure of a single sensor and enhance the system detection rate using the complementary characteristics of the sensors. The radar subsystem provides noise filtering, tracking, and credibility analysis. The two-stage vision detection subsystem can rapidly identify the candidate area form image. The fusion strategy of parallel architecture systems depends on the confidence index of each sensor. Three kinds of scenario conditions (daytime, nighttime, and rainy-day) are implemented in an urban environment to verify the proposed system.

The contributions of this study include the following:

1. In order to solve the shortcomings of each single sensor, by using sensor fusion technology, we integrated the two sensor systems and improved the reliability of the systems.
2. For the fusion architecture of series type, any single sensor failure causes whole system failure. The proposed parallel architecture system depends on the confidence index of each sensor. The system can compensate for each other's sensors and avoid the limitations of series fusion architecture.
3. Three kinds of scenario conditions (daytime, nighttime, and rainy-day) were implemented in an urban environment to verify the proposed system's viability. The experiment results can provide the baseline of comparison for future research.

2. System Architecture

This study proposed a sensor fusion technology integrating MMW radar and camera for front object detection. The proposed system consists of three subsystems, including a radar-based detection system, vision-based recognition system, and sensor fusion system.

The image captured by the camera can easily be affected by lighting and weather conditions. Furthermore, the estimated distance of the front object derived from the camera image has a low precision. A sufficiently large velocity relative to the front object is necessary for the MMW radar to stably detect it. Accordingly, these two sensor subsystems were combined in a parallel connection to compensate for the limitations of each sensor and improve the robustness of the detection system. The overall architecture of the proposed detection and recognition system is shown in Figure 1.

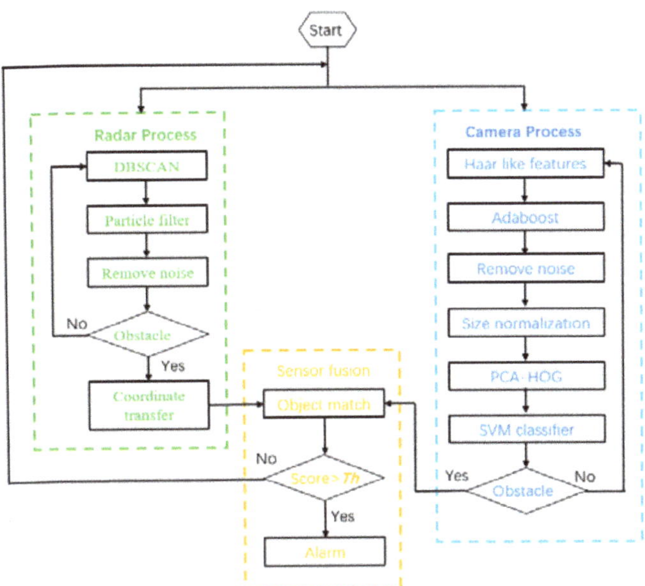

Figure 1. Overall architecture of an on-road obstacle detection system.

A clustering algorithm and particle filter were applied to the MMW radar data to achieve noise removing and multi-object tracking. Then the object detected by the coordinate system of radar sensor

Energies **2020**, 13, 116

were computed from an image coordinate. On the other hand, two stage classifiers were implemented for the foreground segmentation and object recognition for the image data, respectively, then the object information could be obtained. Finally, a radial basis function neural network (RBFNN) was used to fuse the detected object information from the MMW radar and camera.

3. Radar-Based Object Detection

A 24 GHz short-range radar was adopted for front-end environment detection and a multi-object tracking method based on radar was proposed. This method can facilitate tracking multiple object simultaneously and removing noises, which were considered as non-real objects. The flow chart of the proposed radar-based detection subsystem is shown in Figure 2. First, the radar data were divided into different clusters using a clustering algorithm. The particle filter is then used for signal filtering and target tracking. Two kinds of probability scores will be evaluated in the particle filter process. The convergence of the particle swarm can reflect the quality of the tracking. For the stable tracking objects, the particles around the object have a higher weighting in the importance sampling step. Furthermore, these particles have a higher probability of survival in the resampling step. We define the range probability (P_r) as the survival probability of the particles within a radius of 1 m around the object to evaluate the quality of the tracking. On the other hand, the diversity of the particle swarm can cover of all the states of the object. We defined the available probability (P_a) as the survival probability of the particles after the resampling step. During the tracking process, in line with the value of P_a, the system adjusts the particle percentage of resampling to ensure the diversity of the particle swarm. In addition, the confidence index of the target object was derived from the range probability and probability of survival. This confidence index determines the credibility of the actual object. The relative velocity and distance between the vehicle and front object were provided by this subsystem.

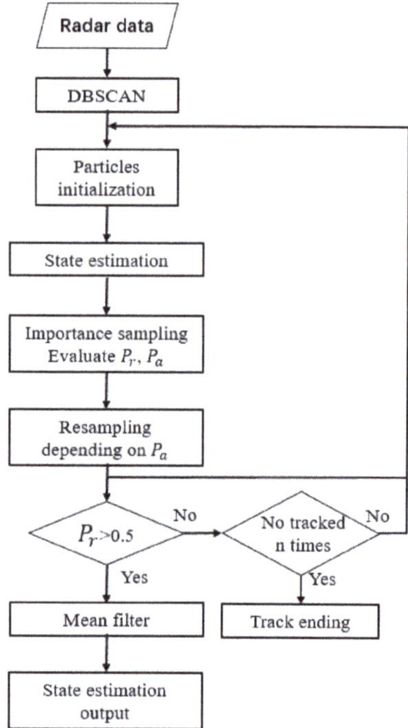

Figure 2. Flow chart of a radar-based object detection subsystem.

3.1. Radar Data Pre-Processing

The MMW radar signals are electromagnetic waves. Both reflection and refraction will occur when the electromagnetic waves occur on the medium. In addition to the reflected wave from the medium itself, some noise signals of non-real objects are also prone to appear. The relationship between relative distance and echo intensity information was statistically analyzed using a vast amount of data collected during experiments. The statistical results are shown in Figure 3. The statistical results of the signal distribution indicate that both real objects and noise show respective concentrations, and only a small part of the distribution of both overlaps. Accordingly, a noise filtering operation was performed. As shown in Figure 3a, after the signal on the left side of red curve was filtered, the subsequent target tracking and particle filter algorithm were performed. Density-based spatial clustering of applications with noise (DBSCAN) algorithm [17] was used to cluster the radar data, and the number of possible front objects was estimated.

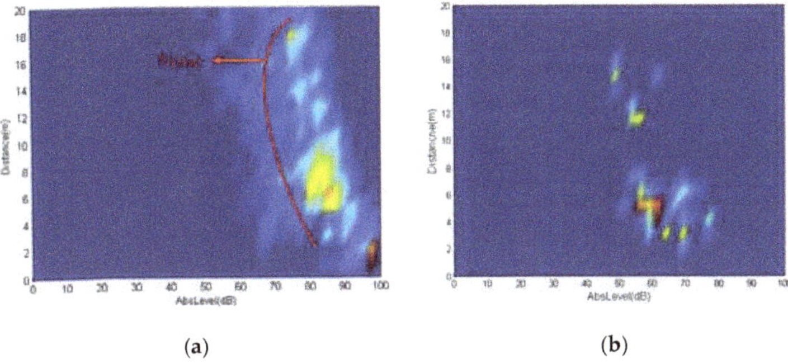

(a) (b)

Figure 3. Statistical results of the radar signal: (**a**) real object and (**b**) non-real object.

3.2. Particle Filter

A particle filter [18] is widely used in many fields, including object tracking, signal processing, and automatic control. In this study, particle filtering was used to filter the radar signal and track the objects in front of a vehicle. The particle filter algorithm uses a finite number of particles to represent the posterior probability of some stochastic process with partial observations. Each particle has the respective weight values that represent the probability of the particle being sampled from the probability density function. The procedure to implement a particle filter algorithm in this study was roughly divided into four steps as follows:

3.2.1. Particle Initialization

To cover all the potential object positions, n pieces of particles were randomly distributed within the radar detection area. Each particle represents a potential position of a real object, where the weight of the particle indicates the probability that the object is at this location.

3.2.2. State Prediction

The state of the object changes over time. Discrete time was used to calculate the object state, and the state of the particle at next moment was predicted by the state and motion model at time $k - 1$. Then the prior probability $P(x_k|x_{k-1})$ was obtained. The equation used to predict the object state is expressed as follows [19]:

$$X_k = FX_{k-1} + GW_k = \begin{bmatrix} 0 & 1 & 0 & T \\ 0 & 0 & 1 & 0 \\ 0 & 0 & 0 & 1 \end{bmatrix} \begin{bmatrix} y_{k-1} \\ \dot{x}_{k-1} \\ \dot{y}_{k-1} \end{bmatrix} + \begin{bmatrix} 0 & \frac{T^2}{2} \\ T & 0 \\ 0 & T \end{bmatrix} \begin{bmatrix} W_x \\ W_y \end{bmatrix} \tag{1}$$

with

$$F = \begin{bmatrix} 1 & 0 & T & 0 \\ 0 & 1 & 0 & T \\ 0 & 0 & 1 & 0 \\ 0 & 0 & 0 & 1 \end{bmatrix}, \quad G = \begin{bmatrix} \frac{T^2}{2} & 0 \\ 0 & \frac{T^2}{2} \\ T & 0 \\ 0 & T \end{bmatrix} \tag{2}$$

where T is the sampling time of the radar sensor, $X_k = \begin{bmatrix} x_k & y_k & \dot{x}_k & \dot{y}_k \end{bmatrix}^T$ denotes the state vector, and x_k and x_{k-1} denote the relative lateral distances between the target object and the sensor at the current time and the previous moment, respectively. y_k and y_{k-1} are the relative longitudinal distances between the target object and the sensor at the current time and the previous moment, respectively. $\dot{x}_k$ and $\dot{y}_k$ represent the lateral and longitudinal relative speeds of the target and the sensor, respectively. W_k is zero-mean Gaussian white noise.

3.2.3. Importance Sampling

This step is based on the concept of a Bayesian filter. The particles that are obtained during the state prediction stage and the information obtained from MMW radar are used to estimate the target position. The Bayesian theorem is used to update the prior probability then obtain the posterior probability. In this step, each particle is assigned a weight. Based on the assumption that the radar measurement area is $M \times N$ blocks, each block unit is 1 m^2. The measurement model of the radar sensor is expressed by Equation (3),

$$z_k^{(i,j)} = h_k^{(i,j)}(x_k) + v_k^{(i,j)} \tag{3}$$

where $v_k^{(i,j)}$ is the measured noise in (i, j) block and its Gaussian white noise with the means equal to 0 and the variance σ^2, while $h_k^{(i,j)}(x_k)$ is the signal strength of the object in the (i, j) block and its point spread function [20] is expressed as follows:

$$h_k^{(i,j)}(x_k) = \frac{\Delta_x \Delta_y I_k}{2\pi \Sigma^2} \cdot \exp\left(\frac{(i\Delta_x - x_k)^2 + (j\Delta_y - y_k)^2}{2\Sigma^2} \right) \tag{4}$$

where Δ_x and Δ_y are the block sizes, I_k is echo strength of the MMW radar, Σ is the blurring degree of the sensor, and the weight value of the particle can be obtained by the following equation:

$$w_k^{\tilde{i}} = \exp\left(\frac{h_k^{(i,j)}(x_k^{\tilde{i}})\left(h_k^{(i,j)}(x_k^{\tilde{i}}) - 2h_k^{(i,j)} \right)}{2\sigma^2} \right). \tag{5}$$

The weight of each particle in the space region is normalized. The normalization method is based on dividing the weight of each particle by the sum of all particle weights, as shown by Equation (6):

$$\hat{w}_k^{\tilde{i}} = \frac{w_k^{\tilde{i}}}{\sum\limits_{i=1}^{n} w_k^{\tilde{i}}}. \tag{6}$$

After the weight of each particle is obtained, the relative position of the object detected by the MMW radar can be estimated. The expected value of the target estimation is expressed as follows:

$$E\left(x_k|y_k\right) = \sum_{i=1}^{n} w_k^{\sim i} f\left(x_k^{\sim i}\right). \tag{7}$$

3.2.4. Resampling

The method of estimating according to the weight of each particle is referred to as the sequential importance sampling (SIS) particle filter [18]. However, this method involves particle degradation, leading to insignificant weight values of most particles after several iterative operations. This triggers the system to perform unnecessary calculations on these particles. Thus, the real target position may not be covered by the remaining particles. The resampling method was used to address this issue. In each iteration process, the particles with smaller weight values were discarded and replaced by particles with larger weight values. After resampling, the weight values of all particles was set at $\frac{1}{n}$, then the next iteration was performed with new particles. The expected value of the target estimation is expressed as follows:

$$E\left(x_k|y_k\right) = \sum_{i=1}^{n} \frac{1}{n} f\left(x_k^{\sim i}\right). \tag{8}$$

3.3. Experimental Verification

A lot of object information was lost while the MMW radar information was processed by internal algorithms. Therefore, the original unprocessed data was obtained from the MMW radar in this study. The proposed particle filter algorithm was used to track the front object and address the issue of losing too much information.

To verify the feasibility of the algorithm proposed in this study, a laser range finder with high precision was used. The measurement error of the adopted lase finder was ±10 mm to record the center position of the frontal object. The experimental equipment installed to verify the radar tracking system is shown in Figure 4. Three verification conditions were set to avoid dark objects and lack of relative speeds, which can lead to losing laser range finder and radar information, as follows: metal and light-colored moving objects, a relative velocity of ±15 km/h or more, and objects moving from far away to nearby.

Figure 4. Equipment setup for radar tracking system verification.

The position of the object measured by the laser range finder is considered as the ground truth, which is illustrated by the blue line seen in Figure 5. The red line represents the tracking result obtained by the proposed particle filtering algorithm. The result of the internal algorithm of the radar sensor is illustrated by the green line. An offset between the detected and actual positions of the object may be observed owing to the characteristics of the radar sensor.

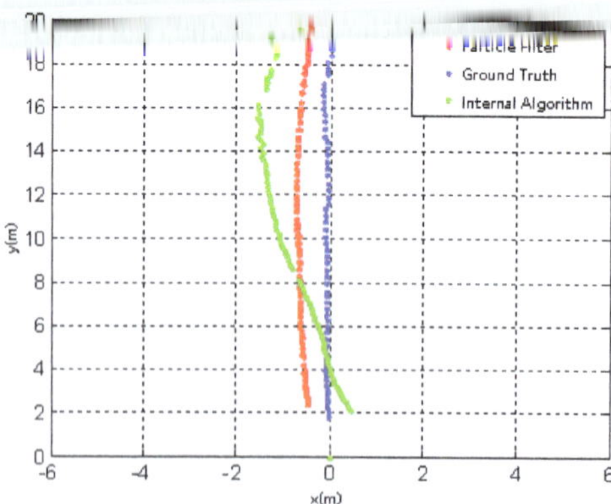

Figure 5. Estimated trajectories of moving objects using different methods.

The error and standard deviation of our proposed particle filter tracking algorithm and the internal algorithm of the radar sensor were compared to the ground truth to verify the tracking results. The error is defined as the absolute value of the estimated position from the algorithm and the ground truth. The average error is the sum of the errors divided by the number of times of detections. As shown in Table 1, the proposed algorithm had better performance considering the average error, the maximum error, and the standard deviation of error of the longitudinal or lateral direction. In addition, the number of times the proposed algorithm effectively detected objects was also greater than that obtained by the sensor internal algorithm.

Table 1. Error of each tracking methods (unit: centimeter).

Method	Average Error		Standard Deviation		Maximum Error	
	Lateral	Longitudinal	Lateral	Longitudinal	Lateral	Longitudinal
The proposed particle filter tracking algorithm	44.48	32.32	18.53	22.35	88.86	99.96
Internal algorithm of the radar sensor	52.66	33.66	42.08	26.47	169.2	116.8

4. Vision-Based Object Recognition

The two-stage vision-based object recognition system was similar to in our earlier work [16]. In the first stage, the Haar-like features algorithm was used to identify the candidate regions of object from foreground segmentation. The second stage is responsible for object recognition. Three kinds of objects (i.e., pedestrians, motorcycles, and cars) can be identified by SVM classifiers. The scheme of the two-stage vision-based object recognition process in shown in Figure 6. The object recognition results are shown in Figure 7.

The distance estimation of image object can be determined by using the polynomial model as expressed in Equation (9):

$$f(y_{im}) = g_0 y_{im}^5 + g_1 y_{im}^4 + g_2 y_{im}^3 + g_3 y_{im}^2 + g_4 y_{im} + g_5 \tag{9}$$

where $f(y_{im})$ is the estimation of distance, while y_{im} denotes the object coordinates v of the image.

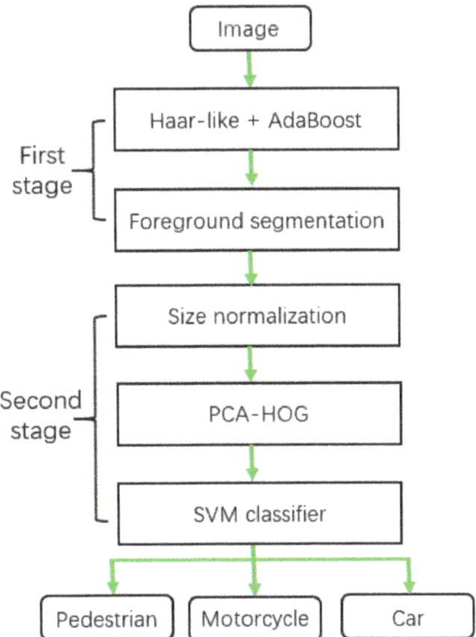

Figure 6. Vision-based object recognition subsystem.

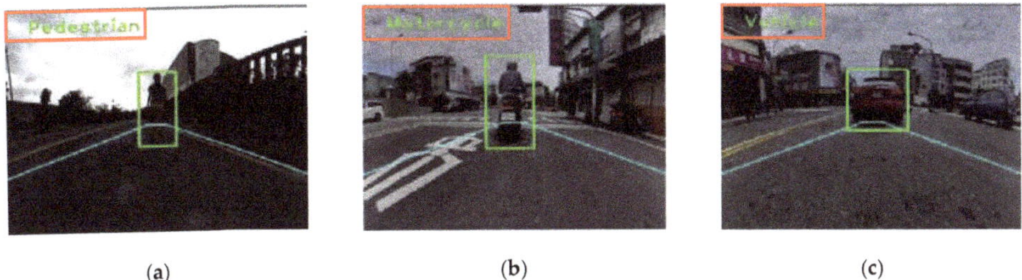

(a)   (b)   (c)

Figure 7. Classification results of vision-based recognition: (**a**) pedestrian, (**b**) motorcycle, and (**c**) car.

5. Sensors Fusion and Decision Mechanism

A single sensor system can operate independently; however, a parallel architecture was adopted in this study to fuse two different sensors. The main purpose of this is to improve the detection rate that can be achieved by a single sensor. The sensor fusion was divided into three parts. First, the two-dimensional coordinate information of the MMW radar was converted into the coordinate of the image. Afterwards, the information obtained by the two sensors was integrated into the same coordinate system. Next, the object information needed to be matched to determine whether the same object information had been obtained by both the MMW radar and camera, and to integrate the detection results of the two systems. Finally, the trusted sensor was determined based on the confidence index of the sensor.

5.1. Coordinate Transformation

The supervised learning algorithms was used to learn the relationship between the MMW radar coordinate and image coordinate system. Before the coordinate transformation, the radar coordinates (x, y) and image coordinate (u, v) needed to be recorded synchronously to be considered as training samples for offline learning. An MMW radar uses electromagnetic waves as a medium, and it exhibits better reflective property to metal objects. Hence, a triangular metal reflector was used as a target object to gather data obtained from the radar and the camera, as shown in Figure 8. A metal reflector were randomly placed in a straight lane at a distance which ranged from 1 m to 12 m in front of the experimental vehicle, and a total of 280 training samples were established.

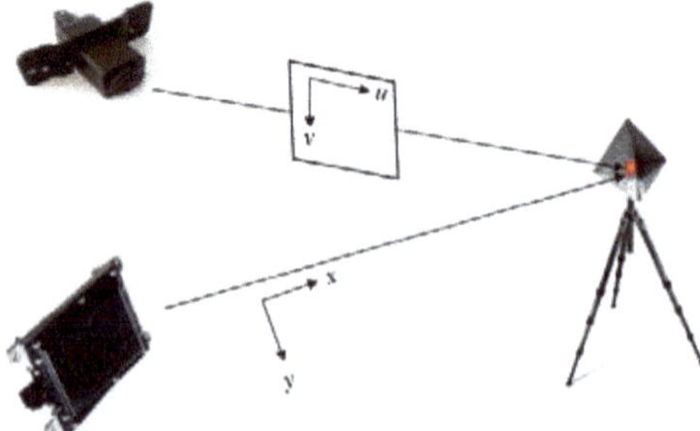

Figure 8. Coordinate transformation of data from the radar and camera.

The camera was installed at an angle parallel to the horizon. When the target object moved from far away to nearby, the position of its center point slightly changed near the center point of the image in the vertical direction. Thus, the variation in the image v-direction coordinate was not obvious. Therefore, the fusion system primarily enabled the neural network to learn the relationship between the MMW radar coordinate (x, y) and the image coordinate (u, v).

From the collected training samples, the longitudinal and lateral distances from the radar were considered as the input of the RBFNN, and the corresponding u coordinate of horizontal direction in the image was considered as an output. This network architecture allows for obtaining the coordinate conversion relationship between these two sensors. The network architecture is shown in Figure 9.

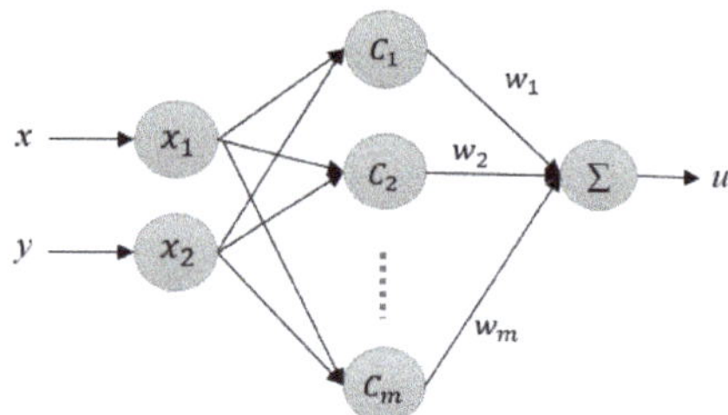

Figure 9. Diagram of radial basis function neural network (RBFNN) architecture.

5.2. Object Match

The MMW radar detection and image recognition systems operate independently, and the two systems obtain information about the detected objects, respectively. To fuse the information of the two systems, the object information must be matched first to determine whether the same object information has been detected by the two sensors. Coordinates shown in the same image may correspond to several different radar coordinate information, as illustrated by the green points shown in Figure 10. In addition, the distance estimated from the image coordinates may be inaccurate owing to the bumpy road surfaces that can cause the vehicle to shake; thus, it is difficult to match the object information and effectively determine whether the same object is detected.

Another RBFNN is used to match the object information and determine whether the same objects are detected by the two sensors. Six factors were entered as the network inputs, which affect the object match, including image coordinate u, object width, object height, object distance estimated from image, object distance measured by the radar, and the u coordinate converted from the radar to the image. Either "match" or "non-match" were obtained as the network output.

(**a**)   (**b**)

Figure 10. Diagram of same image coordinates correspond to different radar coordinates (**a**) Radar detection distance is 2.2 m and (**b**) Radar detection distance is 4 m.

5.3. Decision Strategy

If a single sensor in the sensor fusion of cascade architecture fails, then the entire system will inevitably fail. Meanwhile, the sensor fusion of parallel architecture determines which sensor should be trusted based on the decision mechanism. Although one of the sensors might not detect an object or gives a false alarm, if the other sensor correctly detects the object, then the confidence index of each sensor can be calculated via a scoring mechanism, and a credible subsystem can be determined based on the confidence index.

The confidence index of the radar subsystem was calculated as follows:

$$Score_R = P_r + P_a + A_{rn} \times \eta_r \tag{10}$$

where A_{rn} is the number of times the object tracked by particle filter. η_r is a constant.

The confidence index of the image subsystem was calculated as follows:

$$Score_I = S_d + A_{in} \times \eta_i + \lambda \tag{11}$$

where S_d denotes the distance from the input data point to the SVM hyperplane, A_{in} is the number of times the object tracked in image subsystem, and η_r and λ are constants.

The confidence index of the sensor fusion system was expressed as follows:

$$Score = Score_R + Score_I. \tag{12}$$

When the confidence index *Score* is greater than the set threshold *Th*, the reliability of the system is extremely high, and the output result obtained by the system represents the real situation. If the confidence index of each subsystem is greater than the threshold *Th*, then the subsystem with the highest score is responsible for the entire system decision making process.

6. Experiments

6.1. Experimental Platform and Scenarios

Three kinds of scenario conditions (daytime, nighttime, and rainy-day) were implemented to verify the proposed system. All the scenarios were carried out on urban roads. The MMW radar and camera were mounted on the front bumper of the experimental car, as shown in Figure 11.

Considering the effect of pavement puddles and shadow environment, the daytime scenarios included direct sunlight, pavement puddles, and shadow environments, as shown in Figure 12.

In the nighttime experiment, the scenarios included flashing brake lights of front vehicles, headlight reflections, and poor lighting environments, as shown in Figure 13.

Figure 11. Experimental car and sensors setup.

(a)   (b)   (c)

Figure 12. Daytime scenarios: (**a**) sunlight, (**b**) pavement puddle, and (**c**) shadow.

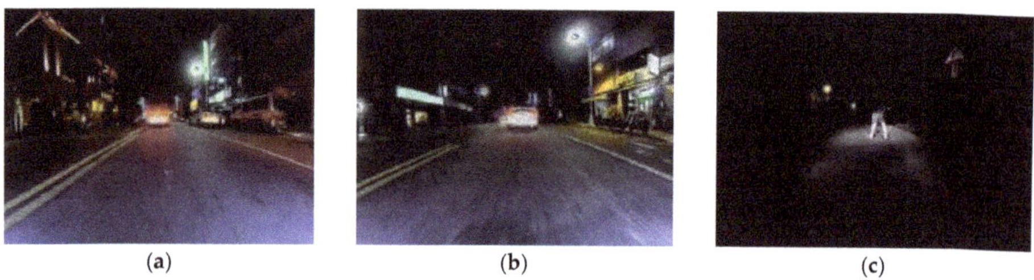

(a)   (b)   (c)

Figure 13. Nighttime scenarios: (**a**) brake light, (**b**) headlight reflection, and (**c**) poor lighting.

In order to reproduce the actual road conditions, we designed a rainy-day scenario too. As the sensors are mounted on the front bumper, the raindrops often adhered to the camera lens during the rainy day experiment, as shown in Figure 14.

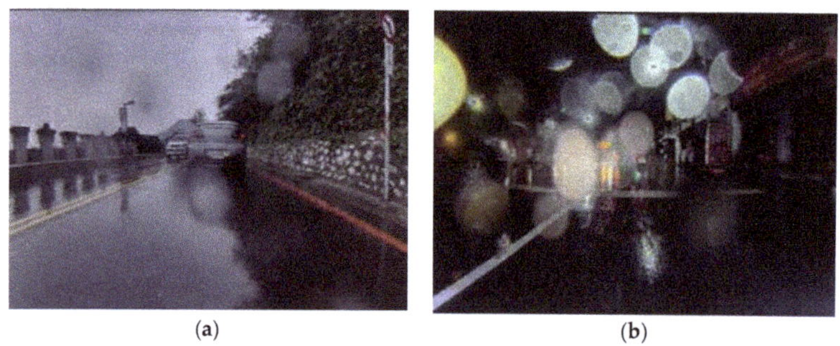

(a)   (b)

Figure 14. Rainy day scenarios: (**a**) daytime and (**b**) nighttime.

6.2. Radar-Based Detection Subsystem

The radar detection subsystem uses MMW radar to perceive the environment ahead. The proposed multi-object tracking algorithm with a particle filter can effectively track the objects in front and remove non-object noise. The radar subsystem experiments tested three different categories of objects under different conditions. The detection results are shown as green circles in Figures 15 and 16. The tests primarily involved a single target in a lane. If there were multiple targets, the alert was reported for closest target to the experimental vehicle. Other targets continued to be tracked.

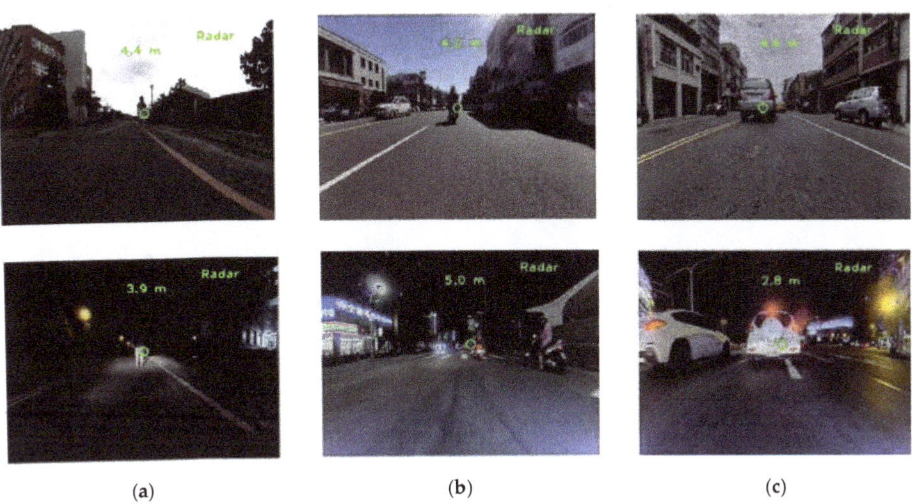

(a)   (b)   (c)

Figure 15. Detection results of radar subsystem, upper row: daytime, lower row: nighttime (**a**) a pedestrian, (**b**) motorcycle, and (**c**) car.

(a)   (b)

Figure 16. Detection results of radar subsystem for a rainy day, (**a**) motorcycle and (**b**) car.

A detection rate exceeding 60% was maintained by the radar detection system during daytime, nighttime, and rainy days. The experimental tests performed under different weather conditions verified that the radar detection system is not affected by weather conditions. The experimental results are listed in Table 2.

Table 2. Detection results of radar systems under different weather conditions.

Condition	Total Frame	Correct Detection	Misinformation	Misjudgment	False Alarm Rate	Detection Rate
Daytime	17,346	11,254	27	6065	0.2%	64.9%
Nighttime	7022	4338	0	2684	0%	61.8%
Rain day	11,193	8135	0	3058	0%	72.6%
Total	35,561	23,727	27	11,807	0.01%	67.0%

3.3. Vision Recognition

The advantages of two-stage vision-based object recognition system are as follows: By using Haar-like features, the first-stage classifier can detect efficiently candidate areas. Unfortunately, the Haar-like algorithm suffers from higher false positive rates (see the purple rectangles in Figure 17). Therefore, the second-stage PCA-HOG algorithm classifier was utilized to compensate for the higher false positive rates of the first-stage result.

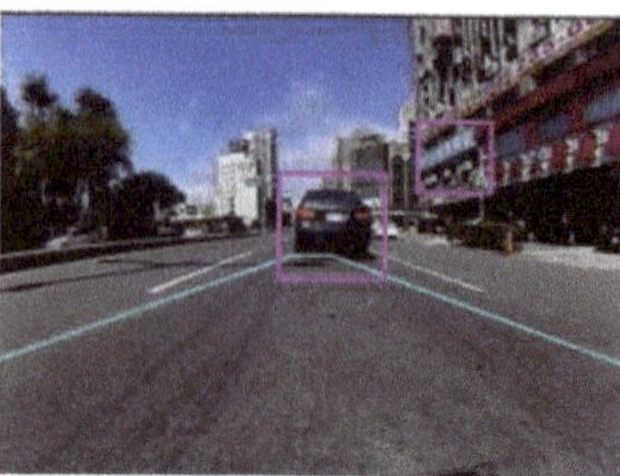

Figure 17. Error detection of the Haar-like algorithm.

The detection results of the vision-based object recognition subsystem are shown as yellow rectangles in Figure 18. The results of the rainy-day experiment are shown as green rectangles in Figure 19.

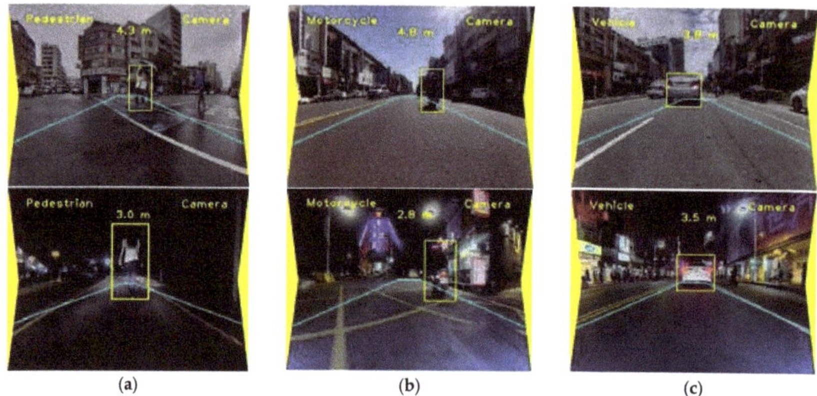

(a)　　　(b)　　　(c)

Figure 18. Detection results of a vision-based subsystem, upper row: daytime, lower row: nighttime (**a**) a pedestrian, (**b**) motorcycle, and (**c**) car.

All the experiments performed under different weather conditions involved three classifications of objects: pedestrians, motorcycles, and cars. The detection results of vision-based systems are listed in Table 3.

Table 3. Detection results of vision-based systems under the different weather conditions.

Condition	Total Frame	Correct Detection	Misinformation	Misjudgment	False Alarm Rate	Detection Rate
Daytime	17,392	14,909	46	2437	0.3%	85.7%
Nighttime	7043	4915	21	2107	0.3%	69.8%
Rain day	11,193	4335	141	6717	1.3%	38.7%
Total	35,628	24,159	208	11,261	0.5%	67.8%

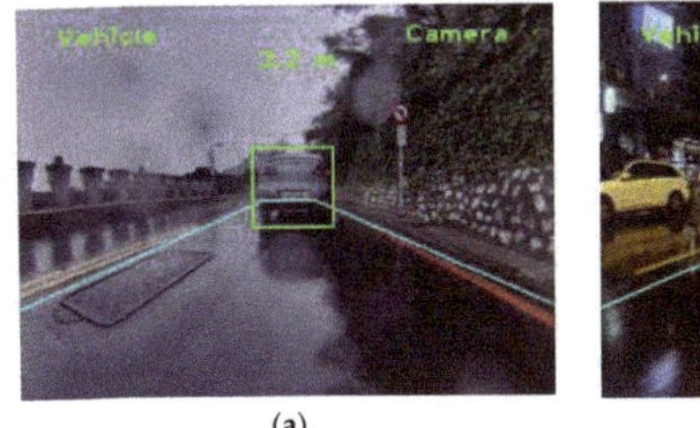
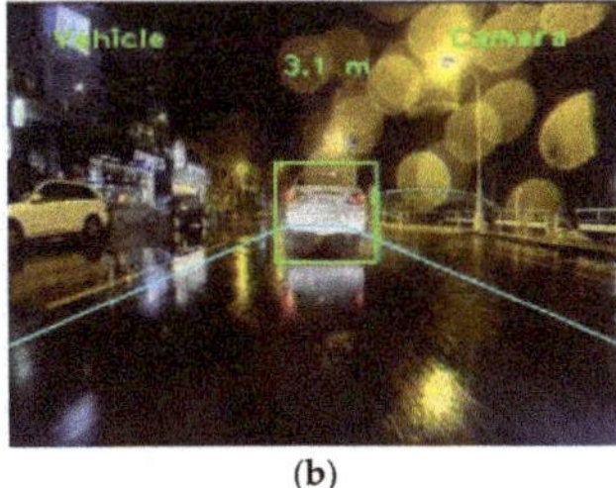

(a)   (b)

Figure 19. Detection results by vision-based subsystem for a rainy day, (**a**) daytime and (**b**) nighttime.

Due to the high sensitivity to light sources, the performance of camera sensor depends on the condition of light sources. For example, suffering in an insufficient light source, the vision-based systems cannot extract completely the features of objects at night. On the other hand, in rainy weather experiments, the raindrops adhering to the camera lens block the object in front of the vehicle. Thus, the system cannot effectively identify the information of the target, leading to the failure of the image subsystem. Therefore, the worst detection rates are achieved at night and on rainy days.

6.4. Sensor Fusion System

This system integrates MMW radar and camera information and improves the scene when one of the detection systems fails by using the sensor fusion of parallel architecture. The system presents complementary characters. For example, as shown in Figure 20, the radar did not detect the front vehicle when the relative speed of the radar and object was relatively small; thus, the camera was used to compensate for the radar failure. On the other hand, when the raindrops adhering to the camera lens blocked the scene, leading to image detection failure, the radar compensated for this situation, as shown in Figure 21.

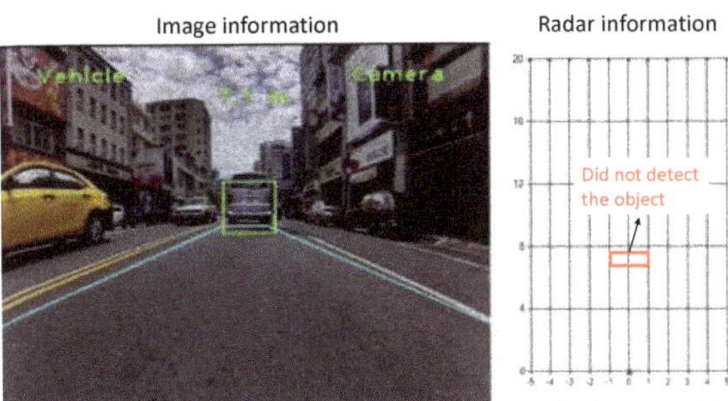

Figure 20. Sensor fusion compensate radar failure.

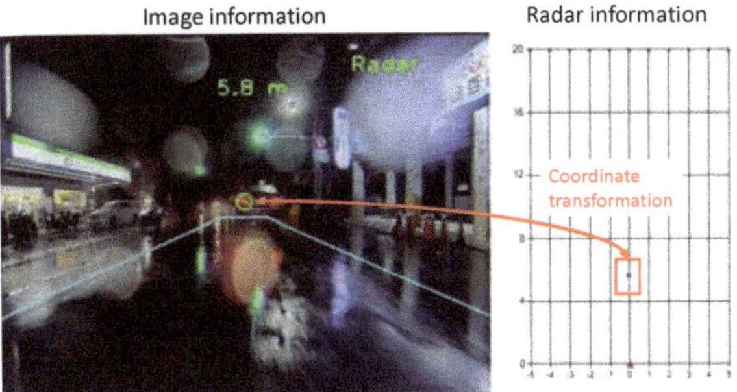

Figure 21. Sensor fusion compensate for image failure.

In addition to compensating for single sensors failures, the system integrates the sensors information when both the radar and camera detect objects simultaneously. The system relies on the coordinate transformation and object matching decision mechanism to determine whether the same objects are detected by the two sensors, as shown in Figure 22.

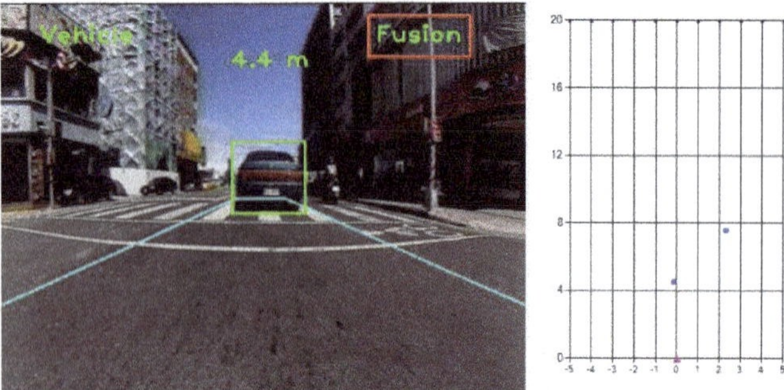

Figure 22. Detection result of sensor fusion system.

The parallel sensor fusion architecture proposed in this study exhibits the advantages of compensating for the disadvantages of relying on a single sensor. It improves the scene in case of subsystem failure and significantly increases the system detection rate and stability, as listed in Table 4. Regardless of the weather conditions, better detection rates were achieved by the sensor fusion system than those obtained when relying on a single subsystem.

Table 4. Detection results of each system under different weather conditions.

Condition	Sensor	Total Frame	Correct Detection	Misinformation	Misjudgment	False Alarm Rate	Detection Rate
	radar	17,392	11,254	27	6065	0.2%	64.7%
Daytime	image	17,392	14,909	46	2437	0.3%	85.7%
	fusion	17,392	16,414	46	978	0.3%	94.3%
	radar	7043	4338	0	2684	0%	61.6%
Nighttime	image	7043	4915	21	2107	0.3%	69.8%
	fusion	7043	6450	21	593	0.3%	91.6%
	radar	11,193	8135	0	3058	0%	72.6%
Rain day	image	11,193	4335	141	6717	1.3%	38.7%
	fusion	11,193	9413	141	9985	1.3%	84.1%

Table 5 lists the detection results of each system for the three object categories under different weather conditions. The sensor fusion system can achieve a detection rate of more than 90%.

Table 5. Detection results of each system.

Sensor	Total Frame	Correct Detection	Misinformation	Misjudgment	False Alarm Rate	Detection Rate
Radar subsystem	35,628	23,727	27	11,807	0.01%	66.6%
Image subsystem	35,628	24,159	208	11,261	0.6%	67.8%
Sensor fusion system	35,628	32,277	208	3143	0.6%	90.5%

We also compared our results with existing related works. The comparison results are listed in Table 6.

Table 6. Comparison with existing related works.

Sensor Type	Object	Fusion Type	Environment	Time Cost	Hardware
Camera [9]	X	X	Daytime	50 ms	Intel i7 3.4 GHz
Camera [12]	Pedestrian	X	Daytime	66–100 ms	Core 2 2.66 GHz
Camera & Lidar [13]	Pedestrian	Series	Daytime	66 ms	Dual-core PC
Camera & Radar [15]	Car	Series	Daytime	16 ms	Intel i7 3.0 GHz
Camera & Radar (the proposed approach)	Car Motor Pedestrian	parallel	Daytime Nighttime Rainy-day	60 ms	Intel i7 2.6 GHz

7. Conclusions

Two types of sensors, an MMW radar and a camera were integrated in this study to develop a frontal object detection system based on sensor fusion using parallel architecture. A particle filter algorithm was employed by the radar detection subsystem to remove noise from non-objects while tracking objects at the same time, and converting the target information into the image coordinates using RBFNN. On the other hand, the image object could be identified as one of three main categories (pedestrians, motorcycles, and cars) by the two-stage vision-based recognition subsystem. The information obtained by the two subsystems was integrated. The sensor with higher credibility was selected as the system output result. Three kinds of experiments (daytime, nighttime, and rainy-days) were performed to verify the proposed system. The experiment results show the detection rates and the false alarm rates of proposed system were approximately 90.5% and 0.6%, respectively. These detection rates are better than those obtained by single sensor systems.

Author Contributions: Conceptualization, Y.-H.L. and J.-W.P.; Data curation, K.-Q.Z.; Formal analysis, J.-W.P. and T.-K.Y.; Investigation, K.-Q.Z.; Methodology, Y.-H.L.; Software, K.-Q.Z.; Supervision, J.-W.P. and T.-K.Y.; Validation, Y.-W.H.; Writing—original draft, Y.-W.H.; Writing—review & editing, Y.-H.L. All authors have read and agreed to the published version of the manuscript.

Funding: This paper was supported by the NSYSU-NUK JOINT RESEARCH PROJECT through National Sun Yat-sen University and National University of Kaohsiung [grant number #NSYSUNUK 107-P003].

Conflicts of Interest: The authors declared no potential conflicts of interest with respect to the research, authorship, and/or publication of this article.

References

1. Mukhtar, A.; Xia, L.; Tang, T.B. Vehicle detection techniques for collision avoidance systems: A Review. *IEEE Trans. Intell. Transp. Syst.* **2015**, *16*, 2318–2338. [CrossRef]
2. Blanc, C.; Aufrere, R.; Malaterre, L.; Gallice, J.; Alizon, J. Obstacle detection and tracking by millimeter wave radar. In Proceedings of the FAC/EURON Symposium on Intelligent Autonomous Vehicles, Lisboa, Portugal, 5–7 July 2004; pp. 322–327.
3. Cho, H.J.; Tseng, M.T. A support vector machine approach to CMOS-based radar signal processing for vehicle classification and speed estimation. *Math. Comput. Model.* **2012**, *58*, 438–448. [CrossRef]
4. Nashashibi, F.; Bargeton, A. Laser-based vehicles tracking and classification using occlusion reasoning and confidence estimation. In Proceedings of the Intelligent Vehicles Symposium, Eindhoven, The Netherlands, 4–6 June 2008; pp. 847–852.
5. Natale, D.J.; Tutwiler, R.L.; Baran, M.S.; Durkin, J.R. Using full motion 3D Flash LIDAR video for target detection, segmentation, and tracking. In Proceedings of the IEEE Southwest Symposium on Image Analysis & Interpretation (SSIAI), Austin, TX, USA, 23–25 May 2010; pp. 21–24.
6. Kmiotek, P.; Ruichek, Y. Multisensor fusion based tracking of coalescing objects in urban environment for an autonomous vehicle navigation. In Proceedings of the IEEE International Conference on Multisensor Fusion and Integration for Intelligent Systems, Seoul, Korea, 20–22 August 2008; pp. 52–57.

7.	Premebida, C.; Monteiro, G.; Nunes, U.; Peixoto, P. A Lidar and vision based approach for pedestrian and vehicle detection and tracking. In Proceedings of the Intelligent Transportation Systems Conference, Seattle, WA, USA, 30 September–3 October 2007; pp. 1044–1049.
8.	Liang, C.W.; Juang, C.F. Moving object classification using local shape and HOG features in wavelet-transformed space with hierarchical SVM classifiers. *Appl. Soft Comput.* **2015**, *28*, 483–497. [CrossRef]
9.	Yang, M.T.; Zheng, J.Y. On-road collision warning based on multiple foe segmentation using a dashboard Camera. *IEEE Trans. Veh. Technol.* **2015**, *64*, 4947–4984. [CrossRef]
10.	Park, S.; Hwang, J.P.; Kim, E.; Kang, H.J. Vehicle tracking using a microwave radar for situation awareness. *Control Eng. Pract.* **2010**, *18*, 383–395. [CrossRef]
11.	Huang, D.Y.; Chen, C.H.; Chen, T.Y.; Hu, W.C.; Feng, K.W. Vehicle detection and inter-vehicle distance estimation using single-lens video camera on urban/suburb roads. *J. Vis. Commun. Image Represent.* **2017**, *46*, 250–259. [CrossRef]
12.	Guo, L.; Ge, P.S.; Zhang, M.H.; Li, L.H.; Zhao, Y.B. Pedestrian detection for intelligent transportation systems combining AdaBoost algorithm and support vector machine. *Expert Syst. Appl.* **2012**, *39*, 4274–4286. [CrossRef]
13.	Oliveira, L.; Nunes, U.; Peixoto, P.; Silva, M.; Moita, F. Semantic fusion of laser and vision in pedestrian detection. *Pattern Recognit.* **2010**, *43*, 3648–3659. [CrossRef]
14.	Wang, T.; Zheng, N.; Xin, J.; Ma, Z. Integrating millimeter wave radar with a monocular vision sensor for on-road obstacle detection applications. *Sensors* **2011**, *11*, 8992–9008. [CrossRef] [PubMed]
15.	Wang, X.; Xu, L.; Sun, H.; Xin, J.; Zheng, N. On-road vehicle detection and tracking using MMW Radar and monovision fusion. *IEEE Trans. Intell. Transp. Syst.* **2016**, *17*, 2075–2084. [CrossRef]
16.	Hsu, Y.W.; Zhong, K.Q.; Perng, J.W.; Yin, T.K.; Chen, C.Y. Developing an on-road obstacle detection system using monovision. In Proceedings of the International Conference on Image and Vision Computing New Zealand (IVCNZ), Auckland, New Zealand, 19–21 November 2018.
17.	Ester, M.; Kriegel, H.; Sander, J.; Xu, X. A density-based algorithm for discovering clusters in large spatial databases with noise. In Proceedings of the Second International Conference on Knowledge Discovery and Data Mining, Portland, OR, USA, 2–4 August 1996; pp. 153–158.
18.	Orhan, E. *Particle Filtering*; Center for Neural Science of New York University Sciences: New York, NY, USA, 2012.
19.	Yu, M.; Oh, H.; Chan, W.H. An improved multiple model particle filtering approach for manoeuvring target tracking using airborne GMTI with geographic information. *Aerosp. Sci. Technol.* **2016**, *52*, 62–69. [CrossRef]
20.	Musso, C.; Champagnat, F. Improvement of the Laplace-based particle filter for track-before-detect. In Proceedings of the International Conference on Information Fusion (FUSION), Heidelberg, Germany, 5–8 July 2016; pp. 153–158.

Article

Direct and Indirect Environmental Aspects of an Electric Bus Fleet under Service

Bogdan Ovidiu Varga, Florin Mariasiu *, Cristian Daniel Miclea, Ioan Szabo, Anamaria Andreea Sirca and Vlad Nicolae

Automotive Engineering and Transports Department, Technical University of Cluj-Napoca, Bdul.Muncii 103-105, Cluj-Napoca, Romania; bogdan.varga@auto.utcluj.ro (B.O.V.); cristymiclea@yahoo.com (C.D.M.); Ioan.Szabo@energom.ro (I.S.); Andreea.Sirca@auto.utcluj.ro (A.A.S.); nicuvlad2@gmail.com (V.N.)
* Correspondence: florin.mariasiu@auto.utcluj.ro

Received: 28 October 2019; Accepted: 3 January 2020; Published: 10 January 2020

Abstract: The reduction of pollutant emissions in the field of transportation can be achieved by developing and implementing electric propulsion technologies across a wider range of transportation types. This solution is seen as the only one that can offer, in areas of urban agglomeration, a reduction of the emissions caused by the urban transport to zero, as well as an increase in the degree of the health of the citizens. This paper presents an analysis of the direct and indirect environmental aspects of a fleet of real electric buses under service in the city of Cluj-Napoca, Romania. The solution of using 41 electric buses to replace Euro-3 diesel buses (with high pollution levels) in the city's transport system eliminates a local amount of 668.45 tons of CO_2 and 6.41 tons of NO_x—pollutant emissions directly associated with harmful effects on human health—annually.

Keywords: electric bus; emissions; urban transportation; energy mix

---

1. Introduction

Contemporary trends of population migration to urban centers and the "metropolization" of urban cities (especially those that are local and regional centers) have been continuously increasing, which has been confirmed by a global increase in population migration. With such an increase in the number of inhabitants in these urban agglomerations, problems related to the public transport of passengers (as an integrated part of the functionality and sustainability of a city) have arisen, which must be solved through the prism of several factors, such as efficiency, versatility, punctuality, modularity, and comfort.

The fact that, in general, the road transport sector is one of the largest net contributors to the generation of NO_x pollutants (about 13% of total pollutant emissions) and 27% of total greenhouse gas (GHG) emissions in the atmosphere at the European level presently cannot be ignored [1]. For this reason, the European Union has adopted (and will adopt) numerous laws and regulations related to the substantial reduction of pollutant emissions caused by transport using internal combustion engines as an energy source.

At the present time, most urban public transportation systems use buses equipped with internal combustion engines, which use fossil fuels as an energy source. Even in the short term, the use of renewable sources (biofuels) has been accepted worldwide as an immediate solution to completely or partially replace fossil fuels (as they can contribute to reducing global GHG emissions); however, there are many limitations to their use, regarding the protection of the environment.

Pollutant emissions and greenhouse gas emissions caused by functioning of internal combustion engines, such as CO_2, NO_x, and PM (particulates), have been shown to lead to serious problems, with negative impacts on the environment and human health [2–5].

Thus, the electrification of urban passenger transport systems (in any form of electric vehicles) is currently seen (and has been implemented, in places) as a relevant/potential solution for a massive/total reduction of local pollution and greenhouse gas emissions.

Especially in Europe, urban centers are often characterized by unique architectural structures, with development occurring around an old town center populated with historic and public buildings. These buildings are permanently exposed to the corrosive effects of emissions and vibration caused by road traffic and, accordingly, several limitations have been imposed on vehicular access in such areas. Primary initiatives in this regard have been made in big cities, such as Vienna, Winchester, Madrid, Berlin, and Cluj-Napoca [6–9]. As a conclusion, studies have generally shown that small and medium cities could successfully adopt sustainable urban transport technologies based on the use of electric transport vehicles [10,11].

It is worth mentioning that emissions reduction is done locally, using electric vehicles—the so called zero emission vehicle (ZEV) concept—and, in the global mode, the intensity of the emissions depends directly on the energy mix used in the production of electricity [12].

The recent massive addition of the electric powertrain in vehicles has been mainly implemented in the construction of cars used as a means of personal travel and in the construction of buses for passenger transport. Conceptually, the construction of an electric powertrain does not differ between a car and an electric bus; the general structures are shown in the Figure 1.

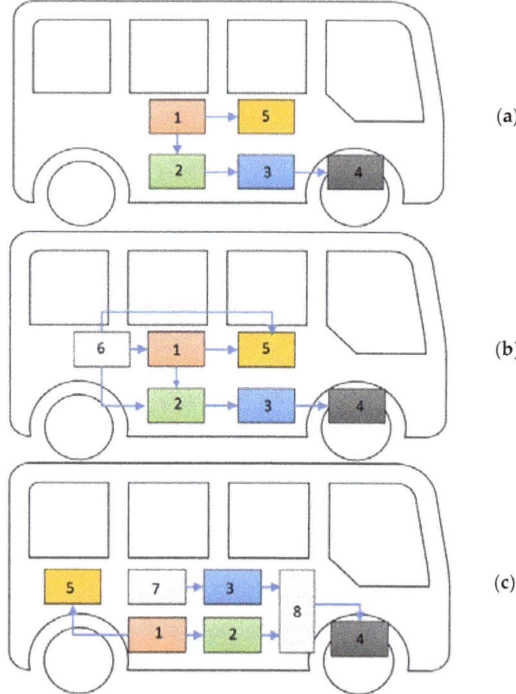

Figure 1. General structures of electric and hybrid powertrains used in the construction of buses: (a) electric bus; (b) hybrid series; and (c) hybrid parallel. 1, battery; 2, electric motor; 3, transmission; 4, final drive; 5, auxiliaries; 6, generator; 7, engine; and 8, torque converter/coupler.

Electric buses have different constructive solutions for the powertrains, these differences relating to the power source for the electric engine. Generally speaking, most common construction solutions for contemporary electric buses are: battery electric, hybrid series, and hybrid parallel; their basic structures (configurations) are shown in Figure 1.

The hybrid technology (both series and parallel types) uses both an internal combustion (IC) engine and an electric motor to produce the required traction power. In the serial configuration of a hybrid propulsion unit, the IC engine is used only to generate the electricity (through a generator) required by the electric traction motor(s) or to be stored in the bus batteries. In the parallel configuration of a hybrid propulsion unit, both the internal combustion engine and the electric motor provide the traction force necessary to move the bus. The traction force can be supplied by both engines (IC engine and/or electric motor) through a torque converter, as well as by independent driving of the traction wheels. In general, the storage capacity of the batteries is much higher in the case of electric buses (being the only energy source available), compared to the buses with hybrid propulsion, where the electric motor operation is performed only in well-defined situations (i.e., starting, slope climbing, increasing the acceleration in the case of overtaking, and so on), the rest of the operations using the IC engine. The hybrid type construction of powertrain greatly helps to significantly reduce the pollutant emissions caused by transport equipped with only an IC engine (using petrol, diesel, gas, or renewable fuels); but does not totally eliminate emissions, as battery electric buses do.

The main barrier that must be overcome for the massive penetration of electric means of transport is that related to the autonomy or range (distance in km) that the vehicles have, compared to the autonomy of those vehicles powered by internal combustion engines. At the beginning of the development of electric vehicles (EVs), the autonomy was relatively low (40–60 km) and the electric vehicles were intended only for urban use and for small trips within the urban surroundings; however, at present, the autonomy achieved by certain electric vehicles has approached the threshold of traveling 400 km with only a single battery charge [13,14].

This has been made possible due to the exponential development of technologies related to increasing the storage capacity of electricity in batteries (as well as the energy source of the electric vehicle), Li-Ion technology being one of the most versatile and efficient technologies from this point of view.

Li-Ion technology is not a new technology just emerging into the market; research into the energy performance of Li-Ion batteries has been carried out since the 1970s, with their primary application being in the field of mobile electronic equipment. However, their application in the electrification of vehicles has been delayed, due to their high manufacturing costs and the safety restrictions, which had to be overcome by manufacturers [15,16]. Due to the high cost of these batteries, the purchase price of EVs is expected to remain higher than that of gasoline or diesel vehicles; this condition will be a key determinant of further massive penetration in the automotive market (despite EV's savings in fuel and maintenance costs) [17–20]. Once these barriers (along with many others [21]) in the market have been overcome, an increasing number of electric vehicles will be available in the automotive market, both for personal use and as a means of transport for passengers in urban agglomerations.

Studies on the possibilities of implementing electric buses in urban transport systems have been carried out by numerous researchers. Most of them analyzed computer simulation methods for different operating scenarios and proposed various algorithms for calculating and estimating the energy efficiency of electric buses. Stempien and Chan [22] presented a comparative study of different bus powertrain designs by a comparison that included such factors as powertrain technologies (i.e., fuel cell, fuel cell electric, battery electric, hybrid electric, IC diesel, and compressed natural gas buses), capital and operating costs, fuel consumption, and fuel cycle emissions. In their study, they used the data presented by Erkkilae et al. [23], which considered bus energetic consumption values (used also further to calculate the amount of indirect emissions) of 0.66–1.23 kWh/km for a lower load and 0.7–1.45 kWh/km for a higher load. By compiling the existent data in the literature, the authors affirmed, as a major conclusion, that the battery electric bus technology is one of the most competitive options for advanced public transport systems in constrained urban areas in the future. Lajunen A. [24] presented an analysis regarding the energy consumption and cost-benefits of hybrid and electric city buses. The analysis was made by simulation (using the ADVISOR vehicle simulation program) to define the energy efficiency of buses. The author did not use real data from exploitation as data sources for models, stating that " … there are no available, comparable results of the energy consumption in the literature for the plug-in hybrid and electric city buses because their commercialization is in early stages". Moreover, an important remark (conclusion) of the author was that the exploitation condition (operation schedule and route planning) of a hybrid or electric bus is a condition that must be considered before introduction in an urban transport system for efficient energy use. A recent study of Vepsalainen et al. [25] studied the energy efficiency of an electric bus using a computationally efficient model for energy demand prediction. This study represented a novel approach to predict energy consumption variation with a wide range of uncertain factors (i.e., temperature, battery technical and functioning parameters, rolling resistance, and payload) and the simulation results gave values of 0.43–2.30 kWh/km (1.20 kWh/km average value with standard deviation of 0.32 kWh/km) net energy consumption. No data about real routes were applied as computational data in this study, however.

Data from the real exploitation of three types of electric buses in certain traffic conditions over a particular route in Macao (8.8 km long) was presented by Zhou et al. [26]. The net energetic consumption measured was 1.38–1.75 kWh/km for a 12 m e-bus type and 0.79 kWh/km for an 8 m e-bus type, resulting a reduction in CO_2 emissions, from a life-cycle perspective, of 19%–35% (compared with a diesel bus). Based on the same exploitation condition previously presented, Song et al. [27] continued the study, regarding the benefits of the introduction of electric buses into Macao's urban transportation system by calculating the reduction of GHG emissions. The results were situated between 56.47–133.76 kgCO_2eq/100 km (average value of 127.99 kgCO_2eq/100 km), taking into consideration the particularities of the energy mix for electricity production.

The aim of this paper is to quantify the direct and indirect CO_2 and NO_x pollutant emissions due to real urban exploitation of an electric bus fleet under service, in the particular case of Cluj-Napoca city. Real technical data related to the energy efficiencies of the electric buses, which have been integrated into the urban transport system of Cluj-Napoca, are presented and analyzed (from direct and indirect pollutant emissions emission point of view), in order to show that there are major differences between data obtained by computer simulation and the real ones (taking into account the particular operating conditions).

2. Materials and Methods

2.1. Cluj-Napoca City's Urban Passenger Transportation System

The city of Cluj-Napoca is a large urban agglomeration located in Transylvania, in the northwestern area of Romania, with a stable population of approx. 400,000 inhabitants. Besides the stable population, the city is an important university (academic) center and, so there is also a permanent fluctuation of approx. 100,000 students that study and live in the city. Thus, the city of Cluj-Napoca can be seen as a large urban agglomeration with an architectural mix between the old central area (medieval) and the peripheral neighborhoods (contemporary), featuring all of the typical problems related to the optimal operation of a public passenger transport system. From the point of view of the main characteristics of the urban transport system, the total length of the routes served for the urban public transport of passengers in the city of Cluj-Napoca is 355.3 km, of which buses are used on 279.4 km (47 lines), trolleybuses are used on 51.95 km (seven lines), and trams are used on 23.95 km (four lines). In addition to these, there are also routes in the metropolitan area, which total 278.45 km.

The totality of the urban means of transport for passengers of Cluj-Napoca city consists of 297 buses (of which 41 are electric buses and 256 are diesel buses Euro 4–Euro 6), 81 trolleybuses, 27 trams, and nine diesel minibuses. The 41 electric buses represent 13.8% of the total buses and 9.90% of the total means of the urban transport fleet. Furthermore, 149 (36% of the total) vehicles out of the total means of transport use electric powertrains. From the point of view of the number of passengers transported, 78.6% are transported by buses, 14.6% by trolleybuses, and 6.8% by trams.

The need to reduce local pollutant emissions has become a stringent contemporary demand and, from this point of view, the administration of the City of Cluj-Napoca has decided to purchase and put under service a fleet of 41 electric buses. The model that won the international tender was the Solaris 12E, a bus model that meets all the standards and requirements of a passenger transport in terms of energy efficiency, security, and comfort.

In the construction of the Solaris 12E electric buses (Figure 2, Table 1), which are used for urban passenger transport, two types of Li-Ion type battery are used. A total of 11 buses use $LiFePO_4$-type batteries and the other 30 are equipped with NMC ($LiNiMnCoO_2$)-type batteries (arranged in five separate packs). The predicted range is approx. 140 km for a single (full) battery charge, a charging process that can be performed in both slow and fast modes.

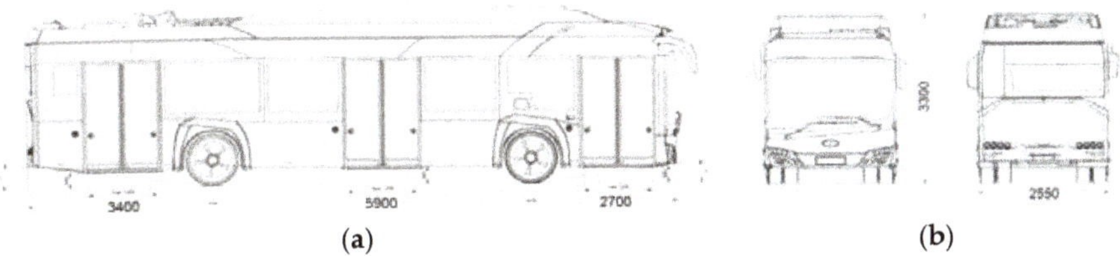

(a)　　　　　　　　(b)

Figure 2. Overall dimensions of the Solaris 12 E bus. (**a**) side view (**b**) front view.

Table 1. Main technical data of the Solaris 12 E bus.

Parameter	Value
Engine	Electric portal axle ZF AVE130 2×110 kW
Traction battery technology	$LiFePO_4$ technology: 58.8 kW pack nominal energy; 687.02 V nominal voltage NMC technology: 50.7 kW pack nominal energy; 651.20 V nominal voltage
Charging system	Plug-in (optional pantograph)
Front axle	ZF independent suspension
Rear (drive) axle	ZF portal axle with integrated electric motors
Suspension leveling system	ECAS air suspension with lowering/raising function: lowering and raising the bus, lowering right side by 70 mm, raising by approx. 60 mm.
Passenger capacity seated	Max. 37 + 1 (depending on door arrangement and batteries)

2.2. Research Methodology

As presented above, the aim of this paper was to quantify the direct and indirect CO_2 and NO_x emissions due to urban exploitation of an electric bus fleet, in the particular case of Cluj-Napoca city.

There were two directions we might take to estimate the environmental effects of exploitation of the electric bus fleet: calculation of the direct reduction of pollutant emissions as a result of replacing IC Euro-3 norm buses (diesel) with electric buses, and calculation of the indirect reduction of pollutant emission by considering the energy mix for production of electrical energy.

The intensity of CO_2 and NO_x emissions for a Euro 3 bus were derived as the average value from different specialized studies on urban bus emissions [28–32]; from which, we arrived at the values of 1259 gCO_2/km and 12.08 gNO_x/km.

Also, a scenario was considered in which the 41 Euro 3 buses would have been replaced with new Euro 6 buses. In this case, for the Euro 6 buses the emissions values were considered to be 1133 g/km for CO_2 and 1.11 g/km for NO_x (based on reference [33]). It can be observed that the difference in of the CO_2 emissions is approx. 10% lower for Euro 6 buses, but the use of selective catalytic reduction (SCR) systems using aqueous urea solutions for Euro 6 bus's exhaust emissions control, results in a reduction of NO_x emissions by more than 10 times.

The fleet of electric buses is permanently monitored by the management center for urban transportation of Cluj-Napoca city, which is achieved by direct internet connection between each electric bus and the management center. The technical parameters for the operation of the electric buses (e.g., energy consumed—related to battery-out current and voltage, energy recovered—related to brake regeneration, battery state of charge, temperature inside the bus, speed, operation of auxiliary systems, temperature of electric motor, operating errors, and so on) and specific data taking into account the routes on which the buses are under service (e.g., route served, GPS position, number of passengers ascended, number of passengers descended, and so on) are provided in real time. Thus, these data have been stored and can be accessed for the continuous monitoring of the electric bus fleet from the energetic and economic efficiency points of view. According to the main data of the electric bus fleet over one year, a total distance of 530,944 km had been traveled, an average load of 3089 passengers per month per bus (1,519,788 passengers per year for the 41-bus fleet), and with an average energy consumption of 0.96 kWh/km and 0.38 kWh/km energy recovered/generated (i.e., net energy consumption was 0.58 kWh/km). The energy recovered/generated by electric buses was obtained due to the regenerative braking process (converting the kinetic energy of the bus into electric energy, which is stored in its batteries). This process improves the overall efficiency of a bus by increasing its operational range (autonomy).

It should be mentioned that there were no dedicated (preferred) exploitation routes for each electric bus, since they served passenger transport routes depending only on the management and transport needs at that time. For the fluidization of traffic in the city of Cluj-Napoca, there are traffic lanes dedicated to buses and trolleybuses, in which their average speed is approx. 15 km/h (compared to the average city traffic speed of 13.7 km/h) with a peak speed of 50 km/h, which means that the energy recovered by the electric buses by the regenerative braking process had high values. Aspects related to the exploitation and operational data of the electric bus fleet under service are presented in Figures 3–7.

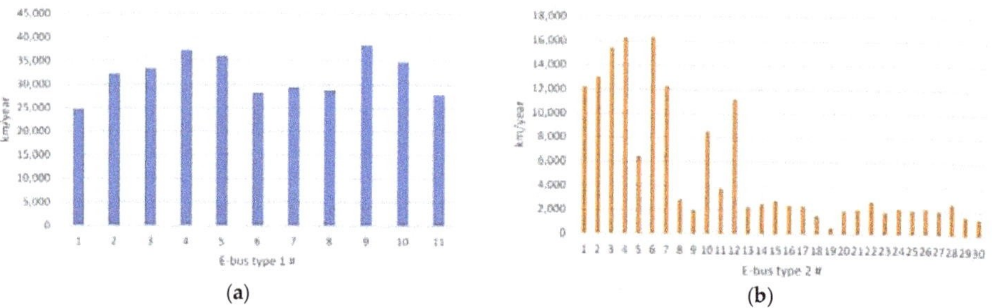

Figure 3. Number of kilometers traveled annually per each e-bus: (**a**) LiFePO$_4$ battery bus type; and (**b**) NMC battery bus type.

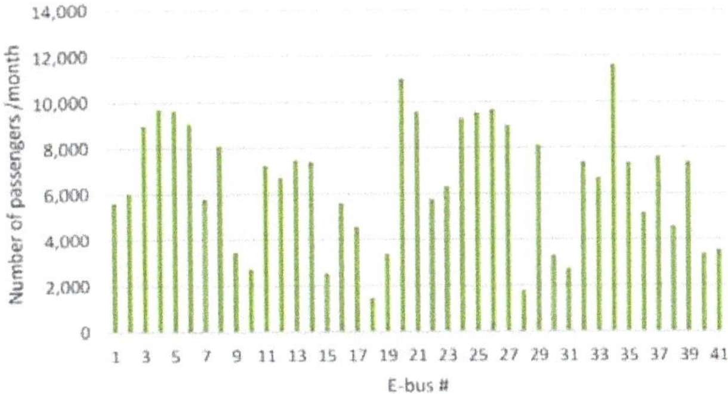

Figure 4. Number of passengers carried per month.

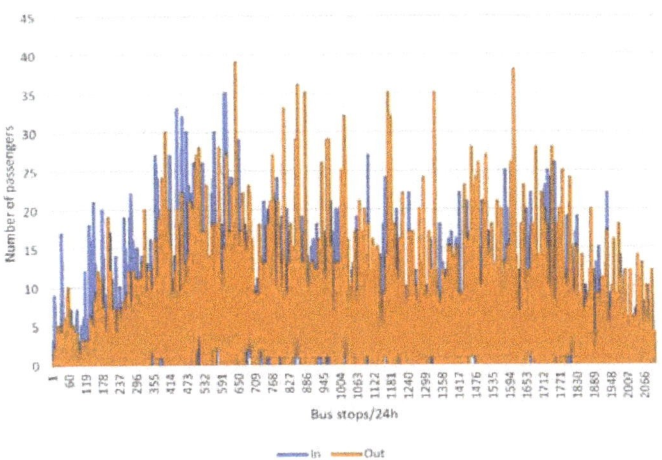

Figure 5. Passenger number dynamics for one day's operations of an electric bus.

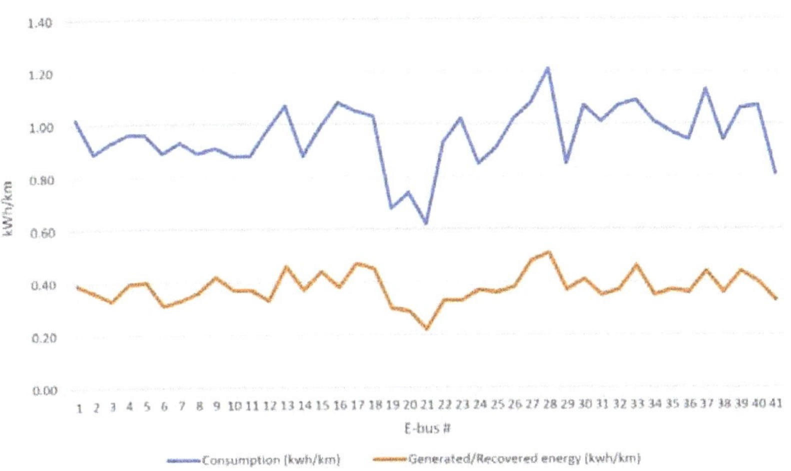

Figure 6. Energy balance of the fleet of electric buses (kWh/km).

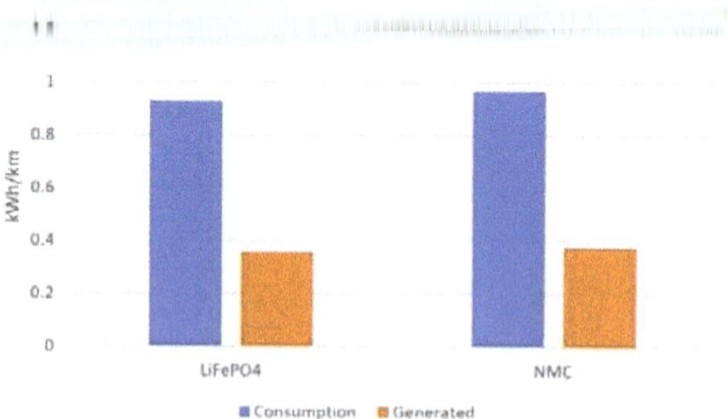

Figure 7. Energy balance (net energy consumption), according to battery type.

3. Results and Discussions

3.1. Direct Emissions Reduction

The direct emissions reduction due to replacing 41 diesel buses (Euro-3 pollution norm) with electric buses, considering the emissions intensity of the considered pollutants and the exploitation data presented previously, was found to be 668.45 tons of CO_2 and 6.41 tons of NO_x per year. Furthermore, taking into account the number of passengers transported by the electric bus fleet, it can be said that each passenger contributed to reducing the local pollution caused by urban traffic by 439.83 gCO_2 and 4.22 gNO_x per year.

In the case of the scenario of replacing Euro 3 buses with new Euro 6 buses, the direct reduction of CO_2 emissions would be only 66.9 tons and in the case of NOx emissions of 5.83 tons, per year. Under these conditions, taking into account the intensity of the use of buses as a means of urban transport from the point of view of the transported passengers, the reduction of local pollution would be 44 gCO_2/passenger/ year and 3.8 gNO_x/passenger/year. It can be observed that when using Euro 6 buses instead of Euro 3 buses, the difference between CO_2 emissions does not show great differences, but the reduction of NO_x emissions is large, sensitive equal to the reduced amount by using electric buses.

3.2. Indirect Emissions Reduction

To estimate the indirect pollutant emissions reduction by exploitation of the electric bus fleet, it is necessary to analyze the energetic mix of energy production used in charging the batteries of the fleet.

The method used to calculate the amount of emission of each considered pollutant is presented in Equation (1) (in the case of CO_2, but applicable also for the NO_x pollutant emission calculation), considering the energy consumption of the buses (kWh/km) and the intensity of pollutant emission function of energy source mix:

$$EVB_{CO2\ (g/km)} = TD_{km} \times CO_{2\ (emissivity)} \times E_{balance} \times P_{losses}, \tag{1}$$

where $EVB_{CO2\ (g/km)}$ represents the amount of CO_2 emissions, TD_{km} is traveled distance, $CO_{2\ (emissivity)}$ is the pollutant emissivity due to the energy source mix used for electric energy production, $E_{balance}$ is the effective energy consumption of an e-bus, and $P_{losses} = 1.15$–1.22 are the losses caused by the electric distribution grid and the EVs internal electrical circuits. Since there are no official final data for Romania regarding the emission intensity for the electricity production in 2019 year (on the basis of the used fuel), data were taken from the references [12,34] and Figure 8 [35].

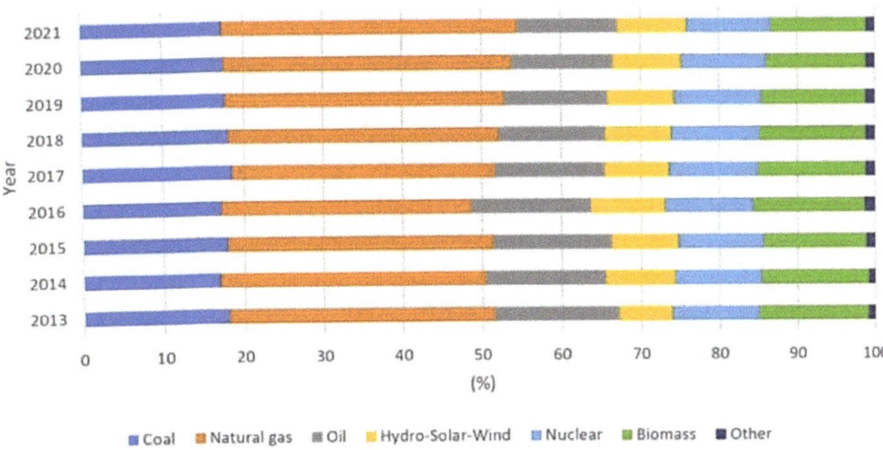

Figure 8. The energy mix of the electricity production in Romania (historical, current, and forecast) [35].

Figure 8 shows the Romanian energy mix over a period of 9 years (both historical and that forecast for 2020 and 2021) [35]. It can be noted that, within this energy mix, the "green energy" in 2019 represented a share of 34.2% of the total (renewable + nuclear), from which the average value of the intensity of CO_2 production of 434.38 g/kWh and the average value of the intensity of NO_x emissions production of 2.17 g/kWh for 2019 were derived (Table 2).

Table 2. CO_2 and NOx emissions intensities, depending on the energy mix for the year 2019 [11,34].

Electricity Generation Sources	Energy Mix (%)	CO_2 Emission Factor (gCO_{2eq}/kWh)	Direct CO_2 Emission by Fuel (gCO_2/kWh)	NO_X Emission Factor (gNO_{Xeq}/kWh)	Direct NO_X Emission by Fuel (gNO_X/kWh)
Coal	17.6	1000	170	6	1.056
Oil	12.9	650	83.85	4	0.516
Natural Gas	35.3	500	176.5	1.7	0.6001
Renewable	23.2	15	3.48	0.006	0.001392
Nuclear	11	5	0.55	0	0

Based on this data, it is possible to calculate the amount of pollutant emissions indirectly eliminated in the atmosphere by using the fleet of 41 electric buses: 153.83–163.19 tons of CO_2 per year (average value of 158.51 tons of CO_2 per year) and 0.770–0.815 tons of NO_x per year (average value of 0.792 tons of NO_x per year). If considering the total number of kilometers traveled annually by all 41 electric buses under service, then the reduction of the considered emissions was 298.54 gCO_2/km and 1.49 gNO_x/km per year.

3.3. Effect of Battery Technologies

A separate discussion is related to the battery technologies used in the bus powertrain designs. As mentioned before, there were two types of batteries used: $LiFePO_4$-type and NMC-type.

Li-Ion technology has been used in the construction of many types of batteries, the difference being the material used in the construction of the cathode (generally, graphite is used as the anode material). Thus, different acronyms have been used to identify this, such as: LFP, lithium iron phosphate; LCO, lithium cobalt oxide; LMO, lithium manganese oxide; NMC, lithium nickel manganese cobalt oxide; and NCA, lithium nickel cobalt aluminum oxide [15]. This particularity in the construction of the cathode causes differences in battery properties, in terms of specific energy, specific power, energy density, performance, voltage level, safety behavior, life span, and cost.

The major differences between the $LiFePO_4$ and NMC batteries used for the electric propulsion of the Solaris 12E can be identified according to the following main operating characteristics:

- Cycle life: $LiFePO_4$—10,000 (80% retained capacity), NMC—10,000 (60% retained capacity);
- Recommended C-rate: $LiFePO_4$—C/2, NMC—C/5;
- Ability to work in high-temperature environments: $LiFePO_4$—YES (up to 120–140 °C), NMC—NO;
- Danger of thermal runaway and fire hazard: $LiFePO_4$—NO, NMC—YES; and
- Thermal management equipment: $LiFePO_4$—NO, NMC—YES.

Based on these main characteristics (among others), it has been considered that NMC-type batteries can cover a large variety of different applications (from small and home electronics to industrial energy storage facilities) and show great potential for the future automotive industry, mainly due to their higher energy density, as compared to $LiFePO_4$ technology.

Nevertheless, it can be seen, from Figure 7, that the differences regarding the net energy consumption are relatively negligible and do not significantly influence the values of pollutant emissions by e-bus exploitation; however, they can be helpful in managing the fleet, by allocating routes in which the possibility of energy recovery (by the regenerative braking process) is greater. The energetic balance between energy consumption and energy generated (recovered) is 0.55 kWh/km for $LiFePO_4$-type batteries and 0.58 kWh/km for NMC-type batteries (5.45%), which indicates that the influence of battery type on bus exploitation parameters does not have a major influence on the overall energetic efficiency of an electric bus.

4. Conclusions

Even though, at present, the costs associated with the introduction of electric buses into urban passenger transport systems are comparatively high, compared to buses equipped with modern internal combustion engines, their immediate utility is given by the local reduction of polluting emissions; appealing to the zero emission vehicle (ZEV) concept. In the particular case of the city of Cluj-Napoca, defined as a city with a medieval structure and infrastructure (old and historic buildings, narrow streets, multiple markets place, and so on), the introduction and operation of a fleet of 41 electric buses managed to eliminate a local amount of 668.45 tons of CO_2 and 6.41 tons of NO_x—pollutant emissions directly associated with harmful effects on human health—annually.

The emission balance was positive in both the cases of CO_2 and NO_x pollutants. The exploitation of 41 electric buses (in the particular traffic conditions of Cluj-Napoca city) had managed to reduce global pollution emissions by 509.95 tons of CO_2 and 5.618 tons of NO_x each year.

These values could be improved further, in order to increase the values of the indirect pollutant and GHG emissions eliminated through local energy management, whereby the energy mix used by the city contained a higher percentage of green energy (such as wind, solar, and/or energy from renewable sources).

The operation of a fleet of urban electric buses was shown to be feasible, taking into account the small average speeds of traffic in urban agglomeration areas and the fact that, when stopped in traffic, an electric bus consumed a minimal amount of energy (only that needed for supplying its auxiliary systems).

The obtained values of the net energy consumption (lower than those found in the specialized literature) were directly influenced by the type and power of the electric propulsion group of the bus (162 kW power of the Solaris 12 E electric motor, comparative to 180 kW for BYD 9K and 240 kW for e-CITARO), the traffic conditions (the existence of dedicated/exclusive lines for buses and which by their nature increase the efficiency of the braking energy regeneration process) and the traffic conditions and the behavior of driver.

Furthermore, local policies to reduce polluting emissions need to be developed and supported further (taxes, preferential parking places, dedicated commuting lanes, education, and so on) in the personal and passenger transportation fields, in order to increase the share of electric and hybrid vehicles in the urban transport system.

Author Contributions: Conceptualization and supervision, B.O.V. and F.M.; methodology and resources, C.D.M. and V.N.; data curation, I.S. and A.A.S.; writing—original draft preparation, F.M.; writing—review B.O.V. and F.M. All authors have read and agreed to the published version of the manuscript.

Funding: This research received no external funding.

Conflicts of Interest: The authors declare no conflict of interest.

References

1. Sabine, G.; Michael, G.; Nicole, M.; Katarina, M. *European Environmental Agency (2019) European Union Emission Inventory Report 1990–2017 under the UNECE Convention on Long-Range Transboundary Air Pollution (LRTAP)*; EEA Report No 08/2019; EEA: Copenhagen, Denmark, 2019; ISSN 1977–8449. [CrossRef]
2. Crouse, D.L.; Goldberg, M.S.; Ross, N.; Chen, H.; Labreche, F. Postmenopausal breast cancer is associated with exposure to traffic-related air pollution in Montreal, Canada: A case-control study. *Environ. Health Perspect.* **2010**, *118*, 1578–1583. [CrossRef]
3. Hung, L.J.; Tsai, S.; Chen, P.; Yang, Y.; Liou, S.; Yang, C. Traffic air pollution and risk of death from breast cancer in Taiwan: Fine particulate matter (PM 2.5) as a proxy marker. *Aerosol Air Qual. Res.* **2011**, *12*, 275–282. [CrossRef]
4. Manzetti, S. Ecotoxicity of polycyclic aromatic hydrocarbons, aromatic amines and nitroarenes from molecular properties. *Environ. Chem. Lett.* **2012**, *10*, 349–361. [CrossRef]
5. Manzetti, S. Polycyclic aromatic hydrocarbons in the environment: Environmental fate and transformation. *Polycycl. Aromat. Compd.* **2013**, *33*, 311–330. [CrossRef]
6. Wall, G.; Felstead, T.; Richards, A.; McDonald, M. Cleaner vehicle buses in Winchester. *Transp. Policy* **2008**, *15*, 55–68. [CrossRef]
7. Garcia Sanchez, J.A.; Lopez Martinez, J.M.; Martin, J.L.; Flores Holgado, M.N.; Morales, H.A. Impact of Spanish electricity mix, over period 2008-2030, on the Life Cycle energy consumption and GHG emissions of Electric, Hybrid Diesel-Electric, Fuel Cell Hybrid and Diesel Bus of the Madrid Transportation System. *Energy Convers. Manag.* **2013**, *74*, 332–343. [CrossRef]
8. Varga, B.; Mariasiu, F.; Moldovanu, D.; Iclodean, C. *Electric and Plug-In Hybrid Vehicles–Advanced Simulation Methodologies*; Springer: Basel, Switzerland, 2015; pp. 477–524.
9. Varga, B.; Iclodean, C.; Mariasiu, F. *Electric and Hybrid Buses for Urban Transport-Energy Efficiency Strategies*; Springer: Basel, Switzerland, 2016; pp. 85–103.
10. Kuhne, R. Electric buses–An energy efficient urban transportation means. *Energy* **2010**, *35*, 4510–4513. [CrossRef]
11. Nurhadi, L.; Boren, S.; Ny, H. A sensitivity analysis of total cost of ownership for electric public bus transport system in Swedish medium size cities. *Transp. Res. Procedia* **2014**, *3*, 818–827. [CrossRef]
12. Varga, B.; Mariasiu, F. Indirect environment-related effects of electric car vehicles use. *Environ. Eng. Manag. J.* **2018**, *17*, 1591–1597.
13. Junquera, B.; Moreno, B.; Álvarez, R. Analyzing consumer attitudes towards electric vehicle purchasing intentions in Spain: Technological limitations and vehicle confidence. *Technol. Forecast. Soc. Chang.* **2016**, *109*, 6–14. [CrossRef]
14. Un-Noor, F.S.; Padmanaban, L.; Mihet-Popa, M.N.; Mollah, E.; Hossain, A. Comprehensive Study of Key Electric Vehicle (EV) Components, Technologies, Challenges, Impacts, and Future Direction of Development. *Energies* **2017**, *10*, 1217. [CrossRef]
15. Shabbir, A.; Trask, S.E.; Dees, D.W.; Nelson, P.A.; Lu, W.; Dunlop, A.R.; Polzin, B.J.; Jansen, A.N. Cost of automotive lithium-ion batteries operating at high upper cutoff voltages. *J. Power Sources* **2018**, *403*, 56–65.
16. Brand, M.; Gläser, S.; Geder, J.; Menacher, S.; Obpacher, S.; Jossen, A.; Quinger, D. Electrical safety of commercial Li-ion cells based on NMC and NCA technology compared to LFP technology. In Proceedings of the EVS27 International Battery, Hybrid and Fuel Cell Electric Vehicle Symposium, Barcelona, Spain, 17–20 November 2013.

17.   Tzeng, G.H.; Lin, C.W.; Opricovic, S. Multi-criteria analysis of alternative-fuel buses for public transportation. *Energy Policy* **2005**, *33*, 1373–1383. [CrossRef]

18.   Li, Y.; Bao, L.; Deng, H.; Lu, D. Promoting the Marketization of Battery-Electric-Bus (BEB) based on the Satisfaction Model. *Procedia-Soc. Behav. Sci.* **2013**, *96*, 1146–1155. [CrossRef]

19.   Mirchandani, P.; Adler, J.; Madsen, O.B.G. New logistical issues in using electric vehicle fleets with battery exchange infrastructure. *Procedia-Soc. Behav. Sci.* **2014**, *108*, 3–14. [CrossRef]

20.   Christidis, P.; Focas, C. Factors Affecting the Uptake of Hybrid and Electric Vehicles in the European Union. *Energies* **2019**, *12*, 3414. [CrossRef]

21.   Vassileva, I.; Campillo, J. Adoption barriers for electric vehicles: Experiences from early adopters in Sweden. *Energy* **2017**, *120*, 632–641. [CrossRef]

22.   Stempien, J.P.; Chan, S.H. Comparative study of fuel cell, battery and hybrid buses for renewable energy constrained areas. *J. Power Sources* **2017**, *340*, 347–355. [CrossRef]

23.   Erkkilae, K.; Nylund, N.O.; Pellikka, A.P.; Kallio, M.; Kallonen, S.; Kallio, M.; Ojamo, S.; Ruotsalainen, S.; Pietikäinen, O.; Lajunen, A. eBUS-Electric bus test platform in Finland. In Proceedings of the eBUS-Electric Bus Test Platform in Finland, EVS27 International Conference, Barcelona, Spain, 17–20 November 2013.

24.   Lajunen, A. Energy consumption and cost-benefit analysis of hybrid and electric city buses. *Transp. Res. Part C* **2014**, *38*, 1–15. [CrossRef]

25.   Vepsalainen, J.; Otto, K.; Lajunen, A.; Tammi, K. Computationally efficient model for energy demand prediction of electric city bus in varying operating conditions. *Energy* **2019**, *169*, 433–443. [CrossRef]

26.   Zhou, B.; Wu, Y.; Zhiu, B.; Wang, R.; Ke, W.; Zhang, S.; Hao, J. Real-world performance of battery electric buses and their life-cycle benefits with respect to energy consumption and carbon dioxide emissions. *Energy* **2016**, *96*, 603–613. [CrossRef]

27.   Song, Q.; Wang, Z.; Wu, Y.; Li, J.; Yu, D.; Duan, H.; Yuan, W. Could urban electric public bus really reduce the GHG emissions: A case study in Macau? *J. Clean. Prod.* **2018**, *172*, 2133–2142. [CrossRef]

28.   Mahmoud, M.; Garnett, R.; Ferguson, M.; Kanaroglou, P. Electric buses: A review of alternative powertrains. *Renew. Sustain. Energy Rev.* **2016**, *62*, 673–684. [CrossRef]

29.   Tucki, K.; Orynycz, O.; Świć, A.; Mitoraj-Wojtanek, M. The Development of Electromobility in Poland and EU States as a Tool for Management of CO_2 Emissions. *Energies* **2019**, *12*, 2942. [CrossRef]

30.   Oprešnik, S.R.; Seljak, T.; Vihar, R.; Gerbec, M.; Katrašnik, T. Real-World Fuel Consumption, Fuel Cost and Exhaust Emissions of Different Bus Powertrain Technologies. *Energies* **2018**, *11*, 2160. [CrossRef]

31.   Arcentales, D.; Silva, C. Exploring the Introduction of Plug-In Hybrid Flex-Fuel Vehicles in Ecuador. *Energies* **2019**, *12*, 2244. [CrossRef]

32.   Nylund, N.O.; Erkkilä, K.K.; Lappi, M.; Ikonen, M. Transit Bus Emission Study: Comparison of Emissions from Diesel and Natural Gas Buses. Research Report PRO3/P5150/04. 2004. Available online: https://www.cti2000.it/Bionett/BioG-2004001%20Transit%20Bus%20Emission%20Study.pdf (accessed on 31 July 2019).

33.   Maciej, G. Comparative studies exhaust emissions of the Euro VI buses with diesel engine and spark-ignition engine CNG fuelled in real traffic conditions. In *MATEC Web of Conferences (VII International Congress on Combustion Engines)*; EDP Sciences: Les Ulis, France, 2017; Volume 118, p. 00007.

34.   POST-Parliamentary Office of Science and Technology. Carbon Footprint of Electricity Generation, Report Number 268. 2006. Available online: http://www.parliament.uk/documents/post/postpn268.pdf (accessed on 20 August 2019).

35.   Romanian National Institute for Statistics. Statistica Energiei. Available online: http://www.insse.ro/cms/ro/content/statistica-energiei (accessed on 20 August 2019). (In Romanian)

Article

Multi-Objective Shark Smell Optimization Algorithm Using Incorporated Composite Angle Cosine for Automatic Train Operation

Longda Wang [1][iD]**, Xingcheng Wang** [1,*]**, Zhao Sheng** [2] **and Senkui Lu** [1]

[1] School of Marine Electrical Engineering, Dalian Maritime University,
Dalian 116026, China; wanglongda@dlmu.edu.cn (L.W.); karsenlu@163.com (S.L.)
[2] School of Electronic and Information Engineering, Beijing Jiaotong University,
Beijing 100044, China; zhaosheng@bjtu.edu.cn
* Correspondence: dmuwxc@dlmu.edu.cn; Tel.: +86-411-8472-7801

Received: 16 December 2019; Accepted: 30 January 2020; Published: 6 February 2020

Abstract: In this paper, an improved multi-objective shark smell optimization algorithm using composite angle cosine is proposed for automatic train operation (ATO). Specifically, when solving the problem that the automatic train operation velocity trajectory optimization easily falls into local optimum, the shark smell optimization algorithm with strong searching ability is adopted, and composite angle cosine is incorporated. In addition, the dual-population evolution mechanism is adopted to restrain the aggregation phenomenon in shark population at the end of the iteration to suppress the local convergence. Correspondingly, the composite angle cosine, considering the numerical difference and preference difference, is used as the evaluation index, which ameliorates the shortcoming that the traditional evaluation index is not objective and reasonable. Finally, the Matlab/simulation and hardware-in-the-loop simulation (HILS) results for automatic train operation show that the improved optimization algorithm proposed in this paper has better optimization performance.

Keywords: automatic train operation; multi-objective optimization; shark smell optimization algorithm; composite angle cosine; dual-population evolution mechanism; hardware-in-the-loop simulation

---

1. Introduction

Railway transportation is an essential means of transportation; it cannot be replaced by others owing to its own superiorities such as safety, energy efficiency, comfortable nature, punctuality, large volume of transport, convenient, accurate parking, etc. [1]. Automatic train operation (ATO) target velocity trajectory optimization is a practical multiple optimization problem for railway transportation, the multiple performance indicators such as energy consumption, parking punctuality, comfort, accurate parking, and so on. Of particular note is the increasing energy demand, thus energy-saving is becoming a research hot spot in automatic train operation [2]. Many researchers' studies about energy efficiency or energy-saving have been proposed in recent literatures [3–6]. The function of automatic train operation (ATO) target velocity trajectory optimization is crucial, it could make the train in the real-time optimal state as much as possible with multi-objective comprehensive performance index satisfied, so as to reduce energy consumption in automatic train operation. Therefore, improving the optimization and tracking control effect involves using the corresponding optimization algorithm effectively by incorporating improvement strategies appropriately.

Multi-objective train operation optimization has been a hot issue in the field of railway research in recent years. To obtain more satisfactory optimization results, a multi-objective optimization model of the speed trajectory for the high-speed train is established and an improved algorithm based on differential evolution and simulated annealing algorithms is designed [7]. A genetic algorithm with the binary encoding method is designed for obtain the high-quality timetables of urban rail transit systems based on two-objective (energy-saving strategies and service quality levels) model formulated [8]. A method about design speed profiles to be programmed robustly and efficiently is proposed in automatic train operation equipment for one metro line based on the two indicators of the running time and the energy consumption [9]. The balance between saving energy and running faster has been investigated, and an improved genetic algorithm is used to search the ideal optimal train target trajectories [10]. A novel multiple optimization model based on switching optimization framework for moving block signal (MBS) system is proposed [11]. A predictive train rescheduling model incorporating the model predictive control (MPC) mechanism and the non-analytical prediction model are proposed to be taken into consideration synergistic safe and efficient operations in high speed trains [12]. A microsimulation system about train timetable evaluation from the viewpoint of passengers to simulate both train operation and passengers' train choice behavior is developed [13].

Further research is necessary based on some of the above research results, and three crucial factors about multi-objective automatic train operation (ATO) target velocity trajectory optimization should be taken seriously. First, there are many uncertainties and complex relations in actual automatic train operation (ATO), and it is difficult to obtain ideal optimization results only by using traditional optimization algorithm. Many literatures (e.g., [7–10], etc.) have studied and improved traditional optimization algorithms about automatic train operation (ATO) optimization, so as to obtain more ideal optimization results. It is easy for traditional optimization algorithms to fall into the local optimum, and there are also problems of blind searching, premature stagnation, and slow convergence, in the end of iteration especially. Compared with improved traditional algorithms, the improved shark smell optimization (SSO) algorithm has more powerful searching capabilities, even in the end of the iteration. To improve the global optimization performance of shark smell optimization (SSO) algorithm, relevant researchers have proposed quite a few improvement strategies and relevant experiments showing that the improved SSO algorithm has more effective performance than other algorithms contrasted. The intrinsic mechanism of SSO algorithm is introduced in detail [14]. To solve the optimal capacitor placement problem satisfying the operating constraints, a new shark smell optimization (SSO) algorithm is proposed [15]. A new model for multiyear expansion planning of distribution networks (MEPDN) is proposed, and, to solve the above MEPDN model optimization problem, a new evolutionary algorithm-based solution method called Binary Chaotic Shark Smell Optimization (BCSSO) is presented [16]. A novel forecasting algorithm based on neural network (NN) and a novel chaotic shark smell optimization (CSSO) algorithm are proposed [17].

Meanwhile, the driving experience for automatic train operation (ATO) target velocity trajectory optimization should not be ignored. A considerable number of researchers are interested in researching the affect of driving experience for automatic train operation (ATO) optimization, such as in [3,4,9,10], etc. In addition, a series of manual driving strategies that will minimize energy consumption for high-speed trains have been researched [18]; an expert system that contains expert rules and a heuristic expert inference method about intelligent train operation optimization for subway has been developed [19]; an intelligent safe driving method (ISDMs) is proposed to obtain better speed–distance curves [20]. Note that preference indices of driving experience are applied in automatic train operation algorithm [21]. Preference information is widely used in multiple decision-making (MPDM) problems such as multi-objective optimization problems, plenty of research findings show that the optimization performance of multi-objective optimization algorithm can be significantly improved using incorporated appropriate preference information. A new method to solve multiple decision-making (MPDM) problems based on the preference information is proposed [22]. A preference information based on the weighted sum aggregation is proposed to better solve the multi-objective optimization problem, and the numerical experiments show that the method has obvious advantages in both calculation accuracy and computation time [23]. A multi-criteria selection method that the preference scale changes with the change of multi-criteria decision-making problem is proposed, and Monte Carlo method is used to verify the feasibility of the algorithm [24].

In fact, during automatic train operation, there are extensive problems need to be considered such as real-time velocity sampled inaccurately, signal delay, and packet loss in transmission and a certain degree of unstable in tracking control system, so a certain proportion of literatures use real vehicle experiments and actual driving data (Ning'xi line, Yizhuang line, Shanghai Railway Transit in China) to verify the effectiveness of the algorithm [4,19,21]. Due to the situation of the actual automatic train operation experiment, it is difficult to implement, and traditional simulation based on pure software environment cannot truly reflect the actual automatic train operation process; hardware-in-the-loop simulation (HILS) is often used in automatic train operation due to its characteristics [25,26]. At present, plenty of HILS-related products are researched, developed, and applied in various fields of rail vehicles, traction control system, hybrid electric vehicles, and electric cars, and numerous relative research findings have been achieved [27–30].

Based on the above research findings, an effective automatic train operation velocity trajectory optimization algorithm that can give full play to the role of driving preference information is needed, and multi-objective shark smell optimization algorithm with powerful optimization should also be emphasized, so as to achieve more satisfactory optimization results for the automatic train operation (ATO). An improved multi-objective shark smell optimization algorithm using incorporated composite angle cosine for automatic train operation is proposed in this paper on the basis of literatures [14,23,28] and several similar literatures. The following summarizes the main contributions of this paper.

(I) An improved multi-objective shark smell optimization algorithm (ISSO) based on incorporated composite angle cosine, dual-population mechanism and fusion distance measurement index is proposed to solve practical automatic train operation (ATO) target velocity trajectory optimization effectively.

(II) To verify the effectiveness of ISSO, two scenarios about rail transit line No.12 and Jinpu line No.1 in Dalian, China are chosen for simulation test. The results of Matlab/simulation and hardware-in-the-loop simulation (HILS) show that the ISSO proposed in this paper (ISSO) (I) has good performance in optimization precision, (II) has fast optimization speed, and (III) can obtain the smooth target velocity trajectory tracked control by "Controller" easily achievable.

The paper is organized as follows. Section 2 introduces the optimization model of automatic train operation. Section 3 illustrates the improved multi-objective shark smell optimization algorithm (ISSO) using incorporated composite angle cosine for automatic train operation proposed in this paper. Section 4 provides the Matlab/simulation results and hardware-in-the-loop simulation (HILS) results to illustrate the proposed method. Section 5 concludes this article.

2. Automatic Train Operation Target Velocity Trajectory Optimization Model

2.1. Constraint Model of Automatic Train Operation

For ensuring the automatic train operation is secure, stable, and accurate, many constraints such as the dynamic equation of automatic train operation, position variable constraint, velocity limitation and so on should be considered [31].

The dynamic equation of automatic train operation is described as follows,

$$\begin{cases} Mv\frac{dv}{ds} = u_f F_t\left(u,v\right) - R\left(v,s\right) - u_b B_r\left(u,v\right) \\ \frac{dt}{ds} = \frac{1}{v} \\ v(s) \leq v_{\lim}(s) \\ s_S = 0, \Delta s = |s_E - D| < \Delta s_{\max} \end{cases} \tag{1}$$

where t is the present running time of the train; s is the present running position of the train; M is the train mass, $M = (1 + r_m)M_T$; r_m is the rotating mass factor; M_T is the weight of the train; $u_f F_t\left(u,v\right)$ and $u_b B_r\left(u,v\right)$ are the traction force and braking force of the current velocity, respectively; $R\left(v,s\right)$ is the additional resistance determined by the current speed and position of the train, s_S and s_E are the positions of starting point and terminal point respectively; D is the actual running distance; Δs_{max} represents the allowable maximum parking error; Δs represents the actual parking error; $v(s)$ is the actual velocity of the position s; $v_{lim}(s)$ are the limit velocity of the position s; and u represents the control sequence of automatic train operation [32,33]. The control modes for the above control sequence of full traction, partial traction for cruising, coasting, and partial braking for cruising and full braking are adopted in the paper, which are represented by $[1, 0.5, 0, -0.5 - 1]$. u_f and u_b are the traction and braking coefficients which needs to satisfy the following constraints, respectively.

$$\begin{cases} u_f, u_b \in [0, 1] \\ u_f \cdot u_b = 0 \end{cases} \tag{2}$$

The inflection position corresponding to each control modes should keep increasing order [34].

$$0 < S_1 < S_2 < \cdots < S_j < \cdots < S_k < D_{max} \tag{3}$$

where S_j represents the j th inflection point position for corresponding control mode and k represents the size of control sequence.

For ensuring the automatic train operation is secure and prevent accidents such as derailment, the velocity limitation should be set up in advance.

$$V_x = \begin{cases} 0 \leq v \leq V_x \\ V_{x1} \ (0 \leq S < Sp_1) \\ V_{x2} \ (Sp_1 \leq S < Sp_2) \\ V_{x3} \ (Sp_2 \leq S < Sp_3) \\ \cdots \\ V_{xkx} \ (Sp_{kx-1} \leq S \leq Sp_{kx}) \\ V_{xkx+1} \ (Sp_{kx} \leq S \leq D) \end{cases} \tag{4}$$

where V_x represents the maximum train velocity allowed in each subinterval, Sp_j represents the starting point of the $j + 1$ th subinterval (also the terminal point of the j th subinterval), and $kx + 1$ represents the number of the subintervals.

2.2. Multi-Objective Index for ATO Target Velocity Trajectory Optimization

Automatic train operation (ATO) target velocity trajectory optimization is a practical optimization problem that needs to meet multiple performance indicators such as energy consumption, parking punctuality, comfort, accurate parking, and so on.

The train energy consumption is expressed as the energy consumed by overcoming resistance during the whole process, and the specific calculation formula is described as follows,

$$K_E = \int_0^D f(u, v) \, ds \approx \sum_{i=2}^n (Ma_i + R_i)(s_i - s_{i-1}) \tag{5}$$

where K_E is the energy consumption, a_i is the acceleration of the i th condition, s_i is the position of the i th condition, and R_i is the resistance of the i th condition [19].

The comfort level is expressed by the sum of the absolute value of the difference of the acceleration of the adjacent working conditions in the running process, and the specific calculation formula is described as follows,

$$K_{Jerk} = \frac{\int_0^D |\Delta a| \, ds}{D} \approx \frac{\sum_{ia=2}^n |a_{ia} - a_{ia-1}|}{D} \tag{6}$$

where K_{Jerk} is the measure of comfort, a_{ia} is the acceleration of the *ia* th condition, $\overline{T}$ represents the actual running time, and $|\Delta a|$ is the absolute value of the acceleration changing rate [35].

The train punctuality is the absolute value of the difference between the actual running time and the prescribed running time, and the specific calculation formula is described as follows,

$$K_T = |\overline{T} - \overline{T}_{Exp}| \tag{7}$$

where K_T is the measure of punctuality and $\overline{T}_{Exp}$ represents the prescribed running time [35].

The target vector of the multi-objective optimization problem is $F(x) = (f_1(x), f_2(x), \cdots, f_k(x))$, and the optimization model is described as follows,

$$\begin{cases} & \min\{F(x)\} \\ \text{s.t.} & g(x) \leq 0, \ i = 1, 2, \cdots, m \\ & x = (x_1, x_2, \cdots, x_n), \ x \in D^* \end{cases} \tag{8}$$

where k is the number of optimization targets, x is the decision variable, and $g(x)$ is the inequality constraint for x. x'' is the absolute optimal solution, and if and only if any $x' \in D^*$, $F(x'')$ is superior to $F(x')$.

The multi-objective comprehensive performance index $F(u)$ is composed of energy consumption, comfort, and running time.

$$\begin{cases} F(u) = (K_E(u), K_{Jerk}(u), K_T(u)) \\ \min\{F(u)\} \end{cases} \tag{9}$$

where *min* denotes the minimum value of $F(u)$, which is the minimum value of each sub-objective function of $F(u)$.

2.3. Linear Weighted Target Method

Compared with the multi-objective optimization problem, the single objective optimization problem is easier to solve. It is a practical and effective way to transform the original multi-objective optimization problem into a single objective optimization problem. For multi-objective optimization problems, there is a degree of unfair measures caused by units and magnitude orders difference of various evaluation indexes. Therefore, index importance and difference between units and magnitude orders need to be considered, so as to give the appropriate weight factors. To eliminate the negative influence caused by the difference between units and magnitude orders, the data needs to be normalized. The calculation formula of the normalized linear weighted target $F'(x)$ can be expressed as follows,

$$\begin{cases} F'(x) = \sum_{i=1}^{k} w'_i w''_i f_i(x) \\ w''_i = \dfrac{f_i(x) - \min(f_i(x))}{\max(f_i(x)) - \min(f_i(x))} \end{cases} \tag{10}$$

where w'_i represents the index importance weight factor ($\sum_{i=1}^{k} w'_i = 1$), which reflects the relative importance of the *i* th optimization index. w'' denotes the correction weight factor, which eliminates the negative influence caused by the difference of dimensions and orders of magnitude for *i* th optimization index. *max* and *min*, respectively, represent the maximum and minimum values of the function [35,36].

As can be seen from the calculation Formula (10), normalization adopting can reduce the difficulty setting weight factors, index importance need to be considered exclusively, the other factor (difference between units and magnitude orders) have been solved by normalization effectively. Yet, the selection of the index importance weight factor by the linear weighted target method lacks the specific theoretical basis, so there is certain subjective limitation in this method.

2.4. Angle Cosine Method

For multi-objective optimization problems, there is an angle between any solution vector and the target demand vector in the solution space, and the angle cosine is less than or equal to 1. The target demand vector is the target vector of the desired optimal solution which may not be the final optimization solution. However, the target demand vector plays an active role in guiding the global convergence of the optimization algorithm in the process of iterative optimization. The specific angle cosine of the target vector and the target demand vector is shown in Figure 1, where the two axes represent two optimization objectives; the three solid lines in the axes, respectively, denote the solution vector $T1$, the solution vector $T2$, and the target demand vector C; the arc expresses the solution space; the two dotted lines in the axis represent the boundary of the solution space; the angles between the solution vector $T1$, $T2$, and the target demand vector C are, respectively, represented by the angles $\angle 1$ and $\angle 2$; and the angle cosines are expressed by γ_1 and γ_2 [37].

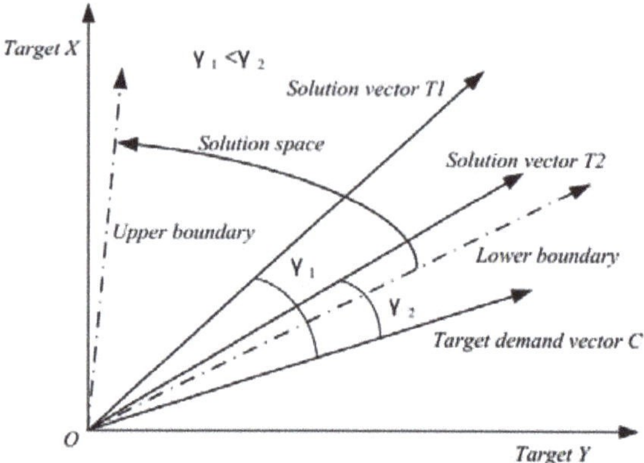

Figure 1. The angle cosine diagram of the solution vector and the target demand vector.

The calculation formula of the angle cosine γ_1 and γ_2 is expressed as follows,

$$
\gamma_1 = \frac{(T1,C)}{\|T1\| \bullet \|C\|} = \frac{\sum\limits_{i=1}^{ni} t_{1,i} \bullet c_i}{\sqrt{\sum\limits_{i=1}^{ni} t_{2,i}^2} \bullet \sqrt{\sum\limits_{i=1}^{k} c_i^2}}
$$
$$
\gamma_2 = \frac{(T2,C)}{\|T2\| \bullet \|C\|} = \frac{\sum\limits_{i=1}^{ni} t_{2,i} \bullet c_i}{\sqrt{\sum\limits_{i=1}^{ni} t_{2,i}^2} \bullet \sqrt{\sum\limits_{i=1}^{k} c_i^2}}
$$

$$(11)$$

where $(T1, C)$ and $(T2, C)$ represent the dot product between the solution vector $T1$, $T2$, and the target demand vector C; $\|A\|$ is the length of vector A; $\bullet$ represents the numerical multiplication; $t_{1,i}$, $t_{2,i}$, and c_i express the normalized value of the i th optimization index of the solution target vector $T1$, $T2$, and the target demand vector C; and ni represents the optimization index number.

As can be seen from Figure 1, the solution vector $T1$ is worse than the solution vector $T2$ caused by the solution vector $T2$ closer to the target demand vector C; meanwhile, $\angle 1 > \angle 2$ and $\gamma_1 < \gamma_2$. Thus, angle cosine can be used as the multi-objective optimization evaluation index. The target demand vector can be estimated and calculated reasonable in practical, so the angle cosine method is more objective and reasonable.

3. Improved Shark Smell Optimization Algorithm for ATO Target Velocity Trajectory Optimization

3.1. Shark Smell Optimization Algorithm

As the best hunter in nature, the shark has foraging behavior that goes forward and rotates, which can be extremely efficient in finding prey [16]. The optimization algorithm for simulating shark foraging is a highly efficient optimization algorithm [18]. For any given location, the sharks move at a speed to the particle that has the more intense scent, so the initial velocity vectors are defined as follows.

$$\left[V_1^1 , V_2^1, \ldots\ldots\ldots, V_{NP}^1 \right] \tag{12}$$

The shark has inertia when it swims, so the velocity formula of each dimension is defined as follows,

$$V_{i,j}^k = \eta_k \cdot R_1 \cdot \left. \frac{\partial(OF)}{\partial x_j} \right|_{x_{i,j}^k} + \alpha_k \cdot R_2 \cdot v_{i,j}^{k-1} \tag{13}$$

where $j = (1, 2, \cdots, ND), i = (1, 2, \cdots, NP)$, and $k = (1, 2, \cdots, k_{max})$; ND represent the number of dimension; NP represents the number of velocity vectors (size of shark population); k_{max} represents the number of iteration; OF represents the objective function; $\eta_k \in [0 , 1]$ represents the gradient coefficient; a_k represents the weight coefficient, it is also a random number between $[0, 1]$; and R_1 and R_2 are two random numbers between $[0, 1]$.

The speed of the shark is necessary to avoid over the boundary, and the specific speed limitation formula is described as

$$\left| v_{i,j}^k \right| = \min \left[\left| v_{i,j}^k \right|, \left| \beta_k \cdot v_{i,j}^{k-1} \right| \right] \tag{14}$$

where β_k represents the speed limit factor of the k th iteration.

The shark has a new position Y_i^{k+1} due to moving forward, and Y_i^{k+1} is determined by the previous speed and position, which is expressed as

$$Y_i^{k+1} = X_i^k + V_i^k \cdot \Delta t_k \tag{15}$$

where Δt_k represents the time interval of the k th iteration. In addition to moving forward, sharks usually rotate along their path to look for stronger odor particles and improve their direction of movement, which is a real way of moving.

The rotating shark moves in a closed interval which is not necessarily a circle. From the point of view of optimization, sharks implement local search at each stage to find better candidate solutions. The search formula for this position is as follows,

$$Z_i^{k+1,m} = Y_i^{k+1} + R3 \cdot Y_i^{k+1} \tag{16}$$

where $m = (1, 2, \cdots, M)$ presents the number of points at each stage of the location search; R_3 is the random number between $[-1, 1]$. If the shark finds a stronger scent point in the rotation, it moves toward the point and continues the search path. The location search formula is described as follows,

$$X_i^{k+1} = \arg\ \max \left\{ OF(Y_i^{k+1}), OF(Z_i^{k+1,1}), \\ \cdots, OF(Z_i^{k+1,M}) \right\} \tag{17}$$

As can be seen from the above formula, Y_i^{k+1} is obtained from the linear movement and $Z_i^{k+1,M}$ is obtained from the rotation movement. Sharks will choose the candidate solution with higher evaluation index value as shark's next location X_i^{k+1}.

3.2. Preference Information and Composite Angle Cosine

To improve the optimization effect, the preference phenomenon should be considered, and the data or information used to quantify the impact of preferences on evaluations is called preference information [38]. The preference vector angle is a kind of preference information, which is used to reflect the degree of preference between the solution vector and the target demand vector. The schematic diagram of the preference phenomenon classification and the preference angle is shown in Figure 2.

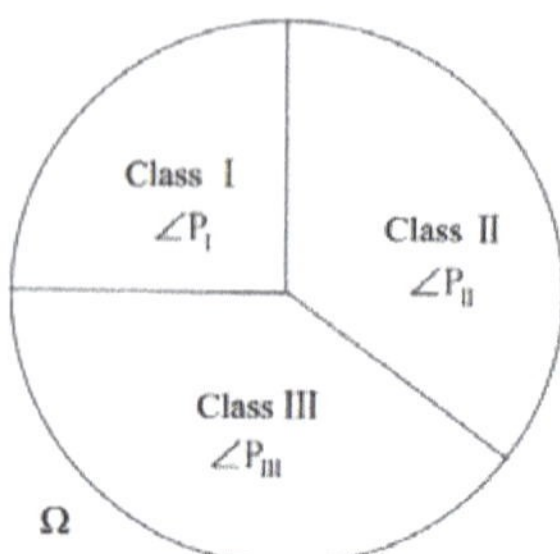

Figure 2. Schematic diagram of the preference phenomenon classification and the preference angle.

As can be seen from Figure 2, according to the preference phenomenon, preference event Ω is divided into three categories: ($ClassI$, $ClassII$, and $ClassIII$), and the corresponding preference angles are ($\angle P_I$, $\angle P_{II}$, and $\angle P_{III}$).

If only the numerical angle cosine is used as evaluation index in optimization, the preference phenomenon will be ignored, and it is easy to lead the evaluation result unreasonable. To take account of the numerical difference and preference difference between the solution vector and the target demand vector, a new evaluation index is proposed in this paper, that is, the compound angle cosine. The compound angle cosine is the cosine of the sum of the numerical angle and the preference angle. The schematic diagram of the compound angle cosine is shown in Figure 3.

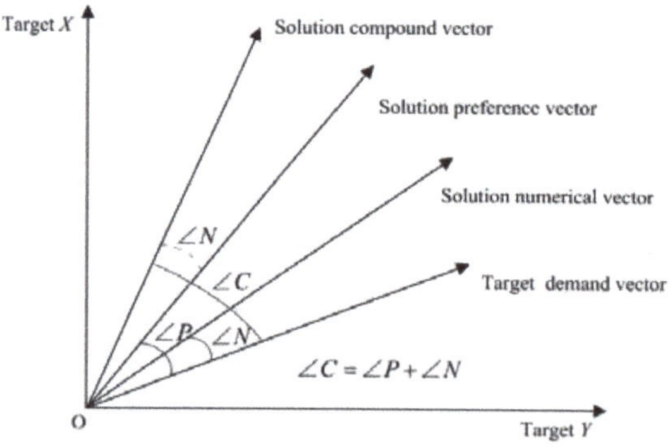

Figure 3. The schematic diagram of the compound angle cosine.

In Figure 3, the angle $\angle N$ represents the numerical angle; it is used to reflect the numerical calculated result between solution vector and the target demand vector. The angle $\angle N$ represents the preference angle; it is used to reflect the preference value between solution vector and the target demand vector; the angle $\angle C$ represents the composite angle, $\angle C = \angle P + \angle N$. The composite angle cosine is used as the evaluation index for the optimization algorithm proposed in this paper. The specific calculation formula of the composite angle cosine is as follows.

$$\angle C = \angle P + \angle C$$
$$\cos C = \cos P \cos N - \sin P \sin N \tag{18}$$

In some optimization problems, there are several the preference events. The specific calculation formula of the composite angle is as follows,

$$\angle P = \sum_{ip=1}^{np} \angle P_{ip} \tag{19}$$

where ip represents the preference event index, np represents the preference event number, and P_{ip} represents the preference angle of the i th preference event.

There are three preference events in the automatic train operation velocity trajectory optimization with the multi-objective comprehensive performance index $F(u)$,$(F(u) = (K_E(u), K_{Jerk}(u), K_T(u)))$. The specific preference categories categories circumstances in automatic train operation velocity trajectory optimization are shown in Table 1. Where, "Prefect", "Qualified", and " Unqualified" represents the valuation level of preference event according to preference phenomenon, the preference angle for "Prefect", "Qualified", and "Unqualified" are 0, $\pi/24$, $\pi/12$, respectively; E_1, E_2, and E_3 represent the boundary values of valuation level for energy consumption, which are decided by train parameters, line conditions, prescribed running time, and running distance; the boundary values of valuation level for "Comfort level" and "Time error" are decided by train operation regulation.

Table 1. The preference categories circumstances in automatic train operation velocity trajectory optimization.

Categories & Valuation Level	Prefect	Qualified	Unqualified
Energy Consumption	$K_E(u) < E_1$ KJ	$K_E(u) < E_2$ KJ	$K_E(u) < E_3$ KJ
Comfort level	$K_{Jerk}(u) < 4.2\,\mathrm{m/s^3}$	$4.2\,\mathrm{m/s^3} \leq K_{Jerk}(u) < 7.5\,\mathrm{m/s^3}$	$7.5\,\mathrm{m/s^3} \leq K_{Jerk}(u) < 13.4\,\mathrm{m/s^3}$
Time error	$K_T(u)) < 0.16\,\mathrm{s}$	$0.16\,\mathrm{m/s} \leq K_T(u)) < 0.20\,\mathrm{m/s}$	$0.20\,\mathrm{m/s} \leq K_T(u)) < 0.50\,\mathrm{m/s}$

If a certain performance index of the solution vector T cannot reach the valuation level "Unqualified", solution vector T is impermissible. At present, only valuation level "Prefect" and "Qualified" are permitted for most of automatic train operation scenarios.

3.3. Fusion Distance

Mahalanobis distance is the data covariance distance defined by Mahalanobis, which can accurately calculate the covariance distance between two samples. The formula of Mahalanobis distance between the sample X to be examined and the basic space set Y is expressed as

$$
\begin{aligned}
\Sigma = Cov(X,Y) &= E\left[(X - E(X))(Y - E(Y))\right] \\
&= \begin{bmatrix}
Cov(x_1, y_1) & Cov(x_1, y_2) \cdots Cov(x_1, y_j) \\
Cov(x_2, y_1) & Cov(x_2, y_2) \cdots Cov(x_2, y_j) \\
\vdots & \vdots \quad \ddots \quad \vdots \\
Cov(x_j, y_1) & Cov(x_j, y_2) \cdots Cov(x_j, y_j)
\end{bmatrix}
\end{aligned}
\tag{20}
$$

where Σ is the expected matrix of the covariance matrix for the basic space set Y.

The fusion distance is the combination of Mahalanobis distance and Euclidean distance, taking into account the independence and relevance of the characteristic variables, which can effectively improve the accuracy of distance calculation [09]. The specific formula for calculating fusion distance is expressed as

$$
\begin{cases}
d_{Mix} = \omega \times MD(X,Y) + (1-\omega) \times ED(X,Y) \\
C_Y = \begin{bmatrix}
\rho_{Y_1 Y_1} & \rho_{Y_1 Y_2} & \cdots & \rho_{Y_1 Y_n} \\
\rho_{Y_2 Y_1} & \rho_{Y_2 Y_2} & \cdots & \rho_{Y_2 Y_n} \\
\vdots & \vdots & \ddots & \vdots \\
\rho_{Y_n Y_1} & \rho_{Y_n Y_2} & \cdots & \rho_{Y_n Y_n}
\end{bmatrix} \\
\omega = \sqrt{1 - |C_Y|}
\end{cases}
\tag{21}
$$

where d_{Mix} represents the fusion distance, MD represents the Mahalanobis distance, C_Y represents the correlation coefficient matrix for the sample set Y, n is the size of sample set Y, Y_i $(i = 1, \cdots, n)$ is the corresponding elements of sample set Y, and ρ is the correlation coefficient. The fusion distance is fused by the weight ω with the relevant information, and the Euclidean distance is fused by $1 - \omega$.

3.4. Particle Swarm Optimization

Particle swarm optimization (PSO) is an optimization algorithm proposed by American psychologist Kennedy and electrical engineer Eberhart in 1995. The update formula of the velocity and position of the particle in dimensional space is described as follows,

$$
\begin{cases}
v_{ip,t+1}^d = \omega \times v_{ip,t}^d + c1 \times rand \times (p_{ip,t}^d - x_{ip,t}^d) \\
\quad + c2 \times rand \times (p_t^g - x_{ip,t}^d) \\
x_{ip,t+1}^d = x_{ip,t}^d + v_{ip,t+1}^d
\end{cases}
\tag{22}
$$

where $ip \in [1,2,...,N]$ is ip th particle of the particle population; N represents the size of particle population; $d \in [1,2,...,D]$ is the d th dimension of the particle; D represents the number of dimension; $t \in [1,2,...,T]$ is the t th iteration; T represents the number of iteration; $c1$ and $c2$ represents the acceleration constants; $rand$ is the random real number of the interval $(0,1)$; ω is the weight coefficient, which is used to balance the degree of global search and local search; the position vector is represented as $\overrightarrow{X_{ip}} = (x_{ip,1}, x_{ip,2}, \cdots, x_{ip,d}, \cdots, x_{ip,D})$; the velocity vector is expressed as $\overrightarrow{V_{ip}} = (v_{ip,1}, v_{ip,2}, \cdots, v_{ip,d}, \cdots, v_{ip,D})$; the optimal location of the particles' individual is recorded as $\overrightarrow{P_{ip,t}} = (p_{ip,t}^1, p_{ip,t}^2, \cdots, p_{ip,t}^d, \cdots, p_{ip,t}^D)$; the best position of the swarm is denoted as $\overrightarrow{P_{g,t}} = (p_{g,t}^1, p_{g,t}^2, \cdots, p_{g,t}^d, \cdots, p_{g,t}^D)$.

3.5. Dual-Population Evolution Mechanism

Due to the limitations of both the evolutionary environment and the initial population, the problem of slow evolution and even stagnant evolution will appear during late iteration [40]. In the long process of iteration, the optimal individual will dominate all the population to some extent, making it difficult to converge globally. The proposal of dual-population strategy makes the optimal individuals of two populations exchange with each other, and the long-term dominance of the optimal individual in the original population is easily lost due to the change of the population environment [41,42].

To improve the optimization performance of SSO algorithm, an improved strategy combining genetic algorithm, particle swarm algorithm, and SSO algorithm based on dual-population Evolution Mechanism is proposed in this paper. In the process of iteration, the SSO algorithm brings each individual of the shark population into the optimal position rapidly. Particle swarm algorithm has the same defect of local convergence as SSO algorithm due to its fixed foraging behavior. At the same time, the crossover and mutation of genetic algorithm can prevent the SSO algorithm based on evolutionary from stalling immediately when it falls into the local optimum, and it can cause certain disturbance and mutation, which can help the SSO algorithm to jump out of the local optimum dilemma. In addition, the composite angle cosine is used as the evaluation index and the parallel evolutionary mechanism of dual-population is adopted to prevent the population from being dominated by extreme individuals. The flowchart of improved shark smell optimization algorithm proposed in this paper is shown in Figure 4.

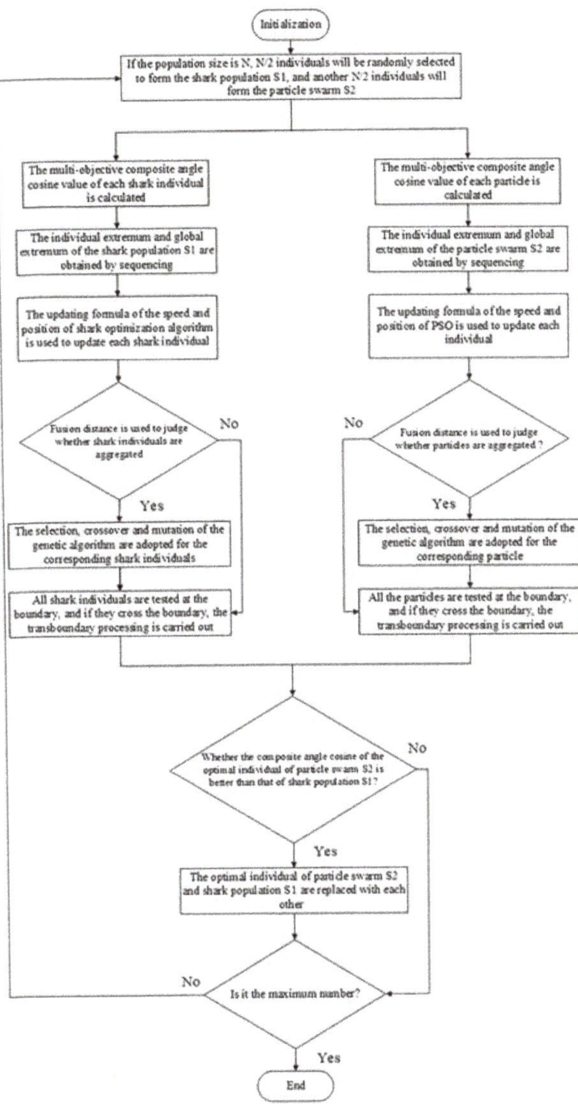

Figure 4. The flowchart of improved shark smell optimization algorithm proposed in this paper.

As can be seen form Figure 4, the dual-population strategy uses two populations (shark population S_1 and particle swarm S_2) to evolve at the same time, and compares the optimal individuals of the two populations, so as to break the equilibrium state in the population and to jump out of the local optimum.

In the optimization process, some generated particle may be beyond the boundary, the specific truncated formula is described as follows,

$$\neg x_i^k > x_{max} \Rightarrow x_i^k = x_{max}$$
$$\neg x_i^k < x_{min} \Rightarrow x_i^k = x_{min}$$

(23)

where $\neg ... \Rightarrow ...$ represents the sign of "if ... then ..." ; x_i^k is the updated particle; x_{max} is the maximum value of particle boundary; x_{min} is the minimum value of particle boundary.

4. Experimental Simulation

4.1. Data and Parameters for Experimental Simulation

In this paper, the scenarios about Jinpu line No.1 and rail transit line No.12 in Dalian, China are selected as the research objects. Jinpu line No.1 is the intercity railway line that is 46.76 km long and has 11 stations in the initial stage, which extend from Jiuli light rail station to the terminal of Zhenxing road in Dalian under construction. Rail transit line No.12 is an urban rail transit line that is 40.38 km long and has 8 stations, which extend from Hekou station to terminal of Lvshun New Port. The simulation running line of scenario about Jinpu line No.1 is from the Jiuli to the 19th bureau, and the interval length between the above two station is 2.74 km. The running simulation line of scenario about rail transit line No.12 is from Lvshun New Port to Tieshan Town, and the interval length between the above two station is 2.94 km, with two long downhill and a long uphill ramps. The main parameters of the above scenarios are shown in Tables 2 and 3, ramp parameters and velocity limit for automatic train operation are shown in Figure 5.

Table 2. The main parameters of the scenario about Jinpu line No.1 in Dalian.

Parameter Name	Parameter Characteristics
Train weight (t)	209
Maximum running speed (km/h)	80
Formation plan	2 motor 2 trail
Mean starting acceleration (m/s^2)	(0~35 km/h) $\geq$ 1.0
Mean acceleration (m/s^2)	(0~80 km/h) $\geq$ 0.6
Mean braking deceleration (m/s^2)	(80~0 km/h) $\geq$ 1.0

Table 3. The main parameters of the scenario about rail transit line No.12 in Dalian.

Parameter Name	Parameter Characteristics
Train weight (t)	211
Maximum running speed (km/h)	80
Formation plan	2 motor 2 trail
Mean starting acceleration (m/s^2)	(0~35 km/h) $\geq$ 1.0
Mean acceleration (m/s^2)	(0~80 km/h) $\geq$ 0.6
Mean braking deceleration (m/s^2)	(80~0 km/h) $\geq$ 1.0

The calculation formula of traction characteristics is expressed as follows,

$$\begin{cases} F(v) = F_{max} & v_q \leq v \leq v_c \\ F(v) = P_{max}/v & v_c \leq v \leq v_d \\ F(v) = P_{max} \times v_d/v^2 & v_d \leq v \leq v_{max} \end{cases}$$

(24)

where $F(v)$ is the instantaneous traction of vehicles, F_{max} is the vehicle's maximum traction, P_{max} is the maximum traction power of the vehicle, v_d is the switching velocity of the constant power zone and the reduced power zone, v_q is the switching velocity of the traction starting region and constant torque region, and v_{max} is the maximum design speed of the vehicle.

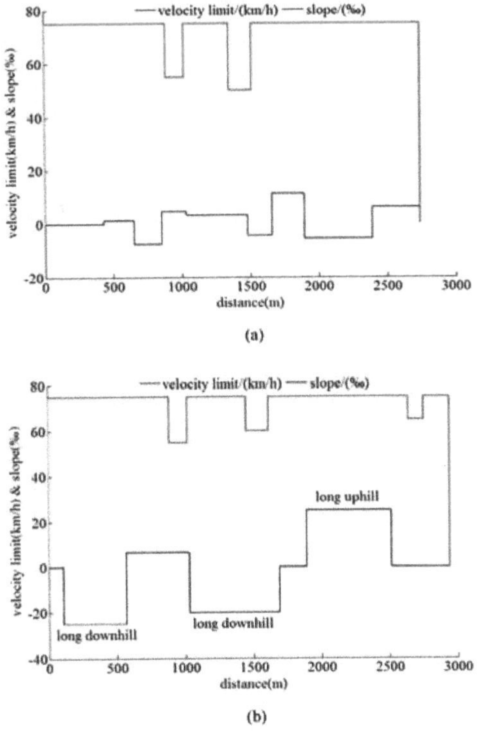

Figure 5. The ramp parameters and the speed limit of experiment simulation for two ATO scenarios. (a) Jinpu line No.1 in Dalian scenario. (b) rail transit line No.12 in Dalian scenario.

Simulation optimization results must satisfy the following conditions; the train instantaneous speed cannot surpass the speed limit, the train must finish the whole process, and the parking error is less than 0.2 m. The improved shark smell optimization algorithm parameters are set as follows; the particle swarm size is 40, the weight coefficient is 0.9, the acceleration coefficients C1 and C2 are 0.5, crossover probability is 0.7, mutation probability is 0.09, selection probability is 0.5, the iteration number is 200, the shark population size is 40, the random number $R1 = 0.4$, $R2 = 0.3$, $R3 = 0.25$, $\eta_k = 0.2$, speed limit factor β_k is 1.3, and the weight coefficient α_k is 0.15. Setting parameters for the optimization algorithm is necessary to consider the convergence speed and optimization effect, such as population size and iteration number, with the increase of population size, the global search ability of the optimization algorithm will be enhanced, but the convergence speed will be reduced; similarly, with the increase of the number of iterations, the finding opportunity of optimal solution will be increase, and the more computing time and resource will be expend. The parameters characteristics and multiple experimental simulation results are taken into account for the above parameters setting. The multi-objective optimization parameters of the ATO scenario simulation of Jinpu line No.1 in Dalian are set as follows; the scheduled running time is 177s; $K_E \in [90,000, 150,000]$; $K_T \in [0, 0.2]$; $K_{Jerk} \in [6, 10]$; the intrinsic weight factors ω_1'; ω_2' and ω_3' are, respectively, 0.5, 0.3, and 0.2; the target demand vector is $[98,000, 6.2, 0.01]$. The multi-objective optimization parameters of the ATO scenario simulation of rail transit line No.12 in Dalian are set as follows; the scheduled running time is 180 s, $K_E \in [80,000, 130,000]$, $K_T \in [0, 0.2]$, $K_{Jerk} \in [5, 10]$, the intrinsic weight factors ω_1', ω_2', and ω_3' are, respectively, 0.5, 0.3, and 0.2, the target demand vector is $[90,000, 5.2, 0.01]$.

4.2. Matlab/simulink Simulation Results for Automatic Train Optimization Scenarios

According to the automatic train operation scenarios of rail transit line No.12 and Jinpu line No.1 in Dalian, China, the approximate optimal solutions are obtained by using the improved algorithm proposed in this paper, traditional improved shark smell optimization algorithm (chaotic shark smell optimization) [17] and traditional improved particle swarm optimization (dynamic multiple populations particle swarm optimization algorithm based on decomposition) [43], the above Matlab/simulink platform is chosen as a simulation platform. The specific configuration of the Matlab/simulink platform used in this paper is described as follow: the Matlab/simulink revision is Matlab GUI 2016b; the major computer configuration is "CPU Core i9-7920X @ 2.9 GHZ" and "Windows 10". The specific Matlab/simulink optimization simulation results are shown in Figures 6 and 7 and Tables 4–7, and three different algorithms are recorded as ISSO, CSSO, and dMOPSO.

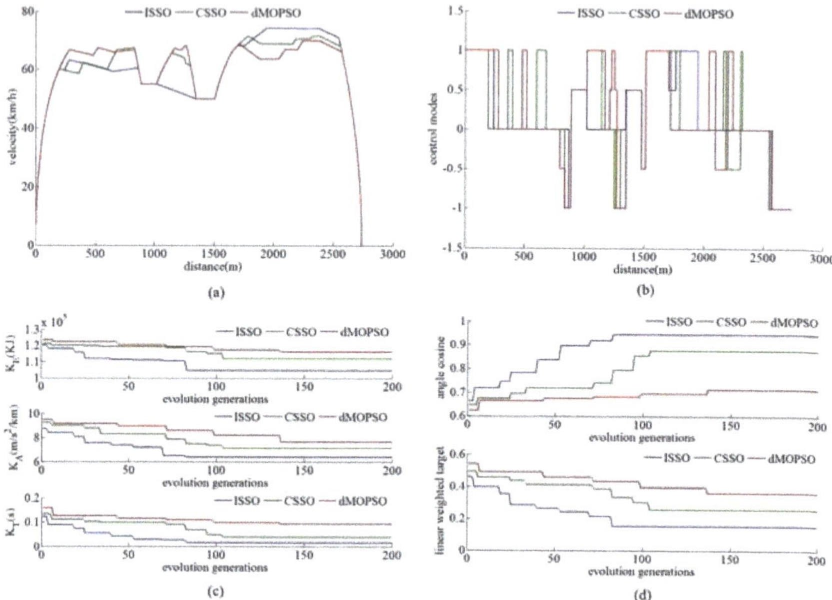

Figure 6. The Matlab/Simulink optimization curves of different algorithms for Jinpu line No.1 in Dalian scenario. (**a**) Jinpu line No.1 in Dalian scenario. (**a**) Target velocity trajectory profiles. (**b**) Operating condition distance curves. (**c**) Iterative convergence curves of each optimization objectives. (**d**) Iterative convergence curves of unified goals.

Table 4. The Matlab/simulink optimization results of different algorithms for Jinpu line No.1 in Dalian scenario.

Algorithm	Energy Consumption	Actual Time	Comfort Level
ISSO	104,739 KJ	177.0153 s	6.392 m/s^2/km
CSSO	110,910 KJ	177.0387 s	7.142 m/s^2/km
dMOPSO	116,157 KJ	177.0924 s	7.655 m/s^2/km

Table 5. The Matlab/simulink evaluate results of different algorithms for Jinpu line No.1 in Dalian scenario.

Algorithm	Angle Cosine	Linear Weighted Target	Computation Time	Convergence Evolution Generations
ISSO	0.9446	0.1498	1722 s	96
CSSO	0.8760	0.2554	1833 s	126
dMOPSO	0.7129	0.3602	2585 s	137

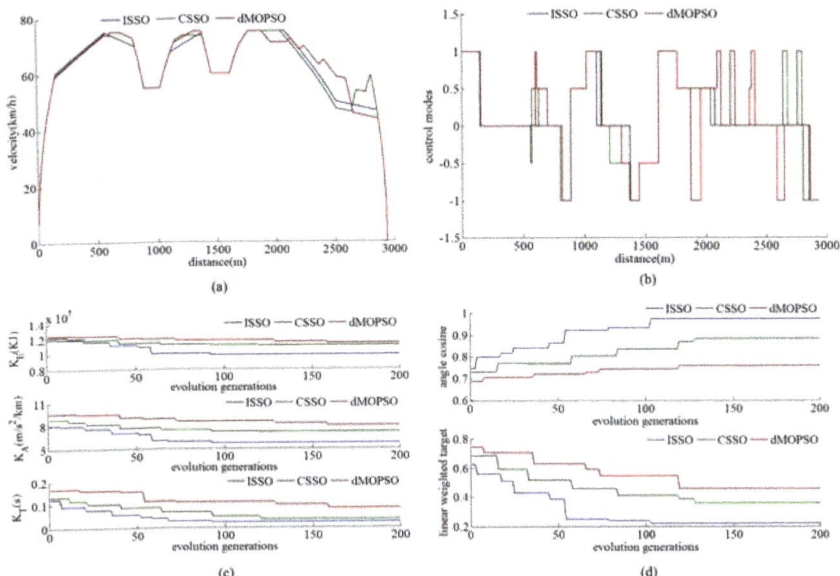

Figure 7. The Matlab/simulink optimization curves of different algorithms for rail transit line No.12 in Dalian scenario. (**a**) Target velocity trajectory profiles. (**b**) Operating condition distance curves. (**c**) Iterative convergence curves of each optimization objectives. (**d**) Iterative convergence curves of unified goals.

Table 6. The Matlab/simulink optimization results of different algorithms for rail transit line No.12 in Dalian scenario.

Algorithm	Energy Consumption	Actual Time	Comfort Level
ISSO	98,749 KJ	180.0139 s	5.638 m/s^2/km
CSSO	107,154 KJ	180.0394 s	6.537 m/s^2/km
dMOPSO	109,469 KJ	180.0878 s	7.408 m/s^2/km

Table 7. The HILS evaluation of the results of different algorithms for Jinpu line No.1 in Dalian scenario.

Algorithm	Angle Cosine	Linear Weighted Target	Computation Time	Convergence Evolution Generations
ISSO	0.9566	0.1776	4476 s	109
CSSO	0.8711	0.2759	4962 s	130
dMOPSO	0.8542	0.3501	7031 s	146

As can be seen in Tables 4–7, the optimization solution obtained by the improved ISSO is superior to that of CSSO and dMOPSO, and three indexes of energy saving, punctuality, and comfort have been improved considerably. The ATO experiment simulation scenario for Jinpu line No.1 from Jiuli to the

19th bureau is a typical slope with ups and downs. It is necessary to keep the train at high speed before driving down the long down ramp. The ATO experiment simulation scenario for rail transit line No.12 from Lvshun New Port to Tieshan Town is in the hilly of Dalian Lvshunkou district, and the hilly region is the typical characteristics of Dalian. When the train is running in such a terrain, the control sequence should be concise. The train speeds up in the long down slope and slows down in the long up slope as much as possible, which can save energy and avoid turbulence. As can be seen from Figures 6a,b and 7a,b (the target velocity trajectory and control sequence), the improved algorithm ISSO can obtain extremely simple control sequence and make the most of long down and up slopes, so as to obtain the target velocity trajectory as smooth as possible. As can be seen from Figures 6c,d and 7c,d (the iterative convergence curves), the convergence rate of ISSO is faster than that of CSSO and dMOPSO. Even in the late iterations, ISSO has a strong ability of global convergence performance.

Compared with other comparison optimization algorithms, ISSO has several obvious advantages in matlab/simulation environment, as there is huge difference between matlab/simulation environment and actual scenario yet, the effectiveness of ISSO is necessary to further test and verify.

4.3. HILS Results for Automatic Train Operation Scenarios

Matlab/Simulink is a simulation technology that is completely separated from the real train operation environment. Therefore, some problems (such as real-time velocity sampled inaccurately, signal delay and packet loss in transmission, a certain degree of unstable in tracking control system, etc.) need to be considered in the actual control process and are cannot be truly reflected. To more effectively test the performance of the optimization algorithm in the actual train operation environment, dSPACE hardware-in-the-loop simulation (HILS) technology is adopted to write the verified optimization algorithm or control algorithm to the chip of the optimizer or controller. Compared to traditional simulation platform based on pure software environment, dSPACE HILS platforms contain the real on-board equipments, which can truly reflect the the real situation for automatic train operation. Due to the restriction of funds and experiment conditions, it is difficult to implement the corresponding actual automatic train operation experiment. Moreover, compared with real actual experiments, dSPACE HILS has the advantages of low experimental cost, implement easily, and high security protection of personal and equipment [33]. Therefore, HILS is highly valued by researchers and developers, abundant HILS-related products are researched, developed, and applied in various fields and numerous relative research results have been achieved [44,45]. The semi-physical simulation equipments in automatic train operation mainly include optimizer, controller, sensors, simulator, conditioning circuit, signal conditioning unit, and connector. The simulation topology diagram of automatic train operation HILS platform, the structure diagram for automatic train operation HILS, and the physical diagram of controller cabinet and simulation cabinet for automatic train operation HILS are shown in Figures 8 and 9.

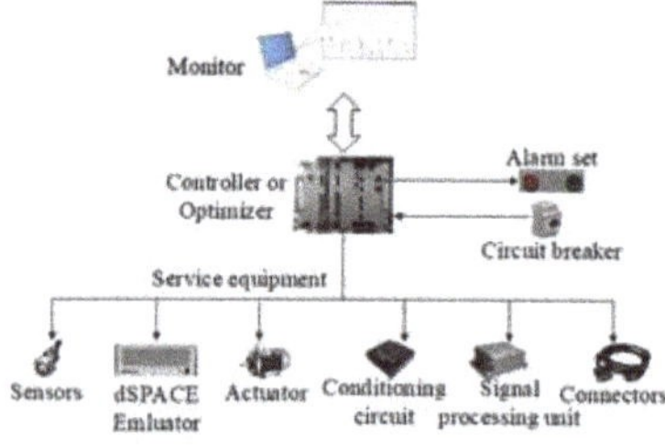

Figure 8. The simulation topology diagram of automatic train operation hardware-in-the-loop simulation (HILS) platform.

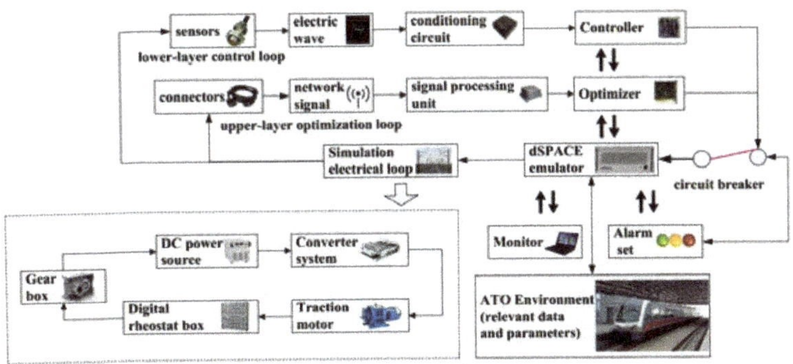

Figure 9. The structure diagram for automatic train operation HILS.

As can be seen from Figure 8, the Controller/Optimizer and its Service equipment make up the automatic train operation HILS platform, the Monitor serves as the monitoring center for human–computer interaction, and the Alarm set and Circuit breaker are the warning and protection center. Alarm set detects the working status of each HILS device in real-time, and the Circuit breaker is used to protect equipment by breaking when a certain kind of exception is detected. As can be seen from Figure 9, the automatic train operation HILS platform contains various actual hardware, and simulation hardware for automatic train operation. The "lower-layer control loop" based on controller and "upper-layer optimization loop" based on optimizer of HILS constitute the two independent communication systems, respectively, and there is a certain connection between the two loops. "Controller" is also named as traction control unit (TCU), which could provide control commands for corresponding equipments in real-time using a proper control algorithm in real-time, enable the the urban rail vehicle to track the ideal profile curve; "Optimizer" is also named as main processor unit (MPU), which could provide the velocity ideal trajectory profile (target speed curve of automatic train operation) for "lower-layer control loop" tracking control based on 'dSPACE emulator'. Second, the "dSPACE emulator", "conditioning circuit", "signal processing unit", "sensors", "connectors", and so on are service equipments for ATO HILS: "dSPACE emulator" provides some correlative simulation environments for the automatic train operation HILS, the related models included such as accurate braking model, traction transformer model, running line model, velocity fluctuation model, etc.; "conditioning circuit" can regulate electrical waves properly for "Controller" appropriately; "signal processing unit" can regulate net signals properly for "Optimizer" appropriately; and the "sensors" and "connectors" are used to feed electrical waves and net signals back to the "Controller" and "Optimizer" in real-time. Third, the "DC power source", "Converter system", "Electric motor", "Digital rheostat box", and "Gear box" are simulation electric hardware equipments of in place of the actual: "DC power source" acts as the actual pantograph; "Converter system" transfer the electric energy from "DC power source" to "Electric motor" through a series of current-voltage conversion processes, it includes AC–DC converter, DC–AC converter, low-pass filters, etc.; "Electric motor" acts as the actual urban rail vehicle motors set; the capacity of simulation circuit loop is smaller than actual. In Figure 11, "train controller cabinet" and "train emulator cabinet" are the vital equipments for automatic train operation HILS, except controller and emulator, conditioning circuit, signal processing unit and corresponding switch groups are included.

Obviously, compared with these automatic train operation scenarios based on pure software (such as Matlab/simulink simulation), their identical scenarios based on HILS are closer to actual. Therefore, based on the same scenarios, tracking control algorithm and HILS platform, the comprehensive performance index for automatic train operation obtained by optimization algorithms could reflect the optimization performance of these algorithms effectively. To further verify the effectiveness of the algorithm, according to the automatic train operation scenarios of rail transit line No.12 and Jinpu line No.1 in Dalian, China, the approximate optimal solutions are obtained by using ISSO, CSSO, and dMOPSO, the above ATO HILS platform is chosen as simulation platform. The specific configuration of the ATO HILS platform used in this paper is described as follow: the Matlab/simulink revision is Matlab GUI 2016b; the major computer configuration is "CPU Core i9-7920X @ 2.9 GHZ" and "Windows 10"; the core chip of "Controller" and "Optimizer" is "TMS320F28335"; the simulation software of "dSPACE emulator" is dSPACE software(revision is control desk 6.1); the communication protocol of the ATO HILS is MVB (multifunction vehicle bus); the MPC (model predictive control) algorithm is adopted as tracking control algorithm; the three optimization algorithms (ISSO, CSSO, and dMOPSO) used are written in the kernel chip of "Optimizer" in "upper-layer optimization loop" for contrasting. The specific HILS optimization results obtained by "Optimizer" in "upper-layer optimization loop" are shown in Figures 10 and 11 and Tables 8–11.

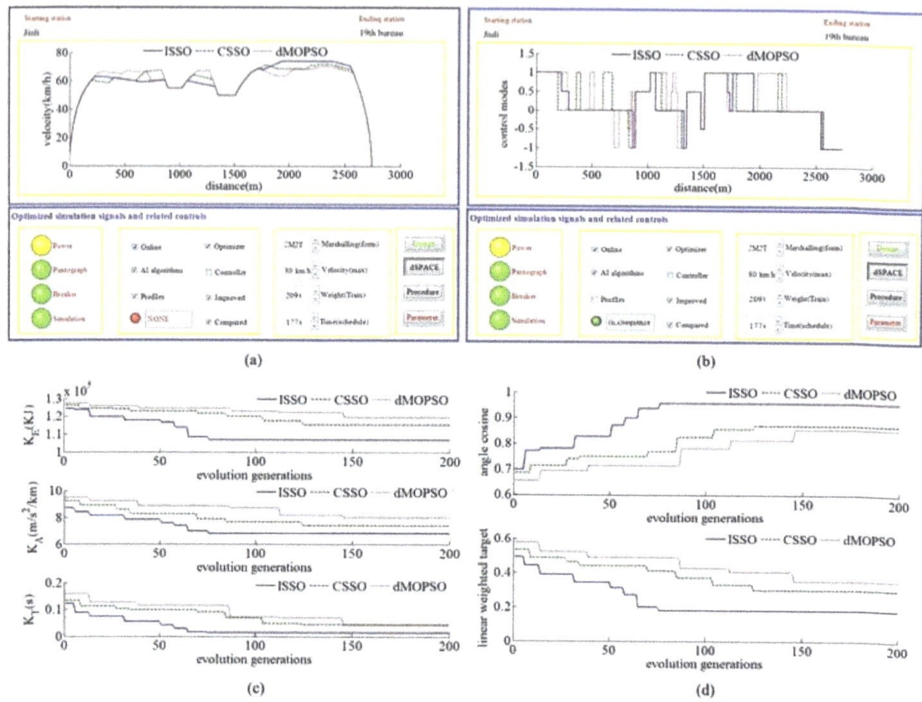

Figure 10. The HILS optimization curves of different algorithms for Jinpu line No.1 in Dalian scenario. (**a**) Target velocity trajectory profiles. (**b**) Operating condition distance curves. (**c**) Iterative convergence curves of each optimization objectives. (**d**) Iterative convergence curves of unified goals.

Table 8. The Matlab/simulink evaluate results of different algorithms for rail transit line No.12 in Dalian scenario.

Algorithm	Angle Cosine	Linear Weighted Target	Computation Time	Convergence Evolution Generations
ISSO	0.9722	0.2205	1935 s	114
CSSO	0.8829	0.3571	2194 s	134
dMOPSO	0.7564	0.4547	2896 s	119

Table 9. The HILS optimization results of different algorithms for Jinpu line No.1 in Dalian scenario.

Algorithm	Energy Consumption	Actual Time	Comfort Level
ISSO	106,741 KJ	177.0135 s	6.824 m/s^2/km
CSSO	115,245 KJ	177.0430 s	7.508 m/s^2/km
dMOPSO	119,569 KJ	177.0428 s	8.029 m/s^2/km

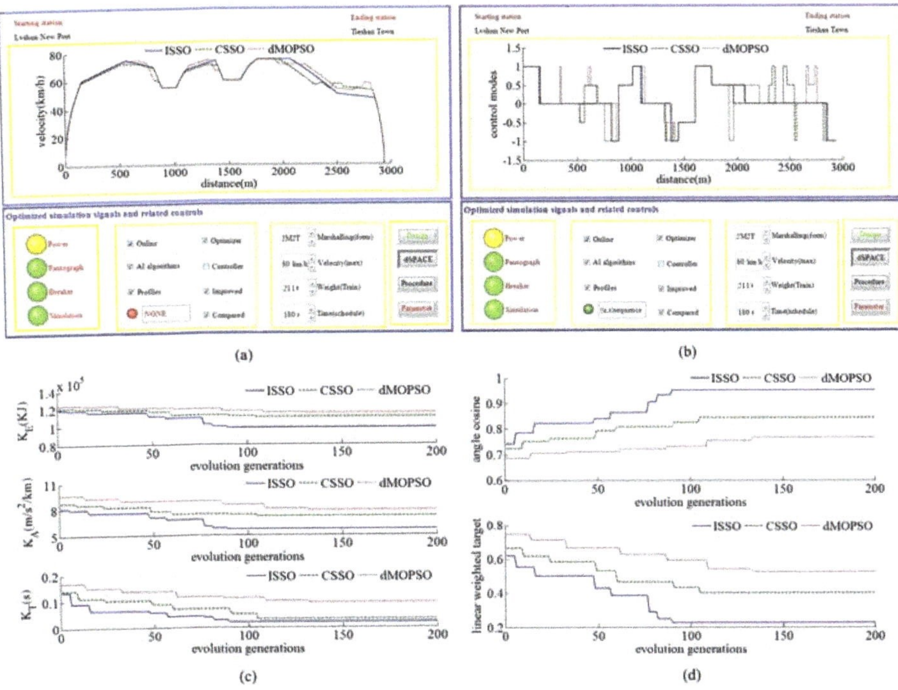

Figure 11. The HILS optimization curves of different algorithms for rail transit line No.12 in Dalian scenario. (**a**) Target velocity trajectory profiles. (**b**) Operating condition distance curves. (**c**) Iterative convergence curves of each optimization objectives. (**d**) Iterative convergence curves of unified goals.

Table 10. The HILS optimization results of different algorithms for rail transit line No.12 in Dalian scenario.

Algorithm	Energy Consumption	Actual Time	Comfort Level
ISSO	98,524 KJ	180.0194 s	5.712 m/s^2/km
CSSO	110,524 KJ	180.0294 s	7.194 m/s^2/km
dMOPSO	114,755 KJ	180.0921 s	7.844 m/s^2/km

Table 11. The HILS evaluate results of different algorithms for rail transit line No.12 in Dalian scenario.

Algorithm	Angle Cosine	Linear Weighted Target	Computation Time	Convergence Evolution Generations
ISSO	0.9515	0.2260	4365 s	99
CSSO	0.8441	0.4009	5027 s	126
dMOPSO	0.7682	0.5251	7885 s	133

In Figures 10a,b and 11a,b (which is related to the optimization effect of automatic train operation process), the power is switched on, the pantograph is raised, the dSPACE simulator is in the working state (the "dSPACE" button is pressed), the human–computer interaction signal is normal (the "Design" button is green), and the circuit breaker is normally closed, which is in a normal optimization simulation state.

According to optimization simulation results of different algorithms in Tables 8–11, compared with the traditional improved particle swarm optimization algorithm (dMOPSO) and traditional improved shark smell optimization algorithm (CSSO), the improved algorithm proposed in this paper (ISSO) has the obvious performance improvement effect, the three indexes of energy saving, punctuality and comfort of target velocity trajectory have been improved considerably; meanwhile, and the convergence evolution generations and computation time have been reduced considerably. As can be seen from Figures 10a,b and 11a,b, the ideal target velocity trajectory obtained by ISSO was the smoothest, compared with the contrasted algorithms (CSSO and dMOPSO), ISSO obtained extremely simple control sequence and made use of the most of long down and up slopes sufficiently, it enables the train to reduce unnecessary traction and braking status and to make full use of coasting status as much as possible. As can be seen from Figures 10c,d and 11c,d, compared with contrasted algorithms (CSSO and dMOPSO), the three optimization objective indexes and two unified goals obtained by ISSO have been improved to a considerable extent not only in the optimization effect but also in the computation efficiency.

Compared with the improving optimization effect of ATO target velocity trajectory optimization, improving the actual tracking control effect is more practical. The optimization effect calculation speed could estimate the performance of the ATO target velocity trajectory optimization algorithms, achievable difficulty for tracking control is also a significant evaluation index. To better verify the effectiveness of the ISSO, the "lower-layer control loop" is used to tracking control according to optimal target velocity trajectory generated by the "upper-layer control loop". The MPC (model predictive control) algorithm has some its own advantages such as good performance in tracking precision, powerful robustness, fast tracking speed, etc. and widely in ATO traction system, so it is selected as tracking control algorithm in this paper. The specific HILS tracking control results obtained by "Controller" in "lower-layer control loop" are shown in Figures 12 and 13 and Tables 12 and 13.

Table 12. The HILS tracking control results of different algorithms for Jinpu line No.1 in Dalian scenario.

Algorithm	Energy Consumption	Actual Time	Comfort Level
ISSO	117,259 KJ	177.0351 s	32.947 m/s^2/km
CSSO	134,845 KJ	177.1045 s	41.859 m/s^2/km
dMOPSO	138,672 KJ	177.1282 s	43.578 m/s^2/km

Table 13. The HILS tracking control results of different algorithms for rail transit line No.12 in Dalian scenario.

Algorithm	Energy Consumption	Actual Time	Comfort Level
ISSO	115,168 KJ	180.0397 s	37.027 m/s^2/km
CSSO	128,122 KJ	180.0823 s	44.254 m/s^2/km
dMOPSO	137,485 KJ	180.1465 s	47.966 m/s^2/km

In Figures 12 and 13 (which relate to the tracking control effect of automatic train operation process according to optimization result), the power is switched on, the pantograph is raised, the dSPACE simulator is in the simulation state (the "dSPACE" button is pressed), the design parameters cannot be changed (the "Design" button is white), and the circuit breaker is normally closed, which is in a normal tracking control simulation state.

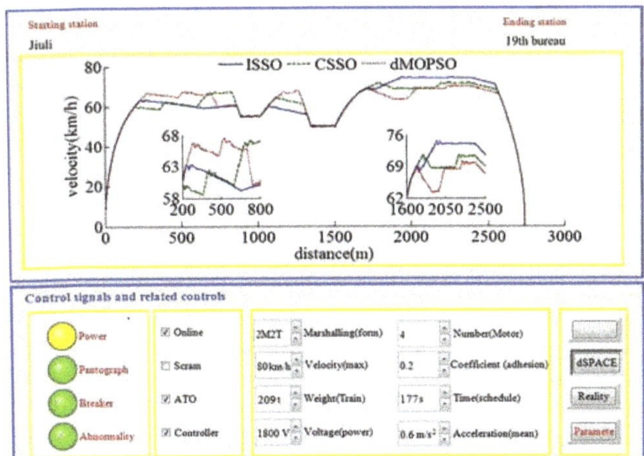

Figure 12. The HILS tracking control curves of different algorithms for Jinpu line No.1 in Dalian scenario.

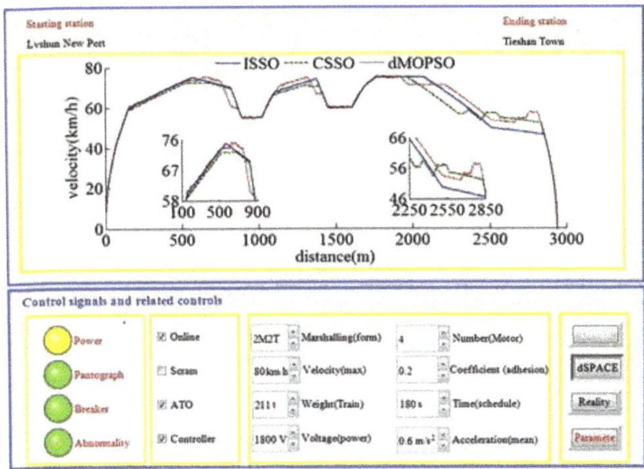

Figure 13. The HILS tracking control curves of different algorithms for rail transit line No.12 in Dalian scenario.

According to optimization simulation results of different algorithms from Tables 12 and 13, compared with the traditional improved particle swarm optimization algorithm (dMOPSO) and traditional improved shark smell optimization algorithm (CSSO), the corresponding tracking control curve tracking the target velocity trajectory, obtained by improved algorithm proposed in this paper (ISSO), has the obvious performance improvement effect, the its three indexes of energy-saving, punctuality, and comfort have been improved considerably. As can be seen from Figures 12 and 13, compared with the contrasted algorithms (CSSO and dMOPSO), the corresponding tracking control curve for ISSO has better tracking control effects. As can be seen from the enlarged areas of Figures 12 and 13, the velocity fluctuation of the velocity distance trajectory curves corresponding to ISSO is weaker, because the target velocity trajectory obtained by ISSO was more ideal and tracking controlled easily in the automatic train operation simulation scenarios.

To verify that the improved algorithm (ISSO) in this paper has more superiorities, such as optimization effect, calculation speed, easily achievable in tracking control, etc., two scenarios about rail transit line No.12 and Jinpu line No.1 in Dalian, China are chosen as optimized objects for

ISSO and some comparison algorithms, the corresponding optimization and tracking control results of matlab simulation and hardware-in-the-loop simulation (HILS) as shown in Tables 4–13 and Figures 5 and 6, and 10–13. Obviously, compared with the contrasted algorithms (CSSO and dMOPSO), the corresponding optimization and tracking control results obtained by ISSO are more ideal. It is indicate that ISSO is a appropriate algorithm with powerful optimization capability, so as to solving practical automatic train operation more effectively.

5. Conclusions

Automatic train operation target velocity trajectory optimization is a very complex issue that needs to take into account multiple objectives, and the ideal optimization solution is not easy to be obtained. An improved multi-objective dual-population shark smell optimization algorithm using incorporated composite angle cosine for automatic train operation velocity trajectory optimization is proposed in this paper, and the specific advantages are described as follows.

For the multi-objective optimization problem, the evaluation index of the solution is very important. Nonetheless, the linear weighted target belongs to the common multi-objective unified target, and there is a problem that subjective parameters are selected blindly. In this paper, the multi-objective angle cosine is used as evaluation index, which can effectively avoid the blind selection of subjective parameters. To make the evaluation index more reasonable, on the basis of the preference information, the composite angle cosine which takes into account both the preference difference and the numerical difference is be used as the evaluation index of the solution in this paper.

Because the updating rules of velocities and positions in the SSO and PSO make all individuals close to the extreme individuals, after a long iteration, the extreme individuals will form a certain degree of dominance over the population, which makes it difficult to converge globally. First, the dual-population evolution mechanism is used to jump out of the local optimum. Second, to suppress the local convergence of optimization algorithm, it is necessary to determine accurately whether the individual aggregation occurs in shark population. In this paper, the fusion distance can be used to take into account the relativity and independence of speed and time, which can detect whether there is the phenomenon of individual aggregation preciously, so the local convergence is better suppressed. At the same time, this paper also introduces the dual-population collaborative optimization mechanism of SSO algorithm and particle swarm algorithm to further improve the optimization performance of the algorithm.

According to the the Matlab/simulink results and HILS results about automatic train operation scenarios, compared with the conventional improved shark smell optimization algorithm and the conventional improved particle swarm optimization algorithm, the improved algorithm (ISSO) proposed in this paper improves the calculation accuracy and optimization ability of the optimization algorithm to some extent, so that more ideal target velocity trajectory can be obtained. This clearly shows that ISSO have been improved to a considerable extent for the automatic train operation target velocity trajectory optimization not only in the pure software scenarios but also in the semi-physical scenarios. As the automatic train operation HILS is close to its actual experiment, the problem that the ISSO has worse improvement optimization effect in the actual than anticipatory could be avoid to certain extent.

Author Contributions: The work presented here was performed in collaboration among all authors. L.W. designed, analyzed, and wrote the paper, and completed the simulation experiment. X.W. guided the full text and provided simulation conditions. Z.S. conceived idea and involved simulation experiment. S.L. involved simulation experiment and analyzed the data. All authors have read and agree to the published version of the manuscript.

Funding: This research was funded by The Nature Science Foundation of China (grand number 51609033 and 61773049).

Conflicts of Interest: The authors declare no conflicts of interest.

Abbreviations

The following abbreviations are used in this manuscript:

ATO	automatic train operation
HILS	hardware-in-the-loop simulation
ISSO	improved shark smell optimization
CSSO	chaotic shark smell optimization
dMOPSO	multi-objective particle swarm optimization based on decomposition

References

1. Wang, G.; Xiao, S.; Chen, X. Application of Genetic Algorithm in Automatic Train Operation. *Wirel. Pers. Commun.* **2018**, *102*, 1695–1704. [CrossRef]
2. Tan, Z.; Lu, S.; Bao, K.; Zhang, S.; Wu, C.; Yang, J.; Xue, F. Adaptive Partial Train Speed Trajectory Optimization. *Energies* **2018**, *11*, 3302. [CrossRef]
3. Gu, Q.; Tang, T.; Cao, F.; Song, Y. Energy-Efficient Train Operation in Urban Rail Transit Using Real-Time Traffic Information. *IEEE Trans. Intell. Transp. Syst.* **2014**, *15*, 1216–1233. [CrossRef]
4. Bai, Y.; Ho, T.K.; Mao, B.; Ding, Y.; Chen, S. Energy-Efficient Locomotive Operation for Chinese Mainline Railways by Fuzzy Predictive Control. *IEEE Trans. Intell. Transp. Syst.* **2014**, *15*, 938–948. [CrossRef]
5. Song, Y.; Song, W. A Novel Dual Speed-Curve Optimization Based Approach for Energy-Saving Operation of High-Speed Trains. *IEEE Trans. Intell. Transp. Syst.* **2016**, *17*, 1564–1575. [CrossRef]
6. Domínguez, M.; Fernández-Cardador, A.; Cucala, A.P.; Pecharromán, R.R. Energy Savings in Metropolitan Railway Substations Through Regenerative Energy Recovery and Optimal Design of ATO Speed Profiles. *IEEE Trans. Autom. Sci. Eng.* **2012**, *9*, 496–504. [CrossRef]
7. Shangguan, W.; Yan, X.; Cai, B.; Wang, J. Multiobjective Optimization for Train Speed Trajectory in CTCS High-Speed Railway With Hybrid Evolutionary Algorithm. *IEEE Trans. Intell. Transp. Syst.* **2015**, *16*, 2215–2225. [CrossRef]
8. Huang, Y.; Yang, L.; Tang, T.; Cao, F.; Gao Z. Saving Energy and Improving Service Quality: Bicriteria Train Scheduling in Urban Rail Transit Systems. *IEEE Trans. Intell. Transp. Syst.* **2016**, *17*, 3364–3379. [CrossRef]
9. Fernandez-Rodriguez, A.; Fernández-Cardador, A.; Cucala, A.P.; Domínguez, M.; Gonsalves, T. Design of Robust and Energy-Efficient ATO Speed Profiles of Metropolitan Lines Considering Train Load Variations and Delays. *IEEE Trans. Autom. Sci. Eng.* **2015**, *16*, 2061–2071. [CrossRef]
10. Bocharnikov, Y.V.; Tobias, A.M.; Roberts, C.; Hillmansen, S.; Goodman, C.J.; Optimal driving strategy for traction energy saving on DC suburban railways. *IET Electr. Power Appl.* **2012**, *181*, 51–62. [CrossRef]
11. Gu, Q.; Tang, T.; Ma, F. Energy-Efficient Train Tracking Operation Based on Multiple Optimization Models. *IEEE Trans. Intell. Transp. Syst.* **2016**, *17*, 882–892. [CrossRef]
12. Zhou, Y.; Tao, X. Robust Safety Monitoring and Synergistic Operation Planning Between Time and Energy-Efficient Movements of High-Speed Trains Based on MPC. *IEEE Access* **2018**, *6*, 17377–17390. [CrossRef]
13. Kunimatsu, T.; Hirai, C.; Tomii, N. Train timetable evaluation from the viewpoint of passengers by microsimulation of train operation and passenger flow. *Electr. Eng. Jpn.* **2012**, *181*, 51–62. [CrossRef]
14. Abedinia, O.; Amjady, N.; Ghasemi, A. A new metaheuristic algorithm based on shark smell optimization. *Complexity* **2014**, *21*, 97–116. [CrossRef]
15. Gnanasekaran, N.; Chandramohan, S.; Kumar, P. S.; Imran, A.M. Optimal placement of capacitors in radial distribution system using shark smell optimization algorithm. *Ain Shams Eng. J.* **2016**, *7*, 907–916. [CrossRef]
16. Ahmadigorji, M.; Amjady, N. A multiyear DG-incorporated framework for expansion planning of distribution networks using binary chaotic shark smell optimization algorithm. *Energy* **2016**, *102*, 199–215. [CrossRef]
17. Abedinia, O.; Amjady, N. Short-term wind power prediction based on Hybrid Neural Network and chaotic shark smell optimization. *Int. J. Precis. Eng. Manuf.-Green Technol.* **2017**, *5*, 13062–13076. [CrossRef]
18. Sicre, C.; Asunción, P.C.; Antonio, F.; Lukaszewicz, P. Modeling and optimizing energy-efficient manual driving on high-speed lines. *IEEJ Trans. Electr. Electron. Eng.* **2012**, *7*, 633–640. [CrossRef]

19. Liang, Y.; Chen, D.; Li, L. Intelligent Train Operation Algorithms for Subway by Expert System and Reinforcement Learning. *IEEE Trans. Intell. Transp. Syst.* **2014**, *15*, 2561–2571.

20. Cheng, R.; Yu, W.; Song, Y.; Chen, D.; Ma, X.; Cheng, Y. Intelligent Safe Driving Methods Based on Hybrid Automata and Ensemble CART Algorithms for Multihigh-Speed Trains. *IEEE Trans. Cybern.* **2019**, *49*, 3816–3826. [CrossRef]

21. Meng, J.; Xu, R.; Li, D.; Chen, X. Combining the Matter-Element Model With the Associated Function of Performance Indices for Automatic Train Operation Algorithm. *IEEE Trans. Intell. Transp. Syst.* **2019**, *20*, 253–263. [CrossRef]

22. Ma, J.; Fan, Z.; Jiang, Y.; Mao, J. An optimization approach to multiperson decision-making based on different formats of preference information. *IEEE Trans. Syst. Man, Cybern. Part A Syst. Hum.* **2006**, *36*, 876–889.

23. Kaddani, S.; Vanderpooten, D.; Vanpeperstraete, J.M.; Aissi, H. Weighted sum model with partial preference information: Application to Multi-Objective Optimization. *Eur. J. Oper. Res.* **2017**, *26*, 665–679. [CrossRef]

24. Nelyubin, A.P.; Podinovski, V.V. Multicriteria choice based on criteria importance methods with uncertain preference information. *Comput. Math. Math. Phys.* **2016**, *57*, 1475–1483. [CrossRef]

25. Mayet, C.; Delarue, P.; Bouscayrol, A.; Chattot, E. Hardware-In-the-Loop Simulation of Traction Power Supply for Power Flows Analysis of Multi-Train Subway Lines. *IEEE Trans. Veh. Technol.* **2017**, *66*, 5564–5571. [CrossRef]

26. De Farias, A.B.C.; Rodrigues, R.S.; Murilo, A.; Lopes, R.V.; Avila, S. Low-Cost Hardware-in-the-Loop Platform for Embedded Control Strategies Simulation. *IEEE Access* **2019**, *7*, 111499–111512. [CrossRef]

27. Terwiesch, P.; Keller, T.; Scheiben, E. Rail vehicle control system integration testing using digital hardware-in-the-loop simulation. *IEEE Trans. Control Syst. Technol.* **1999**, *7*, 352–362. [CrossRef]

28. Yang, X.; Yang, C.; Peng, T.; Liu, B.; Gui, W. Hardware-in-the-Loop Fault Injection for Traction Control System. *IEEE J. Emerg. Sel. Top. Power Electron.* **2018**, *6*, 696–706. [CrossRef]

29. Zhang, H.; Zhang, Y.; Yin, C. Hardware-in-the-Loop Simulation of Robust Mode Transition Control for a Series-Parallel Hybrid Electric Vehicle. *IEEE Trans. Veh. Technol.* **2016**, *63*, 1059–1069. [CrossRef]

30. Abdelrahman, A.; Algarny, K.; Youssef, M. A Novel Platform for Powertrain Modeling of Electric Cars With Experimental Validation Using Real-Time Hardware in the Loop (HIL): A Case Study of GM Second Generation Chevrolet Volt. *IEEE Trans. Power Electron.* **2018**, *33*, 9762–9771. [CrossRef]

31. Cheng, J; Howlett, P. Application of critical velocities to the minimisation of fuel consumption in the control of trains. *Automatica* **1992**, *28*, 165–169. [CrossRef]

32. Wang, P.; Goverde, R.M.P. Multiple-phase train trajectory optimization with signalling and operational constraints. *Transp. Res. Part C Emerg. Technol.* **2016**, *69*, 255–275. [CrossRef]

33. Chen, D.; Chen, R.; Li, Y.; Tang, T. Online Learning Algorithms for Train Automatic Stop Control Using Precise Location Data of Balises. *IEEE Trans. Intell. Transp. Syst.* **2013**, *14*, 1526–1535. [CrossRef]

34. Howlett, P.; Cheng, J. Optimal driving strategies for a train on a track with continuously varying gradient. *J. Aust. Math. Soc.* **1997**, *38*, 388–410. [CrossRef]

35. Liang, Y.; Liu, H.; Qian, C. A Modified Genetic Algorithm for Multi-Objective Optimization on Running Curve of Automatic Train Operation System Using Penalty Function Method. *Int. J. Intell. Transp. Syst. Res.* **2019**, *17*, 74–87. [CrossRef]

36. Cheng, H.; Huang, W.; Zhou, Q.; Cai, J. Solving fuzzy multi-objective linear programming problems using deviation degree measures and weighted max-min method. *Appl. Math. Model.* **2013**, *37*, 6855–6869. [CrossRef]

37. Fan, Z.; Liu, Z.; Li, X.; Gao, Z.; Su, X. Cross-study validation and combined analysis of microarray data for cancer using vector cosine angle method. In Proceedings of the 2005 IEEE Engineering in Medicine and Biology 27th Annual Conference, Shanghai, China, 17–18 January 2006; pp. 4810–4813.

38. Xu, Z. Multiple-attribute group decision-making with different formats of preference information on attributes. *IEEE Trans. Syst. Man Cybern. Part B* **2007**, *37*, 1500–1511.

39. Liu, G.; Wang, X. Fault diagnosis of diesel engine based on fusion distance calculation. In Proceedings of the Advanced Information Management, Communicates, Electronic and Automation Control Conference, Xi'an, China, 3–5 October 2016; pp. 1621–1627.

40. Zhou, J.; Wang, C.; Li, Y.; Wang, P.; Li, C.; Lu, P.; Mo, L. A multi-objective multi-population ant colony optimization for economic emission dispatch considering power system security. *Appl. Math. Model.* **2017**, *45*, 684–704. [CrossRef]

41. Park, T.; Ryu, K.R. A Dual-Population Genetic Algorithm for Adaptive Diversity Control. *IEEE Trans. Evol. Comput.* **2010**, *14*, 865–884. [CrossRef]

42. Xu. M.; You, X.; Liu, S. A novel Heuristic communication Heterogeneous dual population Ant Colony Optimization algorithm. *IEEE Access* **2017**, *5*, 18506–18515. [CrossRef]

43. Peng, H.; Li, R.; Cao, L.; Li, L. Multiple Swarms Multi-Objective Particle Swarm Optimization Based on Decomposition. *Procedia Eng.* **2011**, *15*, 3371–3375.

44. Hasanzadeh, A.; Edrington, C.S.; Stroupe, N.; Bevis, T. Real-Time Emulation of a High-Speed Microturbine Permanent-Magnet Synchronous Generator Using Multiplatform Hardware-in-the-Loop Realization. *IEEE Trans. Ind. Electron.* **2014**, *61*, 3109–3118. [CrossRef]

45. Nariman, R.T.; Venkata, D. A General Framework for FPGA-Based Real-Time Emulation of Electrical Machines for HIL Applications. *IEEE Trans. Ind. Electron.* **2015**, *62*, 2041–2053.

Article

A Method for the Optimization of Daily Activity Chains Including Electric Vehicles

Dimitrios Rizopoulos * and **Domokos Esztergár-Kiss**

Department of Transport Technology and Economics (KUKG), Faculty of Transportation Engineering and Vehicle Engineering (KJK), Budapest University of Technology and Economics (BME), 1111 Budapest, Hungary; esztergar@mail.bme.hu
* Correspondence: dimrizopoulos@gmail.com; Tel.: +30-697-865-6250

Received: 29 January 2020; Accepted: 14 February 2020; Published: 18 February 2020

Abstract: The focus of this article is to introduce a method for the optimization of daily activity chains of travelers who use Electric Vehicles (EVs) in an urban environment. An approach has been developed based on activity-based modeling and the Genetic Algorithm (GA) framework to calculate a suitable schedule of activities, taking into account the locations of activities, modes of transport, and the time of attendance to each activity. The priorities of the travelers concerning the spatial and temporal flexibility were considered, as well as the constraints that are related to the limited range of the EVs, the availability of Charging Stations (CS), and the elevation of the road network. In order to model real travel behavior, two charging scenarios were realized. In the first case, the traveler stays in the EV at the CS, and in the second case, the traveler leaves the EV to charge at the CS while conducting another activity at a nearby location. Through a series of tests on synthetic activity chain data, we proved the suitability of the method elaborated for addressing the needs of travelers and being utilized as an optimization method for a modern Intelligent Transportation System (ITS).

Keywords: daily activity chains; electric vehicles; optimization; charging stations; intelligent transportation systems; ITS

---

1. Introduction

In most urban environments, there is an ever-growing need for navigation through transportation networks. Although more and more mobility services are offered to travelers, the complexity of their use has risen alongisde the offer and demand for such services. Thus, the utilization of Intelligent Transportation Systems (ITS) and the development of travel-related services have become an immense need for the facilitation of the everyday life of citizens.

With the expected rise of the use of Electric Vehicles (EVs) all over the world [1] and the positive impacts that they are expected to bring [2], advanced EV travel planning algorithms are now essential more than ever [3]. Alongside the growth of EVs on global markets and the impact that they will have on power grids, transportation networks, and the environment [4], the study for intelligent driver assistance solutions for EV users is critical. Not only their development and availability are vital, but also the user experience and the satisfaction of specific user preferences are two crucial points when creating such services [5]. Nonetheless, those solutions can help to mitigate the risk of electrification of transportation and deal with one of the enormous obstacles in EV adoption that is called range anxiety, a term used to describe the psychological worry of EV drivers that the remaining electric energy in the vehicle is not enough to attend their destination. While the battery capacity of the EVs is growing at a satisfactory rate and is even expected to double until 2030 [6], it is proven that range anxiety is still one of the most significant concerns of EV drivers and is one of the most critical obstacles towards universal adoption and mainstream use [7]. While it has been shown that EV users that have a charger available at home mainly cover their needs at a satisfactory rate, for users that solely depend on public charging infrastructure, charging can be a more challenging task [8].

Alongside the ability of those methods to address problems of EV drivers on an individual level, they can improve the situation for the system-level [3], too. Mitigation strategies on electrification on a regional level, as their importance as highlighted in recent articles [9], can benefit from driver assistance ITS, since the underlying methods can be independent of the spatial-context and are reproducible over several regions according to needs of travelers of those specific regions.

Apart from the work of Cuchý et al. [3], little research has been conducted to study the flexibility or willingness to use travel planning services for EV drivers. However, the authors believe that based on other previous research on other modes of transport, and mainly conventional automobiles [10], a driver assistance system customized for EV users can be of crucial importance. This observation can be further highlighted by other literature studies that have shown that travel patterns can vary further when influenced by technologic systems [11]. While traveler groups' flexibility can differ according to age groups [12] and social characteristics [13], if the benefits acquired from the planning process are perceived as worthy [14], benefits emerge for planning trips in pre-trip phase [15] or in real-time [16].

The goal of this research is to provide a realistic approach to the Daily Activity Chain Optimization (DACO-EV) problem by exploiting activity-based modeling and the Genetic Algorithm (GA) framework. While the incorporation of the EV range constraints is a significant part of our work, a particular focus is also given to mechanisms that enable the implementation of such a system in a real-world ITS system that will serve the travelers. Section 2 presents a literature review on related topics. Section 3 provides a detailed description of the DACO-EV method, its parameters, and its attributes. Section 4 discusses the implementation of the method and to provide realistic solutions to travelers. Section 5 presents the results, including use cases. Section 6 serves as a conclusion to the article and its contribution.

2. Literature Review

To the best of our knowledge, an approach to the DACO-EV method, including enough parameters to render its solution useful in realistic settings, does not exist in the scientific literature prior to the publication of this article. However, insights and valuable research directions can be extracted from articles regarding the incorporation of electromobility into transportation problems.

The activity-based analysis of transportation systems stems from the idea that the demand for transportation between locations is interconnected with the demand for participation in some activity. As indicated by a series of works in the literature [17–19], it has been extensively studied and used for the analysis of travel behavior and the planning of transportation systems. Our work on the optimization of daily activity chains falls under the umbrella of activity-based research in transportation. Recent notable articles [20,21] show that activity-based modeling has been successfully applied to cases where electromobility is incorporated into transportation systems. In the first paper by Dong et al. [20], the impact of different types of deployment of charging infrastructure is analyzed in regard to the range anxiety of EV users. An activity-based method is used to evaluate the deployments according to travel patterns. In the work by Kontou et al. [21], activity-travel patterns from a National Household Travel Survey data are used, in order to assess two schemes for centralized charging management of EVs, one scheme that regards preferences of individuals and another that considers the government perspective. The results indicate the differing nature of the interests of those two stakeholders, and confirm that the availability of charging stations at the workplace can greatly affect the charging profiles of EV travelers. The latter observation is one of great importance that can be found in other studies [8], too.

In regard to the problem of Daily Activity Chains Optimization (DACO), the basic instance of DACO-EV, it has been previously addressed in the literature, and a GA approach has been introduced [22,23]. Charypar et al. [24] have addressed a similar problem and introduced a similar solution approach. The authors have considered several simplifications, such as the computation of distances according to geometric distance and not based on the real transportation network. Another similar work was conducted by Abbaspour et al. [25], where the authors tackled the same problem but had a specific focus on the touristic aspect of the optimization of activity chains. Although the aforementioned papers regard heuristic optimization methods that have been developed to solve the DACO problem, different modelings could also be considered. A modern approach was presented by Liao et al. [26], who modeled the problem as a graph super-network. They included space-time constraints to enable the valid selection of locations for activities in time, multi-modal transport options and modeled the parking choice. While their activity-travel scheduling algorithm computes the final solutions to the problem in minimal time, in the scheduling problem and the modeling, the priorities and the flexibilities of the travelers are not included, like in the case of DACO.

There is a lack of exact optimization approaches for the DACO or DACO-EV in the literature. Exact optimization methods, where the gap between the solution calculated and the globally optimal solution of the problem is zero, are based on mathematical models of the transportation problem. Although exact approaches tend to provide better solutions in terms of reduction of cost according to criteria, they cannot yield solutions in the amount of time that heuristics can, and usually take much more time to complete the calculations. Works can be found in the literature where routing in multi-modal networks has been addressed with exact approaches [27,28].

One motivation for the extension of our method was to provide an efficient tool dealing with the range anxiety of EV drivers in urban environments. Range anxiety is one of the main bottlenecks of EV adoption and affects both inner-city and inter-city trips. Many studies [29] exist in the literature that addressed this topic and provided useful insights regarding key factors that can evaluate its effect and help deal with it. Although technology, and its advancement, is one of the main ways to deal with range anxiety, other key factors are considered as important for the medium-term horizon, such as battery costs, coverage of the charging stations network and CO_2 vehicle standards [7]. Range anxiety has also been studied in the context of daily schedules by Neubauer et al. [8]. Significant findings in their studies showed that workplace charging could play an essential role in promoting the utility of EVs for high mileage commuters. Additionally, the broadly available charging infrastructure has been shown to be important to both high and low mileage commuters. Range anxiety importance has also been shown in computational studies that discussed its relationship with the CS network. In the work by Guo et al. [30], the battery charging station location problem was addressed where range anxiety and distance convenience were taken into account. While the KSIGALNS algorithm that they developed to solve their bi-level integer programming model was proven to be very effective compared to standard and previous solutions approaches, the analysis of range anxiety as a parameter shows that it is an essential factor to the location strategy of the CS network. Other interesting computational studies exist in the context of range anxiety, such as the one by Esmaili et al. [31], where the authors examine EVs as distributed energy storage units and their potential contribution to microgrids when vehicle-to-grid service is considered. An interesting result that emerges from the solution of their Mixed-Integer Linear Programming (MILP) model is that when range anxiety, as modelled in their mathematical optimization program, is in higher levels, then the average State of Charge of the EV drivers rises and, as a result, the total cost of the microgrid is higher, too.

Range anxiety, however, is a phenomenon that is being progressively dealt with on both a personal and systemic level. It is more evident than ever that not all EVs are the same regarding experiencing range anxiety [32]. There are observations that range anxiety is minimally experienced by users of specific types of EVs in specific regions, such as in the study by Gorenflo et al. [33], in which, their analysis of the e-bike usage and battery charging data supports that range anxiety does not exist among the participants of the survey. Finally, in order to underline the importance of guidance systems, like the one presented in this paper, we further present the study of Cuchý et al. [3]. In their work, they were able to evaluate multi-destination transportation scenarios according to single-user perspective and infrastructure perspective by using AgentPolis simulation framework. Except for the obvious benefits for EV users, they showed that when users plan their mobility up-front, the total utilization of infrastructure can increase by more than 100%.

Particular focus was given to articles dealing with EV routing and the parameters that were considered when it comes to the range that vehicles can cover. Extensive work on the subject has been conducted by the team of Baum. Notably, in work conducted in 2013 [34], they extended their algorithm, named Customizable Route Planning (CRP), to calculate fast queries on graphs that are suited to Electric Vehicles (EV). Their goal was to calculate energy optimal routes, where several parameters can be included, such as the recuperation of the electric vehicle, the battery capacity constraints, and the dynamic behavior in energy consumption of the vehicle. In their next work conducted in 2016 [35], they proposed a framework with a more holistic approach to the problem, where parameters are included, such as the use and location of charging stations, turn costs, and the adjustment of speed in order to save energy. In their work conducted in 2015 [36], they introduced an approach named CHArge, which solves the EV routing problem in realistic settings. They discussed the properties of charging functions, which are used to map an initial State of Charge (SoC) and the duration of charging sessions with a resulting SoC. The works by Baum et al. [34–36], are very close to our approach in terms of methods and calculation mechanisms that were used to deal with the incorporation of electromobility. Additionally, useful information about the charging network, charging sessions, plugs, and strategies were also extracted from the work of Moghaddass et al. [37].

There is a series of articles that are connected to our work since they address the optimization of activity chains considering the use of EVs. The work of Liao et al. [38] is very close to our approach, but they do not solve the same problem. The authors provide modeling and solution to the EV shortest travel time path problem and the fixed tour EV touring problem. They consider a battery swap system for the touring problem, where two cases are discussed: the on-site station model, where each city is considered as a swapping station, and the off-site model, where the swapping station is further away from the city. Although their approach addresses a similar problem to ours, the presented work is not directly linked to a real-world application. In the works of Cuchý et al. [39,40], the authors tackled the Whole Day Mobility Planning with Electric Vehicles (WDMEV) problem and modeled it according to graph theory. The definition of the problem is similar to DACO-EV, but the authors included fewer parameters than in our problem definition. They did not include the priorities of the travelers, and they are neglecting the time windows of the desired attendance of the traveler. Furthermore, the authors did not include in the models the consumption according to Worldwide Harmonized Light Vehicle Test Procedure (WLTP), different EV models, and it is not clear if they consider the charging connector types. In the modeling, the overall weight of the vehicle and the option to have a final desired amount of energy in the battery when the tour ends were not included. Finally, the availability of the chargers was not taken into account in the calculation of the routes, but they considered waiting time at Charging Station (CS) according to randomly parameterized models. The aforementioned parameters may lead to very different solutions spaces and computational results, so the two problems and methods cannot be directly compared.

Overall, the main contribution of this article lies in the elaboration of a method that addresses the DACO-EV problem, including all the necessary parameters that allow the solution of the problem to be personalized to the specific needs of travelers, and enabling the use of the method in realistic settings. The method includes:

- Consumption calculation mechanisms of the EV according to the features of the road network and the EV market model (battery capacity, weight, charging rate, charging plug, WLTP ratings).
- The Starting State of Charge (SSoC) and the desired Final Stage of Charge (FSoC) of the EV used by the traveler.
- A real charging stations network and the availability of those charging stations based on past usage data.
- Two charging scenarios in order to further model realistic travel behavior.

3. Method Description

3.1. Parameters of the DACO-EV

The DACO-EV method incorporates the use of an EV; therefore, it requires modeling of the constraints regarding the energy equilibrium that takes place in the energy storage mechanism of the car while visiting several locations. Because EVs are still in an early phase in their adoption by the public, the available charging facilities are not always optimally distributed, and the minimum distance between available charging points can reach up to several kilometers depending on the type of road network [41]; this situation renders a demanding reality for EV drivers that need to utilize the public charging network to cover their charging needs. The extra parameters that affect this energy equilibrium can be grouped into three major categories, such as vehicle design, driver behavior, and the environment [42]. In our attempt to provide a realistic and practical method for the elaboration of meaningful solutions to the EV drivers, we have included several of those parameters, while we ignored others of lesser importance.

The problem and the elaborated method are based on three sets of parameters. The first set is defined by the traveler and refers to his or her schedule and overall preferences concerning the activities. The second group of parameters is about the constraints related to electromobility when the traveler utilizes an EV. The third set consists of the parameters that do not depend on the user and refer to the existing network, the timeframe of operation of facilities and services. Table 1 refers to the parameters that are set by the traveler.

Table 1. The first group of parameters.

Parameter	Description
Starting time of the tour	Earliest time in the day when the first trip to the first activity can begin. Solutions that start earlier than this time are considered infeasible.
Ending time of the tour	Latest time in the day when the traveler can arrive at the ending (final) activity.
The starting position of the traveler	The starting location of the tour. Commonly it is also the ending position of the tour and is associated with the traveler's home.
The ending position of the traveler	The ending location of the tour.
Activities and their type	This parameter refers to the types of activities that need to be conducted. The type can be chosen from a predefined list (e.g., bank, restaurant, hairdresser, bakery).
Locations of the activities	Defines which are the locations that the traveler would like to conduct each activity.
Processing time at each activity	This parameter is used to specify how much time is needed for the traveler to conduct each activity.
Priorities of the activities	The traveler specifies the level of flexibility for each activity. It can be spatially and temporally fixed, only spatially flexible, only temporally flexible, or flexible.
Desired earliest start time for each activity	The traveler must specify when the preferred time for each activity to start is.
Desired latest end time for each activity	The traveler must specify when the preferred time for each activity to end is.
Used EV model	By having specified the manufacturer and the model of EV that is used by the traveler, we can utilize the information about several aspects of the vehicle. The range and the possible charging locations of the EV are the most important of those aspects (if the mode of transport used is EV).
Starting State of Charge (SSoC)	The energy level of the battery at the start of the tour (if the mode of transport used is EV).
Final State of Charge (FSoC)	The desired energy level of the battery at the end of the tour (if the mode of transport used is EV).

Particular emphasis should be put on the second group of parameters about the EV and its use. Since the traveler specifies the used EV model, the parameters of this second group can be considered as sub-parameters of the first group (Table 2).

Table 2. The second group of parameters concerning the characteristics of the Electric Vehicle (EV).

Parameter	Description
The capacity of the battery of the EV	Also referred to as usable battery capacity, which is the maximum amount of energy that can be stored in the vehicle.
Plug type of the EV	The type of plug that the vehicle uses for charging.
The consumption rate of the EV	The consumption rate of energy is a function of the vehicle speed and the use of auxiliaries. This rate can be extracted from an EV database.
Charging rate of the EV	The time needed to reach the desired energy levels while charging. This rate can also be extracted from an EV database.
Energy recuperation	The EV can recharge its battery while braking, especially when the path to be followed is a downhill road.
Vehicle weight	The weight of the vehicle plays a vital role in how much energy is consumed, especially in cases where the tour includes gains or losses of elevation.

Finally, there is the third group of parameters that are needed for the problem to be fully described. Those parameters are not defined for each passenger, but instead, they are specified for the network, the modes of transport, and the types of activities available. They can also be described as the static parameters of the DACO-EV problem (Table 3).

Table 3. The third group of parameters that are independent of the user's input.

Parameter	Description
Network topology	The transportation network that is connecting the available locations for each type of activity. The elevation of the road segments is considered part of this parameter.
Alternative locations and Time-windows	Locations of facilities of activities, time-windows of operation of the facilities, and types of activities that can be conducted at each location.
Charging stations network	Locations of charging points and types of chargers available at each point.
Availability of charging points	Availability of charging stations as calculated by past usage data.

3.2. Degrees of Freedom of the Problem and Decision Variables of the Solution

At this point, we should make a reference to the decision variables of the DACO-EV that fully define a solution for the travelers. Those decision variables are essential for the modeling as they will form the chromosomes of the solution encoding of the Genetic Algorithm. In Table 4, the decision variables and their descriptions are provided.

All of the aforementioned variables (ones included in Table 4) are the independent variables of the problem. By deciding on the order, the locations, the time windows of attendance, the modes of transport, and the charging time, we decide on a specific way that the first tour changes and gets optimized. Of course, that information is not enough in order to fully articulate a meaningful solution for a traveler in a human-readable format. From the independent variables, we can extract a series of information, such as exact paths to be followed in between activities, battery power after the visits at each of the locations, starting and ending time of the whole tour, travel time of the tour, starting and final state of charge of the EV.

Table 4. Decision variables of DACO-EV.

Decision Variables	Description
Order of the activities in the chain	The order of the activities in the tour in relation to their initial order.
Locations of activities	The traveler can conduct the same activity at multiple locations. This variable refers to one of the available locations that have been chosen for a specific solution.
Time windows of attendance to each activity	This variable is used to specify the arrival and departure of the traveler to the activity.
Charging time, charging station and subtour information	This variable is a set of quantities that contains all the information that regards the subtour for charging if one is included.

3.3. Charging Scenarios

To fully cover the description of the DACO-EV problem, we also need to deal with the possible types of charging. As a first charging scenario, we consider the case where the EV users want to charge their vehicle at a CS, and they want to conduct one of their activities at another location. This is called the *en passant* charging scenario. Usually, it can occur when a charger and a location for an activity are close to each other, and the user can just walk a few hundred meters to the facility and back to the charging station after completing the activity. As a second charging scenario, the classic scenario of charging is included, where the traveler again deviates from the first tour and embarks on a subtour to a CS, but in this case, he or she just waits in the car until the charging is finished.

First, we provide a graphical example of the classic charging scenario of the DACO-EV (Figure 1). Given a set of five initial activities, the activity chain is optimized. For simplicity reasons, we only provide an optimized tour where nothing changes in the spatial context except for the inclusion of a charging session. The charging station is indicated by the blue circle, which is a new activity in the chain.

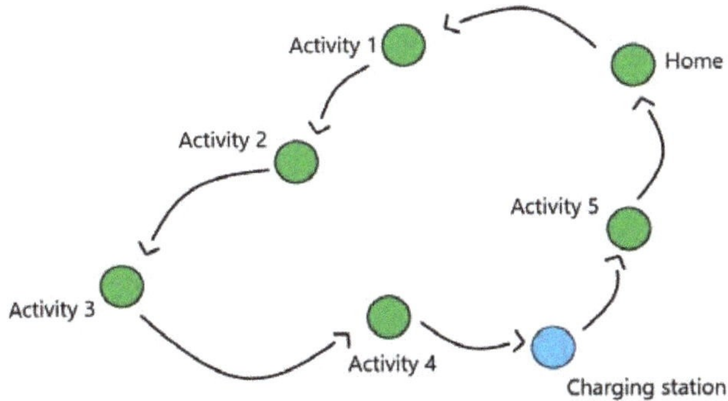

Figure 1. The classic charging scenario of the DACO-EV problem.

In Figure 2, present the *en passant* charging scenario of the DACO-EV. The charging detour happens between Activities 4 and 5, and the transfers from the CS to the location of Activity 5 are represented by the dotted lines. In this case, the EV remains at the CS, while the user processes Activity 5 using another mode of transportation (e.g., walking).

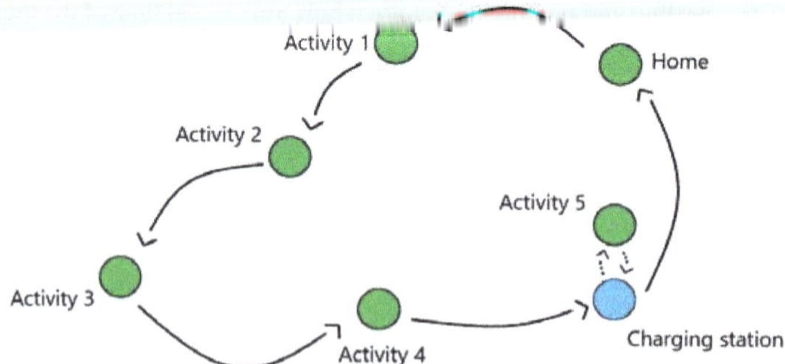

Figure 2. The *en passant* charging scenario of the DACO-EV problem.

3.4. Optimization Method

The method is based on a Genetic Algorithm (GA) that iteratively solves the Travelling Salesman Problem with Time-Windows (TSP-TW) for different combinations of the characteristics of the possible solutions and evaluates each of them according to several criteria. The efficiency of the proposed method has been previously tested and proved for the primary instance of the problem in the articles by Esztergár-Kiss et al. [22,23]. After running the algorithm for several iterations, a more optimal solution for the initial schedule and the preferences of the traveler is derived. In the algorithm, we make sure that after the traveler executes the tour suggested, and given the EV usage constraints, the EV has the desired FSoC when it reaches home.

By defining the properties (i.e., decision variables) of an individual (i.e., candidate solution) of the problem, we can form the populations of solutions in the solution space. Then the genetic operators of selection, mutation, and crossover are utilized to search the available space of solutions and derive the ones that are more suitable for each case of traveler. Except for the genetic operators that have been used, and according to GA literature [43], the method is characterized by a set of parameters, such as:

- Population size—Number of solutions initially created at the solution space initialization phase and kept by the selection operator at the end of each iteration of the algorithm. The population size that was selected for the solution of the DACO-EV is 30 individuals.
- Base mutation probability—Defines how much we want to search for solutions with totally new attributes compared to an initial population of solutions. The mutation probability that was selected is 20%.
- Crossover probability—Defines how many new solutions (i.e., individuals) are produced at each iteration of the algorithm based on previous populations. The crossover probability that was selected for the efficient solution of the problem is 10%.
- Generations—Defines the iterations of the GA algorithm that run until we get a final solution. No other termination criterion is used for the algorithm, which means the number of generations, will define the final optimality gap of the solution calculated by the run of the algorithm. The number of generations until a more optimal solution is calculated for the traveler is 30. The number of generations for this algorithm can vary according to if pre-optimization techniques are applied to the solution approach or not.

3.4.1. The Genetic Algorithm

Although the GA framework contains more or less standard data structures, functions, and operations utilized for each application, it was considered appropriate that a more formal specification of the algorithm is presented. This algorithm describes how temporal and spatial flexibility were handled and includes several technical details in order to enable its reproduction.

Algorithm 1	A genetic algorithm for addressing the DACO-EV problem
Input:	Parameters for the execution of the GA, Parameters of the problem
Output:	The solution of the problem: an optimal schedule for the traveler
Steps:	
1:	Read the data and the parameters, create appropriate data structures;
2:	Initialize a population of feasible solutions (i.e., individuals) for the DACO-EV problem;
3:	For generation i in the range [0, generations):
3.1:	Apply the crossover operator between the individuals of generation *i*;
3.2:	Apply the mutation operator to the individuals of *generation i*;
3.2.1:	Stochastically swap the order of two activities in any individual;
3.2.2:	If there is a spatially flexible activity, then stochastically change its location to a new suitable location;
3.2.3:	If an EV is used and if the SoC at the final location is not at the desired level, or a negative number at any point in the tour, add a detour at a random CS. If the detour exists, stochastically change the location of the CS in the tour;
3.3:	Calculate the distances matrix (and the kilometric distances, elevation gained, and elevation lost matrices if needed) for the locations involved in the individuals of generation *i*; Normalize the attributes (calculate time-windows, battery power at each activity/location) of the individuals according to the order of the activities and the detour;
3.4:	Evaluate all individuals according to the fitness function;
3.4.1:	Label as infeasible all individuals who do not satisfy the *temporal constraints*;
3.4.2:	Label as infeasible all individuals who do not satisfy the *priorities' constraints*;
3.4.3:	Label as infeasible all individuals who do not satisfy the *energy equilibrium*. Apply *Scaled Ranking* of infeasible solutions;
3.4.4:	If an EV is used, label infeasible all individuals who include CS which are unlikely to be available (As described in Section 4.3. Availability of Charging Stations) ;
3.5:	Select the individuals with the best fitness to form a population and reproduce in the next generation;
4:	Select the individual with the best fitness from the population of the final generation;
5:	End of the algorithm.

3.4.2. Fitness Function and Optimization Criteria

For the efficient solution of the DACO-EV, several criteria have been considered. While most of them regard travel time, whose reduction has been our primary aim, a few more criteria were considered as important to include. The method is capable of handling five criteria. They were weighted by the corresponding parameters a, b, c, d, e to create an efficient multi-criteria fitness function for the problem.

$$U(X) = (a * TT) + (b * TTST) + (c * AT) + (d * WT) + (e * CT) \tag{1}$$

Fitness function U has been applied to every candidate solution X that has been calculated by the genetic algorithm. This evaluation, as indicated in the algorithm at *Step 3.4*, was applied to each potential solution, when not infeasible, and has been utilized to label and prioritize according to the potential needs of travelers. For the first criterion, TT, its actual value is given in regard to the time that the user has spent traveling within his or her tour. The second criterion, the Travel Time in the Subtour (TTST), regards travel time only in the charging subtour. This specific type of travel time has been calculated as an extra variable, in order to be able to be separately included in cases, where the effect of the subtour needs to be minimized in the overall solutions. Arrival Time (AT) at the final destination is also another criterion considered that can be essential for some users, as well as, the Walking Time(WT), when it occurs in the sub-tour for charging, and the en-passant scenario. Finally, Charging Time (CT) has been additionally considered as an extra criterion, and has been added to the function as a possibility for optimization. The reason for including all five criteria was to provide a robust optimization approach that is applicable to a wide set of real-world applications. While usage of the method is independent of spatial context, the potential needs of travelers vary across different cities, countries, and the available charging stations network.

For the values of the parameters a, b, c, d, e, different solutions are derived. The proper choice of the values is of great importance to the type of solutions that will be calculated. A sensitivity analysis of the effect of the parameters on the solutions produced is one of the future steps to take in order to assure suitable performance for each real-world case (i.e., region of application) and understand the effect of each criterion on the activity chain.

4. Implementation of the Method

For the efficient functioning of a modern application that will support the travelers in their daily commutes, the implementation of the method has been realized. Because of the complexity of the problem addressed and the vast amount of calculations needed, a series of heuristic rules were applied to the method in order to direct the search of the solution space and yield meaningful results within realistic computation times. This addition to the method is called the pre-optimization phase, which creates a smaller solution space for the algorithm to search. However, in this paper, we focus on the charging related functions of the method.

For the implementation of the method, Python programming language was utilized. Data from OpenStreetMap and Geofabrik [44] was used for establishing the road network, OpenTripPlanner (OTP) engine [45], was the primary tool for the calculation of road distances and routes, and calculations for the availability of CS were conducted based on data from e-MOBI [46] about the charger network. The chargers based on which the availability was calculated were 89. Elevation data are taken from NASA's Shuttle Radar Topography Mission (SRTM). Regarding the EV model used, the traveler is allowed to enter the vehicle that he or she is going to use, and we were able to extract the properties (battery capacity, available plug types, consumption according to WLTP, vehicle weight) of each vehicle from an electric vehicle database [47]. A simplification was made regarding the charging rate, which was considered according to the available plug types at both the charging station and the EV. A matchmaking mechanism was developed to match EVs with fast charging plugs if they are likely to be available, or with standard charging plugs, otherwise. Regarding the priorities of the travelers, we included them in four levels which are described in Table 5:

Table 5. Priorities of the travelers considered.

Priority Label	Priority Label Description	Spatial Flexibility	Temporal Flexibility
1	Totally fixed	×	×
2	Spatially flexible	√	×
3	Temporally flexible	×	√
4	Totally flexible	√	√

When an activity is assigned with Priority Label 1, it means that the traveler considers this activity as both spatially and temporally fixed. Priority Label 2 is assigned by the travelers to activities that are spatially flexible, which means they can be conducted at several locations, but temporally not, meaning that the algorithm is bound to calculate the visit within the desired time windows. With Priority Label 3, the traveler indicates that the activity must be conducted at the provided location, but it can happen in any time-window within the day. Oppositely to Priority Label 1, Priority Label 4, is the case where the traveler indicates that the activity is totally flexible, which means that it can be both conducted in another location (providing the same type of service) and at any time (within the overall timeframe of the user and the operating hours of venues and shops).

Before actually providing information about the implementation of the GA and its mechanisms that allow the effective solution of the DACO-EV, an overview of the work is presented in Figure 3 that depicts the method, its parameters, external tools, and data used. On the one hand, the traveler provides an initial activity chain and information about his or her preferences by setting the first, and thus the second, group of parameters. On the other hand, the third group of parameters provide external information, such as the network topology (road network and elevation), the electric vehicle database for consumption calculations, and the alternative activities' location database for calculating the optimized activity chain by examining the full spectrum of the spatial possibilities for the activity chain. Finally, based on the charging station usage data and other related information to the CS (i.e., charging stations database), the availability of charging stations was calculated using data mining techniques. Before using the heuristic optimization model, a pre-optimization is performed, and the routing engine is used to calculate optimal routes between activity locations. As a result, the genetic algorithm calculates optimized activity chains for users.

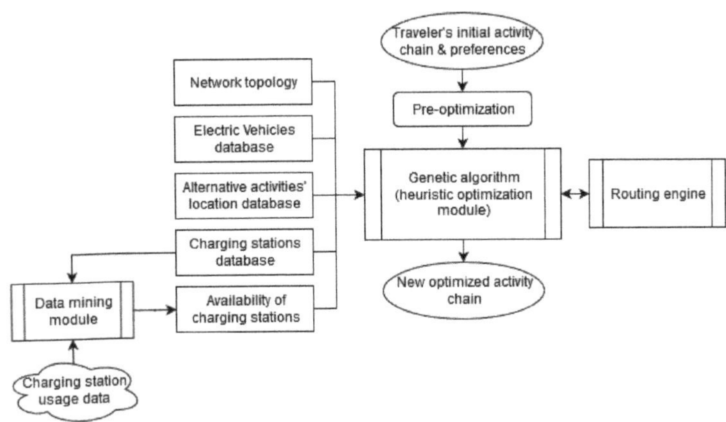

Figure 3. The method elaborated.

4.1. Explanation of the Pre-Optimization Phase

The pre-optimization of each solution is a useful tool that can help make the GA approach more efficient. In the context of this work, several techniques were implemented in order to make the GA compute better results faster. The pre-optimization phase aims at reducing the initial solution space of the algorithm, which is created every time that an instance of the problem is created according to the static and dynamic parameters of the problem, from a generalized vast solution space to, what one can call, a personalized solution space.

The personalized network is achieved by deciding on some aspects of the solution before triggering the actual search for the optimal solution, thus creating a smaller solution space. It considers the activities with spatially fixed priorities to serve as gravity points for the optimization procedure. More specifically, after the traveler provides the input of the preferences, we have an initial set of choices that the traveler initially intends to follow. This cutting of the solution space happens around the locations of activities that the travelers consider as totally or spatially fixed. This creates a subset of alternative locations of activities and charging stations. Then this subset of the initial space is searched with the GA in the main optimization phase.

For example, if an activity is fixed (Priority Label 1) in the search for an optimal solution, all other temporal and spatial choices concerning this activity can be ignored. In the same way, if the activity is temporally fixed, we include all alternative facilities (locations), but we do not permit the algorithm to search outside the time windows specified by the traveler. If the activity is spatially fixed, then we cut out the parts of the solution space that has to do with other location of facilities for the activity, and we only allow the algorithm to search for different time windows. If we have the fourth case of total flexibility, then we have to search both for alternative facilities and time windows of attendance, and we cannot pre-optimize.

4.2. Calculation of the Charging Time and Deciding the Charging Scenario

Charging time is a deterministic variable and a function of the rest of the attributes. The amount of time spent at the CS is calculated with a forward lookup of the energy requirements of the EV to cover the full tour. This calculation occurs by comparing the *SSoC*, the *FSoC*, and the remaining energy in the battery of the vehicle at every location. The candidate solutions (i.e., individuals) to the problem fully describe the locations to visit and the order of attendance to the activities. From those two attributes, we can extract the exact routes between locations of activities, as well as the kilometric distance, elevation gained and lost, and charging time at the CS.

Several aspects of the detour are modeled as independent variables and can stochastically change. Thus, we included a heuristic rule in the solution process that further directs our search. This rule defines that if the CS is within a direct distance K from the location of the activity after the trip to the CS, then the charging scenario is set to *en passant* charging scenario. When the CS is at a direct distance greater than K, then the charging scenario is set to a classic charging scenario. By setting K to the appropriate value, the algorithm can lead to better solutions by avoiding assigning the *en passant* scenario to cases where the location of the CS is too far away from the next location of the activity. The values of K can vary according to what modes are included and are available in the detour when the EV is left for charging. In our implementation, we only added walking mode for the *en passant* scenario, and we have noticed that the implementation yields satisfying results for the values of 800 to 1000 m for urban environments.

In the cases where the solution gets mutated to include the detour right at the end of the tour before the traveler goes to the desired final destination, the charging scenario is always set to the classic charging scenario. In Figure 4, a combination is given, where the heuristic rule is applied, and charging station A is utilized with the *en passant* scenario, while charging station B is utilized with the classic charging scenario. For the optimization of this initial activity chain (i.e., green markers), charging stations that fall within a distance K from the location of Activity 5, can be utilized under the en passant scenario (dotted lines) since the walking distance is considered walkable. Otherwise, the algorithm sets the charging method to the classic charging scenario.

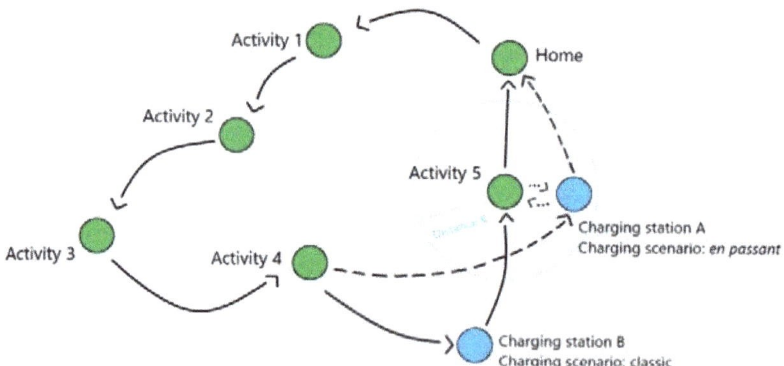

Figure 4. The difference between the classic and the *en passant* charging scenarios, and the role of distance K.

4.3. Availability of Charging Stations

Regarding the availability of charging stations, a method has been elaborated in order to incorporate this parameter of the problem. We used the data provided by e-Mobi [46], which is the organization responsible for e-mobility investments (especially for deploying charging stations) in Hungary, to derive a suitable metrics that could be included in the optimization of the DACO-EV problem for travelers in Budapest. The dataset provided information about the names, locations, and plug types of chargers in Budapest. Furthermore, we had a list of charging sessions for those chargers that happened between the 23rd of February 2018 and the 12th of February 2019. For this period, we applied data mining techniques to derive average statistical properties that lead to a percentage of the occupation, in similar terms, probability for the charger to be occupied for each hour of a day. If the charger was occupied for 20 min on average in a specific hour in the examined period, then it was assumed that this charger was occupied for 33.33% on average.

These statistics were included in the method for the optimization of the DACO-EV. The main component of the method that was utilized for including the availability of CS was the evaluation operator of the GA. Every time we needed to evaluate a solution that emerged from the mutation and crossover operators of the GA, we applied the evaluation operator and checked whether the CS included in the solution is likely to be occupied. We aimed to prioritize solutions in the population that include chargers that are less likely to be occupied. The prioritization occurs only by labeling the solutions, and the corresponding charger as feasible or infeasible. Other than that, the percentage to be occupied did not have any impact on the fitness function and did not change its value. In order to avoid empirical bias in how the solutions get labeled, we implemented a labeling mechanism based on stochasticity. The method produced a random percentage in the positive continuous set of [0,100], and if that random percentage was lower than the calculated percentage of the occupation, then the solution was marked infeasible. In that regard, CS, whose rate of occupation was high were more likely to be labeled as infeasible and were not included in the population of solutions.

In Figure 5, we provide example graphs for the availability of Charging Stations (CS) as calculated by the method. The percentage of occupation for four random CS in Budapest can be seen for 24 h. The percentage of hourly occupation can significantly vary between the chargers. In the case of CS 1 and CS 3, nearly 0% occupancy rate can be derived, while in the case of CS 2 and CS 4, the occupancy rate reaches almost 50% in the morning hours.

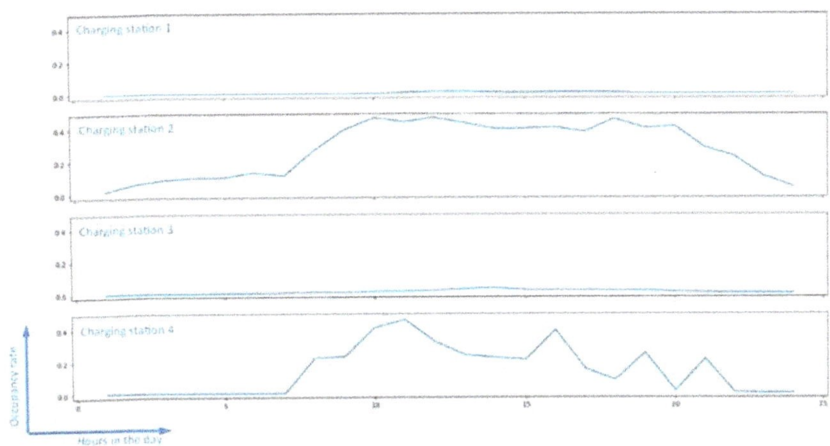

Figure 5. Examples of the availability of charging stations throughout the day.

As a future step of the data mining method for the calculation of the availability of charging stations, we would include a calculation of the rate of occupation also according to the day of the week. A more enormous dataset than the one available would be more suitable for this kind of extension. For example, this extension would help the method performing better on special days like weekends, where the availability may be different from weekdays.

4.4. Energy Consumption Model and the Incorporation of Elevation

The incorporation of elevation of the road segments was a significant part of our work, and it was based on several articles [48–51], with the work of Fišer [52] being a vital starting point to our work. In his thesis, he proposed an Approximated Consumption Function:

$$g(e,\ m) = \begin{cases} \kappa\cdot\delta(e) + \lambda\cdot\Delta elev_e, & if\ \Delta elev_e \geq 0 \\ \kappa\cdot\delta(e) + a\cdot\Delta elev_e, & otherwise \end{cases} \tag{2}$$

where g is the consumption function, δ is a function that assigns a distance to an edge, e and m are the edge traversed and the EV model correspondingly, $\Delta elev_e$ is the difference in elevation from the start of the edge u to end of it, v. Furthermore, κ, λ, and a are tuning constraints through which we can control the values of function g in order to yield realistic results. However, to cover the needs of our more practical approach, we moved on to introduce the Practical Consumption Function, which can be described by the following equation:

$$g(e,\ m) = \begin{cases} ConsWLTP_{e,m} + coeff_{uphill} * (m*g*\Delta elev_e), & if\ \Delta elev_e \geq 0 \\ ConsWLTP_{e,m} - coeff_{downhill} * (m*g*\Delta elev_e), & otherwise \end{cases} \tag{3}$$

where *ConsWLTP* is the consumption according to EV model and the *WLTP* standard, and the Electric Vehicle database [47]. The reason for using $coeff_{uphill}$ and $coeff_{downhill}$ is to emulate the performance of the motor of the vehicle when going uphill and the performance of the recuperation system going downwards. The energy recuperated while braking on a flat road is ignored. In comparison to other approximation functions used for considering the consumption for EV routing, our Practical Consumption Function avoids using further tuning parameters and includes the actual interchange between the potential and kinetic energy of the vehicle. Nonetheless, the results derived from our Practical Consumption Function are realistic and very close to findings reported in the literature of EV routing. The incorporation of more advanced formulas like ones presented in [53,54] our method can be considered as the next step in our research.

In order to provide the necessary background for the use of the Practical Consumption Function, the data for the distance and elevation are derived from the OpenTripPlanner API, which bases its calculations on OpenStreetMap data for the road network and SRTM data for the elevation. The aim of this model is not to introduce an EV consumption calculation model as compared to recent approaches [53,54], but to provide a mechanism for our method calculating meaningful results considering limited computation times. The model itself is based on WLTP standard for the consumption of EV models that already includes several parameters and laboratory tests and is only extended to include the effect of elevation, for the road segments traversed in the tour of the traveler. The consideration of more complex approaches are not necessary and not within the scope of this research.

5. Results

5.1. Use Cases

The experiments were run on ten instances of problems, where the solution space can significantly differ according to different preferences of the user. While one traveler may need to attend a set of fixed activities, another may be flexible concerning the locations of the activities or the temporal schedule of the activities. This means that the first traveler's initial schedule and preferences would not allow the method to search for alternatives since there is no flexibility concerning the schedule, while the second traveler's preferences would enable the method to search through a vast solution space for other options.

For all of the cases used for our tests, the activity chains contained at minimum one out-of-the-house activity and at maximum four out-of-the-house activities. For the first six instances of problems, cases 1 to 6, we considered synthetic cases of the problem, where the locations, types of activities, and flexibility were based on real activity chains of travelers, but also randomized in some of their attributes to a certain degree. For the rest of the examined cases, cases 7 to 10, we kept the set of locations and types of activities the same as in case 6, and then we included potential combinations of flexibility flags. We consider cases 1 to 6 the balanced cases, which could be input from real-world travelers, and cases 7 to 10, the extreme cases where the computational limits of the algorithm are tested.

In our use cases for the DACO-EV, the same vehicle was considered, and the $SSoC$ was set to a medium-range percentage while the $FSoC$ was required to be at a higher percentage than the $SSoC$, which means that a detour for charging is needed in all cases of problems displayed. For the results presented, and regarding case-specific characteristics, and the results presented below, for the provided instance of tests, for cases 1, 2, and 5 to 10 we have four out-of-the-house activities, whereas for cases 3 and 4, three out-of-the-house activities were conducted. Finally, let us state that the results provided here are calculated based on one of the criteria that are available in the method described and in Section 3.4.2, that of travel time. Thus, the parameters in the fitness function are all set to *0* except for parameter *a* that is set equal to *1*.

5.2. DACO-EV Results

The results for the evaluation of the performance of the method in addressing the DACO-EV problems are presented. The performance of the algorithm is compared in regard to the travel time in the initial activity chain of the travelers and the resulting activity chain in both cases when precomputation is applied and when not, are given. The columns of the tables are the following:

- Initial time: Average travel time of the initial tour of the traveler (min)
- Opt. time (no pre): Average travel time for an optimized tour (without precomputation applied to the solution process) (min)
- Subtour (no pre): Average travel time in the subtour for a DACO-EV tour (without precomputation applied to the solution process) (min)
- Comp. time (no pre): Average computation time for the optimization of a tour (without precomputation applied to the solution process) (s)
- Opt. time (pre): Average travel time for an optimized tour (with precomputation applied to the solution process) (min)
- Subtour (pre): Average travel time in the subtour for a DACO-EV tour (with precomputation applied to the solution process) (min)
- Comp. time (pre): Average computation time for the optimization of a tour (with precomputation applied to the solution process) (min)

In Table 6, the results addressing the several cases of experiments are presented.

Table 6. Comparisons for the runs addressing the DACO-EV—effect of precomputation on solutions and computation time.

	Initial Time (Min)	Opt. Time (No Pre) (Min)	Subtour (No Pre) (Min)	Comp. Time (No Pre) (s)	Opt. Time (Pre) (Min)	Subtour (Pre) (Min)	Comp. Time (Pre) (s)
Case 1	32.61	57.79	30.19	59.25	38.11	23.57	14.28
Case 2	79.60	85.62	33.5	72.56	75.79	21.37	24.41
Case 3	70.75	66.73	27.23	85.53	55.37	13.83	17.59
Case 4	67.28	65.73	28.81	25.49	59.27	22.94	13.94
Case 5	81.28	85.33	39.82	110.43	75.64	33.95	19.89
Case 6	79.98	74.82	28.71	105.05	62.35	10.36	26.95
Case 7	79.98	81.15	26.38	35.48	80.72	24.34	20.50
Case 8	79.98	87.32	35.04	67.42	31.57	15.6	41.85
Case 9	79.98	81.67	30.86	37.65	76.03	25.19	18.81
Case 10	79.98	77.77	31.52	303.78	28.43	9.74	51.68

It can be stated that the developed approach provides appropriate results that can help EV drivers in their daily commutes. On the one hand, the overall travel time gets optimized(travel time for attending activities plus travel time in the introduced subtour), while the computation times remain at acceptable levels for a driver assistance system, that can be run by the driver for a few seconds in his or her pre-trip planning phase.

The optimized overall travel time is not reduced in all cases. This is because for the tests that we have run, in which charging is required because of the SSoC and FSoC levels, an extra charging activity or charging session was added. Although activity chains are optimized to reduce travel time, in some cases (e.g., cases 1 and 2), the travel time is raised. When precomputation is applied, the algorithm still finds only a solution with more travel time than the initial one (e.g., case 1), but it can calculate solutions that have less travel time than the initial one (e.g., case 2). It can be noticed that the travel time for the subtour, which, in most cases, is a big part of the overall travel time. A positive observation is that the computation time in balanced cases was 19.51 seconds(s) on average, which is a reasonable solution time if the method can be possibly reproduced to serve real travelers in their daily lives.

In Figures 6 and 7, and Tables 7 and 8, we present an instance of case 3 in more detail to show how an activity chain of the traveler is initially set and how it can change after applying our algorithm. In Figure 6, we can see the initial activity chain, and in Figure 7, we can see how the algorithm produces an optimal activity chain. The travel time of the tour depicted in Figure 6 is 70.75 minutes (min), while the tour depicted in Figure 7 takes 55.37 min. The figures present the spatial aspects of the initial and optimized activity chain, and the tables present the temporal attributes of the two activity chains.

Figure 6. Initial activity chain of the traveler for case 3.

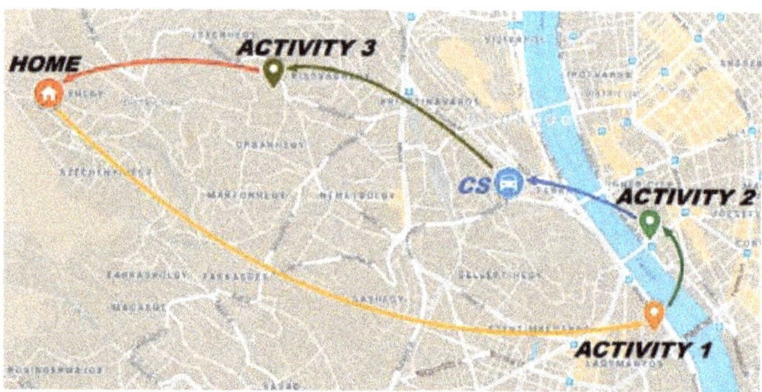

Figure 7. Optimized activity chain of the traveler for case 3.

Table 7. Part of the input of the traveler regarding the spatial and temporal aspect of activities.

Activity Order	Type of Activity	Processing Time	Priority Label	Demand Time Start	Demand Time Close
Start	Home	0	1	480	480
1	University	480	1	540	1020
2	Burger Place	30	4	1060	1090
3	Coffee Shop	20	2	1170	1190
End	Home	0	3	1440	1440

Table 8. Attributes of the solution regarding the visits to the activities.

Activity Order	Type of Activity	Processing Time	Priority Label	Time of Attendance Start	Time of Attendance Close
Start	Home	0	1	480	480
1	University	480	1	540	1020
2	Burger Place	30	4	1024.37	1054.37
CS	Charging Station	97.2	–	1064.57	1161.77
3	Coffee Shop	20	2	1170	1190
End	Home	0	1	1201.3	1201.3

Regarding the initial activity chain of the traveler, it can be noticed that it includes three out-of-the-house activities, each with different levels of flexibility. Except for the first visit, that is conducted at the university, the other two activities are spatially flexible. This fact, in combination with the significant difference between the SSoC and the FSoC that forces the algorithm to add a visit to a CS, changes the spatial attributes of the initial activity chain to a great extent. In other words, the method proposes different locations to be visited by the traveler. In this specific example, the order of the visits to the activities does not change, but a charging session is added between the visits to the locations for the second and the third activities. Finally, we can notice that while the optimized activity chain includes an extra activity that corresponds to the detour for charging, the selection of alternative locations and suitable CS enables the reduction of the overall travel time.

5.3. Research Implications for EV Usage and Energy Systems

Apart from the obvious benefits that the method presents on an individual level for travelers, such as decreasing the travel time in their daily activity chains and dealing with range anxiety, there are two main takeaways regarding the use of EVs. Firstly, on an individual level, the method can be an effective supporting tool that helps minimize consumption, distributing drivers to the suitable CS according to availability and allowing them to find possible charging opportunities according to two realistic charging scenarios. The travelers that do not own chargers at their homes can greatly benefit from this method since it allows the calculation of tours and routes, that guarantee a FSoC for the drivers when reaching their home. On a system-level, the use of such a method, although it is aimed for personal usage, can provide benefits for cities. The better distribution of drivers to charging stations is a major possible benefit, as well as, the more efficient tours that are to be followed will minimize the consumed energy and the load on the charging stations network as a whole. As shown in work by Cuchý et al. [3], and their simulation efforts based on multi-destination planning tools like the one provided in this article, those methods can allow the better alignment of supply and demand and can help to make the overall transition to electromobility smoother.

6. Conclusions

A method has been elaborated for the solution DACO-EV problem, which many EV drivers face in their daily lives. Given the initial schedule of activities of the traveler, their locations, and priorities concerning each activity, our method calculates an optimal tour that guides the travelers in a better way.

At the same time, travelers do not need to face range anxiety since the solution guarantees that the EV battery level remains above the desired energy level. In the modeling, two charging scenarios were included, where the traveler can charge the EV either while staying in the EV at the CS or leaving it for charging at a CS while conducting another activity. This article describes the following contributions of the method:

- Consumption calculation mechanisms based on previous literature and real-world EV usage attributes.
- Calculation of an optimal tour based on the Starting State of Charge (SSoC) and the desired Final Stage of Charge (FSoC) of the EV used.
- A real charging stations network for Budapest, Hungary, and the availability of those charging stations based on past usage data.
- Two real-world charging scenarios for the detour of travelers, when they need to charge their vehicles.

Our research efforts aimed to develop an optimization model that is capable of modeling the real-world problem of optimization of daily activity chains for EV users and can serve as a driver assistance solution that reduces the range anxiety of EV users. The model itself can be used for studying the utilization of CS networks in urban environments and the study of the expected behavior of EV drivers. As discussed earlier in this article, although those methods address the problem on an individual level, they can be beneficial for the system-level, the CS network infrastructure, and the road network.

As a few next steps in this research, we propose further investigation of the incorporation of the availability of charging stations. It should be studied how solution times and the quality of solutions delivered by the algorithm change when the availability is considered according to different data mining methods. A worth mentioning research direction is the inclusion of the effect of traffic on the tours that happen on rush-hours and may get delayed travel times in regard to traffic. The utilization of this algorithm as part of a simulation framework based on activity-based modeling is a long-term research goal.

Author Contributions: Conceptualization, D.R. and D.E.-K.; methodology, D.R. and D.E.-K.; writing—original draft, D.R. and D.E.-K.; data curation, D.R. and D.E.-K.; software, D.R.; supervision, D.E.-K. All authors have read and agreed to the published version of the manuscript.

Funding: Electric Travelling project has received funding from the ERA-NET COFUND Electric Mobility Europe (EMEurope), on the basis of grant contract No. ZFF24/2018-ITM_SZERZ concluded with the Ministry for Innovation and Technology.

Acknowledgments: The research reported in this paper was supported by the Higher Education Excellence Program in the frame of the Artificial Intelligence research area of Budapest University of Technology and Economics (BME FIKP-MI/SC).

Conflicts of Interest: The authors declare no conflict of interest.

References

1. Dua, R.; White, K.; Lindland, R. Understanding potential for battery electric vehicle adoption using large-scale consumer profile data. *Energy Reports* **2019**, *5*, 515–524. [CrossRef]
2. U. S. D. of Energy, Electric Vehicle Benefits. Available online: https://www.energy.gov/eere/electricvehicles/electric-vehicle-benefits (accessed on 25 June 2019).
3. Cuchý, M.; Štolba, M.; Jakob, M. Benefits of Multi-Destination Travel Planning for Electric Vehicles. Proceedings of 2018 21st International Conference on Intelligent Transportation Systems (ITSC), Maui, HI, USA, 4–7 November 2018; pp. 327–332.
4. Vokony, I.; Hartmann, B.; Kiss, J.; Sőrés, P.; Farkas, C. Business Models to Exploit Possibilities of E-mobility: An Electricity Distribution System Operator Perspective. *Period. Polytech. Transp. Eng.* **2019**, *48*, 1–10. [CrossRef]

5. Orbulov, V.; Lógó, E. Assessment of Applicability of the Service Design Method on Electric Vehicles. *Period. Polytech. Transp. Eng.* **2019**, *48*, 52–56. [CrossRef]

6. I. E. Agency, Global EV Outlook 2019. Available online: https://www.iea.org/reports/global-ev-outlook-2019 (accessed on 1 July 2019).

7. Statharas, S.; Moysoglou, Y.; Siskos, P.; Zazias, G.; Capros, P. Factors Influencing Electric Vehicle Penetration in the EU by 2030: A Model-Based Policy Assessment. *Energies* **2019**, *12*, 2739. [CrossRef]

8. Neubauer, J.; Wood, E. The impact of range anxiety and home, workplace, and public charging infrastructure on simulated battery electric vehicle lifetime utility. *J. Power Sources* **2014**, *257*, 12–20. [CrossRef]

9. Ou, S.; Yu, R.; Lin, Z.; Ren, H.; He, X.; Przesmitzki, S.; Bouchard, J. Intensity and daily pattern of passenger vehicle use by region and class in China: Estimation and implications for energy use and electrification. *Mitig. Adapt. Strateg. Glob. Chang.* **2019**, 1–21.

10. Bast, H.; Delling, D.; Goldberg, A.; Müller-Hannemann, M.; Pajor, T.; Sanders, P.; Wagner, D.; Werneck, R.F. Route planning in transportation networks. In *Algorithm Engineering*; Springer: Cham, Switzerland, 2016; Volume 9220, pp. 19–80.

11. Streit, T.; Allier, C.-E.; Weiss, C.; Chlond, B.; Vortisch, P. Changes in Variability and Flexibility of Individual Travel in Germany. *Transp. Res. Rec. J. Transp. Res. Board* **2015**, *2496*, 10–19. [CrossRef]

12. Kim, H.; Xiang, Z.; Fesenmaier, D.R. Use of The Internet for Trip Planning: A Generational Analysis. *J. Travel Tour. Mark.* **2015**, *32*, 276–289. [CrossRef]

13. Tarigan, A.; Fujii, S.; Kitamura, R. Intrapersonal variability in leisure activity-travel patterns: The case of one-worker and two-worker households. *Transp. Lett.* **2012**, *4*, 1–13. [CrossRef]

14. Pell, A.; Nyamdzawo, P.; Schauer, O. Intelligent transportation system for traffic and road infrastructure-related data. *Int. J. Adv. Logist.* **2016**, *5*, 19–29. [CrossRef]

15. Farag, S.; Lyons, G. What Affects Use of Pretrip Public Transport Information? *Transp. Res. Rec. J. Transp. Res. Board* **2008**, *2069*, 85–92. [CrossRef]

16. Cats, O.; Koutsopoulos, H.N.; Burghout, W.; Toledo, T. Effect of Real-Time Transit Information on Dynamic Path Choice of Passengers. *Transp. Res. Rec. J. Transp. Res. Board* **2011**, *2217*, 46–54. [CrossRef]

17. Kitamura, R. An evaluation of activity-based travel analysis. *Transportation (Amst.)* **1988**, *15*. [CrossRef]

18. Axhausen, K.W.; Gärling, T. Activity-based approaches to travel analysis: Conceptual frameworks, models, and research problems. *Transp. Rev.* **1992**, *12*, 323–341. [CrossRef]

19. Algers, S.; Eliasson, J.; Mattsson, L.-G. Is it time to use activity-based urban transport models? A discussion of planning needs and modelling possibilities. *Ann. Reg. Sci.* **2005**, *39*, 767–789. [CrossRef]

20. Dong, J.; Liu, C.; Lin, Z. Charging infrastructure planning for promoting battery electric vehicles: An activity-based approach using multiday travel data. *Transp. Res. Part C Emerg. Technol.* **2014**, *38*, 44–55. [CrossRef]

21. Kontou, E.; Yin, Y.; Ge, Y.-E. Cost-Effective and Ecofriendly Plug-In Hybrid Electric Vehicle Charging Management. *Transp. Res. Rec. J. Transp. Res. Board* **2017**, *2628*, 87–98. [CrossRef]

22. Esztergár-Kiss, D.; Rózsa, Z.; Tettamanti, T. Extensions of the Activity Chain Optimization Method. *J. Urban Technol.* **2018**, *25*, 125–142. [CrossRef]

23. Esztergár-Kiss, D.; Rózsa, Z.; Tettamanti, T. An activity chain optimization method with comparison of test cases for different transportation modes. *Transp. A Transp. Sci.* **2019**, *9935*, 1–23. [CrossRef]

24. Charypar, D.; Nagel, K. Generating complete all-day activity plans with genetic algorithms. *Transportation (Amst.)* **2005**, *32*, 369–397. [CrossRef]

25. Abbaspour, R.A.; Samadzadegan, F. Time-dependent personal tour planning and scheduling in metropolises. *Expert Syst. Appl.* **2011**, *38*, 12439–12452. [CrossRef]

26. Liao, F.; Arentze, T.; Timmermans, H. Incorporating space–time constraints and activity-travel time profiles in a multi-state supernetwork approach to individual activity-travel scheduling. *Transp. Res. Part B Methodol.* **2013**, *55*, 41–58. [CrossRef]

27. Saharidis, G.K.D.; Rizopoulos, D.; Fragkogios, A.; Chatzigeorgiou, C. A hybrid approach to the problem of journey planning with the use of mathematical programming and modern techniques. *Transp. Res. Procedia* **2017**, *24*, 401–409. [CrossRef]

28. Aifadopoulou, G.; Ziliaskopoulos, A.; Chrisohoou, E. Multiobjective Optimum Path Algorithm for Passenger Pretrip Planning in Multimodal Transportation Networks. *Transp. Res. Rec. J. Transp. Res. Board* **2007**, *2032*, 26–34. [CrossRef]

29. Xie, C.; Boyles, S.D.; Ding, J.; Wu, X. Travel Behavior and Transportation Systems Analysis of Electric Vehicles. *J. Adv. Transpor.* **2018**, *2018*, 1–2. [CrossRef]

30. Guo, F.; Yang, J.; Lu, J. The battery charging station location problem: Impact of users' range anxiety and distance convenience. *Transp. Res. Part E Logist. Transp. Rev.* **2018**. [CrossRef]

31. Esmaili, M.; Shafiee, H.; Aghaei, J. Range anxiety of electric vehicles in energy management of microgrids with controllable loads. *J. Energy Storage* **2018**. [CrossRef]

32. Lane, B.W.; Dumortier, J.; Carley, S.; Siddiki, S.; Clark-Sutton, K.; Graham, J.D. All plug-in electric vehicles are not the same: Predictors of preference for a plug-in hybrid versus a battery-electric vehicle. *Transp. Res. Part D Transp. Environ.* **2018**. [CrossRef]

33. Gorenflo, C.; Rios, I.; Golab, L.; Keshav, S. Usage Patterns of Electric Bicycles: An Analysis of the WeBike Project. *J. Adv. Transp.* **2017**. [CrossRef]

34. Baum, M.; Dibbelt, J.; Pajor, T.; Wagner, D. Energy-optimal routes for electric vehicles. In Proceedings of the 21st ACM SIGSPATIAL International Conference on Advances in Geographic Information Systems—SIGSPATIAL'13, Orlando, FL, USA, 5–8 November 2013; pp. 54–63.

35. Baum, M.; Dibbelt, J.; Gemsa, A.; Wagner, D. Towards route planning algorithms for electric vehicles with realistic constraints. *Comput. Sci.-Res. Dev.* **2016**, *31*, 105–109. [CrossRef]

36. Baum, M.; Dibbelt, J.; Gemsa, A.; Wagner, D.; Zündorf, T. Shortest feasible paths with charging stops for battery electric vehicles. In Proceedings of the 23rd SIGSPATIAL International Conference on Advances in Geographic Information Systems—GIS '15, Seattle, WA, USA, 3–6 November 2015; pp. 1–10.

37. Moghaddass, R.; Mohammed, O.A.; Skordilis, E.; Asfour, S. Smart Control of Fleets of Electric Vehicles in Smart and Connected Communities. *IEEE Trans. Smart Grid* **2019**, *10*, 6883–6897. [CrossRef]

38. Liao, C.-S.; Lu, S.-H.; Shen, Z.-J.M. The electric vehicle touring problem. *Transp. Res. Part B Methodol.* **2016**, *86*, 163–180. [CrossRef]

39. Cuchý, M.; Štolba, M.; Jakob, M. Whole Day Mobility Planning with Electric Vehicles. In Proceedings of the 10th International Conference on Agents and Artificial Intelligence, Madeira, Portugal, 16–18 January 2018; pp. 154–164.

40. Cuchý, M.; Štolba, M.; Jakob, M. *Integrated Route, Charging and Activity Planning for Whole Day Mobility with Electric Vehicles*; Filipe, J., Fred, A., Eds.; Springer: Berlin/Heidelberg, Germany, 2019; Volume 358, pp. 274–289.

41. Lucas, A.; Prettico, G.; Flammini, M.; Kotsakis, E.; Fulli, G.; Masera, M. Indicator-Based Methodology for Assessing EV Charging Infrastructure Using Exploratory Data Analysis. *Energies* **2018**, *11*, 1869. [CrossRef]

42. Varga, B.; Sagoian, A.; Mariasiu, F. Prediction of Electric Vehicle Range: A Comprehensive Review of Current Issues and Challenges. *Energies* **2019**, *12*, 946. [CrossRef]

43. Goldberg, D.E. *Genetic Algorithms in Search, Optimization & Machine Learning*; Addison-Wesley Publishing Company, Inc.: Reading, MA, USA, 1989.

44. Geofabrik Website. Available online: https://www.geofabrik.de/ (accessed on 8 August 2019).

45. McHugh, B. The OpenTripPlanner Project. Available online: https://www.opentripplanner.org/ (accessed on 8 August 2019).

46. e-Mobi. Available online: https://e-mobi.hu/ (accessed on 8 August 2019).

47. Electric Vehicle Database. Available online: https://ev-database.org/ (accessed on 1 August 2019).

48. Mehar, S.; Senouci, S.M.; Remy, G. EV-planning: Electric vehicle itinerary planning. In Proceedings of the 2013 International Conference on Smart Communications in Network Technologies (SaCoNeT), Paris, France, 17–19 2013; pp. 1–5.

49. Sachenbacher, M.; Leucker, M.; Artmeier, A.; Haselmayr, J. Efficient Energy-Optimal Routing for Electric Vehicles. In Proceedings of the Twenty-Fifth AAAI Conference Artificial Intelligence, San Francisco, CA, USA, 7–11 August 2011.

50. Wu, X.; Freese, D.; Cabrera, A.; Kitch, W.A. Electric vehicles' energy consumption measurement and estimation. *Transp. Res. Part D Transp. Environ.* **2015**, *34*, 52–67. [CrossRef]

51. Musardo, C.; Rizzoni, G.; Staccia, B. A-ECMS: An Adaptive Algorithm for Hybrid Electric Vehicle Energy Management. *Europ. J. Cont.* **2005**, *11*, 1816–1823.

52. Fišer, T. Integrated Route and Charging Planning for Electric Vehicles. Bachelor's Thesis, Czech Technical University, Prague, Czech Republic, 2017.

53. De Cauwer, C.; Van Mierlo, J.; Coosemans, T. Energy Consumption Prediction for Electric Vehicles Based on Real-World Data. *Energies* **2015**, *8*, 8573–8593. [CrossRef]
54. De Cauwer, C.; Verbeke, W.; Coosemans, T.; Faid, S.; Van Mierlo, J. A Data-Driven Method for Energy Consumption Prediction and Energy-Efficient Routing of Electric Vehicles in Real-World Conditions. *Energies* **2017**, *10*, 608. [CrossRef]

Article

New Dispatching Paradigm in Power Systems Including EV Charging Stations and Dispersed Generation: A Real Test Case

Fabio Cazzato [1], Marco Di Clerico [1], Maria Carmen Falvo [2],*, Simone Ferrero [1] and Marco Vivian [2]

[1] E-Distribuzione, Enel Group, 00198 Rome, Italy; fabio.cazzato@e-distribuzione.com (F.C.); marco.diclerico@e-distribuzione.com (M.D.C.); simone.ferrero@e-distribuzione.com (S.F.)

[2] DIAEE-Department of Astronautics, Energy and Electrical Engineering, University of Rome Sapienza, 00184 Rome, Italy; marco94vivian@gmail.com

* Correspondence: mariacarmen.falvo@uniroma1.it

Received: 22 January 2020; Accepted: 18 February 2020; Published: 20 February 2020

Abstract: Electric Vehicles (EVs) are becoming one of the main answers to the decarbonization of the transport sector and Renewable Energy Sources (RES) to the decarbonization of the electricity production sector. Nevertheless, their impact on the electric grids cannot be neglected. New paradigms for the management of the grids where they are connected, which are typically distribution grids in Medium Voltage (MV) and Low Voltage (LV), are necessary. A reform of dispatching rules, including the management of distribution grids and the resources there connected, is in progress in Europe. In this paper, a new paradigm linked to the design of reform is proposed and then tested, in reference to a real distribution grid, operated by the main Italian Distribution System Operator (DSO), *e-distribuzione*. First, in reference to suitable future scenarios of spread of RES-based power plants and EVs charging stations (EVCS), using Power Flow (PF) models, a check of the operation of the distribution grid, in reference to the usual rules of management, is made. Second, a new dispatching model, involving DSO and the resources connected to its grids, is tested, using an Optimal Power Flow (OPF) algorithm. Results show that the new paradigm of dispatching can effectively be useful for preventing some operation problems of the distribution grids.

Keywords: Dispatching Service; Dispersed Generation; Distribution Grid; Electrical Vehicle; Optimal Power Flow; Power Flow; Power System; Simulation Models

1. Introduction

Stopping climate change requires a revolution in the way we produce and consume energy. On the production side, the most important step is to reduce the generation from carbon and other fossil fuel to near-zero, and to increase the amount of the Renewable Energy Sources (RES). On the consumption side, the revolution requires an improvement of the efficiency and in the electrification of the energy demand. In the transport sector, these two objectives can be reached thanks to the spread of Electric Vehicles (EVs), whose Charging Stations (EVCS) power is enhancing (up to 250 kW) to allow a faster recharge [1–4]. All these changes affects the operation of the power systems, at transmission grids level in High Voltage (HV) and distribution grids level in Medium Voltage (MV) and Low Voltage (LV) [5,6]. The issues are different in the two stages of the power system: most of the RES-based power plants are connected to the distribution grids (Dispersed Generation, DG) and all the EVCS are connected to the distribution grids, thus, the first impact is on the distribution level. However, at the same time, the DG, typically RES-based, also has indirect impacts on the transmission level. some problems related to the operation of the transmission grids, due to the DG, are:

- Reduction of the equivalent load on the HV grid, with reactive power management problems in light-load hours and thus high voltage problem;
- Reduction or turning off of conventional power plants generation, reducing the reserves for the control of the whole power system, and in particular causing a reduction of the inertia and of the spinning reserve and thus frequency control issues,
- Inverse power flows on the primary substations in some hours,
- Dispatching reserves and rules for controlling the frequency, inappropriate for the reliability of the grid, increasing costs of the dispatching service for the Transmission System Operators (TSOs).

On the other hand, some problems related to the operation of the distribution grid, due to DG and EVCS are:

- Congestions on some branches,
- Voltage drops and critical voltage control,
- Inverse power flows on some lines through the primary substations,
- Increasing of the power losses,
- Power quality issues.

Many of the listed issues are due to the fact that the RES-based DG and EVCS are still managed with a "fit and forget" approach: the amount of electricity available in production or needed in consumption is put into or delivered by the system, without any limits. This type of approach is no longer feasible, especially when the rate of DG and EVs become high, as is the case in Europe.

A solution of most of the listed issues could be an active management of distribution grids and of the sources, there connected, in charge to the Distributor System Operators (DSOs), similar to the management done by the TSOs with the HV grids. For all these reasons, a reform of the electrical dispatching is in progress in Europe, including Italy [7]. The common objective of the review process in all the European Countries is to test new ways for getting the necessary resources to guarantee the reliability of the whole power system, through a new form of dispatching service. It also includes taking in new entities, such as DG, MV and LV flexible end-users (such as EVCS) and their combinations, and thus involving in the dispatching functions the DSOs.

In literature, many recent works are present on the impact on the transmission and distribution grids of RES-based DG and EVCS. Some paper are dedicated only to the impacts and possible management solution for the RES-based DG on the distribution grids [8–11]. Others are related only to the influence and possible control strategies of EVCS, in distribution grids, according to a Vehicle-to-Grid approach, in reference to a smart grid environment [11–21]. Other papers are dedicated to analyze the combination of the effect of the two (DG and EVCS) and of their suitable management in the distribution grids operation and planning [22–25]. Few papers deal with the problems related both to the distribution and transmission grids, linked to the DG and EVCS [26,27].

In this context, the originality of the present work is to propose and test a new dispatching model on a real distribution system, in line with the reform that is in progress in Europe and in Italy [7], that implies a smart management of EVCSs, coordinated with DG, mainly RES-based. Analyses and simulations are done on the new model in reference to a real distribution system, in reference to future scenarios (2030), owned and operated by the most important DSO in Italy, *e-distribuzione*.

The work is so organized. Section 2 includes a short description of new dispatching models proposed by the reform in progress. The main features of the real grid object of the analysis are in Section 3. Section 4 shows the methodologies and the assumptions for the simulations. Section 5 describes and discusses the obtained results. In Section 6, conclusions are drawn.

2. Dispatching Models in the European Reform

The Italian TSO (TERNA) is moving forward in the dispatching reform guided by the Italian Authority for electricity (ARERA) in accordance with the European policies and rules [7]. In 2013, the debate concerning new dispatch models started. The investigation was through three possible models managing the dispatching sources that exploit RES-based generators, which can be summarized as follows:

- Model 1, called "Extended Central Dispatching": all the sources connected to the transmission and distribution grids, including generators and flexible end-users, must provide dispatching services to the TSO, through the ancillary service market, and only local system services to the DSO for solving grid problems at distribution level, with a direct call.
- Model 2, called "DSO Local Dispatching": the TSO can purchase system services from traditional generation power plants, RES-based power plants or flexible end-users connected to the High Voltage (HV) grids, and directly from DSO, through the ancillary service market. The DSO itself can indeed purchase dispatching services or local system services, from the DG and flexible end-users, connected to its medium voltage (MV) and low voltage (LV) grids, through a new ancillary service market at distribution level.
- Model 3, called "Planned Profile Exchange at HV/MV Electrical Substation (ESS)": all the sources connected to the transmission grids, including generators and flexible end-users, must provide dispatching services to the TSO. The DSO is responsible only for the planned energy exchange in a single primary ESS, or in a zone that includes several primary ESS, but it does not provide any dispatching service to the TSO. This profile is guaranteed by DSO, using the DG and flexible end-users, connected to its MV and LV grids; they can be also called for system services on the distribution grids with a direct call by DSO.

Nowadays the reforms are heading toward Model 1, where the TSO can directly draw new dispatching sources from entities connected to the distribution system. The requests of the TSO do not consider the issues that may arise in the distribution grids, such as congestions, under-voltage and over-voltage phenomena and so on; thus, the DSOs are not only suffering the unavoidable spread of DG, but also an external control, which may lead to further problems. The control of the DG and of the flexible loads, by the DSO over its system, could be the best solution to avoid technical issues (Model 2 or 3). In the meantime, to guarantee a predictable exchange profile at primary ESS, reducing the global dispatching cost could be easy in this case for the DSO and could be useful for TSO (Model 3). Obviously, this can be done in a smart grid, where suitable communication [15] and control technologies are available between the DSO and the dispersed resources. This is the reason why the most important Italian DSO, *e-distribuzione*, is investigating the opportunity to apply Model 2 or 3 instead of Model 1, which is currently tested by the Italian TSO.

3. *E-Distribuzione* Test Case

The study has been carried out on a real distribution grid belonging to *e-distribuzione*, which is the most important DSO in Italy, and is involved in many studies on the impact of EVCS and DG on distribution grids [28–32]. The main features of the distribution grid are summarized in Figure 1.

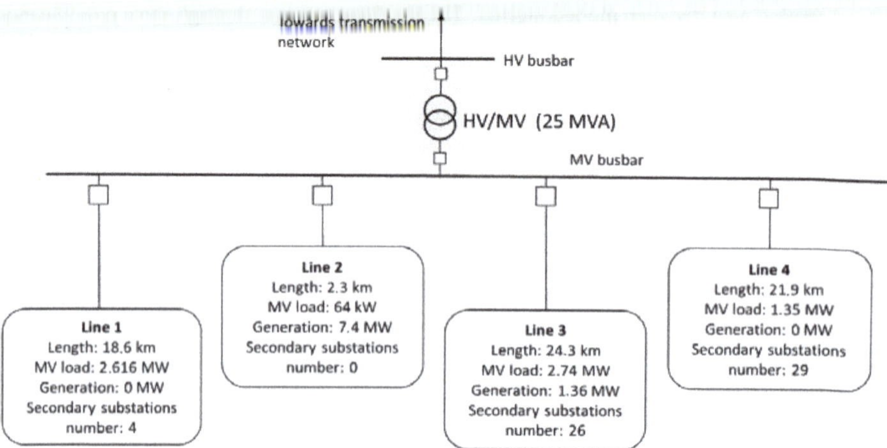

Figure 1. *e-distribuzione* distribution grid.

The system is located in Southern Italy and it is subtended to a primary Electrical Sub-station (ESS) with a 25 MVA power transformer and includes four lines. The main data on the lines are summarized in Table 1.

Table 1. MV lines characteristics.

Line	Length [km]	% in Overhead Line Naked	% in Underground	% in Overhead Line Insulated
1	18.6	59%	0%	41%
2	2.3	0%	0%	100%
3	24.3	1%	0%	99%
4	21.9	85%	4%	11%
Tot	67.1	45%	1%	54%

All the MV generators are Photovoltaic (PV) systems with a total peak power of 8.76 MW. The powers and the lines to which they are connected are detailed in Table 2.

Table 2. MV PV Generators.

MV Generation Plants	Power [kW]	Line
G1	7983	2
G2	48	3
G3	1250	3
G4	60	3

Big consumers connected directly to MV lines are detailed in Table 3 with their available power and delivery line.

For each passive customer (industrial and residential) in MV, *e-distribuzione* has provided data about the active and reactive power profile in time, every 15 min, in reference to the four seasons and the two typical days of the week (workdays and weekends). The LV loads and generators are represented, in their respective secondary ESS, as aggregate profiles of active and reactive power, because the information about the LV network are not given. The production profile of the PV systems have been calculated in reference to the irradiation values in the analyzed area through System Advisor Model (SAM) software [33,34]. In addition, *e-distribuzione* provides the technical features of the MV grid components.

All these data, useful for the simulations, are not reported here, both for lack of space and privacy policies.

Table 3. MV passive customers.

MV Customers	Power [kW]	Line
C1	20	3
C2	550	3
C3	350	4
C4	64	2
C5	500	4
C6	866	1
C7	505	4
C8	501	1
C9	567	1
C10	682	1
C11	300	3
C12	120	3
C13	250	3
C14	1500	3
TOT	6775	-

4. Models for Simulations

The system has been implemented in a MATLAB environment and simulated through MATPOWER Power Flow (PF) and Optimal Power Flow (OPF) algorithms [35,36]. Twelve simulations have been analyzed according to season and the day of the week was considered. The PF and OPF run every 15 min. The simulations take into account only the active powers as control variables, because it is assumed to be a dispatching model that deals only with the control of the active power of the sources involved with a different objective, according to the reform proposals in progress in Europe and in Italy.

A first PF analysis in the present scenario was performed, to check the operation of the grid in the present and without any type of dispatching action implemented. The results show that the peak demand is reached during the summer evening. Nevertheless, the grid constraints, such as voltage and current limits, are widely respected all year long and the grid turns out to be discharged and highly suitable for new sources integration.

4.1. Assumptions for Simulations

An evaluation was carried out about possible future scenarios in 2030 considering the targets imposed by the European Commission and Italian Authority [1–3]. The future scenarios have been chosen based on the experience on the field of *e-distribuzione* authors and their expertise coming out by the management and operation of the Italian distribution grid [28]. It is well known that the forecasts are dependent on many different factors, such as investments, policies and politics of the selected nation. For these reasons, three different scenarios have been deduced according to the speed of the growth, as shown in Table 4.

Table 4. Percentage of growth for 2030.

Growth	PV Increase at LV Level	PV Share at MV Level	EVs Share
	[%]	[%]	[%]
Slow	+100	50	10
Moderate	+150	70	12
Fast	+200	90	15

Since 31 GW of PV power has to be installed at distribution level to reach 51 GW, for the LV size, the installed power of domestic PV plants should at least double and in the fast growth scenario, it should triple. Thus, the variation will be considered within 100 and 200%. Concerning the PV installed at MV, the current plants are left as they are, while new PV plants are supposed to rise in the proximity of the MV customers, affecting between 50% and 90% of them. Most of the studies highlight a sudden boom in the sale of EV between 2020 and 2025, when the technology would be mature to be worldwide adopted. Unfortunately, nowadays, Italy holds one of the last positions in the European ranking of EV readiness, hence, the forecasts are discouraging; even in the best scenario the percentage goal to contain the emission is not achieved. The Italian percentages of EV share in 2030 according to *e-distribuzione* fluctuate between 10 and 15%.

The integrated EVCSs are assumed to be:

- Home station, with a rated power of 7.2 kW;
- Pole station, with a rated power of 22 kW;
- Super-fast station, with a rated power of 350 kW + 2150 Kw;

The number of every kind of EVCS is proportional to the EVs share and the number of vehicles circulating in the area, but it is obtained following different methodologies and considering a battery capacity of 80 kWh. It is worth to note that the study is focused on how the distribution grid could be affected by the increase of installed PV power, EVCSs and their management. Thus, any kind of technical information, such as kind of socket connection, details of the recharge and battery properties, is avoided, because they do not concern this dissertation.

From the DSO point of view, the EVCS could be seen as a flexible load. Indeed, the charging stations can not only withdraw energy from the grid, as is pretty much common today, but they will be able to inject power, as is shown in many pilot projects, with so-called Vehicle to Grid (V2G) technology [13–21]. In this framework, an assumption for simulations is to consider that 25% of the batteries are not accessible, because they need to be charged.

4.2. OPF Objetive Functions

In reference to these scenarios and assumptions, different OPF analyses have been performed in reference to different objective function. In particular, to minimize the losses, to get a peak shaving function and to get a voltage drop containment.

The higher the power crossing the primary ESS, the higher are the losses in the system, and consequently, the operation cost. This is the reason why the DSOs are often concerned about the minimization of the active power flow in the primary ESS. The peak shaving concept lies on the assumption that some loads or generator can be shifted in time to avoid high losses on the grid. Any storage available in the grid could be a source to fulfil this need, since it can inject and withdraw energy whenever it is needed, respecting the capacity and maximum power constraints. The batteries of the EV connected to the distribution grid can be a useful storage for this task. In particular, the peak shaving corresponds to an energy injection (or withdraw in case of negative peak) that has to be withdrawn (or injected) within the day to allow the daily cycle of an EV battery. The proposed EV battery cycle considers that the EV drivers change the range setting of the lower limit of the State of Charge (SOC) to 50%. Furthermore, the hypothetic daily trip (9% of discharge) is divided into the route from home to office and vice versa (Figure 2).

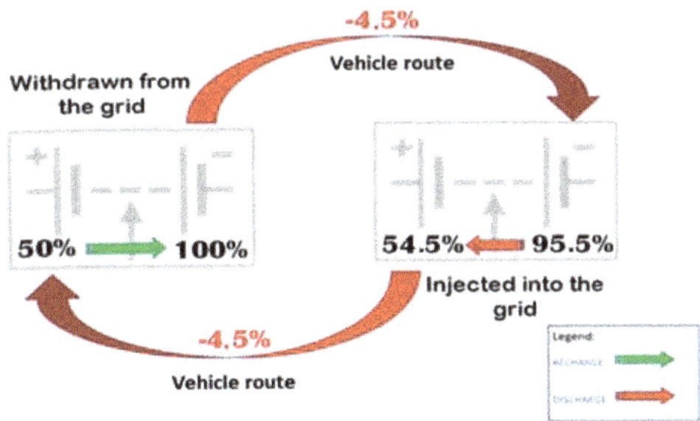

Figure 2. EV battery cycle of use hypnotized in the simulations.

The energy injected to the grid has to be the 82% of the energy withdrawn during the day to have a perfect charging cycle. The energy spent during the route is seen as a loss from the grid point of view, but it is in fact the energy that makes the EVs move, thus, the most useful from the driver's point of view. Two algorithms have been carried on with two different primary goals: to reduce the evening peak and to reduce the reverse power flow. These two purposes cannot be completely decoupled from each other; indeed, as an EV battery is going to be discharge in the evening, it will unavoidably recharge during the day, flatting the negative peak and vice versa. The goal of the analysis is to evaluate the power, the energy and the percentage of EVs involved guaranteeing a given peak shaving service. This evaluation of peak shaving potential is based on thresholds that limit the positive active power, which can flow through the primary ESS; once this limit is broken, the algorithm is set to exploit the charging station to balance the grid. The evening peak set point is 4.7 MW, because it allows a realistic exploitation of the EVs batteries without requiring more power than the available one. The minimum active power threshold is set to −2.7 MW.

As the voltage drop is remarkable during summer evening, the home stations are programmed to inject power to the grid when violation occurs. An important assumption is made: the stations belong to the DSO or the EVs users are inclined to accept less power availability, the available power indeed can be modified to provide local services, only if there is a will from the battery side. The percentage of the availability of the battery to be charged or discharged and the presence of the vehicles have been considered equal to 45% of the total installed power. This analysis aims to verify whether the defined percentage of dispatchable power can guarantee the containment of the voltage drop within ±6.3%.

5. Results and Discussion

5.1. PF Analysis

A first PF analysis has been made in reference to the future scenarios, for checking the performance of the grid, without any dispatching model implemented. The results are here reported in reference to the case more stressing, which is the one for the fast growth scenario in spring and summer. The impact of increased PV systems, home stations and pole stations on the grid in terms of active power at the MV bus-bar is shown, for the fast growth scenario, in Figure 3a for the spring season and in Figure 3b for the summer season, compared to the real profile of nowadays. It is evident that the presence of the EVCS has as results an increasing of the amount of active power at MV bus-bar, only during evening hours, linked to the evening use of the home stations. During the noon hours, the prevalent effect is of increasing of the negative active power at MV bus-bar (up to 5 MW), due to the higher production by new PV plants.

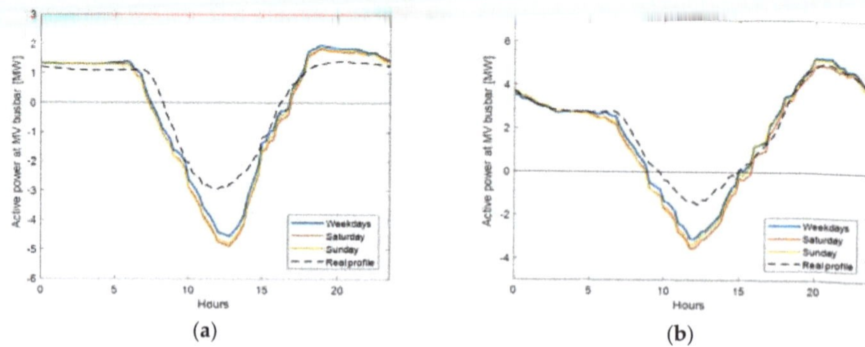

Figure 3. Active power at MV side of the primary ESS for fast growth scenario during spring (**a**) and summer (**b**) considering the impact of Home stations and pole stations.

Figure 4 shows the trend of the active power in the primary ESS adding also the super-fast EVCS, which could require up to 750 kW, during spring season (a) and summer season (b). The effect is the same described for the case in Figure 3.

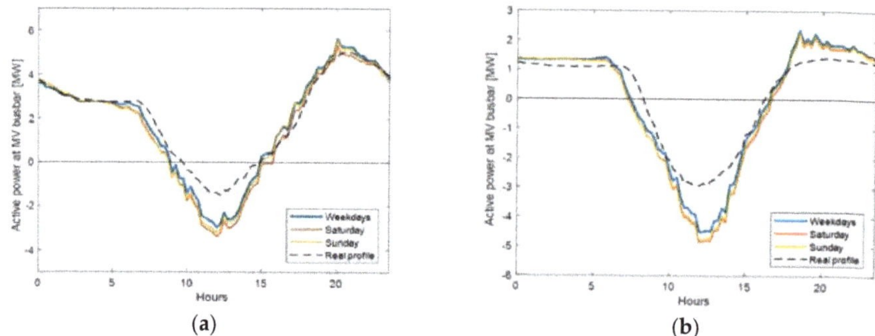

Figure 4. Active power at MV side of the primary ESS for fast growth scenario during spring (**a**) and summer (**b**) considering the impact of Home stations, pole stations and super-fast stations.

From the simulation results, the voltage was also checked: a drop of 6.71% was tested during summer, but it was confined within 5% during the rest of the year. Nevertheless, the grid still did not experience any congestion and it is highly suitable for further installations.

5.2. OPF Analysis In Case Of Evening Peak Shaving Service

The most interesting scenario analyzed is the most stressful, that is, the fast growth one during a summer weekday. A threshold assumed to be of 4.7 MW implicates an energy injection of 760 kWh, which stands for an exploitation of 29 home stations more than the foreseen ones. The peak power indeed reaches 800 kW (Figure 5), which means a contemporary recharge of about 111 EVs. The recharges consider the energy spent for the daily trip and they occur during the rest of the day, preferably during the time of reverse power flow. The 111 vehicles are 60% of the total EVs fleet, but 79% of the available EVs.

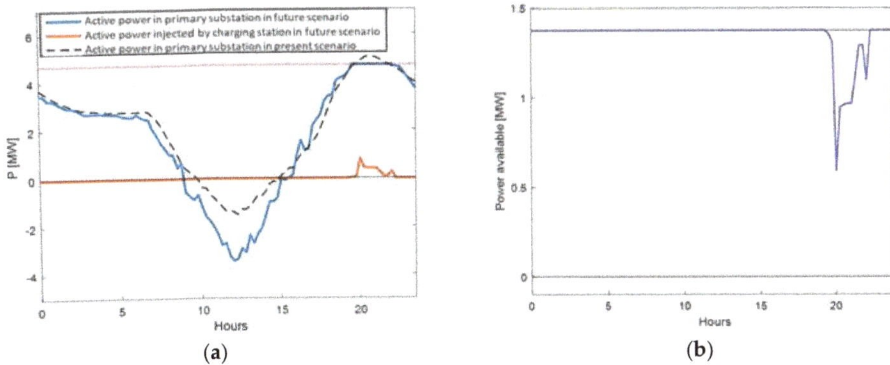

Figure 5. Active power at the MV side of the primary ESS and power injected by the EV home stations for the evening peak shaving (**a**); power available by the home stations (**b**).

5.3. OPF Analysis In Case Of Reverse Power Flow Shaving

From the energy point of view, the system will need to charge 22 EVs more to increase the energy demand of 880 kWh. On the other hand, 949 kW of peak power (Figure 6) requires 21 EVCS more, with the strong assumption that they are used at the same time. In this case, the peak power dictates the right spread of EVs which is equal to 49%. Nevertheless, this percentage is not actually a needed condition for the right functioning of the system, it should need just 42 EVs plugged at the same time. The recharge implies the discharge of part of the battery during the evening; this helps to smooth out the peak power and the voltage.

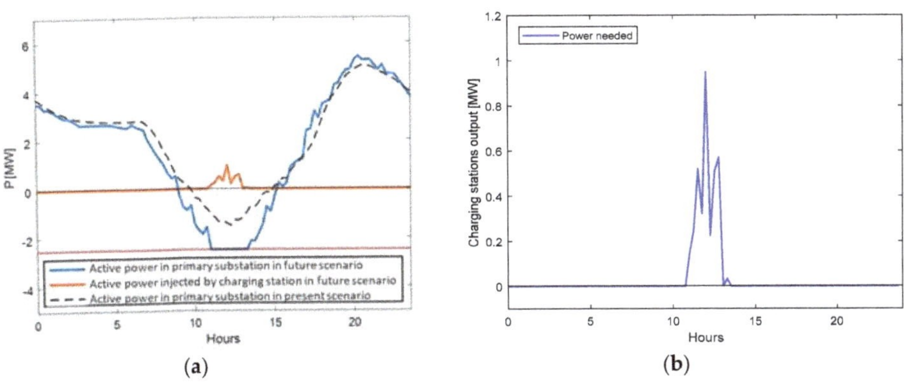

Figure 6. Active power at MV side of the primary ESS and power injected by the EV pole station for the reverse power flow shaving (**a**); power withdrawn by the EV pole stations (**b**).

Table 5 shows the results of the peak shaving analysis: the power and energy required by the DSO vary according to the goal set. The EV drivers' awareness is calculated as the number of EV drivers that allow the grid to use their own battery over the total number of EV drivers in the area. The easier objective to implement appears to be the noon negative peak shaving, which requires only 22% of drivers' awareness. Nevertheless, this solution implies that installing 21 EVCS is too many compared to the circulating fleet (420 more EVs should be bought).

Table 5. Power and energy required for peak shaving services.

Goal	Set-Point	Withdrawal		Injection		EV Drivers Awareness
Evening peak shaving	4.7 MW	926	kWh	760	kWh	60%
		-	kW	800	kW	
Noon negative peak shaving	−2.7 MW	880	kWh	721	kWh	22%
		949	kW	100	kW	

The noon shaving goal instead could be reach without any economic effort; it simply depends on the EVS drivers' availability, which should reach a quite high percentage.

5.4. OPF Analysis In Case Of Voltage Drop Containment

The active power through the primary ESS has not changed a lot, except for the evening peak between 8 pm and 10 pm, when the home stations inject active power to compensate the voltage violation (Figure 7). The home stations are indeed the only source that can be exploited because it is the only one available during those hours.

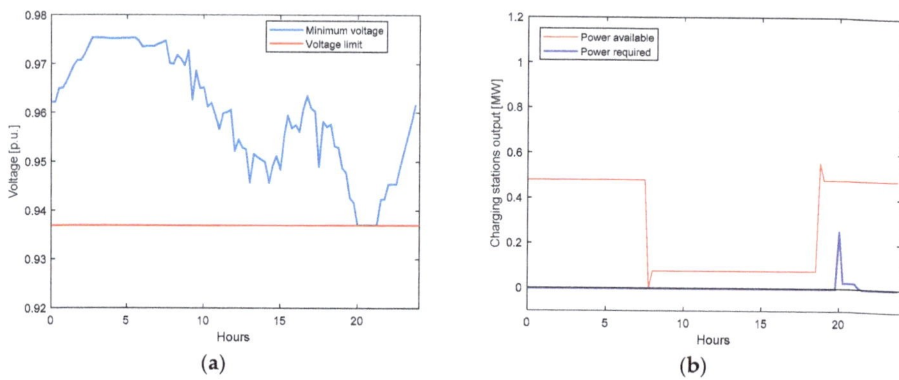

(a) (b)

Figure 7. Profile of the minimum voltage in the whole network (**a**); power injected by the EVs home station to contain the voltage violation (**b**).

The total power requested by the grid is 273 kW, thus, 57% of the total dispatchable power. The violation is highly contained within the threshold; this scenario requires only 37 available EVs batteries for a total of 160 kWh. The energy needed is much lower than the one that the involved EVs can provide; thus, the reduction of energy in each car will be almost derisory.

6. Conclusions

A new paradigm for the management of distribution grids, with big amount of EVCS and RES-based DG, is investigated in reference to a real distribution grid, belonging to the main Italian DSO, *e-distribuzione*. First, in reference to future scenarios of spread of RES-based power plants and EVCS, some PF analyses have been made to check the performance of the distribution grid in reference to the usual operation. Second, considering a new dispatching model involving DSO and EVCS, the opportunity to provide different local and system services are tested, using OPF algorithms.

The impact of EVCS and DG power growth, in 2030, is modelled and simulated in MATLAB environment, for three future scenarios, according to the intensity of the diffusion. After running a first PF simulation, the network detected a high peak power, given by the evening loads and a negative peak power provided by the DG plants. Thus, with OPF algorithms, the flexible management of the installed EVCS for providing services to the DSO was tested. The first algorithm was intended to cut off the voltage violations that occur if the voltage limits are at ±6.3%, as local service. The OPF algorithm tends to maximize the available power at the EVCS and acts only if a violation occurs. The study proved that it is possible to successfully contain the voltage drop within 6.3%, assuming an EVs availability of 45%. Another OPF algorithm had the goal to restrict the active power flow in the primary ESS within an upper and lower limit. This is effectively done to lessen the grid losses and provides a given value of maximum input/output power to the transmission grid, as system service. The tests have shown that the availability of EVs drivers to provide this kind of services is fundamental rather than high spread of EVs. In general, all the results show that the proposed paradigm of dispatching involving DSO and their resources (EVCS and DG) can effectively improve the distribution grid performance, to prevent its operation problems (such as under-voltages or congestions) and to support the whole power system management with system services.

Author Contributions: Conceptualization, M.C.F., F.C., M.D.C.; methodology, M.V., M.D.C. and S.F.; software, M.V.; validation, M.V., M.D.C. and S.F.; formal analysis, M.V. and M.C.F.; investigation, M.V.; resources, M.V. and M.C.F.; data curation, M.V., M.D.C. and S.F.; writing-original draft preparation, M.V.; writing, review and editing, M.V. and M.C.F.; visualization, M.V.; supervision, M.C.F. and F.C.; project administration, M.C.F. and F.C. All authors have read and agreed to the published version of the manuscript.

Funding: This research received no external funding.

Conflicts of Interest: The authors declare no conflict of interest. The funders had no role in the design of the study; in the collection, analyses, or interpretation of data; in the writing of the manuscript, or in the decision to publish the results.

References

1. International Energy Agency. Global EV Outlook 2019. Technical Report. 2019. Available online: https://webstore.iea.org/download/direct/2807?fileName=Global_EV_Outlook_2019.pdf (accessed on 14 November 2019).
2. European Commission. Communication from the Commission to the European Parliament, the Council, the European Economic and Social Committee and the Committee of the Regions; A Policy Framework for Climate and Energy in the Period from 2020 to 2030. 2014. Available online: https://eur-lex.europa.eu/legal-content/EN/ALL/?uri=CELEX%3A52014DC0015 (accessed on 19 February 2020).
3. MISE, Italian Minister Ministry of Economic Development. Proposta di Piano Nazionale Integrato per L'energia e il Clima. 2018; p. 46. Available online: https://www.mise.gov.it/images/stories/documenti/Proposta_di_Piano_Nazionale_Integrato_per_Energia_e_il_Clima_Italiano.pdf (accessed on 19 February 2020).
4. Transport & Environment. Electric Surge: Carmakers' Electric Car Plans across Europe 2019–2025. 2019, p. 30. Available online: https://www.transportenvironment.org/sites/te/files/publications/2019_07_TE_electric_cars_report_final.pdf (accessed on 19 February 2020).
5. U.S. Department of Energy, Office of Electricity Delivery & Energy Reliability. Modern Distribution Grid: Advanced Technology Maturity Assessment. 2017. Available online: https://gridarchitecture.pnnl.gov/modern-grid-distribution-project.aspx (accessed on 19 February 2020).
6. Vijay, A.; Fouquet, N.; Staffell, I.; Hawkes, A. The value of electricity and reserve services in low carbon electricity systems. *Appl. Energy* **2017**, *113*, 111–123. [CrossRef]
7. ARERA. Italian Autorithy for Energy and Environment. 393/2015/R/eel-Riforma Organica Della Regolazione del Servizio di Dispacciamento Dell'energia Elettrica e Conseguente Attivazione del Progetto Interdirezionale RDE (Riforma del Dispacciamento Elettrico). 2015. Available online: https://www.arera.it/allegati/docs/15/393-15.pdf (accessed on 19 February 2020).
8. Liang, X. Emerging Power Quality Challenges Due to Integration of Renewable Energy Sources. *IEEE Trans. Ind. Appl.* **2017**, *53*, 855–866. [CrossRef]
9. Paduchuri, C.B.; Dash, S.S.; Bayındır, R.; Behera, R.K.; Subramani, C. Analysis and experimental investigation for grid-connected 10 kW solar PV system in distribution networks. In Proceedings of the 2016 IEEE International Conference on Renewable Energy Research and Applications (ICRERA), Birmingham, UK, 20–23 November 2016.
10. Tamura, J.; Kondo, K.; Baba, J. Study of the LFC Signal Dispatching Method to Compensate Short Period Power Fluctuation by Use of Many Photovoltaics. In Proceedings of the 2018 IEEE PES Asia-Pacific Power and Energy Engineering Conference (APPEEC), Kota Kinabalu, Malaysia, 7–10 October 2018; pp. 1–6.

11. Dreidy, M.; Mokhlis, H.; Mekhilef, S. Inertia response and frequency control techniques for renewable energy sources: A review. *Renew. Sustain. Energy Rev.* **2017**, *69*, 144–155. [CrossRef]
12. Boudina, R.; Wang, J.; Benbouzid, M.; Khoucha, F.; Boudour, M. Impact evaluation of large-scale integration of electric vehicles on power grid. *Front. Energy.* **2017**. [CrossRef]
13. Omran, N.G.; Filizadeh, S. A semi-cooperative decentralized scheduling scheme for plug-in electric vehicle charging demand. *Int. J. Electr. Power Energy Syst.* **2017**, *88*, 119–132. [CrossRef]
14. Min, Z.; Qiuyu, C.; Jiajia, X.; Weiwei, Y.; Shu, N. Study on influence of large-scale electric vehicle charging and discharging load on distribution system. In Proceedings of the 2016 China International Conference on Electricity Distribution (CICED), Xi'an, China, 10–13 August 2016; pp. 1–4.
15. Qiao, Z.; Yang, J. Low-voltage distribution network reconfiguration considering high penetration of electric vehicles: A UK case study. In Proceedings of the 2016 IEEE International Conference on Renewable Energy Research and Applications (ICRERA), Birmingham, UK, 20–23 November 2016.
16. Turker, H.; Pirsan, V.; Bacha, S.; Frey, D.; Richer, J.; Lebrusq, P. Heuristic strategy for smart charging of Plug-In Electric Vehicle in residential areas: Variable charge power. In Proceedings of the 2014 International Conference on Renewable Energy Research and Application (ICRERA), Milwaukee, WI, USA, 19–22 October 2014; pp. 938–944.
17. Gaur, P.; Soren, N.; Bhowmik, D. Impact Assessment of Vehicle-to-grid Technology in LFC of Multi-area Solar-thermal Power System. *IJRER* **2018**, *8*, 1580–1590.
18. Brusaglino, G. Integration of road electric vehicles into the smart grid system. In Proceedings of the 2013 International Conference on Clean Electrical Power (ICCEP), Alghero, Italy, 1–13 June 2013; pp. 177–182.
19. Vinayak, P.; Sindhu, M. An intelligent control strategy for vehicle-to-grid and grid-to vehicle energy transfer. In Proceedings of the First International Conference on Materials Science and Manufacturing Technology, Tamil Nadu, India, 12–13 April 2019.
20. Eajal, A.A.; Shaaban, M.F.; El-Saadany, E.F.; Ponnambalam, K. Fuzzy logic-based charging strategy for Electric Vehicles plugged into a smart grid. In Proceedings of the 2015 IEEE International Conference on Smart Energy Grid Engineering (SEGE), Oshawa, ON, Canada, 17–19 August 2015; pp. 1–6.
21. Chukwu, U.C.; Mahajan, S.M. Modeling of V2G net energy injection into the grid. In Proceedings of the 2017 6th International Conference on Clean Electrical Power (ICCEP), Santa Margherita Ligure, Italy, 27–29 June 2017.
22. Rodriguez-Calvo, A.; Cossent, R.; Frias, P. Integration of PV and EVs in unbalanced residential LV networks and implications for the smart grid and advanced metering infrastructure deployment. *Int. J. Electr. Power Energy Syst.* **2017**, *91*, 121–134. [CrossRef]
23. Ahmadian, A.; Sedghi, M.; Aliakbar-Golkar, M. Fuzzy Load Modeling of Plug-in Electric Vehicles for Optimal Storage and DG Planning in Active Distribution Network. *IEEE Trans. Veh. Technol.* **2016**, *66*, 99. [CrossRef]
24. Coninx, K.; Moradzadeh, M.; Holvoet, T. Combining DSM and Storage to Alleviate Current Congestion in Distribution Grids. In Proceedings of the 2016 IEEE PES Innovative Smart Grid Technologies Conference Europe (ISGT-Europe), Ljubljana, Slovenia, 9–12 October 2016; pp. 1–2.
25. Masotti, L.; Dolara, A.; Leva, S. Vehicle-to-Grid for peak shaving in a Medium Voltage Grid with PV plants. In Proceedings of the 2019 International Conference on Clean Electrical Power (ICCEP), Otranto, Italy, 2–4 July 2019.
26. Hadush, S.Y.; Meeus, L. DSO-TSO cooperation issues and solutions for distribution grid congestion management. *Energy Policy* **2018**, *120*, 610–621. [CrossRef]
27. Shashank, N.G.; Tianyang, Z.; Choung, J.K.; Rajit, G. Transmission, Distribution deferral and Congestion relief services by Electric Vehicles. In Proceedings of the 2019 IEEE Power & Energy Society Innovative Smart Grid Technologies Conference (ISGT), Washington, DC, USA, 18–21 February 2019.
28. Di Clerico, M.; Cazzato, F.; Di Martino, D.; Marmeggi, F. The increase of distributed generation on Enel Distribuzione's network: State of the art, actions and strategies for integration. In Integration of Renewables into the Distribution Grid. In Proceedings of the CIRED 2012 Workshop: Integration of Renewables into the Distribution Grid, Lisbon, Portugal, 29–30 May 2012; pp. 1–4.
29. Di Clerico, M.; Caneponi, G.; Cazzato, F.; Cochi, S.; Falvo, M.C.; Manganelli, M. EVs Charging Stations in Active Distribution Grids: A Real Case-Study of Smart Integration. In Proceedings of the 2016 International Symposium on Power Electronics, Electrical Drives, Automation and Motion (SPEEDAM), Anacapri, Italy, 22–24 June 2016.

30. Di Clerico, M.; Caneponi, G.; Cazzato, F.; Cochi, S.; Falvo, M.C.; Manganelli, M. Planning studies for active distribution grids in presence of EVs charging stations: Simulation on a real test network. In Proceedings of the CIGRE 2016, Paris, France, 22–26 August 2016.

31. Di Clerico, M.; Caneponi, G.; Cazzato, F.; Cochi, S.; Falvo, M.C.; Manganelli, M. Active distribution grids and EV charging stations: A centralized approach for their integration. In Proceedings of the 2018 7th International Conference on Renewable Energy Research and Applications (ICRERA), Paris, France, 14–17 October 2018; pp. 1466–1471.

32. Di Clerico, M.; Caneponi, G.; Cazzato, F.; Cochi, S.; Falvo, M.C.; Manganelli, M. EV charging stations and RES-based DG: A centralized approach for smart integration in active distribution grids. *Int. J. Renew. Energy Res.* **2019**, *9*, 605–612.

33. Photovoltaic Geographical Information System. Available online: https://re.jrc.ec.europa.eu/pvg_tools/en/tools.html#TMY (accessed on 26 November 2019).

34. Blair, N.; DiOrio, N.; Freeman, J.; Gilman, P.; Janzou, S.; Neises, T.; Wagner, M. *System Advisor Model (SAM) General Description (Version 2017.9.5)*; National Renewable Energy Laboratory: Golden, CO, USA, 2018; pp. 1–19.

35. Zimmerman, R.D.; Murillo-Sánchez, C.E. *MATPOWER: User's Manual*; Version 7.0b1; Power System Engineering Research Center, 2018; pp. 23–37. Available online: https://matpower.org/docs/manual.pdf (accessed on 19 February 2020).

36. Majidi Nezhad, M.; Groppi, D.; Laneve, G.; Marzialetti, P.; Piras, G. Oil Spill Detection Analyzing "Sentinel 2" Satellite Images: A Persian Gulf Case Study. *World Congr. Civ. Struct. Environ. Eng.* **2018**. [CrossRef]

Article

Optimal Design for a Shared Swap Charging System Considering the Electric Vehicle Battery Charging Rate

Lingshu Zhong [1] and **Mingyang Pei** [2,3,*]

1   Department of Electric Power, South China University of Technology, Guangzhou 510641, China; z.lingshu@mail.scut.edu.cn
2   Department of Civil and Transportation Engineering, South China University of Technology, Guangzhou 510641, China
3   Department of Civil and Environmental Engineering, University of South Florida, Tampa, FL 33620, USA
*   Correspondence: mingyangpei@mail.usf.edu

Received: 23 February 2020; Accepted: 1 March 2020; Published: 6 March 2020

Abstract: Swap charging (SC) technology offers the possibility of swapping the batteries of electric vehicles (EVs), providing a perfect solution for achieving a long-distance freeway trip. Based on SC technology, a shared SC system (SSCS) concept is proposed to overcome the difficulties in optimal swap battery strategies for a large number of EVs with charging requests and to consider the variance in the battery charging rate simultaneously. To realize the optimal SSCS design, a binary integer programming model is developed to balance the tradeoff between the detour travel cost and the total battery recharge cost in the SSCS. The proposed method is verified with a numerical example of the freeway system in Guangdong Province, China, and can obtain an exact solution using off-the-shelf commercial solvers (e.g., Gurobi).

Keywords: shared swap charging system; electric vehicle; operational design; battery charging rate; binary integer programming

---

1. Introduction

Electric vehicles (EVs) are a promising technology for reducing the environmental impacts of road transport [1] and have increased rapidly in number over the past ten years [2]. However, there are several barriers to overcome for expanding the adoption of EVs. One problem with large-scale EV adoption is the limited maximum driving range [3,4] and range anxiety [5–8], which may make it difficult to complete some long-distance tours. The other problems are high battery purchase cost [9–11] and long charging time [11–13]. To solve the problems above, an increasing number of researchers have focused on deploying EV charging systems, which will significantly shape current EV coverage [10,14]. These infrastructures can generally be divided into three categories [15]: plug-in EVs (PEVs, i.e., slow chargers and fast chargers) [16], wireless charging EVs (WCEVs, i.e., inductive charging during driving) [17,18], and swap charging EVs (SCEVs) [19,20]. Table 1 shows the comparisons between these infrastructures.

Table 1. EV charging infrastructure comparisons.

EV Type	Charging Mode	Power Refill Rate *	Infrastructure Cost	Operation Cost	Usage Scenarios **	Advantages	Disadvantages
PEV	Slow charging	30–75 miles/h	L	L	Home/workplace	Cheap and safe	Long charging time
	Rapid charging	180 miles/h	H	M	Public charging stations	Charge rapidly; not very expensive	Long charging time; powerful cooling systems required
WCEV	Dynamic charging	–	VH	VH	Designed to use on heavy load traffic corridor/ freeway	Charging while moving; without range anxiety	Extremely high investment cost
	Static charging	20 to 100 miles/h	H	H	Public charging stations	Wireless	Long charging time
SCEV	Centralized battery charging	Swap time can be less than a minute	M	M	Public swap charging stations	Shorten charge time sharply; centralized charging	Battery ownership, purchase cost, standardization, and safety issues in the swap and charge process

* The power refill rate is the travel distance that the EV can travel after an hour of charging. The slow charging at home wall box (7 kW) would take 9 h 25 min from 0–100%. A public charger (22 kW) provides 75 miles of driving range in 1 h of charging. The charging rate of the rapid charger (50 kW or more) can provide up to 90 miles of driving range in 30 min. ** Most often used in these proposed scenarios, other scenarios are omitted due to limited space. Cost abbreviations: L—low; M—medium; H—high; VH—very high. Abbreviations: EV—electric vehicle; PEV—plug-in EV; SCEV—swap charging EV; WCEV—wireless charging EV. Data source: the Renault website of https://www.renault.co.uk/electric-vehicles/, the NIO website of https://www.nio.com/nio-power, and the WIRED website of https://www.wired.com/.

Battery swap stations (BSSs) were originally implemented by the company Better Place, which went out of business in 2013 [15]. Then, five cities in China start testing BSS technology, where they serve personal vehicles, and commercial vehicles [21]. Although BSSs best replicate the experience of existing gas stations, there are still issues that prevent their wide-scale implementation. These include battery ownership, high battery purchase cost, complicated battery standardization issues, and safety issues in the swap and charge process. Since not only do an increasing number of EV consumers expect charging approaches that include short charging times (similar to refueling their current fuel vehicles) [22], but also global economic growth means more people can afford high-cost options, the SCEV mode is becoming increasingly popular [4,19,23–25]. SCEVs are good in that they have both fast and economical charging modes [4,26,27]. As shown in Figure 1, a driver can drive into a battery swap station, and a robot replaces the depleted battery with a fully charged spare [28,29]. This swap time could be very short (e.g., less than one minute based on a report from Tesla) with further automation and refinements on the vehicle [30].

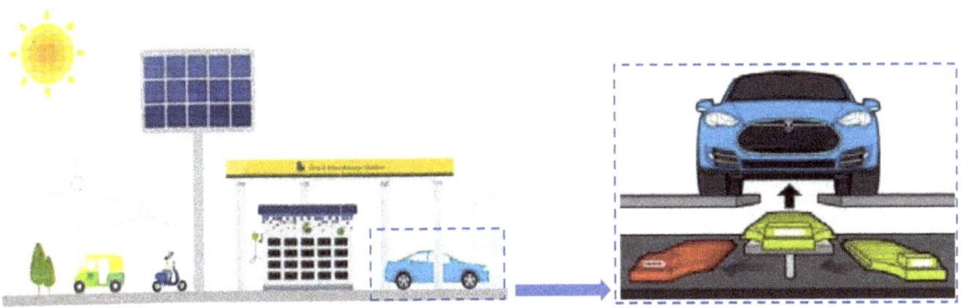

Figure 1. Battery swap automation and refinements on the vehicle (figure source: SUN mobility and Tesla).

Swap charging (SC) can reduce the peak consumption of electricity by centralized charging [30,31] and avoid grid overloading due to mass EV charging [32] because the empty batteries that are swapped out can be charged when electricity is cheap or demand is low. Since SCEVs are considered to be a suitable EV mode, an increasing number of studies on the SC system (SCS) have emerged worldwide [4,19,25,27,33]. The Fluence Z.E. was the first electric car enabled with battery swapping technology and deployed within the Better Place network in Israel and Denmark in 2012 [4,20,27]. Then, with the advanced SC technology, fully automatic battery swapping was even faster than refueling at gas stations. NIO proposed the smallest power swap station in the world which only took up three parking spaces [2,31]. Based on these state-off-the-art battery swapping technologies' tests, some researchers have proposed an advanced concept called shared SCS (SSCS) [31]. The SSCS is an SCS that can provide heterogeneous services and requires online reservations in advance. The SSCS has a lot of differences from the regular SCS mode, and the comparisons are shown in Table 2.

The SCS and the SSCS proposed in this paper are both used for SCEVs, which separate the batteries from the vehicles and allow the SC mode. The SSCS has a few new features, as listed below:

1. **Reserved charging demand**: This feature differs this system from the regular SCS, which can supply service on a come and served basis, as the newly proposed SSCS requires online reservations in advance. All vehicle service strategies (e.g., routing and swapping battery types) can be calculated according to their origins and destinations (ODs), their initial battery power level, etc.

2. **Multi-type battery supplied**: The SCS can only provide fully charged batteries [29], while the SSCS can provide online reservations and allow the BSS to optimally deploy their state of charge (SOC) battery.

Although research focusing on BSS strategies has been ongoing, the results are fragmented. Currently, an integrated way of considering the VRP, battery dispatching, and battery charging efficiency (considering the battery charging rate) has not been fully investigated. To bridge these research gaps and realize the vision of the SSCS, this paper proposes an exact approach to describe the EV routing problem and BSS battery dispatching and determine the optimal SSCS design to minimize the overall system operational cost. We formulate this problem into a binary programming model so that it can deal with the various large-scale strategy issues. This model has a binary decision variable and thus quickly solves an exact solution by off-the-shelf commercial solvers (e.g., Gurobi).

1.2. Contributions

This paper focuses on SC technology and proposes a new structured SSCS to overcome the difficulties in optimal swap battery strategies for a large number of EVs with charging requests and simultaneously considers the varying battery charging rate. The contributions of this paper are mainly three-fold.

- First, we propose an innovative binary programming SSCS model to balance the tradeoff between the vehicle travel cost and battery dispatching cost. This model is a linear integer problem that solves exact solutions by off-the-shelf commercial solvers (e.g., Gurobi).
- Second, we propose an optimal operation strategy for deploying multi-type batteries and simultaneously consider the charging process. In this process, a large number of various charging requests with various initial battery power levels are given various charging strategies (i.e., optimal routes to BSS and battery types). These charging strategies can help improve charging efficiency and minimize the overall system operational cost.
- Finally, a numerical example with real-world freeway data from Guangdong Province, China is conducted to demonstrate the applicability of the proposed model and its effectiveness in reducing construction costs. Overall, this paper provides valuable insights into the future integration of BSSs into long-distance freeway services and offers a numerical method for designing an optimal operational plan for this integrated system.

The remainder of this paper is organized as follows. Section 2 introduces the operation characteristics, notation, and concept of the proposed SSCS. Section 3 formulates the SSCS model with alternative systems. Section 4 tests the proposed model with a numerical example in China and conducts corresponding sensitivity analyses. Finally, Section 5 provides conclusions and recommends future research directions.

2. Model Description

This section introduces the operational process of the SSCS and underlying assumptions. For the convenience of readers, we list some notation frequently used in the paper in Table 4.

Consider a set of vehicle stations $I\{1,\ldots,I\}$ in space. For each vehicle station $i \in I$, there is a BSS. These stations can also be the ODs of vehicles. Consider a set of batteries with varying SOC $Q\{1,\ldots Q\}$ that a shared BSS can provide. Let $q \in Q$ denote the battery SOC. For each station, the number of battery types can be different. Consider a set of the vehicle trip characteristic index $U\{1,\ldots,U\}$ which has a series of various travel demands (i.e., origin station i_u^+, destination station i_u^-, and the initial state of the battery charge q_u^0). Let x_{ujq} denote whether vehicle i heads to station j and replaces the depleted battery with a well-charged battery in the state of $q \in Q$.

To fully understand the operation process of an SSCS, Figure 2 shows an example with shared BSS stations $I = \{1,\ldots,5\}$ and three types of battery SOCs $q = \{1,2,3\}$. In this figure, on each link between two stations, the segment of a different number represents the travel distance between the stations. The different combinations of colors for the stations represent the battery types they provide.

Battery swap stations (BSSs) were originally implemented by the company Better Place, which went out of business in 2013 [15]. Then, five cities in China start testing BSS technology, where they serve personal vehicles, and commercial vehicles [21]. Although BSSs best replicate the experience of existing gas stations, there are still issues that prevent their wide-scale implementation. These include battery ownership, high battery purchase cost, complicated battery standardization issues, and safety issues in the swap and charge process. Since not only do an increasing number of EV consumers expect charging approaches that include short charging times (similar to refueling their current fuel vehicles) [22], but also global economic growth means more people can afford high-cost options, the SCEV mode is becoming increasingly popular [4,19,23–25]. SCEVs are good in that they have both fast and economical charging modes [4,26,27]. As shown in Figure 1, a driver can drive into a battery swap station, and a robot replaces the depleted battery with a fully charged spare [28,29]. This swap time could be very short (e.g., less than one minute based on a report from Tesla) with further automation and refinements on the vehicle [30].

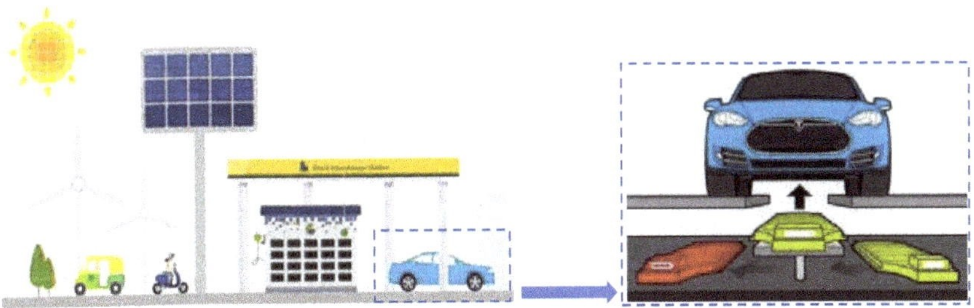

Figure 1. Battery swap automation and refinements on the vehicle (figure source: SUN mobility and Tesla).

Swap charging (SC) can reduce the peak consumption of electricity by centralized charging [30,31] and avoid grid overloading due to mass EV charging [32] because the empty batteries that are swapped out can be charged when electricity is cheap or demand is low. Since SCEVs are considered to be a suitable EV mode, an increasing number of studies on the SC system (SCS) have emerged worldwide [4,19,25,27,33]. The Fluence Z.E. was the first electric car enabled with battery swapping technology and deployed within the Better Place network in Israel and Denmark in 2012 [4,20,27]. Then, with the advanced SC technology, fully automatic battery swapping was even faster than refueling at gas stations. NIO proposed the smallest power swap station in the world which only took up three parking spaces [2,31]. Based on these state-off-the-art battery swapping technologies' tests, some researchers have proposed an advanced concept called shared SCS (SSCS) [31]. The SSCS is an SCS that can provide heterogeneous services and requires online reservations in advance. The SSCS has a lot of differences from the regular SCS mode, and the comparisons are shown in Table 2.

The SCS and the SSCS proposed in this paper are both used for SCEVs, which separate the batteries from the vehicles and allow the SC mode. The SSCS has a few new features, as listed below:

1. **Reserved charging demand**: This feature differs this system from the regular SCS, which can supply service on a come and served basis, as the newly proposed SSCS requires online reservations in advance. All vehicle service strategies (e.g., routing and swapping battery types) can be calculated according to their origins and destinations (ODs), their initial battery power level, etc.

2. **Multi-type battery supplied**: The SCS can only provide fully charged batteries [29], while the SSCS can provide online reservations and allow the BSS to optimally deploy their state of charge (SOC) battery.

3.   **Accurate cost calculated**: Different from a regular SCS, where the economic essence is battery leasing, the SSCS conducts energy leasing. In the pricing strategy, the SCS sets a price for each battery, while the SSCS sets a price for the process of recharging the depleted battery to the same power level as the new battery.

4.   **Charging rate considered**: In this proposed system, the recharge cost of depleted batteries is calculated by considering the battery charging rate curve. The SSCS can help achieve an optimal charging strategy and improve energy usage efficiency.

Table 2. Comparisons between the SCS mode and SSCS mode *.

EV Type	System Mode	Operation Mode	Battery Type **	Average SC Cost	Charging Rate	Average Charging Cost	Residual Value	Capacity of BSS
SCEV	SCS	Come and served	Single	M	–	M	–	–
	SSCS (this paper)	Reserved online in advance	Multiple	L	•	L	•	•

* The symbol • in this table denotes that the factor is considered and symbol – denotes otherwise. ** Battery types: single type—fully charged battery; multiple types—varying state of charge (SOC) batteries. Cost abbreviations: L—low; M—medium. Abbreviations: BSS—battery swap station; SC—swap charging; SCS—swap charging system; SSCS—shared SCS.

1.1. Literature Review

Since public power charging infrastructure plays a critical role in EV systems [7,14,34], an increasing number of researchers have begun to focus on EV routing problems under SC technology and with the battery charging dispatch model [1,4,20,24,25,27,29,30,35], which holds promise to realize long-distance EV travel [4,20]. Here, we summarize some applications and modeling attempts to develop SC in recent years, as shown in Table 3, and the findings can be briefly synthesized as follows.

- The battery charging dispatch model was set up from the grid side to minimize the total cost (e.g., infrastructure deployment cost [4] and sequential decision cost [26]) while satisfying various physical constraints. Later, an increasing number of researchers began focusing on the transportation side due to the massive traffic issue and then dealt with this SCS as a vehicle routing problem (VRP) [27,36,37], location routing problem (LRP) [3,24,38], or battery dispatch management problem [15,26,29,32,39,40]. In this paper, we propose vehicle routing and battery dispatching as two vital indices for optimizing an SSCS.

- Due to technological or application limitations (i.e., an internet-based booking platform; BSS operation information processing center (IPC); centralized vehicle introduction systems) over the past few years, there are only a limited number of recent studies [27] on the BSS online reservation system that focused on various vehicle demands. This study proposes a new operational mode under a new information system (i.e., vehicles require advanced reservations and the IPC gives various service strategies).

- Most previous studies provided only a single battery type (i.e., fully charged battery) [4,36,40], and they only allow depleted batteries to be replaced by a standard SOC battery. However, some researchers have considered providing multi-type batteries, as stated in the references [15,27,29], and the introduction of varying SOC batteries gives more flexibility in optimal applications. Since our SSCS model is based on the battery charging rate, we propose an optimal operation strategy, deploying multi-type batteries simultaneously.

Table 3. Comparison between existing related SC models and our model *.

Authors	Objective Function	Decision Variable	Battery Type	Online Reservation	Various Demands	Battery Charging Rate	Capacity of BSS	Model Approach
Mak et al. (2013) [4]	Building and operating costs	Infrastructure deployment	S	–	–	–	–	MISOCP
Adler and Mirchandani (2014) [27]	Average delay	VRP and battery dispatch	S	•	–	–	•	DP
Yang et al. (2014) [26]	Battery management	Sequential decision	M	–	–	–	–	Simulation
Yang and Sun (2015) [3]	Construction and routing cost	LRP	S	–	–	–	–	MIP
Chen et al. (2017) [37]	Travel distance	VRP	S	–	–	–	–	MIP
Hof et al. (2017) [24]	Construction and routing cost	LRP	S	–	–	–	–	MIP
Amiri et al. (2018) [38]	Total charging cost	LRP	S	–	–	–	•	MINLP
Widrick et al. (2018) [39]	Total reward	Battery dispatch	S	–	–	–	•	DP
Ding et al. (2019) [29]	Total profit	Battery dispatch	M	–	–	•	•	MIP
Jie et al. (2019) [36]	System cost	VRP	S	–	–	–	–	IP
Infante et al. (2019) [40]	Minimum recourse	Battery dispatch	S	–	–	–	•	MILP
Sun et al. (2019) [32]	Battery investment and operating cost	Battery dispatch	S	–	–	–	•	Fluid model
Šepetanc and Pandžić (2020) [15]	Total profit	Battery dispatch and pricing	M	–	–	–	•	MILP
This paper	System operational cost	VRP and battery dispatch	M	•	•	•	•	IP

* The symbol • in this table denotes that the factor is considered, and symbol – denotes otherwise. Battery type abbreviations: M—multiple types (i.e., varying SOC batteries); S—single type (i.e., fully charged battery). Abbreviations: DP—dynamic programming; IP—integer programming; LRP—location routing problem; MINLP—mixed-integer nonlinear programming; MISOCP—mixed-integer second-order cone programming; MIP—mixed-integer programming; VRP—vehicle routing problem.

Although research focusing on BSS strategies has been ongoing, the results are fragmented. Currently, an integrated way of considering the VRP, battery dispatching, and battery charging efficiency (considering the battery charging rate) has not been fully investigated. To bridge these research gaps and realize the vision of the SSCS, this paper proposes an exact approach to describe the EV routing problem and BSS battery dispatching and determine the optimal SSCS design to minimize the overall system operational cost. We formulate this problem into a binary programming model so that it can deal with the various large-scale strategy issues. This model has a binary decision variable and thus quickly solves an exact solution by off-the-shelf commercial solvers (e.g., Gurobi).

1.2. Contributions

This paper focuses on SC technology and proposes a new structured SSCS to overcome the difficulties in optimal swap battery strategies for a large number of EVs with charging requests and simultaneously considers the varying battery charging rate. The contributions of this paper are mainly three-fold.

- First, we propose an innovative binary programming SSCS model to balance the tradeoff between the vehicle travel cost and battery dispatching cost. This model is a linear integer problem that solves exact solutions by off-the-shelf commercial solvers (e.g., Gurobi).
- Second, we propose an optimal operation strategy for deploying multi-type batteries and simultaneously consider the charging process. In this process, a large number of various charging requests with various initial battery power levels are given various charging strategies (i.e., optimal routes to BSS and battery types). These charging strategies can help improve charging efficiency and minimize the overall system operational cost.
- Finally, a numerical example with real-world freeway data from Guangdong Province, China is conducted to demonstrate the applicability of the proposed model and its effectiveness in reducing construction costs. Overall, this paper provides valuable insights into the future integration of BSSs into long-distance freeway services and offers a numerical method for designing an optimal operational plan for this integrated system.

The remainder of this paper is organized as follows. Section 2 introduces the operation characteristics, notation, and concept of the proposed SSCS. Section 3 formulates the SSCS model with alternative systems. Section 4 tests the proposed model with a numerical example in China and conducts corresponding sensitivity analyses. Finally, Section 5 provides conclusions and recommends future research directions.

2. Model Description

This section introduces the operational process of the SSCS and underlying assumptions. For the convenience of readers, we list some notation frequently used in the paper in Table 4.

Consider a set of vehicle stations $I\{1,\ldots,I\}$ in space. For each vehicle station $i \in I$, there is a BSS. These stations can also be the ODs of vehicles. Consider a set of batteries with varying SOC $Q\{1,\ldots Q\}$ that a shared BSS can provide. Let $q \in Q$ denote the battery SOC. For each station, the number of battery types can be different. Consider a set of the vehicle trip characteristic index $\mathcal{U}\{1,\ldots,U\}$ which has a series of various travel demands (i.e., origin station i_u^+, destination station i_u^-, and the initial state of the battery charge q_u^0). Let x_{ujq} denote whether vehicle i heads to station j and replaces the depleted battery with a well-charged battery in the state of $q \in Q$.

To fully understand the operation process of an SSCS, Figure 2 shows an example with shared BSS stations $I = \{1,\ldots,5\}$ and three types of battery SOCs $q = \{1,2,3\}$. In this figure, on each link between two stations, the segment of a different number represents the travel distance between the stations. The different combinations of colors for the stations represent the battery types they provide.

Table 4. Notation.

Sets	
$\mathcal{U}$	Set of the vehicle trip characteristic index, $\mathcal{U}\{1,\ldots,U\}$
$\mathcal{I}$	Set of vehicle stations (i.e., origin stations and destinations), $\mathcal{I}\{1,\ldots,I\}$
$\mathcal{J}$	Set of the shared BSS index, $\mathcal{J}\{1,\ldots,J\}$
Q	Set of varying SOCs that shared the BSS provided, $Q\{1,\ldots Q\}$
Parameters	
u	Index of the vehicle trip characteristics, $u \in \mathcal{U}$
j	Shared BSS index, $j \in \mathcal{J}$
i_u^+	Origin for vehicle trip characteristic index u
i_u^-	Destination for vehicle trip characteristic index u
q_u^0	Initial battery SOC for vehicle trip characteristic index u
q	Battery SOC that shared the BSS provided, $q \in Q$
q_u	Battery capacity of the shared BSS provided for the vehicle trip characteristic index u. $q_u \in Q$
$d_{i,j}$	Travel distance between station i to station j
$\Delta d_{i_u^+,j,i_u^-}$	Distance for charging detour, $\Delta d_{i_u^+,j,i_u^-} d_{i_u^+,j} + d_{j,i_u^-} - d_{i_u^+,i_u^-}$
C_1	Unit detour cost, Yuan/km
C_2	Unit time cost for battery charging process, Yuan/min
C_3	Unit power cost for battery charging process, Yuan/kW
C_4	Unit power salvage value in the battery, Yuan/kW
s	Unit energy consumption per kilometer, kW/km
$f(q)$	Formula of the battery charging time rate with varying SOC
q^L	Lower band of the battery SOC
n_{jq}	Swapping battery supplement at station $j \in \mathcal{J}$, with battery SOC $q \in Q$
Decision variables	
x_{ujq}	Binary variables, $x_{ujq} = 1$ when vehicle i goes to power station j and the battery is replaced by a new battery with power quantity q; $x_{uj} = 0$ otherwise $u \in \mathcal{U}, j \in \mathcal{J}, q \in Q$

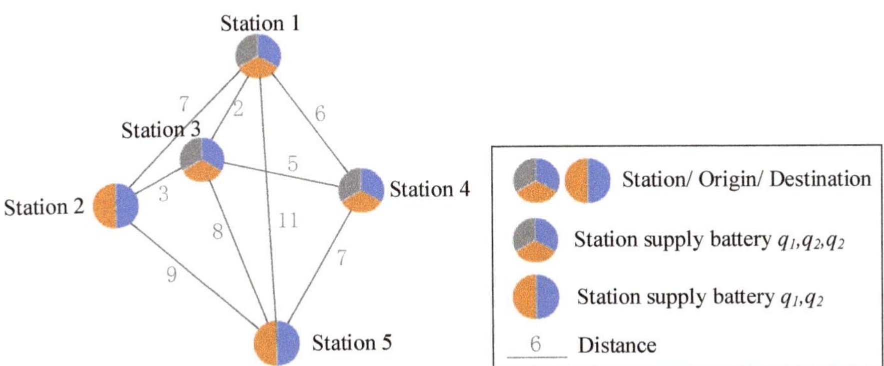

Figure 2. Example network with different battery mode supplies in the SSC station.

In the SSCS, the entire operation process can be divided into three steps, as shown in Figure 3. The vehicle side allows the EV to make online reservations in advance and then follow the instructions from the IPC. The IPC side requires all the vehicles and BSSs to follow centralized guidance, and the BSS side follows the optimal battery replacement and charging strategies. All these system components operate smoothly under the proposed SSCS model.

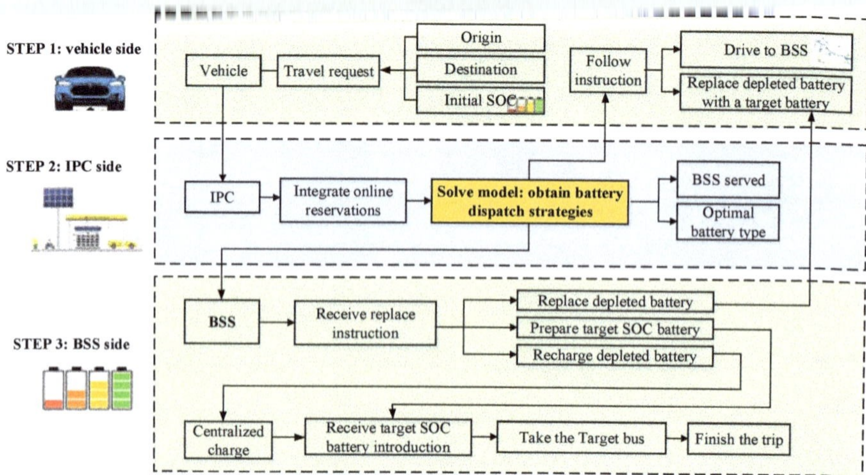

Figure 3. Operation process in the SSCS. Abbreviation: IPC—information processing center.

In previous studies [29,41–43], the battery charging rate is a concave function that satisfies formula $f(q) > 0, f'(q) < 0$. The charging time function can be approximately formulated as a piecewise function $f(q) = \begin{cases} k_1 q + b_1, 0 < q \le q_1 \\ k_2 q + b_2, q_1 < q \le q_2 \\ \cdots \\ k_m q + b_m, q_{m-1} < q \le q_m \end{cases}$. In Figure 4, we plot the varying SOC (q), the battery charging rate ($\frac{dt}{dq}$), and the cumulative time functions of the SOCs of the batteries.

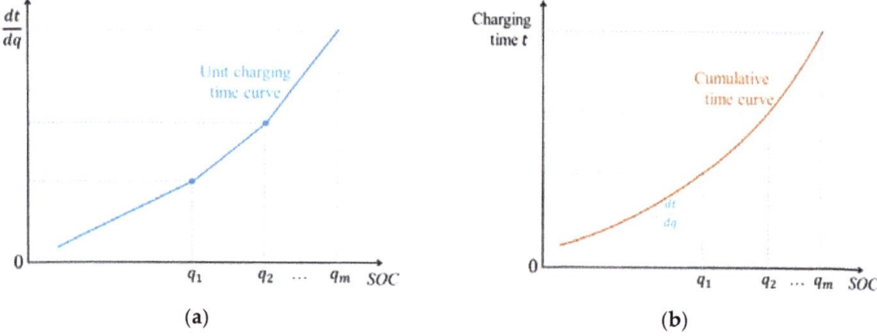

Figure 4. Performances of charging times with current battery power levels: (**a**) curve of the varying SOC (q) and the battery charging rate (dt/dq); (**b**) cumulative time curve of the SOCs of batteries.

To facilitate the model formulation, we introduce the following assumptions in the investigated problem. These assumptions have been used in other studies on operational design for the SC battery system.

Assumption 1. *The battery power consumption of EVs is proportional to the driving distance [37,44]. It is hard to relax this assumption when a battery consumed along a stretch of road is not dependent on the distance; then, the problem becomes an NP-hard problem and appears to be mathematically intractable [14,45,46].*

Assumption 2. *All vehicles in our system share the same battery capacity size. In the previous study, many researchers have already focused on optimizing the battery size to reach a better system income [47].*

Assumption 3. *All vehicles in this system reserve swap batteries online and follow the instructions. This assumption will not be strict in the future because of the connected and autonomous vehicle atmosphere and because it has already been applied in previous studies [27].*

3. Model Formulation

This section provides a model formulation of the investigated problems. Section 3.1 proposes a model to describe the above-defined SSCS problem. Section 3.2 puts forward the physical constraints that make this model applicable in real-world cases. Finally, Section 3.3 compares this proposed system with the benchmark system.

3.1. Objective Function

The objective function formulated in Equation (1) aims to minimize the SSCS system operational costs, which includes three components: the travel cost of the detour in the swapping battery process (F_1), the total battery cost in the battery recharging process (F_2), and the residual value of electricity power in moving EVs (F_3).

$$\min_{x_{ujq}} F_1 + F_2 - F_3 \tag{1}$$

As shown in Equation (2), F_1 denotes the travel cost of the detour in the swapping battery process, and C_1 denotes the unit detour cost. Let $\Delta d_{i_u^+,j,i_u^-}$ denote the distance of the charging detour, $\Delta d_{i_u^+,j,i_u^-} d_{i_u^+,j} + d_{j,i_u^-} - d_{i_u^+,i_u^-}$. The total battery cost in the battery recharging process includes charging time costs and charging energy consumption costs. The total battery recharge cost is cumulative and can be calculated by Equation (3). In this formula, let C_2 denote the unit time cost for the battery charging process, and let C_3 denote the unit power cost for the battery charging process. Equation (4) presents the electricity power residual values of the EVs.

$$F_1 C_1 \sum_{u\in\mathcal{U},j\in\mathcal{J},q\in Q} x_{ujq} \Delta d_{i_u^+,j,i_u^-} \tag{2}$$

$$F_2 \sum_{u\in\mathcal{U},j\in\mathcal{J},q\in Q} x_{ujq} \int_{q_u^0 - d_{i_u^+,j}s}^{q} (C_2 f(r) + C_3)\,dr \tag{3}$$

$$F_3 C_4 \sum_{u\in\mathcal{U},j\in\mathcal{J},q\in Q} x_{ujq}\left(q - d_{j,i_u^-}s\right) \tag{4}$$

3.2. Constraints

The above objective function is subject to a set of constraints, as formulated below.

$$q_u^0 - \sum_{j\in\mathcal{J},q\in Q} d_{i_u^+,j} s x_{ujq} \geq q^L \quad u\in\mathcal{U} \tag{5}$$

$$q_u - \sum_{j\in\mathcal{J},q\in Q} d_{j,i_u^-} s x_{ujq} \geq q^L \quad u\in\mathcal{U} \tag{6}$$

$$\left(q_u^0 - d_{i_u^+,j}s\right) x_{ujq} \leq q_u \quad u\in\mathcal{U}, j\in\mathcal{J}, q\in Q \tag{7}$$

$$\sum_{j\in\mathcal{J},q\in Q} x_{ujq} \leq 1 \quad u\in\mathcal{U} \tag{8}$$

$$\sum_{u\in\mathcal{U},q\in Q} x_{ujq} \leq n_j \quad j\in\mathcal{J} \tag{9}$$

$$x_{ujq} = 0,1 \quad u\in\mathcal{U}, j\in\mathcal{J}, q\in Q \tag{10}$$

Constraints (5) and (6) are related to the safety constraints, which mandates that for each vehicle in the SSCS, the lowest power level value should always exceed the lowest level value (q^L) on the right-hand side (RHS). The left-hand side (LHS) in Constraint (5) denotes the battery power level of vehicle u when it obtains access to a BSS, and the LHS in Constraint (6) denotes the battery power level when vehicle u finishes its trip at its destination. Constraint (7) is a limitation that the SOC of a new swap battery is always higher than the SOC of the depleted battery. Constraint (8) is proposed to limit the EV to only swap the battery once in this model, and the side effects of this constraint can be relieved by multiple inputs and by solving this model. In the future, we will try to put forward a more integrated model. Constraint (9) sets some general constraints of the model, which are related to the network battery power balance, similar to reference [38], which describes the maximum permitted capacity of the battery swapped in each BSS.

3.3. Alternative Systems

A single-type battery system (STBS) is used as an alternative system. The only difference between the STBS and SSCS is that each BSS can only supply a fixed SOC of q_F in an STBS, while the SSCS can supply multiple types of SOCs.

4. Numerical Example

To examine the model performance over different network topologies, we present a numerical example with the designed SSCS over the Guangdong Province freeway network and compare it with the alternative STBS simultaneously. As shown in Figure 5a, the input data included 205,876 records of vehicles passing through 14 key toll stations between 17:00 and 18:00 throughout May 2019. We obtained the corresponding vehicle OD demands, as shown in Figure 5b, and assumed that 50% of the passengers use SCEVs. Then, we assumed that the initial battery SOC of these vehicles followed a random distribution.

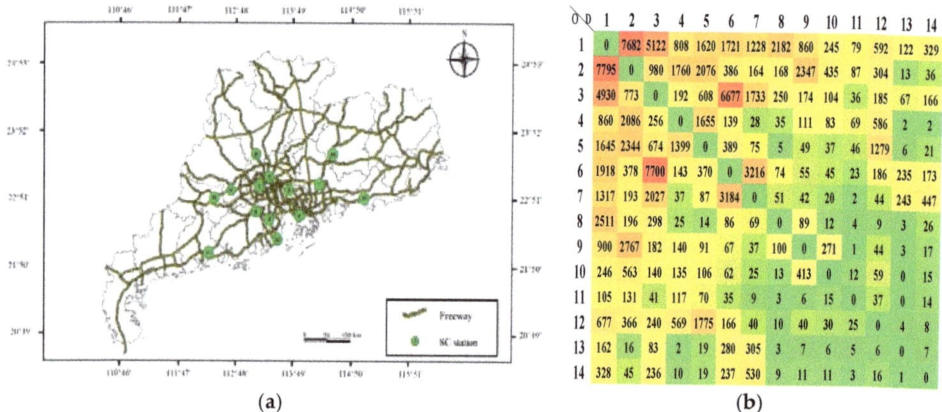

O\D	1	2	3	4	5	6	7	8	9	10	11	12	13	14
1	0	7682	5122	808	1620	1721	1228	2182	860	245	79	592	122	329
2	7795	0	980	1760	2076	386	164	168	2347	435	87	304	13	36
3	4930	773	0	192	608	6677	1733	250	174	104	36	185	67	166
4	860	2086	256	0	1655	139	28	35	111	83	69	586	2	2
5	1645	2344	674	1399	0	389	75	5	49	37	46	1279	6	21
6	1918	378	7700	143	370	0	3216	74	55	45	23	186	235	173
7	1317	193	2027	37	87	3184	0	51	42	20	2	44	243	447
8	2511	196	298	25	14	86	69	0	89	12	4	9	3	26
9	900	2767	182	140	91	67	37	100	0	271	1	44	3	17
10	246	563	140	135	106	62	25	13	413	0	12	59	0	15
11	105	131	41	117	70	35	9	3	6	15	0	37	0	14
12	677	366	240	569	1775	166	40	10	40	30	25	0	4	8
13	162	16	83	2	19	280	305	3	7	6	5	6	0	7
14	328	45	236	10	19	237	530	9	11	11	3	16	1	0

(a)   (b)

Figure 5. The input data for this numerical example: (**a**) designed BSSs in Guangdong Province, China; (**b**) origin and destination (OD) demand distribution.

4.1. Input Parameters

All experiments were performed on a PC with an Intel® Core™ i7-8550U @1.99 GHz CPU and 24 GB RAM. The code was implemented in MATLAB 2019a, calling a commercial solver Gurobi [48–50]. The charging rate we used is normally and approximately fitted to a linear function [41,42], and in this paper, we selected the parameters considering both the vehicle battery characteristics and electric grid characteristics, which are $f(q) = \begin{cases} 1q + 0.2, \, 0 < q < 0.6 \\ 2q - 0.4, 0.6 < q < 0.8 \\ 4q - 2, 0.8 < q < 1 \end{cases}$. Other default parameter values were

stated in Table 5.

Table 5. Default parameter settings.

Parameter	Description	Value	Data Source
C_1	Unit detour cost	1 Yuan/km	EV travel cost (https://afdc.energy.gov/fuels/)
C_2	Unit time cost	1 Yuan/h	Guangzhou Municipal Human Resources and Social Security Bureau reports in 2019 (http://gzrsj.hrssgz.gov.cn/english/)
C_3	Unit power cost	1 Yuan/kW	Electricity price in China (https://www.ceicdata.com/china/electricity-price)
C_4	Unit power salvage value	0.4 Yuan/kW	Related to the PEV charging price (https://afdc.energy.gov/fuels/electricity_charging)
s	Unit energy consumption/km	0.25%/km	Most EVs are currently capable of approximately 100–250 miles of driving before they need to charge (Data source: UC Davis; https://phev.ucdavis.edu/)
v	Average vehicle travel speed in km/h	100 km/h	The operating speed of EVs on the freeway (http://www.0512s.com/lukuang/G94.html)
p^L	Lower battery power limit	20%	Safety suggestion from EV enterprises (e.g., Beijing Automotive Group Co., etc.)

4.2. Optimal Location Result

By solving the proposed SSCS model, the optimal objective value (system operational cost) is 926.3, with a CPU time of 0.6359 s. Figure 6 shows the battery swaps of different OD pairs. In this figure, on each row and column intersection, the different color circles represent the different battery types (i.e., SOC $q = 60\%$, 80%, and 100%), and the circle size represents the type of dispatch frequency. The results show that the total number of batteries swapped for SOC types of 60%, 80%, and 100% are 139, 940, and 352, respectively.

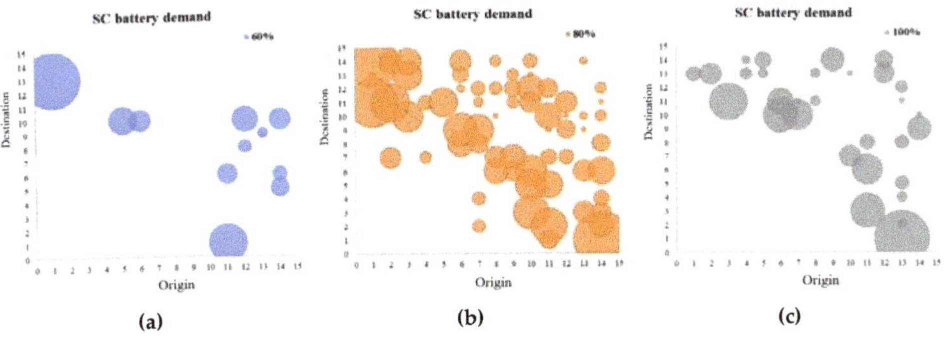

Figure 6. Battery swaps with different OD pairs with battery SOCs of (**a**) $q = 60\%$; (**b**) $q = 80\%$; and (**c**) $q = 100\%$.

We compared the SSCS solutions with the benchmark STBS. In this experiment, we compared the system operational cost and the average battery level before and after SC, with the average energy gap filled, the average battery level at the destination, and the average energy consumption over the traveled distance as the criteria to evaluate the performance of the proposed system. Figure 7 shows the comparison between the SSCS and STBS in a multi-type battery deployment. Most of the batteries deployed in the SSCS and STBS were the same except for stations 4, 5, 9, 10, and 12, which indicates that the introduction of multiple types of batteries does not significantly change the total amount of battery management.

More detailed results are shown in Table 6. As we can determine from the comparison result, the total number of batteries the two systems swapped was the same (i.e., 1431). Since they share different battery types (i.e., SSCS has multi-type batteries, and STBS has single-type batteries), their optimal battery levels are different. Compared to the average battery level before SC, the optimal battery level of the STBS (65%) was much higher than that of the SSCS (32.1%), which is not efficient for energy

image. When compared to the average energy gap the charging process fills, the performance of the SSCS (50.9%) was also better than that of the STBS (35%), which is significantly related to the SC efficiency. Since the average energy consumption for traveled distance was similar (SSCS and STBS are 53% and 54%, respectively), the detour distance did not make a noticeable impact. Overall, the multi-type SC strategies for the SSCS could reduce the system operational cost (54.3%) when compared with the STBS.

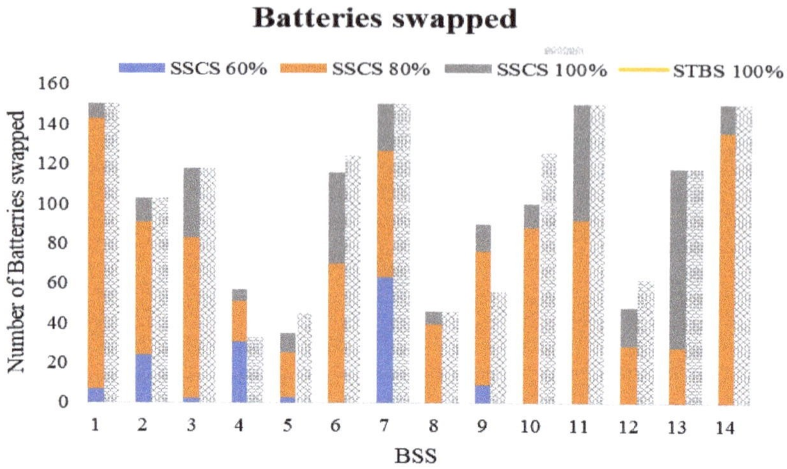

Figure 7. Comparison of the number of swapped batteries.

Table 6. Results comparison with the alternative system.

| | | SSCS | STBS | |
| | | | | |
Evaluation Criteria	SOC	Multi-Type	Single Type $q = 100\%$ *	Rate **
• Total number of batteries swapped	60%	139	–	–
	80%	940	–	–
	100%	352	1431	–
• Average battery level before SC		32.1%	65%	2.025
• Average battery level after SC		83.0%	100%	1.205
• Average energy gap filled		50.9%	35%	0.687
• Average battery level at the destinations (residual energy level)		30.0%	46%	1.533
• Average energy consumption for traveled distance		53%	54%	1.019
• System operational cost		926.3	1428.9	1.543

* $q = 100\%$ indicates that the depleted battery is replaced by a fully charged battery, which is commonly used in the market. ** The rates are calculated by the value of the single-type battery system (STBS) divided by the value of the SSCS.

4.3. Sensitivity Analysis

This section analyzes the sensitivity of critical parameters to the cost components in the SSCS. In each instance, only one parameter is varied, and the other parameters maintain their default values. To evaluate the performances of different parameter combinations, we compared the overall system cost and the multi-type battery combinations. To simplify the sensitivity analysis for vectors C_1, C_2, C_3 and C_4, we varied the values of these parameters and plotted the results in Figure 8. The findings can be briefly summarized as follows.

- We perform a regression analysis of C_1, C_2, C_3, and C_4 with the system operational cost (simplified as F_{SSCS}), as shown in Figure 8a–d and obtain $F_{SSCS} = 541.1C_1 + 304.0C_2 + 276.4C_3 - 436.56C_4 - 78.2$ with $R^2 = 0.995$. This result reveals a high linear correlation with all four critical parameters.
- The optimal result of battery type performance stability with varying values of C_1, C_2, and C_3 is shown in Figure 8. The varying value of C_4 can change the optimal strategy significantly, as shown

in Figure 8d,e. The increased value of C_4 would result in an increased number of vehicles holding more residual energy at the destination.

- Figure 8f shows the performance of the average battery charging time with varying C_2. We learn that C_2 is related to the unit time cost for the battery charging process, and it reaches a plateau period when the value of C_2 is over 1.5.

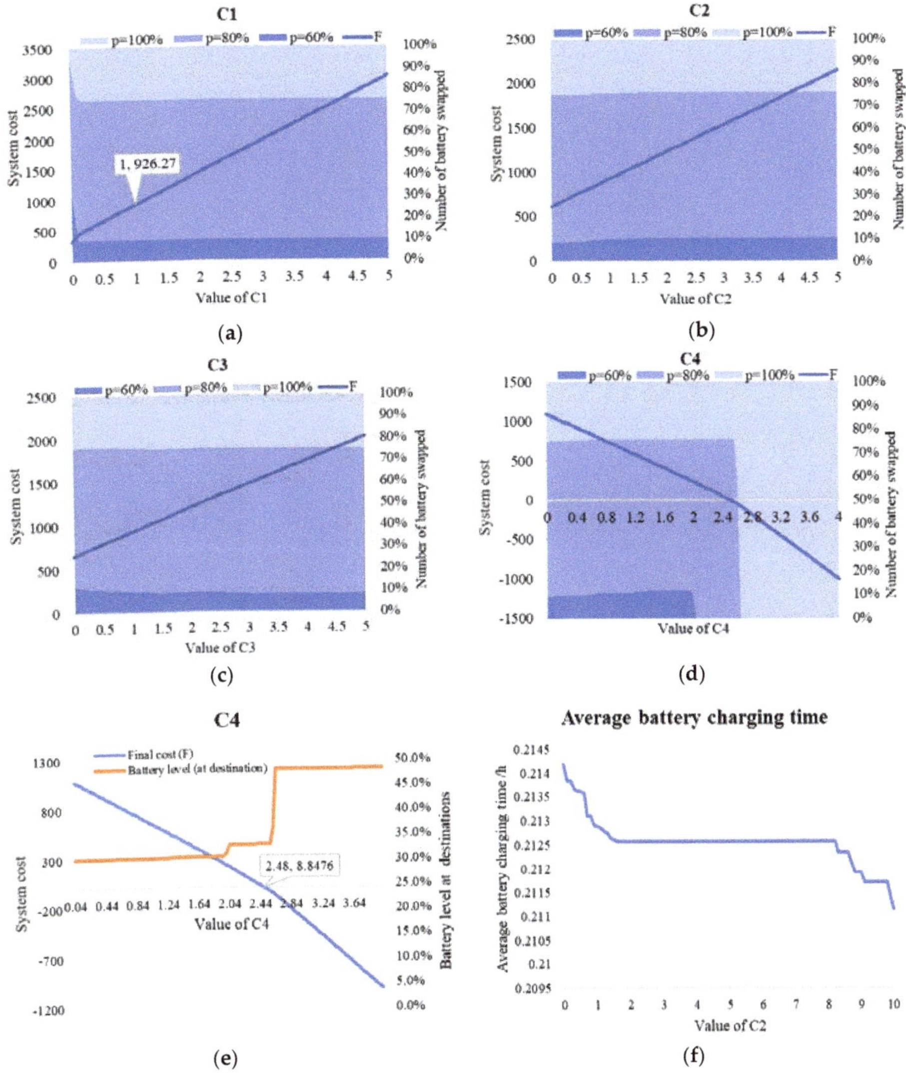

Figure 8. System operational cost performance and number of batteries swapped with varying values of (**a**) C_1; (**b**) C_2; (**c**) C_3; and (**d**) C_4. (**e**) Average battery charging time with varying C_2. (**f**) System operational cost performance and battery level at the destination with varying C_4.

5. Conclusions

SC technology offers the possibility of EVs swapping batteries with other EVs and provides plausible solutions for realizing a long-distance freeway trip. By taking advantage of SC technology, this paper proposes an exact approach to describe SSCS operations and determine the optimal SSCS design (i.e., optimal swap battery strategies for EVs with charging requests and the consideration of varying battery charging rates simultaneously) to minimize the overall system operational cost. In this proposed SSCS system, we formulated this problem into a binary integer programming model that could be solved by off-the-shelf commercial solvers (e.g., Gurobi). We explored a numerical example to illustrate the applications of this model from the freeway system in Guangdong Province, China, and compare it with alternative systems (the STBS). The SSCS was shown to be more effective than the alternative (e.g., a reduction of 54.3% in system operational cost).

This study can be extended in several directions. Future research can be conducted to explore the dynamic and stochastic demands of SCEVs, more variables such as maintenance and service levels of BSSs, variation of electricity prices, more complicated multi-type SC strategy combinations, associated vehicle coordination, more efficient customized solution methodologies, and the allowance of these vehicles to participate in peak shaving and valley filling to improve unreasonable charging and discharging. Moreover, it would be interesting to examine the impact of combinations of autonomous, modular, and EV technologies into this SSCS.

Author Contributions: The authors confirm the contributions to this paper are as follows: study conception and design, L.Z.; data collection, M.P.; analysis and interpretation of results, L.Z. and M.P.; draft manuscript preparation, L.Z. and M.P. All authors have read and agreed to the published version of the manuscript.

Funding: This work was supported by the National Natural Science Foundation of China under Grants U1811463.

Acknowledgments: The China Scholarship Council and its financial support are highly acknowledged.

Conflicts of Interest: The authors declare no conflicts of interest.

References

1. Zheng, Y.; Dong, Z.Y.; Xu, Y.; Meng, K.; Zhao, J.H.; Qiu, J. Electric vehicle battery charging/swap stations in distribution systems: Comparison study and optimal planning. *IEEE Trans. Power Syst.* **2014**, *29*, 221–229. [CrossRef]
2. Mechthild Wörsdörfer. *Global EV Outlook 2019*; International Energy Agency: Paris, France, 2019.
3. Yang, J.; Sun, H. Battery swap station location-routing problem with capacitated electric vehicles. *Comput. Oper. Res.* **2015**, *55*, 217–232. [CrossRef]
4. Mak, H.Y.; Rong, Y.; Shen, Z.J.M. Infrastructure planning for electric vehicles with battery swapping. *Manag. Sci.* **2013**, *59*, 1557–1575. [CrossRef]
5. Sierzchula, W.; Bakker, S.; Maat, K.; Van Wee, B. The influence of financial incentives and other socio-economic factors on electric vehicle adoption. *Energy Policy* **2014**, *68*, 183–194. [CrossRef]
6. Chen, A.; Zhou, Z.; Chootinan, P.; Ryu, S.; Yang, C.; Wong, S.C. Transport Network Design Problem under Uncertainty: A Review and New Developments. *Transp. Rev.* **2011**, *31*, 743–768. [CrossRef]
7. Nie, Y.; Ghamami, M. A corridor-centric approach to planning electric vehicle charging infrastructure. *Transp. Res. Part B Methodol.* **2013**, *57*, 172–190. [CrossRef]
8. Deflorio, F.; Guglielmi, P.; Pinna, I.; Castello, L.; Marfull, S. Modeling and analysis of wireless "charge while driving" operations for fully electric vehicles. *Transp. Res. Procedia* **2015**, *5*, 161–174. [CrossRef]
9. Mouhrim, N.; El Hilali Alaoui, A.; Boukachour, J. Optimal allocation of wireless power transfer system for electric vehicles in a multipath environment. In Proceedings of the 2016 3rd International Conference on Logistics Operations Management (GOL), Fez, Morocco, 23–25 May 2016.
10. Liu, Z.; Song, Z. Robust planning of dynamic wireless charging infrastructure for battery electric buses. *Transp. Res. Part C Emerg. Technol.* **2017**, *83*, 77–103. [CrossRef]
11. Zhao, M.; Li, X.; Yin, J.; Cui, J.; Yang, L.; An, S. An integrated framework for electric vehicle rebalancing and staff relocation in one-way carsharing systems: Model formulation and Lagrangian relaxation-based solution approach. *Transp. Res. Part B Methodol.* **2018**, *117*, 542–572. [CrossRef]
12. Ko, Y.D.; Jang, Y.J.; Lee, M.S. The optimal economic design of the wireless powered intelligent transportation system using genetic algorithm considering nonlinear cost function. *Comput. Ind. Eng.* **2015**, *89*, 67–79. [CrossRef]

13. Yatnalkar, G.; Narman, H. Survey on Wireless Charging and Placement of Stations for Electric Vehicles. In Proceedings of the 2018 IEEE International Symposium on Signal Processing and Information Technology (ISSPIT), Louisville, KY, USA, 6–8 December2018; pp. 526–531.

14. Chen, Z.; He, F.; Yin, Y. Optimal deployment of charging lanes for electric vehicles in transportation networks. *Transp. Res. Part B Methodol.* **2016**, *91*, 344–365. [CrossRef]

15. Sepetanc, K.; Pandzic, H. A Cluster-Based Operation Model of Aggregated Battery Swapping Stations. *IEEE Trans. Power Syst.* **2020**, *1*, 249–260. [CrossRef]

16. Bansal, P. Charging of Electric Vehicles: Technology and Policy Implications. *J. Sci. Policy Gov.* **2015**, *6*, 1–20.

17. Machura, P.; Li, Q. A critical review on wireless charging for electric vehicles. *Renew. Sustain. Energy Rev.* **2019**, *104*, 209–234. [CrossRef]

18. Jang, Y.J. Survey of the operation and system study on wireless charging electric vehicle systems. *Transp. Res. Part C Emerg. Technol.* **2018**, *95*, 844–866. [CrossRef]

19. Yang, J.; Guo, F.; Zhang, M. Optimal planning of swapping/charging station network with customer satisfaction. *Transp. Res. Part E Logist. Transp. Rev.* **2017**, *103*, 174–197. [CrossRef]

20. Schneider, F.; Thonemann, U.W.; Klabjan, D. Optimization of battery charging and purchasing at electric vehicle battery swap stations. *Transp. Sci.* **2018**, *52*, 1211–1234. [CrossRef]

21. Du, J.; Ouyang, M. Review of electric vehicle technologies progress and development prospect in China. In Proceedings of the 2013 World Electric Vehicle Symposium and Exhibition (EVS27), Barcelona, Spain, 17–20 November 2013; pp. 1–8.

22. Pérez, J.M.G. *Emerging Technologies for Electric and Hybrid Vehicles*; MDPI: Basel, Switzerland, 2018; ISBN 3038971901.

23. Mohamed, A.A.S.; Meintz, A.; Zhu, L. System Design and Optimization of In-Route Wireless Charging Infrastructure for Shared Automated Electric Vehicles. *IEEE Access* **2019**, *7*, 79968–79979. [CrossRef]

24. Hof, J.; Schneider, M.; Goeke, D. Solving the battery swap station location-routing problem with capacitated electric vehicles using an AVNS algorithm for vehicle-routing problems with intermediate stops. *Transp. Res. Part B Methodol.* **2017**, *97*, 102–112. [CrossRef]

25. Masmoudi, M.A.; Hosny, M.; Demir, E.; Genikomsakis, K.N.; Cheikhrouhou, N. The dial-a-ride problem with electric vehicles and battery swapping stations. *Transp. Res. Part E Logist. Transp. Rev.* **2018**, *118*, 392–420. [CrossRef]

26. Yang, S.; Yao, J.; Kang, T.; Zhu, X. Dynamic operation model of the battery swapping station for EV (electric vehicle) in electricity market. *Energy* **2014**, *65*, 544–549. [CrossRef]

27. Adler, J.D.; Mirchandani, P.B. Online routing and battery reservations for electric vehicles with swappable batteries. *Transp. Res. Part B Methodol.* **2014**, *70*, 285–302. [CrossRef]

28. Adegbohun, F.; von Jouanne, A.; Lee, K.Y. Autonomous battery swapping system and methodologies of electric vehicles. *Energies* **2019**, *12*, 667. [CrossRef]

29. Ding, T.; Bai, J.; Du, P.; Qin, B.; Li, F.; Ma, J.; Dong, Z. Rectangle packing problem for battery charging dispatch considering uninterrupted discrete charging rate. *IEEE Trans. Power Syst.* **2019**, *34*, 2472–2475. [CrossRef]

30. Kang, Q.; Wang, J.; Zhou, M.; Ammari, A.C. Centralized Charging Strategy and Scheduling Algorithm for Electric Vehicles under a Battery Swapping Scenario. *IEEE Trans. Intell. Transp. Syst.* **2016**, *17*, 659–669. [CrossRef]

31. Yang, J.; Wang, W.; Ma, K.; Yang, B. Optimal Dispatching Strategy for Shared Battery Station of Electric Vehicle by Divisional Battery Control. *IEEE Access* **2019**, *7*, 38224–38235. [CrossRef]

32. Sun, B.; Sun, X.; Tsang, D.H.K.; Whitt, W. Optimal battery purchasing and charging strategy at electric vehicle battery swap stations. *Eur. J. Oper. Res.* **2019**, *279*, 524–539. [CrossRef]

33. He, F.; Yin, Y.; Lawphongpanich, S. Network equilibrium models with battery electric vehicles. *Transp. Res. Part B Methodol.* **2014**, *67*, 306–319. [CrossRef]

34. He, F.; Wu, D.; Yin, Y.; Guan, Y. Optimal deployment of public charging stations for plug-in hybrid electric vehicles. *Transp. Res. Part B Methodol.* **2013**, *47*, 87–101. [CrossRef]

35. Worley, O.; Klabjan, D. Optimization of battery charging and purchasing at electric vehicle battery swap stations. In Proceedings of the 2011 IEEE Vehicle Power and Propulsion Conference, Chicago, IL, USA, 6–9 September 2011; pp. 1–4.

Energies **2020**, *13*, 1213

36.	Jie, W.; Mang, J.; Zhung, M.; Huang, I. The two-echelon capacitated electric vehicle routing problem with battery swapping stations: Formulation and efficient methodology. *Eur. J. Oper. Res.* **2019**, *272*, 879–904. [CrossRef]

37.	Chen, J.; Qi, M.; Miao, L. The Electric Vehicle Routing Problem with Time Windows and Battery Swapping Stations. *IEEE Int. Conf. Ind. Eng. Eng. Manag.* **2016**, *2016*, 712–716.

38.	Amiri, S.S.; Jadid, S.; Saboori, H. Multi-objective optimum charging management of electric vehicles through battery swapping stations. *Energy* **2018**, *165*, 549–562. [CrossRef]

39.	Widrick, R.S.; Nurre, S.G.; Robbins, M.J. Optimal policies for the management of an electric vehicle battery swap station. *Transp. Sci.* **2018**, *52*, 59–79. [CrossRef]

40.	Infante, W.; Ma, J.; Han, X.; Liebman, A. Optimal Recourse Strategy for Battery Swapping Stations Considering Electric Vehicle Uncertainty. *IEEE Trans. Intell. Transp. Syst.* **2019**, 1–11. [CrossRef]

41.	Ouyang, D.; Chen, M.; Liu, J.; Wei, R.; Weng, J.; Wang, J. Investigation of a commercial lithium-ion battery under overcharge/over-discharge failure conditions. *RSC Adv.* **2018**, *8*, 33414–33424. [CrossRef]

42.	Chen, Z.; Liu, W.; Yin, Y. Deployment of stationary and dynamic charging infrastructure for electric vehicles along traffic corridors. *Transp. Res. Part C Emerg. Technol.* **2017**, *77*, 185–206. [CrossRef]

43.	Koç, Ç.; Jabali, O.; Mendoza, J.E.; Laporte, G. The electric vehicle routing problem with shared charging stations. *Int. Trans. Oper. Res.* **2019**, *26*, 1211–1243. [CrossRef]

44.	Wu, X.; Freese, D.; Cabrera, A.; Kitch, W.A. Electric vehicles' energy consumption measurement and estimation. *Transp. Res. Part D Transp. Environ.* **2015**, *34*, 52–67. [CrossRef]

45.	Smith, O.J.; Boland, N.; Waterer, H. Solving shortest path problems with a weight constraint and replenishment arcs. *Comput. Oper. Res.* **2012**, *39*, 964–984. [CrossRef]

46.	Laporte, G.; Pascoal, M.M.B. Minimum cost path problems with relays. *Comput. Oper. Res.* **2011**, *38*, 165–173. [CrossRef]

47.	Hwang, I.; Jang, Y.J.; Ko, Y.D.; Lee, M.S. System Optimization for Dynamic Wireless Charging Electric Vehicles Operating in a Multiple-Route Environment. *IEEE Trans. Intell. Transp. Syst.* **2018**, *19*, 1709–1726. [CrossRef]

48.	Cochran, J.J.; Cox, L.A.; Keskinocak, P.; Kharoufeh, J.P.; Smith, J.C.; Linderoth, J.T.; Lodi, A. MILP Software. In *Wiley Encyclopedia of Operations Research and Management Science*; John Wiley & Sons Inc.: Hoboken, NJ, USA, 2011.

49.	Zhang, Y.; D'Ariano, A.; He, B.; Peng, Q. Microscopic optimization model and algorithm for integrating train timetabling and track maintenance task scheduling. *Transp. Res. Part B Methodol.* **2019**, *127*, 237–278. [CrossRef]

50.	Fuentes, M.; Cadarso, L.; Marín, Á. A hybrid model for crew scheduling in rail rapid transit networks. *Transp. Res. Part B Methodol.* **2019**, *125*, 248–265. [CrossRef]

 *energies* 

Article

Deployment of a Bidirectional MW-Level Electric-Vehicle Extreme Fast Charging Station Enabled by High-Voltage SiC and Intelligent Control

Ziwei Liang, Daniel Merced, Mojtaba Jalalpour and Hua Bai *

Department of Electrical Engineering and Computer Science, University of Tennessee, Knoxville, TN 37996, USA; zliang7@vols.utk.edu (Z.L.); dmercedc@vols.utk.edu (D.M.); mjalalp1@vols.utk.edu (M.J.)
* Correspondence: hbai2@utk.edu

Received: 24 March 2020; Accepted: 7 April 2020; Published: 10 April 2020

Abstract: Considering the fact that electric vehicle battery charging based on the current charging station is time-consuming, the charging technology needs to improve in order to increase charging speed, which could reduce range anxiety and benefit the user experience of electric vehicle (EV). For this reason, a 1 MW battery charging station is presented in this paper to eliminate the drawbacks of utilizing the normal 480 VAC as the system input to supply the 1 MW power, such as the low power density caused by the large volume of the 60 Hz transformer and the low efficiency caused by the high current. The proposed system utilizes the grid input of single-phase 8 kVAC and is capable of charging two electric vehicles with 500 kW each, at the same time. Therefore, this paper details how high-voltage SiC power modules are the key enabler technology, as well as the selection of a resonant-type input-series, output-parallel circuitry candidate to secure high power density and efficiency, while intelligently dealing with the transient processes, e.g., pre-charging process and power balancing among modules, and considering the impact on the grid, are both of importance.

Keywords: silicon carbide; extreme fast charging; DC transformer; electric vehicle

---

1. Technical Challenges of EV Extreme Fast Charging Stations

Transportation is revolutionizing as the world welcomes the benefits of electric transportation. The growing interest in electric vehicle (EV) technology is due to their minimum fuel emissions and air pollution. The demand for this kind of vehicles is growing, and the rate of this growth is also expected to increase in the forthcoming years [1,2]. However, there are concerns to be addressed in this regard, including time of charge, range anxiety, cost of charging, and negative impact of charging demand on the grid. Range anxiety and time of charge are the issues that have been addressed in recent years through developing EV fast chargers. Sixteen hours are needed for level-1 chargers, while only 10 minutes are needed for extreme fast chargers (XFCs). This by itself shows how the charging process can impact the adoption of plug-in hybrid electric vehicles. The concept of a gas-station-like experience for charging EVs has become a reality in recent years due to the advancement of XFCs [3,4], e.g., chargers rated at 400 kW and above. The focus of this work is to design a 1 MW XFC station for two EVs, i.e., 500 kW per car. Take Tesla Model S as an example, which has 75 kWh battery pack. If the state-of-charge (SOC) window is 80%, a 500 kW XFC is then able to charge an EV within 75 kWh × 80%/500 kW = 7 m, presuming the charging loss is ignorable.

While the majority of EV fast charging stations in the United States have a grid connection of three-phase 480 VAC, which is rectified into a DC bus (normally between 600 VDC and 800 VDC), followed by a DC/DC converter with galvanic isolation. Using such an approach to provide 1 MW sees a challenge from high grid current, which in turn challenges the 480 VAC transformer design. As shown in Table 1, the traditional 500 kVA XFC uses a grid transformer of more than 5000 L and more than 2000 kg. To enhance the system power density and reduce cost, elimination of such bulky and heavy grid transformers is a must [5].

Table 1. Comparison of Metrics for a Traditional and Proposed 500 kVA Fast Charging System.

Figures of Merit	Traditional 500 kVA System	HV SiC Enabled 500 kVA System
Power losses (%/kW)	$\eta = \eta_{xfmr} \times \eta_{fast\ charger} = 94.38\%$ @ rated power $P_{losses} = 28.09$ kW	$\eta \geq 97.75\%$ @ rated power $P_{losses} \leq 11.25$ kW $\eta \geq 95\%$ @ 5% rated power $P_{losses} \leq 1.25$ kW
Size–footprint (dm)	$Area_{total} = 3.5$ m^2	$Area_{total} \leq 0.875$ m^2
Size–form factor (dm^3)	$V_{total} = 5190$ liters	$V_{total} \leq 1298$ liters
Weight (kg)	$W_{total} = 3537$ kg	$W_{total} \leq 530$ kg
Specific Power	0.14 kVA/kg	>5 kVA/kg
Power Density	0.09 kVA/L	>9.2 kVA/L
Cooling Method	Air Cooled/Oil Filled	Liquid Cooled
MTTF (Targets)	68,960 h (7.9 years) (PFC + DCX)	75,856 h (8.66 years) (PFC + DCX)

Recently, wide-bandgap (WBG) semiconductor devices have played a role in developing EV chargers [6,7], not only increasing the power density thanks to the higher switching frequency than Si, but also undertaking higher input voltage, such as 6.5 kV silicon carbide (SiC). This potentially allows the medium voltage (MV) transmission lines (2.4–13.4 kVAC) directly to come into the charging station, thereby saving the 480 VAC step-down grid transformers. As compared in Table 1, elimination of the grid transformer can significantly reduce the size and weight of the overall XFC system. Even though the DC/DC converter such as dual active bridge (DAB) and resonant circuits (LLC and CLLC) all need transformers [8–10], such transformers can be operated at a much higher switching frequency, thereby seeing less weight and size penalty while maintaining galvanic isolation.

From Table 1, we can find that the overall efficiency of the proposed 500 kVA system based on the HV SiC switches is improved by 3%. The main reasons for the higher efficiency of the proposed system are as follows. 1) The normal input grid voltage of the traditional 500 kVA system is three-phase 480 VAC, and the input grid voltage of the proposed 500 kVA system is single-phase 8 kVAC, which is more than 17 times that of the traditional system. Given the same power level, higher voltage means lower current, which is beneficial to reduce the power loss of the whole transmission and improve system efficiency. 2) Enabled by 6.5 kV high voltage SiC power modules, the proposed system will connect to MV transmission lines (2.4–13.4 kVAC), thereby directly saving the 480 VAC step-down grid transformers. Therefore, the large power loss of the 60 Hz transformer is eliminated, which helps to achieve the higher efficiency. Additionally, a bidirectional feature is highly demanded, especially when there is a grid outage.

With the above consideration and enabled by high voltage (HV) SiC power modules, an approach shown in Figure 1 is proposed. The input 8 kVAC was split evenly among three H-bridge modules, which act as the power factor controller. The AC is then rectified into 4.3 kVDC, which is forwarded to a resonant-type DC transformer circuit (DCX) made of a primary half-bridge, a three-winding transformer, and two low-voltage H-bridge paralleled to output 1.3 kVDC. Note that this allows the two low-voltage DC (LVDC) buses to be isolated, thereby charging two cars in an isolated manner. The outputs of each of the DCX modules are connected in a parallel output to generate 1 MW power. Each LVDC bus then is connected to a 500 kW buck converter to charge one EV, as shown in Figure 1b.

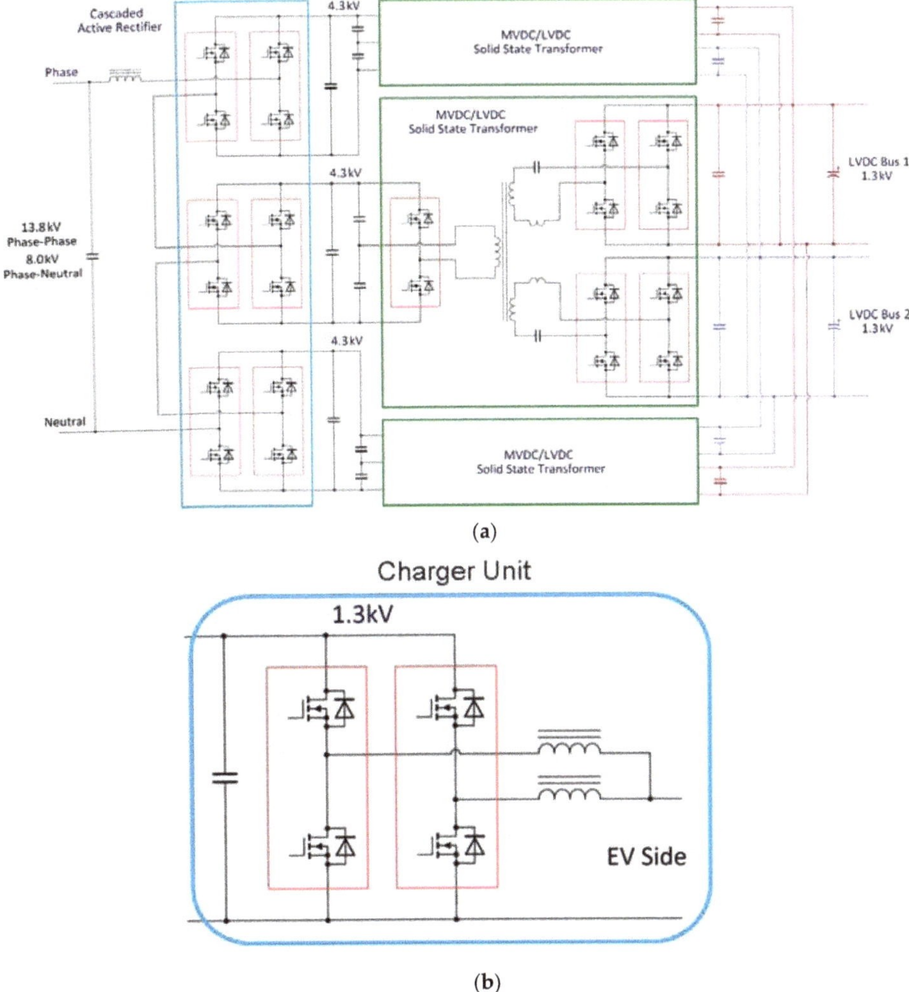

(a)

(b)

Figure 1. (**a**) Block Diagram of proposed design of electric vehicle (EV) extreme fast charger (XFC) station for two vehicles. (**b**) Block diagram of the following 500 kW DC/DC converter.

Essentially, such an input-series, output-parallel (ISOP) design successfully eliminates the 480 VAC/60 Hz transformer, while it also realizes bidirectional power flow. Such an ISOP system, however, brings other challenges. 1) Device electrical ratings. With 8 kVAC single-phase grid, the grid connection of each power factor correction (PFC) module is desired to be at 2.6 kVAC, which is further boosted to 4.3 kVDC. For a 1 MW charging capability, each DC/DC converter module undertakes 333 kW. Selecting the right topology to minimize electrical stress of the switches is key to the success, especially for the DC/DC stage. A detailed comparison of various DC/DC topologies is shown in Table 2, including DAB, DCX, and CLLC. These topologies were chosen because they all provide isolated bidirectional DC/DC conversion by means of a high-frequency transformer and are all widely used [8–12]. As seen in Table 2, even though the peak current and root mean square(RMS) current are high with the DCX converter, the maximum switching off current is close to 0. This is because the DCX topology always operates at the resonant frequency, which is the main reason this paper selects such topology. All SiC devices are switched at 30 kHz in all compared topologies.

Table 2. DCX topology comparison.

Parameter	DAB	CLLC	DCX
Peak Current [A]	112	275	250
RMS Current [A]	93	193	170
Maximum Switching-off Current [A]	112	154	15
C_{DC} (DC-bus cap between PFC and DCDC [mF])	0.8	0.6	0.01
C_{out} (DC-bus cap at output [mF])	0.085	0.085	1×2
C_r (resonant cap [µF])	N/A	0.316	12.665
Transformer [kVA]	420	423	592

The electric stress is then shifted to the resonant circuit and MV transformer. The PFC stage also needs to realize unity power factor and reduce total harmonic distortion (THD), and given the input series structure of the PFCs, the interleaved gate signal is supplied to the PFCs to achieve much smaller current ripple, which could help to reduce the demand of the large bulky DC bus capacitor. This is of importance when reducing the dimensions and cost of the system and improving the power density of the system. Besides, the smart output voltage balancing control is proposed to make sure the three PFC modules have balanced output voltage even under the different load conditions. The balanced output could ensure the same voltage stress among different modules, which is beneficial to the safe operation of the system. This part will be discussed further in Section 2. The DCX topology steps down 4.3 kVDC to a level of ~1.3 kVDC. Considering the importance of the resonance of the DCX, the intelligent resonant frequency tracking method based on the master–slave control is proposed. This control method could adjust the working frequency automatically to reach the resonant frequency, and the master-slave control enables the automatic change of the frequency adjustment operand. Considering that the resonant condition is necessary for the low switching loss of the switches, this novel control method is indispensable for the high-efficiency and safe operation of the system. Given the high-power level and high-voltage level, to avoid the inrush current at the start of the system, the DCX is responsible for pre-charging the output capacitor of the LVDC buses before charging. These control methods will be detailed in Section 3. Section 4 will discuss the impact on the grid. Section 5 is the conclusion.

2. PFC-Stage Control

2.1. Power Balancing

Since the basic control method of the totem-pole PFC has been widely discussed [13,14], this paper will particularly focus on the power balancing among modules. To simplify the analysis, just consider two series PFC stages as shown in Figure 2.

To rectify the AC power from the power grid, the bridge leg of the PFC #n {Sn3, Sn4} is operated with main power frequency, i.e., 50/60 Hz, which could be seen as the rectifier bridge leg. Another bridge leg of the PFC #n {Sn1, Sn2} is pulse width modulation (PWM) controlled to achieve the power factor correction and low THD. In the steady state of this system, because the voltage and current are in phase, the DCX could be considered as the resistive load. Because the switching frequency fs is much higher than the line frequency fg, vin(t) can be assumed as constant within one switching period. Considering the continuous conduction mode (CCM) for both PFCs, steady-state equations can be derived as follows [15]:

$$V_{o1}(1 - d_1) + V_{o2}(1 - d_2) = |V_{in}| \tag{1}$$

$$V_{o1} = |i_{in}| \cdot (1 - d_1) \cdot R_1 \tag{2}$$

$$V_{o2} = |i_{in}| \cdot (1 - d_2) \cdot R_2 \tag{3}$$

where d_1 and d_2 represent the duty cycle of the high-frequency gate signals of PFC#1 and PFC#2.

From Equations (2) and (3), it can be seen that to achieve the output voltage balancing control under the unequal load condition, the duty cycles of two PFCs need to be modulated to accommodate the load difference [15,16]. Figure 3 shows the control block of the PFC weighted output voltage balancing control, composed of three loops, i.e., inner current control loop, outer voltage control loop, and the supplementary voltage balancing loop. The slower outer voltage control loop is to make the PFC stage output voltage follow the reference voltage and supply the reference current to the inner current control loop. The reference current is generated by the averaged voltage loop output V_{voa} and the grid side voltage V_{in}, where V_{voa} supplies the magnitude and vin with its RMS value V_{in} provides the phase of the reference current.

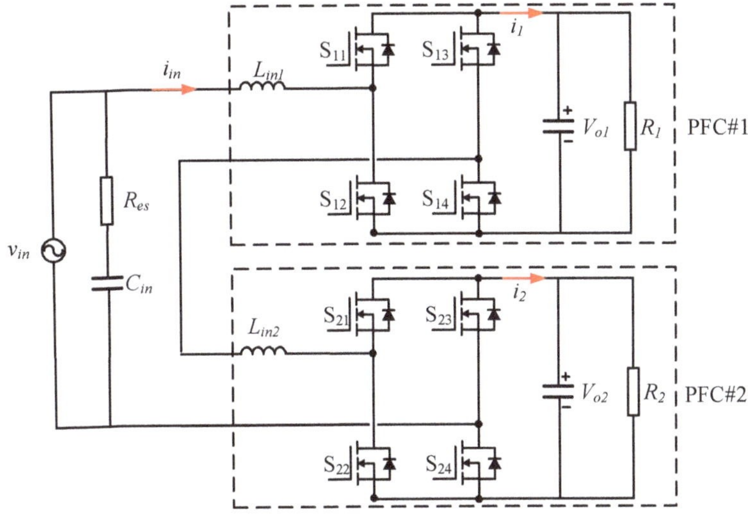

Figure 2. Schematics of two power factor correction (PFC) modules in series.

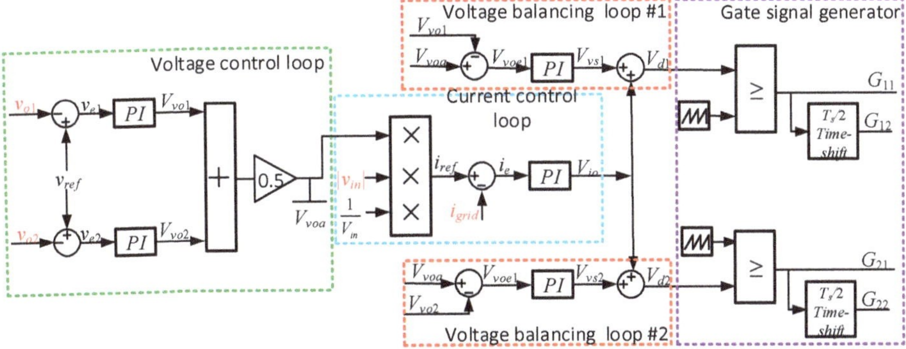

Figure 3. Control scheme for series-connected PFC modules.

The impact of the load difference is thereby eliminated through the voltage balancing control loop, as simulated in Figure 4a,b, where the load of each PFC is not identical, $R_2 = 1.1R_1$. Without balancing control, some diversity of the switch can cause the DC-bus variation. Based on Equations (2) and (3), the same duty cycle will be supplied to two PFCs. Because of the series input structure, the input current will be the same for two PFCs. Therefore, the different load resistance will cause the unbalanced output. The simulation results of the steady-state output voltage waveform without the balancing control is shown in Figure 4a. When the voltage balancing loop is integrated into the control loop, two

PFCs can be controlled individually. Voltage balancing loop could compensate for the deviation of the duty cycle caused by the different loads. After the compensation, the different duty cycle generated based on the output voltages of two PFCs will be supplied to the corresponding PFCs, such that the balanced output could be achieved. Figure 4b shows the voltage balancing control performance based on the simulation model. We then further experimentally validated the voltage balancing control for PFC stages, where we consider the load resistances of two PFCs to be $2R_1 = R_2 = 25\ \Omega$. The experimental setup and the Field Programmable Gata Array (FPGA) control block diagram is shown in Figure 5, and the experimental results are shown in Figure 6. Figure 6a shows the steady-state input and output waveform without the balancing control. It can be seen that the output DC voltage of PFC#1 is almost twice of the output DC voltage of PFC#2. Considering that the load R_2 is twice of the load R_1, the experimental results are consistent with the theoretical analysis. As shown in Figure 6b,c, after integrating the voltage balancing control, the DC bus voltages are kept the same once the system reaches steady state. Therefore, the voltage balancing control method is validated by the simulation and experiments.

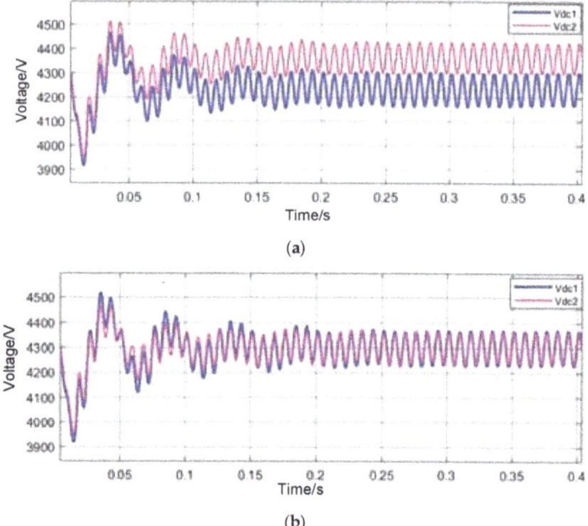

Figure 4. Simulation results of the DC output voltage for series-connected PFC modules: (**a**) without the balancing control and (**b**) with the balancing control.

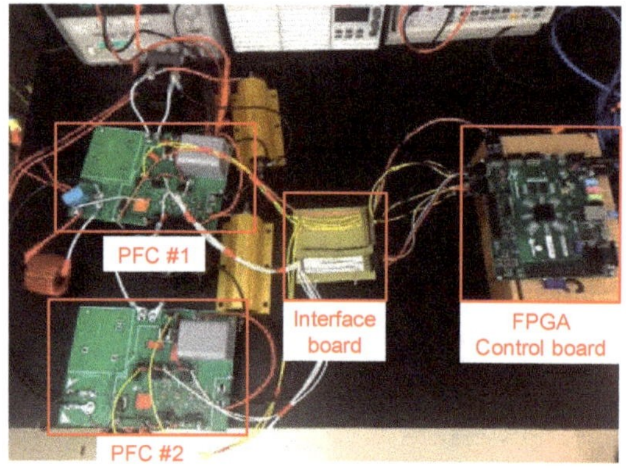

Figure 5. *Cont.*

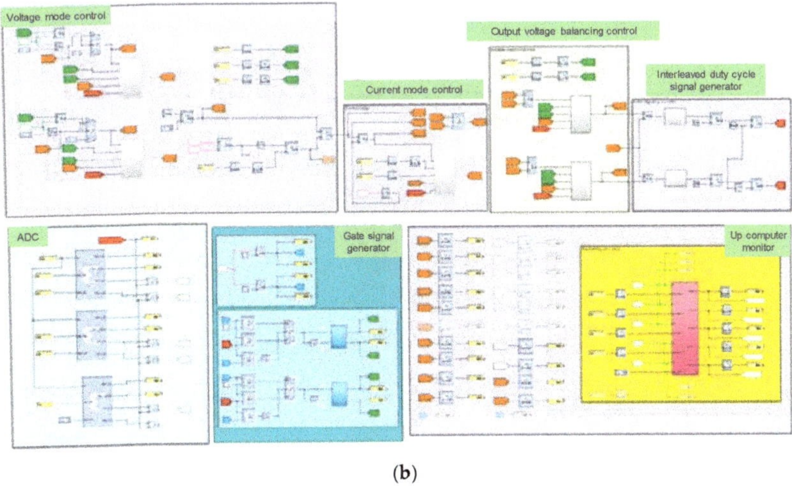

(**b**)

Figure 5. (**a**) Experimental setup of two input series-connected PFC; (**b**) FPGA control block diagram.

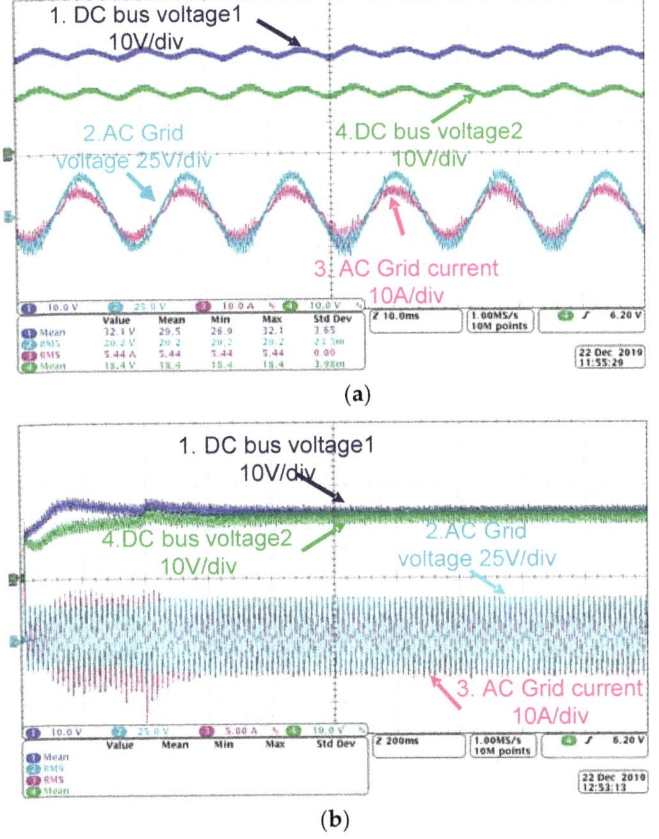

(**a**)

(**b**)

Figure 6. *Cont.*

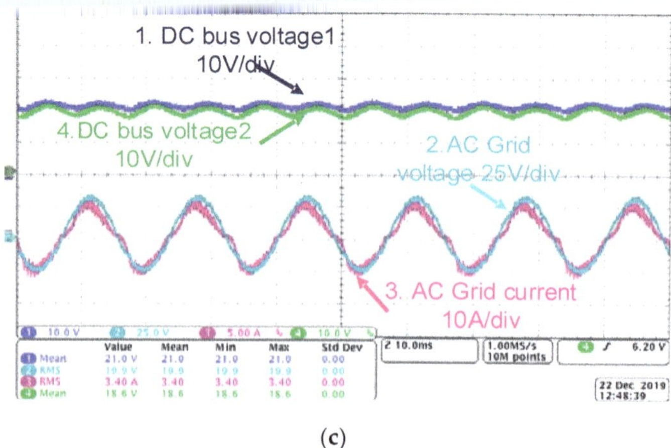

(**c**)

Figure 6. Experimental results for series-connected PFC modules (**a**) without the balancing control; (**b**) with the balancing control and (**c**) Zoom-in view with the balancing control.

2.2. Interleaving Control

Once the three PFC modules balance the power in the modules, we can further employ interleaving control for each PFC, i.e., shifting the gate signals by 120° (1/3Ts). This helps further reduce the grid-side current. As shown in Figure 7a, without the interleaving, the grid side current (I_inductor) sees over 20 A current ripple. With the interleaving control, as shown in Figure 7b, the grid current ripple drops to ~5 A.

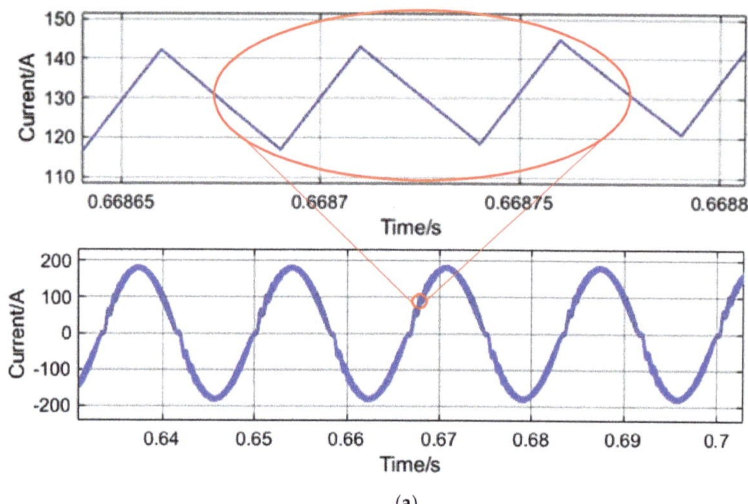

(**a**)

Figure 7. *Cont.*

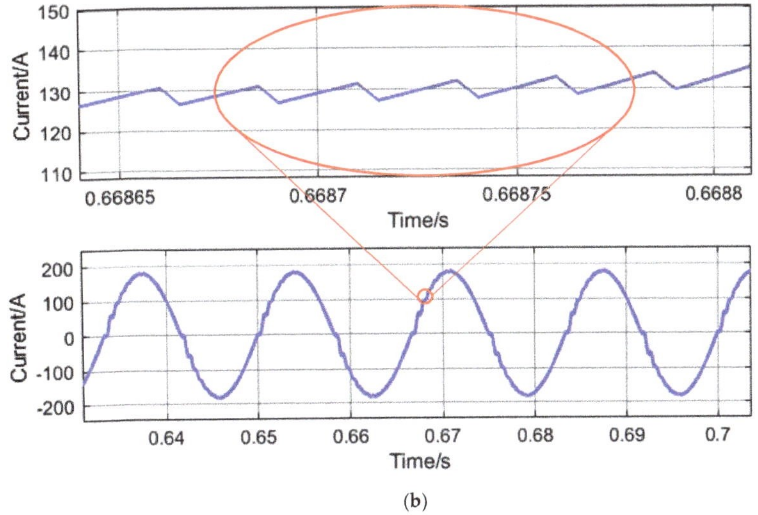

Figure 7. Simulated grid current (**a**) with same gate signal and (**b**) with interleaving gate signal.

3. DCX Stage Control

3.1. Power Balancing Control

The DCX converter topology, as seen in Figure 1a, consists of a MOSFET half-bridge converter on the primary side, a three-winding MV transformer, a resonant capacitor and a MOSFET full-bridge converter on each secondary side. The primary-side and secondary-side MOSFETs share the same complementary PWM gate signals with 50% duty cycle. The relationship of the resonant frequency f_r and the resonant parameters can be expressed by Equation (4), where L_r is the leakage inductance of the transformer and C_r represents the capacitance of the resonant capacitor.

$$2\pi f_r = \frac{1}{\sqrt{L_r C_r}} \tag{4}$$

As previously mentioned, the PFC stage could achieve balanced output voltage, which means that each of the DCX converters will have the same input voltage. However, a slight difference between transformers leakage inductance or the resonant capacitance will lead the converter to deviate away from the resonance point and further creates an unbalanced output current. The main challenge of this unbalanced problem is that when the switching frequency is higher or lower than the actual resonant frequency of the DCX, they both can cause the output current to decrease and unbalance the outputs. Thus, we need to detect the impedance of the resonant tank to determine the changing direction of the switching frequency. To solve this problem intelligently and easily, output current balancing control based on the resonant frequency control is proposed in this paper and shown in Figure 8.

It is noticeable that when the switching frequency of one DCX converter in this paralleled structure is closer to its resonant frequency, its output current will be larger. Thus, in this control method, the DCX converter which has the highest output current will be set as the master converter, and other DCX converters will automatically become the slaves. The frequency of the master will be constant, and the frequency of the slave will be adjusted to match the resonant frequency of the DCX converter. The resonant frequency tracking control is mainly controlling the changing direction of switching frequency based on the changing of the output current. The time step ts and frequency step Δf are constant and can be adjusted according to the system parameters and requirements of the control accuracy. Then, if the output current of DCX #n increased after the last change of the frequency (when time $t = (k-1)t_s)$, the changing direction of the frequency will stay the same as the last time (when time $t = kt_s$). Conversely, the frequency will change in the opposite direction. The control goal is to track the resonant frequency of each DCX automatically and achieve the balanced output currents.

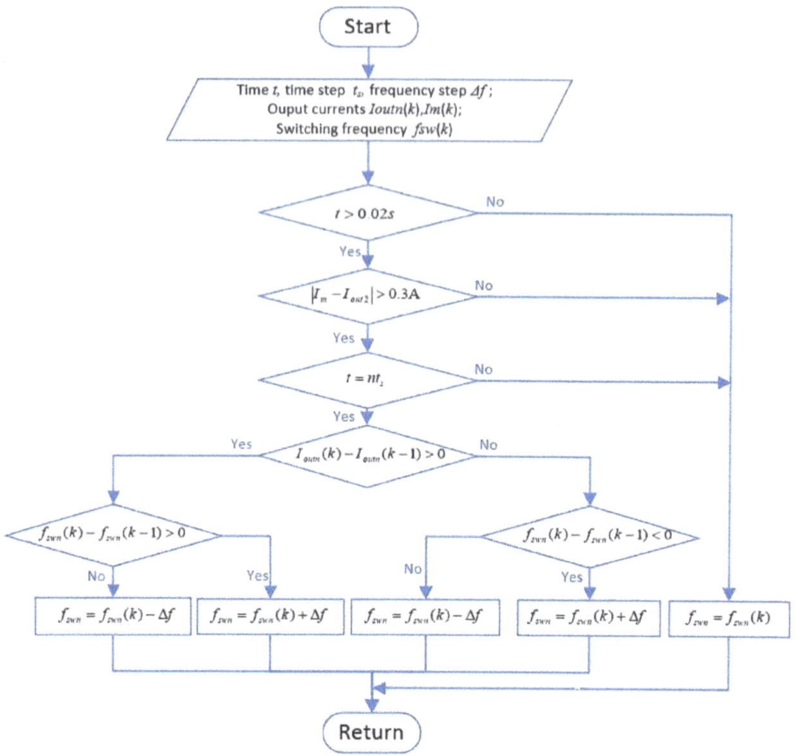

Figure 8. Algorithm depicting output current balancing control of the paralleled DC transformer circuits (DCXs).

Compared with the control method based on phase relationship detection [17] and time measurement of zero diode current [18], among others, the proposed method has three main advantages: 1) changing fs bi-directionally and automatically tracking the resonant condition; 2) setting the master automatically to make all the DCX converters work under the resonant conditions when the system is at the steady state; and 3) having no extra detection and measurement circuit.

Based on three paralleled DCXs, the simulation is conducted to validate the output current balancing control. Three DCXs have different resonant inductors but have the same initial working frequency $f_{initial}$:

$$f_{initial} = \frac{1}{2\pi \sqrt{L_{rd}C_{rd}}} = 20\text{kHz} \tag{5}$$

where L_{rd} and C_{rd} represents the designed resonant inductance and capacitance.

Figure 9a,b shows the output current balancing process of three parallel-connected outputs under two sets of different initial parameters of inductors, 1) $L_{r1} = L_{rd}$, $L_{r2} = 0.9\ L_{rd}$, $L_{r3} = 0.95\ L_{rd}$ and 2) $L_{r1} = 1.05\ L_{rd}$, $L_{r2} = 0.9\ L_{rd}$, $L_{r3} = 0.95\ L_{rd}$. For the first set, based on the control algorithm, there are mainly three stages for the balancing process. First, when $t < 0.02$ s, the system works without the balancing control, and the output currents are unbalanced, which is caused by the different resonant condition. From Figure 9a, when t < 0.02 s, it can be seen that the DCX#1 with the best resonant condition has the largest output current, which is consistent with the theoretical analysis that a better resonant condition will lead to larger output current. Then, when $t > 0.02$ s and before the system reaches the balanced condition (where the deviation of currents are less than 0.3 A), the first DCX is the master controller as it has the largest current and the other two DCXs become the slave automatically. The frequency of the master converter will keep consistent, and two slaves will adjust the frequency to reach the resonant frequency based on the algorithm. Finally, when the output currents reach balance, the algorithm will automatically stop changing the frequency and keep the system working under this steady-state condition. Similar to the first set, the modulation process of the second set also has three modulation stages, which are shown in Figure 9b. The difference is that, for the second set, DCX#1 and DCX#3 have similar frequency deviation, so at first, the output current of these two DCXs are close to each other.

Figure 10 shows the voltage and current waveforms of the transformer primary sides of DCX#2 before and after the DCX reaches the resonant condition. After the DCX reaches the resonant condition, the transformer current waveform changes to a sinusoidal waveform, and the current and voltage are exactly in phase, which indicates zero-current switching-on and -off loss. Figure 11 shows the frequency modulation process of DCX#2 and DCX#3 when the system works under the first set of the parameters of inductance shown in Figure 9a. From Figure 11, we can find that first, the DCX#2 and DCX#3 could modulate the frequency to the real resonant frequency, which is determined by the inductor and capacitor, based on the deviation of the resonant frequency; and second, when the system reaches the resonant condition, the frequency modulation will stop and the controller will keep the frequency constant, such that three DCXs will work on the steady state with balanced outputs.

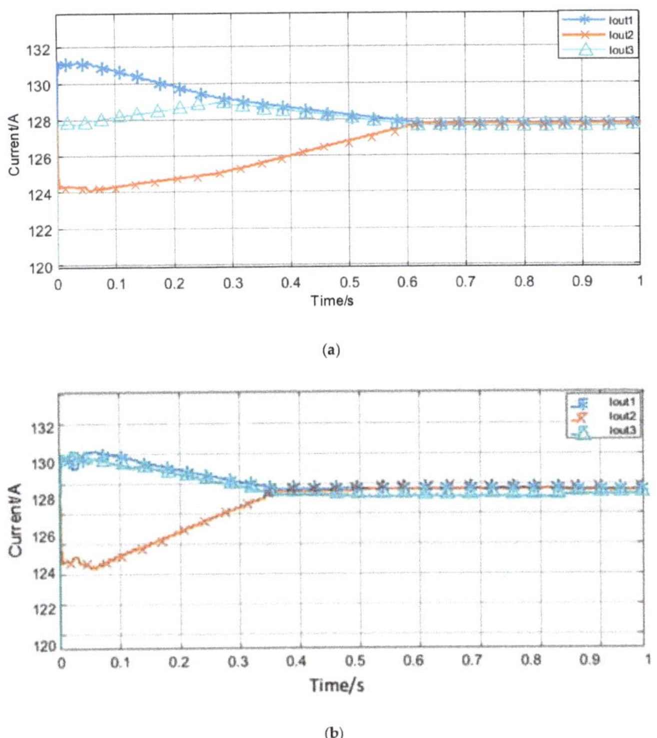

Figure 9. Output current balancing control performance with the resonant inductance of (a) $L_{r1} = L_{rd}$, $L_{r2} = 0.9\,L_{rd}$, $L_{r3} = 0.95\,L_{rd}$ and (b) $L_{r1} = 1.05\,L_{rd}$, $L_{r2} = 0.9\,L_{rd}$, $L_{r3} = 0.95\,L_{rd}$.

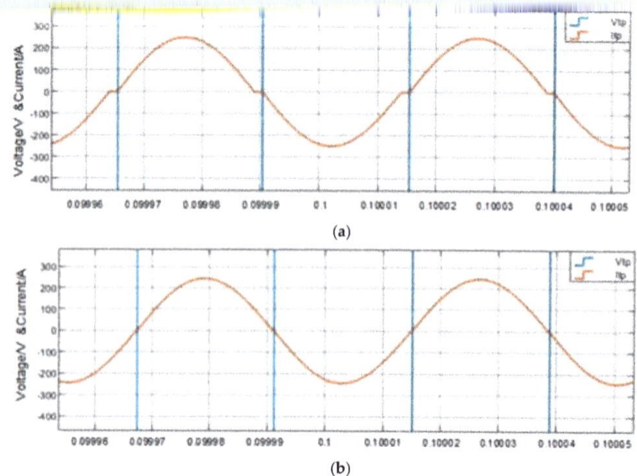

Figure 10. Voltage and current waveforms of the primary side of the DCX transformer (**a**) before the DCX reaches the resonant condition and (**b**) after the DCX reaches the resonant condition.

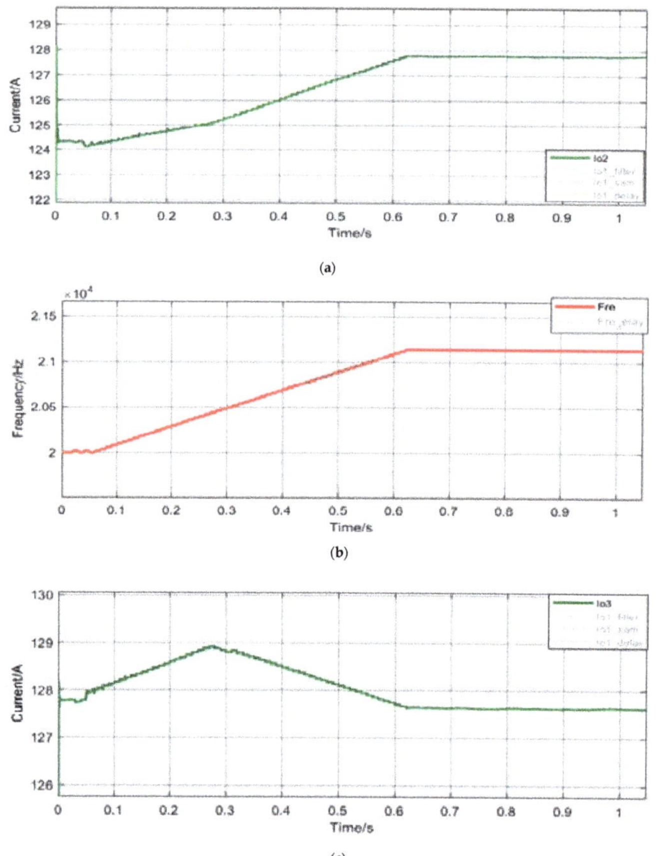

Figure 11. *Cont.*

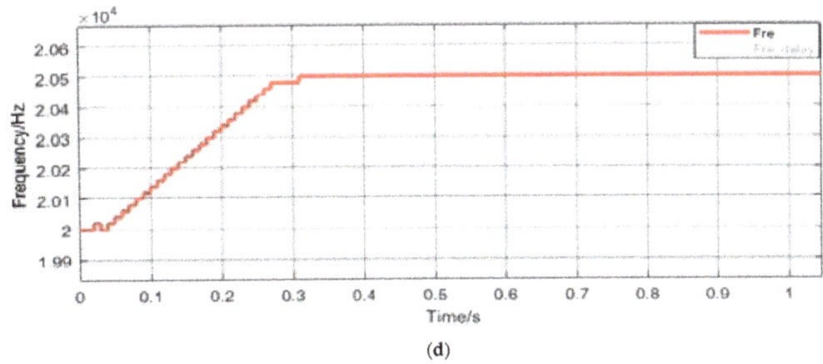

(d)

Figure 11. Output current and switching frequency regulation of the DCX#2 and DCX#3. (**a**) Output current changing with the switching frequency of DCX#2; (**b**) Switching frequency regulation of DCX#2; (**c**) Output current changing with the switching frequency of DCX#3 and (**d**) Switching frequency regulation of DCX#3.

3.2. Pre-Charging Process

Considering high power and voltage level, we need to pre-charge the output capacitor of the LVDC bus before starting to charge the vehicle. Given the nature of its voltage source, directly providing 50% duty cycle for all switches will induce the large inrush current [19]. Instead, we propose to use a smaller duty cycle for the primary-side switch of the DCX in the starting process and prolong the charging time. The simulation results of pre-charging control are shown in Figure 12. Figure 12a shows the pre-charging process; meanwhile, Figure 12b shows the duty cycle of the gate signal changing process. When the system works at steady state, for DCX, the gate signals for the primary and secondary side switches are exactly in phase and with duty cycle of 50%. Differently from the steady state, when the system works on pre-charging stage, the duty cycle of the primary side will increase gradually from 0% to 50%. In this way, we can make sure that the charging power is small at first, which could decrease the dv/dt and charging current of the DC bus capacitor. Figure 12c shows the waveform of the current and voltage during the pre-charging process. From Figure 12c, we can find that the duty cycle smaller than 50% could decrease the current flowing to the capacitor, which is achieved by smaller conduction time. Figure 12d shows the relationship of the pre-charging time and peak value of the current flowing through the primary side switches. The peak current can be seen decreasing as the pre-charging time is increased, which could help to choose the appropriate pre-charging time based on the current rating of devices.

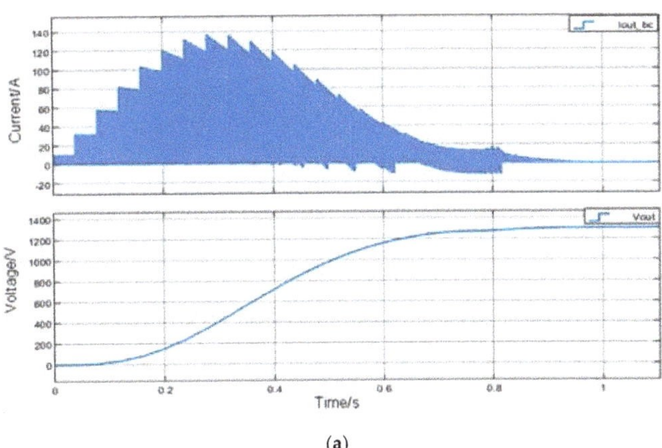

(a)

Figure 12. *Cont.*

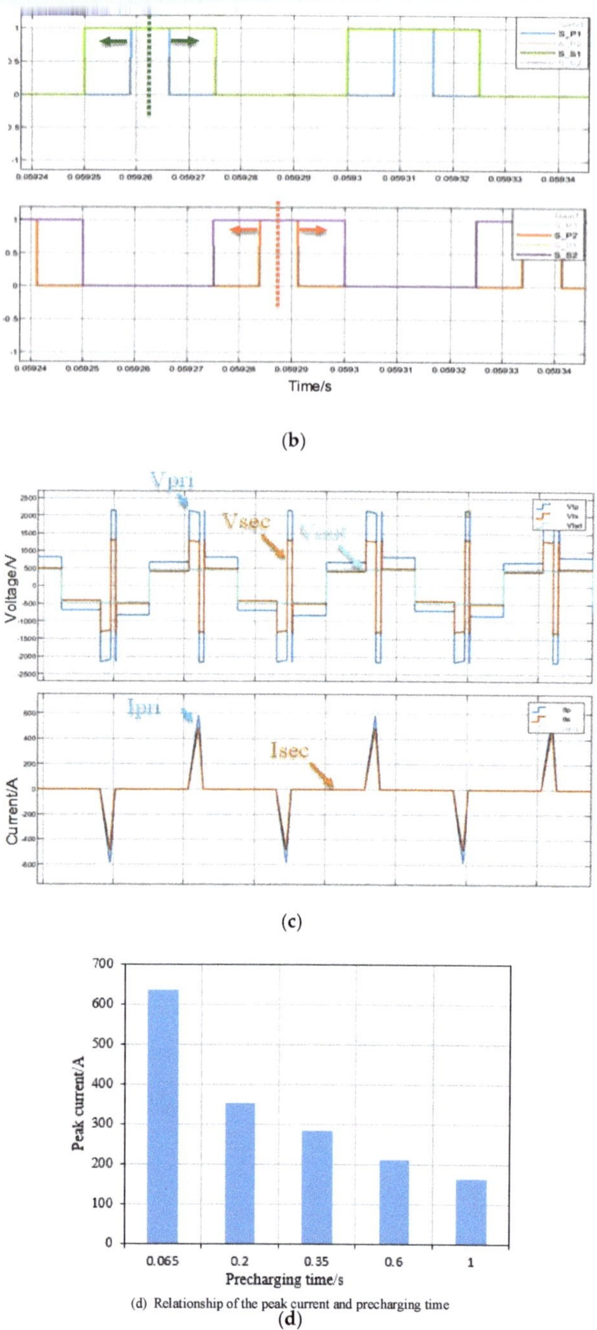

Figure 12. Pre-charging mode of the DCX stage. (**a**) Pre-charging process when the pre-charging time = 1 s; (**b**) Gate signals modulation of the switches; (**c**) Zoom-in waveform of primary and secondary voltage and current; and (**d**) Relationship of the peak current and pre-charging time.

4. Interaction with the Grid

How to fit such a charger to the grid is another concern of this study. In this study, we are using a quasi-steady-state approach with a one-hour resolution to study the effect of installing XFCs on the feeders. For this purpose, the highest load day in 2019 has been picked and the one-hour snapshots of the grid have been selected as the base cases. The snapshots are basically real-time models of the network generated from the real energy management system (EMS) data by state estimators. Topology changes, load variations, generation units' dispatch, and all other grid specifications that are subject to change are included in these models. These snapshots are being generated every ten minutes, but since the demand model for the XFC demand has a one-hour resolution, the snapshots are chosen accordingly. The models are real models of "Utility D".

The XFC demand model that has been chosen for this study is the model that has been suggested in [20]. Authors suggested this demand model based on field data surveys and used statistical methods to generate a MW/time model for the demand for a week, which is shown in Figure 13. What we are using is an hourly demand model for a day which has the highest value during the week.

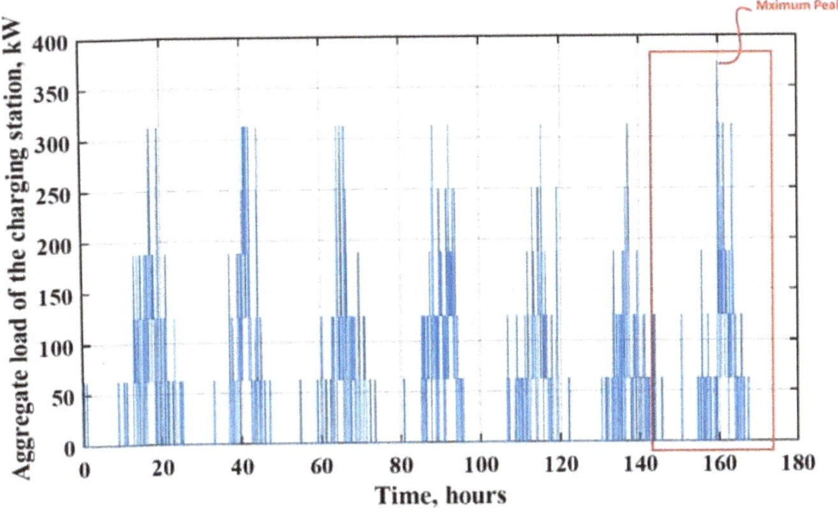

Figure 13. Demand MW/Time model of an XFC station during a week for a ~500 kW station.

In our study, to have a more realistic sense of the effect of chargers on the higher-level grid, we have considered 20% of maximum loadability of the Transmission/Distribution transformer as the aggregated nominal charging power of the stations which are installed at the feeder. The MW demand associated with the time of the day is being added to a certain predetermined substation in the corresponding snapshot. For example, MW demand for 23:00 is added to the substation in the 23:00 snapshot. For each location, it is done for 24 h of a day. Then, the voltage variation, voltage violation, and transformer rating violation are monitored during the day. The results are further investigated for the aforementioned penetration of the EV chargers in the selected locations.

The first step for implementing the study was to choose the substations and geographic areas which make sense in the real world to install XFC stations. Through the field research from operational planning engineers in "Utility D", five different areas with 19 transformers in total have been selected as candidates for the study. These areas are mostly airport, commercial, residential, or touristic areas in the state. All the locations in one area are studied together so that we can have a good understanding of how moving the XFC station might affect the grid locally. In all cases, the power factor has been considered. Therefore, the station is not absorbing any reactive power from the network.

To explain the research findings compactly, only the simulation results of four feeders in Area 2 are shown in Figure 14. Other areas have similar results with Area 2. Area 2 is mostly an important business area with four feeders as the options for connecting the XFC stations. From Figure 14a,b,d, we can find that tx-05, tx-06 and tx-08 are able to supply the demand while the stations are working throughout the day. At feeder tx-07, there is a voltage violation at 3 p.m, which can be found in Figure 14c. Although, the voltage violation also exists in the base case, the 2% voltage drop due to charging load at the time intensifies the problem and might cause a relay action at the location. This case could cause trouble and needs further study. As can be seen from Figure 14, the planner might decide to move the station to another feeder or to take remedial action by installing capacitor banks in the same location.

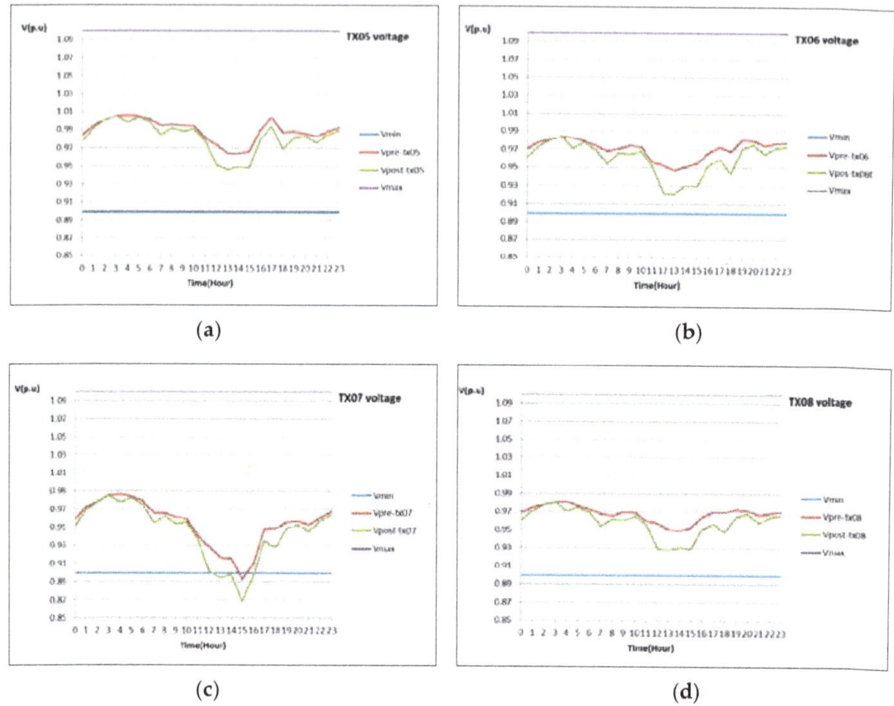

Figure 14. Voltage profile of feeder head before and after XFC installation on 20% penetration in area 2. (a)TX05; (b)TX06; (c)TX07 and (d)TX08.

As has been presented in the last section, we have observed voltage violations on the transformers after XFC integration to the network in four feeders. The voltage problem can be solved using conventional solutions like changing taps or switching in the capacitor banks, if any exist at the location. But in this section, we tend to evaluate the potential use of reactive power injection from the chargers to the grid. This could shed light on the possibility of using the converters as voltage improvement devices so that at least they do not worsen the voltage condition at the feeder at which they are connected. For the purpose of this study, we consider four different scenarios for all five feeders that have a voltage problem. The first scenario is the base case scenario before any station is attached to the grid. The second scenario is the same as we have demonstrated in the previous section. The third scenario is to deploy half the capacity of reactive power (VAR) injection of all stations. Typically, the converters are designed to be able to provide reactive power equal to 1.7 times of their nominal active power. Therefore, the half capacity of the VAR injection will be 0.85 MW. The fourth scenario is applying the full capacity of VAR injection, which is 1.7MW. From Figure 14, we know that the voltage profile of the tx-07 is not ideal even before the installation of the XFC station. To improve the voltage profile, four scenarios are applied to the tr-07, and the results are shown in Figure 15.

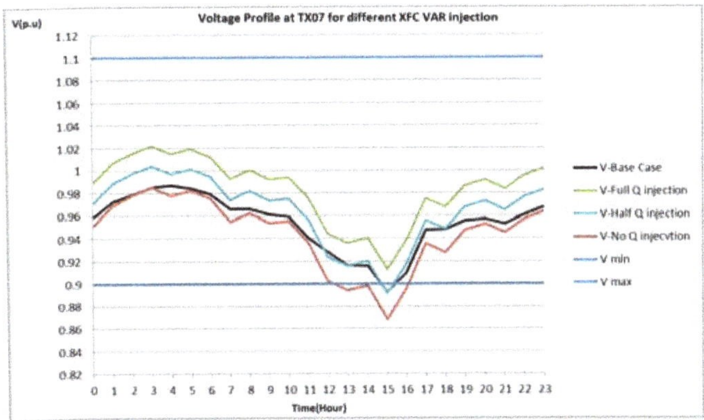

Figure 15. Voltage profile on tx-16 in different scenarios.

At tx-07, when we use half VAR capacity, the voltage profile improves significantly; meanwhile, the daily voltage dip is restored to base case level. At full-capacity VAR injection, there will not be any voltage violation as it can be observed. Therefore, to successfully deploy such XFC, the PFC stage should not always work at unity power factor. When the grid voltage dips, it needs to generate capacitive reactive power to back up the grid.

5. Conclusions and Future Work

A single-phase 1 MW EV XFC station is presented in this paper. Essentially, this is an input-series, output-parallel design that includes a PFC stage and DCX per module. This paper addressed the output voltage balancing control of the series-input PFC considering the unbalanced load condition. The integrated voltage balancing control loop could compensate for the deviation of the duty cycle and achieve the independent control of two PFCs, which could satisfy the output voltage balancing. Considering the paralleled output structure of three DCXs in our model, the output current balancing control is proposed to compensate for the resonant frequency deviation in this paper. Based on the master–slave resonant frequency modulation control, the working frequency of each of the DCXs could be modulated to the corresponding resonant frequency, which could be different from the initially designed value. In addition, the pre-charge mode is investigated to make sure the system could start safely considering the high level and potential large inrush current without the addition of external hardware.

The success of such an XFC is not only determined by the power electronics design and control but also up to the interaction with the grid. This paper finds that such high power can cause the grid-voltage dip at some special moment and location. The proposed approach is to let the PFC stage generate reactive power to support the grid. Future work of this research is to test the system up to 1 MVA once the hardware is ready.

Author Contributions: Conceptualization, Z.L., D.M., H.B., and M.J.; methodology, Z.L. and M.J.; software, Z.L. and M.J.; validation, Z.L. and D.M.; formal analysis, D.M. and H.B.; investigation, Z.L. and M.J.; resources, H.B. and D.M.; writing—original draft preparation, Z.W., H.B., M.J., and D.M.; writing—review and editing, Z.L. and D.M.; supervision, H.B.; project administration, H.B.; funding acquisition, H.B. All authors have read and agreed to the published version of the manuscript.

Funding: This research was funded by ARPA-E CIRCUITS, grant number DE-AR0000901.

Acknowledgments: Authors would like to thank for the support of ARPA-E CIRCUITS program. In addition, authors acknowledge the contribution of Wolfspeed, as the project lead, proposed the topology, constructed the hardware and provided HV SiC modules.

Conflicts of Interest: The authors declare no conflict of interest.

Nomenclature

The following abbreviations are used in this manuscript:

AC	Alternating Current
CCM	Continuous Conduction Mode
C_r	Resonant Capacitance
DAB	Dual Active Bridge
DC	Direct Current
DCX	DC Transformer
DC/DC	DC to DC
d_1, d_2	Duty Cycles for PFC stage
EMS	Energy Management System
EV	Electric Vehicle
FPGA	Field Programmable Gata Array
f_r	Resonant Frequency
f_s	Switching Frequency
$f_{initial}$	Initial working frequency
L_{rd}	Designed resonant inductance
C_{rd}	Designed resonant capacitance
HV	High Voltage
ISOP	Input Series Output Parallel
L_r	Resonant Inductance
LVDC	Low-Voltage DC
MOSFET	Metal Oxide Semiconductor Field Effect Transistor
MTTF	Mean Time To Failure
MV	Medium Voltage
MW	Megawatt
PFC	Power Factor Correction
PWM	Pulse Width Modulation
RMS	Root Mean Square
R_1, R_2	PFC Load Resistances
Si	Silicon
SiC	Silicon Carbide
SOC	State of Charge
T_s	Switching Period
THD	Total Harmonic Distortion
VAC	AC Voltage
VAR	Reactive Power
VDC	DC Voltage
V_{in}	Input Voltage
V_{o1}, V_{o2}	PFC Output Voltages
V_{voa}	Average Output Voltage
WBG	Wide Bandgap
XFC	Extreme Fast Charger
XFMR	Transformer
η	Efficiency

References

1. Hajian, M.; Zareipour, H.; Rosehart, W.D. Environmental benefits of plug-in hybrid electric vehicles: The case of Alberta. In Proceedings of the 2009 IEEE Power & Energy Society General Meeting, Calgary, AB, Canada, 26–30 July 2009; pp. 1–6.
2. Lulhe, A.M.; Date, T.N. A technology review paper for drives used in electrical vehicle (EV) & hybrid electrical vehicles (HEV). In Proceedings of the 2015 International Conference on Control, Instrumentation, Communication and Computational Technologies (ICCICCT), Kumaracoil, India, 18–19 December 2015; pp. 632–636.
3. Ahmed, S.; Bloom, I.; Jansen, A.N.; Tanim, T.; Dufek, E.J.; Pesaran, A.; Burnham, A.; Carlson, R.B.; Dias, F.; Hardy, K.; et al. Enabling fast charging—A battery technology gap assessment. *J. Power Sources* **2017**, *367*, 250–262.
4. Tu, H.; Feng, H.; Srdic, S.; Lukic, S.M. Extreme Fast Charging of Electric Vehicles: A Technology Overview. *IEEE Trans. Transp. Electrif.* **2019**, *5*, 861–878. [CrossRef]
5. Ji, S.; Huang, X.; Zhang, L.; Palmer, J.; Giewont, W.; Wang, F.; Tolbert, L.M. Medium voltage (13.8 kV) transformer-less grid-connected DC/AC converter design and demonstration using 10 kV SiC MOSFETs. In Proceedings of the 2019 IEEE Energy Conversion Congress and Exposition (ECCE), Baltimore, MD, USA, 29 September–3 October 2019; pp. 1953–1959.
6. Chow, T.P. Wide bandgap semiconductor power devices for energy efficient systems. In Proceedings of the 2015 IEEE 3rd Workshop on Wide Bandgap Power Devices and Applications (WiPDA), Blacksburg, VA, USA, 2–4 November 2015; pp. 402–405.
7. Armstrong, K.O.; Das, S.; Cresko, J. Wide bandgap semiconductor opportunities in power electronics. In Proceedings of the 2016 IEEE 4th Workshop on Wide Bandgap Power Devices and Applications (WiPDA), Fayetteville, AR, USA, 7–9 November 2016.
8. Zhao, B.; Song, Q.; Liu, W.; Sun, Y. Overview of dual-active-bridge isolated bidirectional DC–DC converter for high-frequency-link power-conversion system. *IEEE Trans. Power Electron.* **2014**, *29*, 4091–4106. [CrossRef]
9. Wen, H.; Gong, J.; Yeh, C.-S.; Han, Y.; Lai, J. An investigation on fully zero-voltage-switching condition for high-frequency GaN based LLC converter in solid-state-transformer application. In Proceedings of the 2019 IEEE Applied Power Electronics Conference and Exposition (APEC), Anaheim, CA, USA, 17–21 March 2019; pp. 797–801.
10. Li, B.; Li, Q.; Lee, F.C. A WBG based three phase 12.5 kW 500 kHz CLLC resonant converter with integrated PCB winding transformer. In Proceedings of the 2018 IEEE Applied Power Electronics Conference and Exposition (APEC), San Antonio, TX, USA, 4–8 March 2018; pp. 469–475.
11. Everts, J.; Krismer, F.; Keybus, J.V.D.; Driesen, J.; Kolar, J.W. Optimal ZVS modulation of single-phase single-stage bidirectional DAB AC–DC converters. *IEEE Trans. Power Electron.* **2013**, *29*, 3954–3970. [CrossRef]
12. Khalid, U.; Shu, D.; Wang, H. Hybrid modulated reconfigurable bidirectional CLLC converter for V2G enabled PEV charging applications. In Proceedings of the 2019 IEEE Applied Power Electronics Conference and Exposition (APEC), Anaheim, CA, USA, 17–21 March 2019; pp. 3332–3338.
13. Singh, B.; Chandra, A.; Al-Haddad, K.; Pandey, A.; Kothari, D.P. A review of single-phase improved power quality ac~dc converters. *IEEE Trans. Ind. Electron.* **2003**, *50*, 962–981. [CrossRef]
14. Jovanovic, M.; Jang, Y. State-of-the-art, single-phase, active power-factor-correction techniques for high-power applications—an overview. *IEEE Trans. Ind. Electron.* **2005**, *52*, 701–708. [CrossRef]
15. Sankar, U.A.; Mishra, S.; Viswanathan, K.; Naik, R. Control of a series input boost pre-regulator with unbalanced load. In Proceedings of the IECON 2014 - 40th Annual Conference of the IEEE Industrial Electronics Society, Dallas, TX, USA, 29 October–1 November 2014; pp. 4582–4588.
16. Mukhopadhyay, A.; Mishra, S. FPGA based novel voltage balancing technique for series input boost pre-regulator. In Proceedings of the IECON 2017 - 43rd Annual Conference of the IEEE Industrial Electronics Society, Beijing, China, 29 October–1 November 2017; pp. 1160–1165.
17. Li, H.; Jiang, Z. On automatic resonant frequency tracking in LLC series resonant converter based on zero-current duration time of secondary diode. *IEEE Trans. Power Electron.* **2015**, *31*, 4956–4962. [CrossRef]
18. Kundu, U.; Chakraborty, S.; Sensarma, P. Automatic resonant frequency tracking in parallel LLC boost DC–DC converter. *IEEE Trans. Power Electron.* **2014**, *30*, 3925–3933. [CrossRef]

19. Xiang, X.; Chen, H.; Shu, J.; Tu, W.; Wang, L.; Liu, J.; Hu, P.; Xiaonan, Y.; Hongkun, C.; Jing, S.; et al. A novel precharge control strategy for modular multilevel converter. In Proceedings of the 2015 IEEE PES Asia-Pacific Power and Energy Engineering Conference (APPEEC), Brisbane, QLD, Australia, 15–18 November 2015; pp. 1–5.

20. Deng, Q.; Tripathy, S.; Tylavsky, D.; Stowers, T.; Loehr, J. Demand Modeling of a dc Fast Charging Station. In Proceedings of the 2018 North American Power Symposium (NAPS), Fargo, ND, USA, 9–11 September 2018; pp. 1–6.

Article

A Study on Electric Vehicles Participating in the Load Regulation of Urban Complexes

Qiwei Xu [1], Jianshu Huang [1], Yue Han [2], Yun Yang [1] and Lingyan Luo [1,*]

1   State Key Laboratory of Power Transmission Equipment & System Security and New Technology, Chongqing University, Chongqing 400044, China; huangjs@cqu.edu.cn (Q.X.); 20132160@cqu.edu.cn (J.H.); 201734131008@cqu.edu.cn (Y.Y.)
2   Northeast Branch of State Grid Corporation of China, Shenyang 110180, China; yuehanying@163.com
*   Correspondence: luoly@cqu.edu.cn

Received: 6 April 2020; Accepted: 4 June 2020; Published: 8 June 2020

Abstract: Urban complex (UC) is the main place of citizens' life and work. The construction of an UC often needs to expand the capacity of the power equipment. This paper proposes to use electric vehicles (EVs) in an UC to reduce the power load of the UC during peak periods, so that lower capacity power equipment can be used to reduce the construction costs of the UC and the transformation of electrical facilities. In order to find the relationship between parking and power load in the UC, the UC is decomposed into different functional areas for research. Then, we build a parking information database for clustering and calculation. Divide the load peak into adjustment intervals of equal duration. The EVs parked in the UC for each regulation interval (RI) are grouped according to parking characteristics. Establish an objective function with the minimum load variance during peak hours. The discharge capacity of each group in each RI is obtained and distributed to each EV to realize peak load reduction of UC. Finally, the results of case analysis show that the strategy can reduce the peak load effectively thus save the cost of UC construction.

Keywords: electric vehicle; V2G; urban complex; peak shaving; smart grid

---

1. Introduction

The consumption of fossil fuels has brought about serious ecological issues such as global warming and air pollution. The promotion of electric vehicles (EVs) can effectively solve these problems. Therefore, many countries have promulgated various policies to support the development of EVs. It can be predicted that the number of EVs will increase rapidly in the future. The large-scale popularization of EVs will require huge demand for electricity. As electricity consumers, uncontrolled charging of many EVs may cause unsafe operations in the power grid, but EVs can also be used as mobile energy storage for load regulation. Hence, proper control of EVs' charging and discharging will have an important impact on the construction and operation of the power grid. As the busiest place for people and cars gathering in the city, the urban complex (UC) has high electricity demand. The construction of UCs will cause the reconstruction and expansion of the surrounding distribution power grid. It will cost lots of money for reconstruction and expansion of the power grid. To solve this problem, we will mainly study EVs with the function of participating in the load regulation of UC.

There are many studies on charging load mode of EVs at present. In order to establish the charging load model of EV cluster in charging stations and residential areas, a queuing theory modeling method was employed in [1]. Based on the National Household Travel Survey (NHTS) database of the United States Department of Transportation in 2009, the statistical rules of EV trips, ending time and driving distance are studied and analyzed. On this basis, the probable charging model of EVs was completed in [2]. Considering the price and tax rate, fuel price, policy support and charging safety of EVs, a forecasting method based on an agent model was proposed and used to predict the development scale of EVs in Iceland from 2012 to 2030 in [3]. A statistical analysis of the 24-hour travel of EVs was carried out, and the load curve of charging at home was calculated in [4]. An Autoregressive Integrated Moving Average model (ARIMA) method for demand forecasts of conventional electrical load and charging demand of EVs' parking lots was presented in [5], and the parameters of the ARIMA model should be tuned so that the mean square error of the forecaster could be minimized. Researchers in [6] pointed out that in the United Kingdom, EVs with 10% permeability would lead to a 17.9% increase in peak power demand per day, and a 35.8% increase if permeability reaches 20%.

The strategies of peak-shaving and valley-filling are one of the research focuses. A time-sharing charging control strategy for EVs based on the predicted load curve is presented in [7]. By increasing the tariff during the peak load period and reducing the tariff during low load period, users can choose to charge during low load period to realize peak shift [8]. Researchers in [9–11] established the simulation model of EV participating in load frequency adjustment and study its mechanism and control effect. They demonstrated the feasibility of EV participating in grid frequency regulation. When the grid frequency is high, controlling a large number of EVs charging promotes the frequency of the grid to decline; and when the grid frequency is low, controlling a large number of EVs discharge promotes the frequency of power grid to rise. Scholars simulate and calculate the potential benefits of 250 EVs participating in grid regulation in New England. The calculation results show that when EV only provides downward FM service (the EV only charges), each EV can bring about an annual revenue of $700–900. When the EV provides both upward and downward FM service (EVs both charge and discharge), each EV can annually generate revenue of $1250 to $1400 for its users [12].

The authors in [13] proposed a stochastic unit commitment for isolated power systems in which the risk of high operating costs is limited using the conditional value-at-risk risk measure. A unit commitment formulation that accounts for the requirements of spinning reserves was presented in [14]. A procedure to determine the optimal design of an isolated system with a high penetration of renewable energy sources was proposed in [15]. Several electric vehicle charging algorithms were proposed explicitly considering their negative impacts on the transmission and distribution grids in [16]. An agent-based coordinated dispatch strategy for electric vehicles and renewable units at distribution level was presented in [17]. A reserve contract optimization model designed for electric vehicles with vehicle-to-grid capability was proposed in [18]. The authors in [19] investigated the optimal planning of the Nordic transmission system in 2050 for a 100% electric vehicle penetration.

Researchers in [20] found that V2G strategy effectively reduced contamination emissions and cut down the investment of variable load plants. Researchers in [21] also considered that the participation of V2G into power systems could reduce peak power generation costs. The authors in [22] believed that V2G application of heavy-duty electric vehicles was not only possible but also necessary for more profits while electric buses without V2G revenue would not be cost-effective compared with traditional diesel buses. The results in [23] showed that electric trucks could generate an additional enormous income for owners if they provided V2G regulation service. The authors in [24] proposed a two-stage approximate dynamic programming framework for the optimal charging strategy in a commercial building parking lot and simulates a number of scenarios where the vehicle arrival behavior is modeled as a Poisson process.

The coordination of EVs and distributed energy sources is another research focus. When there are enough intelligent generators, power electronic devices and interactive chargers in the power grid, EVs can be used as energy storage devices and standby power supply in case of accidental power outages [25–28]. EVs can help integrate intermittent renewable energy such as wind and solar energy into the grid more efficiently. V2G system can provide more help for grid operators, including reactive power support, active power regulation, and load balancing through peak shaving and valley filling, current harmonic filtering. These systems can provide auxiliary services such as frequency control and rotating reserve, which can significantly improve the efficiency, stability and reliability of power grid [29–32]. According to the Demand Side Management (DSM) method, the charging and discharging status of EVs will change with the load, which can not only improve the reliability of power grid operation, but also reduce the impact on power grid if a large number of EVs are connected to the power grid [33]. A strategy to optimize the control of EV group, not only considering the cost of EV users, but also considering the benefits of EVs accessing the grid to participate in auxiliary peak shaving, frequency modulation and standby [34,35]. EV clusters and a double-level optimal strategy are used according to different priorities in [36], and a real-time optimal strategy from EV clusters to individuals is completed.

We can find that most of studies have considered that EVs should charge or discharge at the charging station. In fact, users prefer to charge near home because the charging stations are always located in remote places. The smart city is a new urban construction concept. This means the integration of urban function and services to optimize efficiency of urban management and services. It will improve the quality of citizens' life. The UC is generally located in busy areas with business, office, residential functions and so on. This aligns with the smart cities concept. Many people live near the UC. Obviously, it is a better choice for users to charge and discharge in the UC. It is convenient for people if we build the garage of an UC as a charging and discharging power station. This design can not only meet the charging needs of users, but also let EVs participate in the load regulation of the UC. Therefore, EVs participating in the load regulation of UC can facilitate the safe and stable operation of the power grid and efficient utilization of power resources in a smart city. This is conducive to establishing the power supply systems of smart cities. However, there are few studies about EVs participating in load regulation of an UC. The vehicle behavior of UCs is very complex. There are many cars of malls, consumers, white-collar workers and residents in an UC, so the relationship between the load characteristics and parking characteristics of different functional areas is obtained by studying the different function areas of an UC in our research. Then, the EVs are clustered into a controllable group and an uncontrollable group. Each group will be regarded as a whole to participate in the load regulation of the UC. In order to make more EVs with different parking times participate in the load regulation, we divide the peak load period into many small time periods defined as the regulation interval (RI). In this way, if the parking time of some vehicles can cover a certain RI, the vehicles can participate in the load regulation of this RI. We set the objective function which minimize the mean square deviation of load and get the total discharge power of each group in each RI and allocate it to each vehicle. Last, the case study proved that the EVs can reduce the peak load of UC effectively. In summary, the main contributions of our research are:

(1) The random charging load model of EVs is established according to the daily travel habits of residents. The influence of the EVs' number and charging power on the charging load is analyzed. It is confirmed that large-scale EV use with random charging will produce a serious charging load, which will influence the safe operation of the power grid.

(2) The construction of UCs is becoming more and more popular among real estate developers. There are few studies about EVs participating in load regulation of UCs. The research carried out in this paper can reduce the peak load of UCs and improve the security of local power grid operation.

(3) It is proposed to divide the UC into different functional areas. Then, we get the relationships between load ratio and parking ratio of these different functional areas. The vehicles parked in the UC are divided into three groups according to the functional areas of the UC. The problem is simplified by this way.

(4) We propose a load shaving strategy involving three types according to their controllability, and the objective function was established to minimize load variance during the peak period of the UC. A case study was done to verify the strategy. The result proves that the strategy proposed in the paper can effectively reduce the peak load through appropriate parameter matching, which is safe to the operation of the power grid.

The rest of the paper is organized as follows: Section 2 describes the problem faced by EV charging. The calculation model of random charging load of EVs is established according to the user's habit.

Then, the relationship between load and parking of the UC is studied. The relationship between load ratio and parking ratio can be obtained by decomposing the UC into different functional areas. In Section 3, the EV discharging strategies in UCs are illustrated and thoroughly described. In Section 4, the effectiveness of the strategy is verified by case study, and the influence of permeability and length of regulation interval on peak shaving is analyzed. Finally, the conclusions are presented in Section 5.

2. Model Description

With the increase of EVs, the concentrated charging of EVs may cause power shortages in local areas. The charge power of EVs during peak load periods will increase the burden of the distribution power network. Since the existing local distribution networks and facilities did not consider the charge demand of EVs during the construction period, they need to invest a lot of money to expand and transform the local distribution power grid for EV charging.

This section will study the influence of stochastic charge on power grid, load characteristics of the UC, the parking patterns in the UC and model the stochastic charge load of EVs. Based on this, our research further proposes a parking garage which can be used in an UC. This parking garage will be built as a charge center with the function of both parking and charge/discharge for EVs. Through the cooperation of flexible charge/discharge strategies, it can solve the problems of grid planning lag in the construction of UCs and the dangers of power grid operation which caused by the stochastic charge of EVs. In this paper, we will propose a discharge strategy for the peak load period.

2.1. Load Model of Stochastic Charging

With the expanded development of EVs, the number of EVs will reach a new scale. Their charging demands will increase significantly. The charge load generation by EVs will represent and ever increasing proportion of the power grid electricity consumption if the number of EVs is increasing. In addition, the charge behavior of EVs users is clustered. The peak load of the power grid will increase dramatically. It is necessary to increase the capacity of the power network to satisfy the demand of EV charging in the future. Therefore, for construction and operation of the power grid, it is particularly important to establish a model of the EVs' charging load.

The factors affecting the charging load can be summarized as battery capacity, charge power and users' behavior. The power battery capacity determines the users' charging frequency. The larger battery capacity means the lower the charging frequency. The charge power is related to the charge time. The greater charge power means the shorter charge time. At the same time, the charge power will influence the peak charge load. Compared with the objective factors mentioned above, users' behavior is the key factor affecting the charging load. The user behaviors have an impact on charging load mainly includes two aspects: the start charge time and State Of Charge (SOC). Concentration of the user's initial charging time will cause the greater the charging load. SOC reflects the user's current battery capacity consumption. If the battery capacity is small, the users generally charge every day. With the increase of the battery capacity of EV, the users' charge frequency will decrease. Our research will consider these factors and establish a power demand model of EV in one day.

According to the National Household Travel Survey 2001 of American, the daily routine mileage s obeys the log-normal distribution [4]:

$$f(s) = \frac{1}{\sqrt{2\pi}s\delta_D} \exp[-\frac{(\ln s - \mu_D)^2}{2\delta_D^2}]$$

(1)

where, μ_D is the expectation of probability density function, δ_D is the standard deviation.

Assume the SOC is full at the first travel in a day. The remain SOC_{end} after the last travel is:

$$SOC_{end} = 1 - k \cdot s$$

(2)

where, k is the SOC consumption per kilometer.

The peak time of off-duty travel of private EVs is typically 17:00–19:00 [4]. Assuming that the starting charge time is the last return time of the trip, combined with the survey results of the United States Department of Transportation on American household vehicles, the final return time of the vehicle is a normal distribution. In Equation (3), t is the final return time of the vehicle, and its probability density function is as follows:

$$f_s(t) = \begin{cases} \frac{1}{\delta_s \sqrt{2\pi}} \exp[-\frac{(t-\mu_s)^2}{2\delta_s^2}], & (\mu_s - 12) < t < 24 \\ \frac{1}{\delta_s \sqrt{2\pi}} \exp[-\frac{(t+24-\mu_s)^2}{2\delta_s^2}], & 0 < t \le (\mu_s - 12) \end{cases} \tag{3}$$

where μ_s is the expectation of probability density function, δ_s is the standard deviation.

The charge power of EVs is related to the characteristics of batteries. The classical three-stage charge mode is usually adopted in the charge process of lithium batteries: pre-charge stage, constant current charge stage and constant voltage charge stage. In the pre-charge stage, lithium batteries are activated to a certain state by a small current, which is usually about 10% of the constant current stage. Constant current charge stage is the main stage of charging. When the battery voltage is higher than 2 V, the activity of the lithium-ion battery is fully activated and the internal resistance is smaller. It can accept a high current charge. Generally, this period is used to charge the battery capacity to about 80%. When the voltage of the lithium battery reaches the predetermined value, the third stage of constant voltage charge is carried out. The current gradually decreases until it drops to 0. The period of first stage and third stage are short, so the charge power can be regarded as a constant.

The charge time t_c can be calculated according to SOC_{end} and charge power.

$$t_c = \frac{(1 - SOC_{end}) \cdot W}{P_c \eta} \tag{4}$$

where P_c is the charge power; η is the charge efficiency, chosen as 0.90 [37], W is the battery capacity of the EVs. Table 1 shows the top 10 most popular EV types in 2019 in China [38]. The battery capacity of the top 10 most popular EVs in China is generally above 40 kWh, except for some mini-cars. According to the data in Table 1, the battery capacity is more than 40 kWh after weighting the market share. With the development of electric vehicle technology, the battery capacity will be larger and larger in the future. In order to simplify the calculation process, the average capacity of power batteries is chosen as $W = 40$ kWh in our research.

Table 1. Battery capacity and market share of the top10 most popular EV types in 2019 of China.

Type	Battery Capacity (kWh)	Market Share
BAIC EU	53.6	13.3%
BYD Yuan EV	40	7.4%
Baojun EV	24	7.2%
Chery eQ	23.6	4.7%
BYD Tang DM	82.8	4.1%
BYD E5	60.48	3.9%
GAC Aion S	49.4	3.9%
Roewe Ei5	52.5	3.7%
GWM R1	28.5	3.4%
Emgrand EV	52	3.4%

According to the travel characteristics, the Monte Carlo method is used to simulate the charging load of EVs. Firstly, we assume the total number of EV is N. For the n-th EV, the daily routine mileage of vehicles s^n and the final return time of the vehicle t^n is randomly generated by Equations (1) and (3). Then, the charge time t_c^n can be calculated by Equation (4). The charging load P_c^n will last from t^n to

$t^n + t_c^n$ for the *n*-th EV. By adding up the charging load of each vehicle in each period *T* we get the total charging load $P(T)$. Algorithm 1 in Table 2 is the calculation process of stochastic charging load of EVs.

Table 2. Calculation process of stochastic charging load of EVs.

Algorithm 1	Calculation process of stochastic charging load of EVs.
Input	probability density function of daily routine mileage and final return time
output	Random charging load $P(t)$ of EVs.
Step1	Generate s^n and t^n by the Monte Carlo method
Step2	Calculate the charge time t_c^n
Step3	When $n \leq N$
	Add the P_c^n to the $P(T)$ from t^n to $t^n + t_c^n$
Step4	Output the random charging load of EVs

According to the calculation process, the stochastic charge load of EVs is established. We take a residential area with 1000 households as an example to calculate the stochastic charge load. Supposing there are 1000 parking spaces, of which 250 parking spaces can provide charging services. We set the charge power as P = 10 kW, and explore the influence of EV number on charge load. As can be observed from the Figure 1, the peak charge load concentrate between 19:00 and 22:00, and with the increase of EV number, the charge load will increase rapidly. When 250 EVs are charged, the peak charge load exceeds 1000 kW. This is a large valve for 1000 residential households.

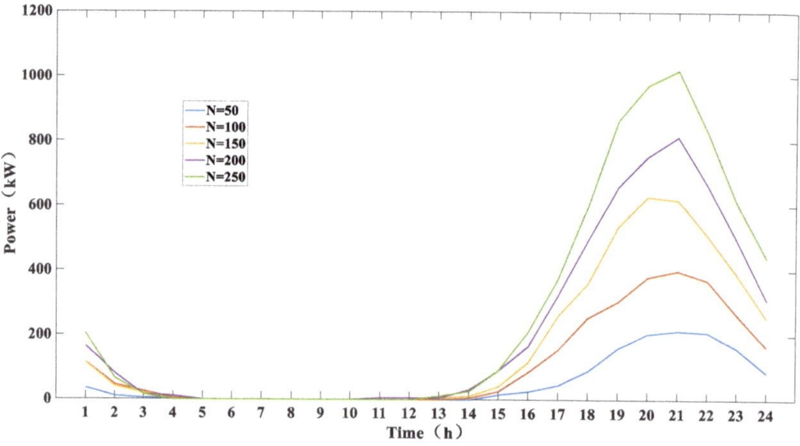

Figure 1. Uncontrolled charge load with different number of EVs when the charge power is 10 kW.

Then, we set the number of EV is N = 100, and find the influence of charge power on charge load. As can be observed from Figure 2 the peak charge load will decrease with the decrease of charge power. It can reduce the charge load impact on power grid effectively. At the same time, the charge time will last long if the charge power decrease. So, if the charge power is too low, the charge period will be too long to satisfy the usage for next time. With the development of battery technology, the battery capacity will be larger and larger. The requirement of charge power will also be greater and greater. There is no point by reducing charge power simply.

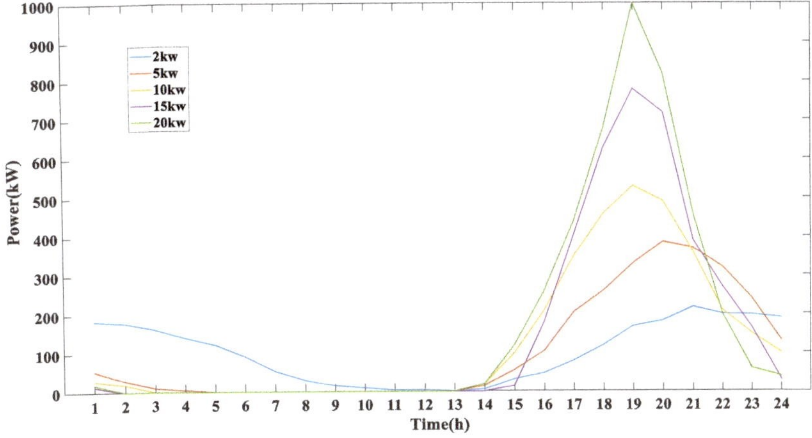

Figure 2. Uncontrolled charge load with different charge powers when the number of EVs is 100.

2.2. Characteristic of UC Load

As shown in Figure 3, an UC is a group of buildings with at least three of a series of functions such as shopping, entertainment, office space, accommodation, residential living and so on. Its area can reach more than hundreds of thousands of square meters or even millions of square meters. It is a multi-functional, efficient and convenient synthetic assembly, which is formed to realize the complete operation system of work and life in smart city. It can provide people with a colorful life. More and more people like to live in or near the UC. Real estate companies will also build more UCs to meet people's needs. However, in order to recoup funds as soon as possible, real estate companies need to take as little time as possible to construct an UC. There is a large demand for electricity, as a result, the original power planning of the region may not meet the needs of the UC.

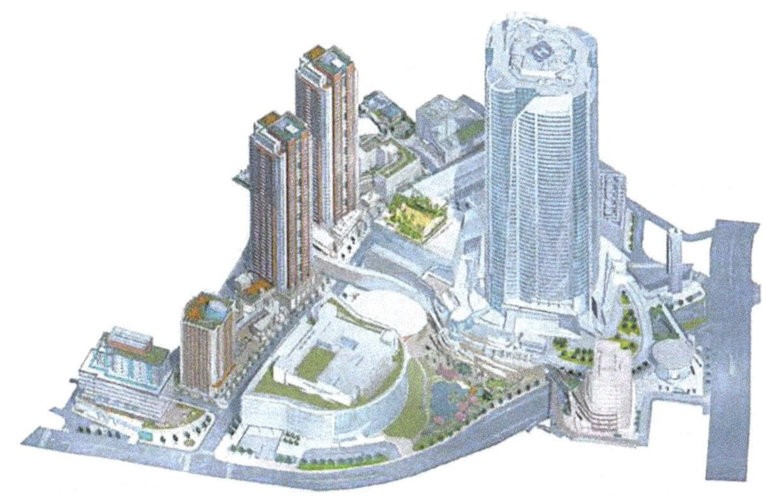

Figure 3. A typical urban complex with a huge underground parking lot.

There are differences between different functional areas of UC in work and rest time. Therefore, the load characteristics of UC are closely related to its functions. Our research divides UC into three types of individual buildings based on the function and running time of UC: commercial type, office type and residential type. Then, the real data of different types is analyzed to get the load and parking characteristics of the different types.

Because the power load of different UC is different in different days, our research uses the load ratio to analyze the load characters at different times of the day, and randomly selects three different types of single buildings on the working day for analysis.

The load ratio α^T can be calculated by:

$$\alpha^T = \frac{P_{con}^T}{p_{max}} \tag{5}$$

Figure 4 shows the load ratio curves at different times for the residential type of building. In the residential type, the load increases significantly from 8:00 a.m., with the peak occurring between 19:00 and 23:00 a.m. This is closely related to residents' living habits. Residents start a new day after waking up at about 8:00 am. Most people will leave home to work during the daytime, but some unemployed residents stay at home for their own business during the daytime on the power load remains at a low level before the other residents come back. Residents will return to the home after work at about 18:00. Then, the demand for electricity will increase. There will then be a peak of electricity consumption until the residents rest at night.

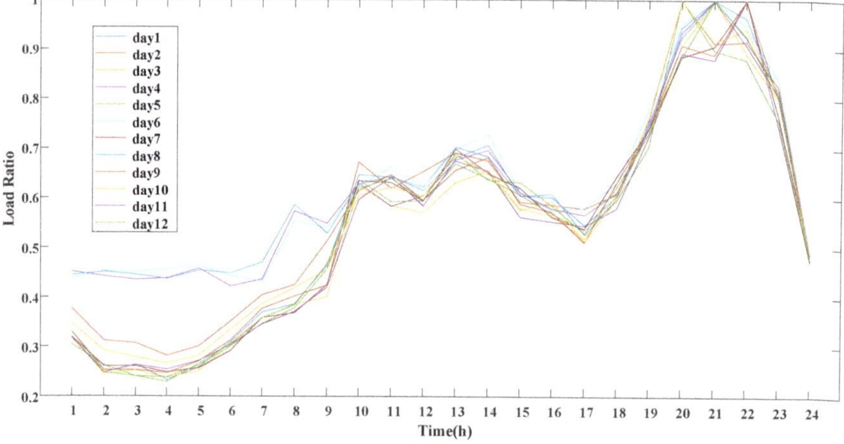

Figure 4. Load ratio curves of residential type buildings at different days.

Figure 5 shows the load ratio curve of office type buildings at different times. It is obvious that the peak load period of office type is mainly concentrated between 9:00 a.m. and 18:00 p.m. The rest of the time load the load is relatively low. The peak load period basically coincides with companies' working hours. The load ratio is relatively stable in peak and valley sections, respectively. The employees' demand for electricity is mainly concentrated on lighting, air conditioning and the consumption of office equipment such as computers during working hours. The demand for electricity is stable. During non-working hours, there is almost nobody in the office, and only a small amount of power is needed, such as emergency channel lighting.

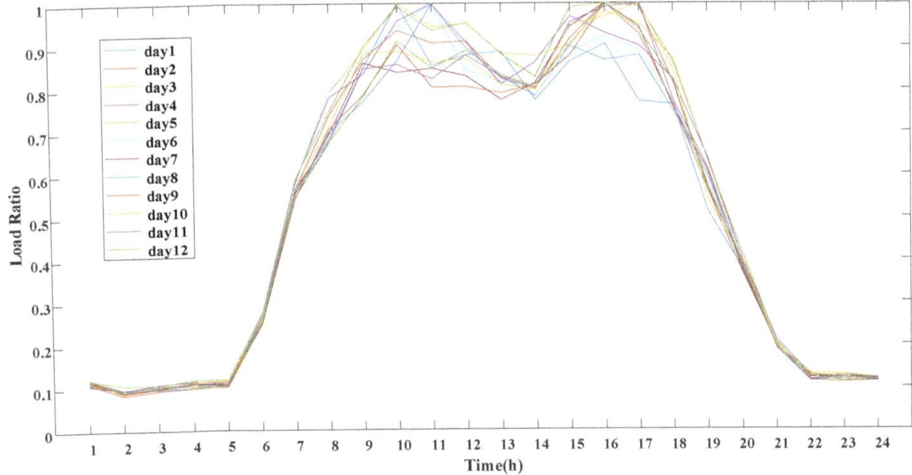

Figure 5. Load ratio curve of office type buildings at different days.

Figure 6 shows the load ratio curves at different times in commercial type buildings. The peak load period of this type mainly concentrated between 9:00 a.m. and 22:00 p.m. which coincides with the usual business hours in the business district. During business hours in business districts, shopping malls, entertainment facilities and restaurants need a large amount of electricity. When business areas are closed, there will be little electricity demand. Therefore, the peak and trough periods of the load ratio curves are obviously the same as the office type.

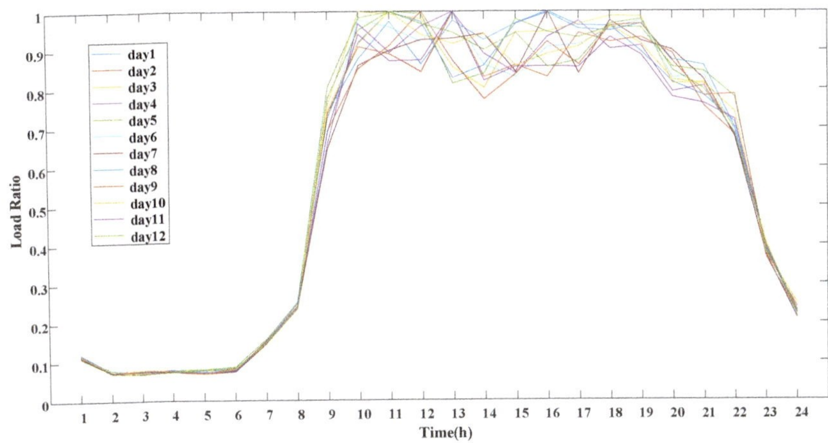

Figure 6. Load ratio curves of commercial type buildings at different days.

The typical load ratio curves of the different types are shown combined in Figure 7. The load ratio is obviously closely related to the activity hours of the different types of building. Especially in commercial and office type buildings, the peak load period is concentrated during business hours. The load ratio of the rest of the time is much lower than the peak period, which is not conducive to the operation of the power grid.

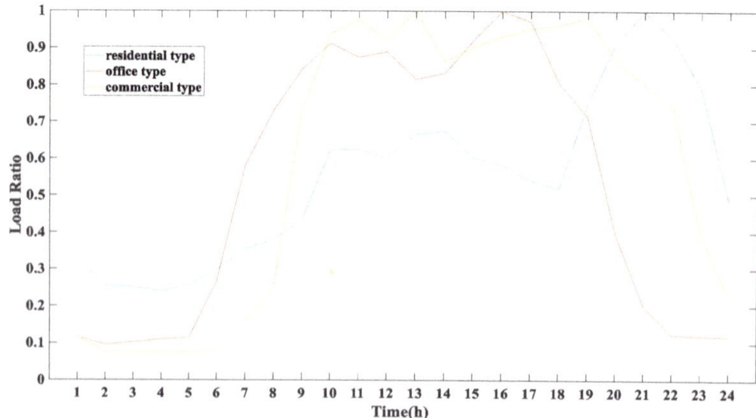

Figure 7. Typical load ratio curve of different building function types.

The overall load ratio α_{UC}^T can be calculated by the following expression:

$$\alpha_{UC}^T = \frac{P_{UC}^T}{P_{UC}^{max}} \tag{6}$$

where P_{UC}^T is the power load at time T, P_{UC}^{max} is the maximum power load in the day.

The overall load ratio of the UC is determined by the load ratio and maximum load of the building types, therefore, the proportion of different building types directly affects the overall load ratio. Developers' positioning to build UCs will determine the load ratio curve of the UC. Figure 8 shows the overall load ratio of a city complex. From Figure 8, it can be observed that the peak load period of the UC is still relatively concentrated, and the peak-valley difference is large.

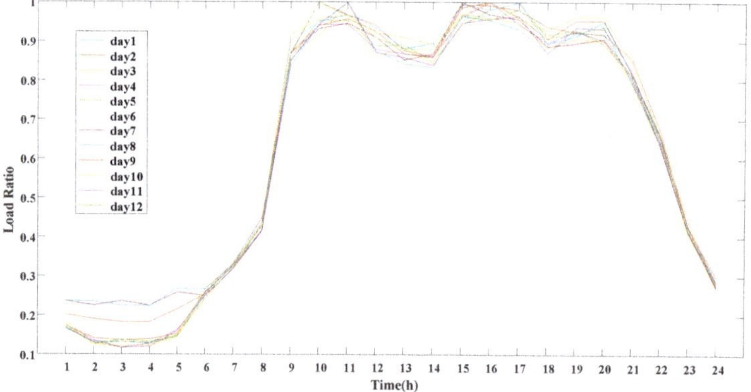

Figure 8. Load ratio curve of UCs at different days.

2.3. Parking Behavior of UC

The parking characteristics of commercial type, office type and residential type building in an UC are quite different. The peak parking time and parking turnover ratio are different in different UC building functional types. In order to analyze the parking characteristics more conveniently and accurately, we should find out the rules relating different parking characteristics and load characteristics [39]. There are three types of UC, so the parking ratios under these three types of UC are discussed separately.

The parking ratio α_{park} is defined as the ratio of the number of parking lots to the number of parking spaces provided by the garage per period.

$$\alpha_{\text{park}} = \frac{N_t}{N_{\max}} \tag{7}$$

where N_t is the number of parking lots available at time t and $N_{\max}$ is the total number of parking spaces in the garage

Figure 9 is a typical residential type parking ratio curve. The parking ratio of residential type buildings mainly depends on the habits of residents. Some residents drive to work during the day. After work, these residents drive home. Therefore, the probability of parking during the day is relatively low, while the ratio of parking at night is higher. Some families have more than one car, and some cars are used for other temporary trips besides work. These cars will be parked in the garage for long time during a day. Therefore, the parking ratio in residential areas is always high.

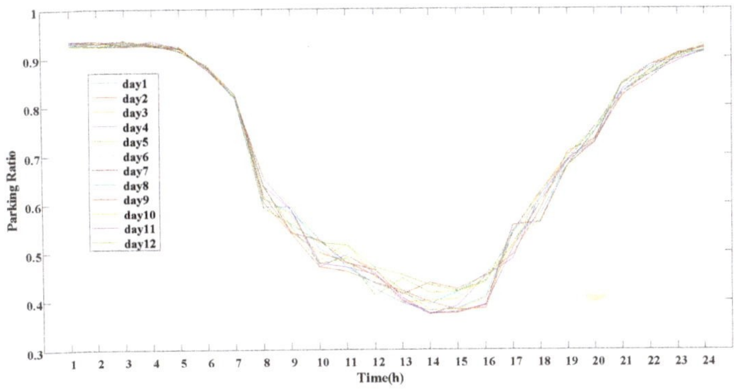

Figure 9. Typical parking ratio curves of residential type buildings at different days.

Figure 10 is a typical parking ratio curve of an office type building. Employees drive from home to work and leave their cars in the garage. Some of the employees go home for a rest or out for lunch, so there is a slight drop in parking at noon. The utilization ratio of the garage maintains at a high level during the working hours, and utilization ratio of the garage is very low during other periods.

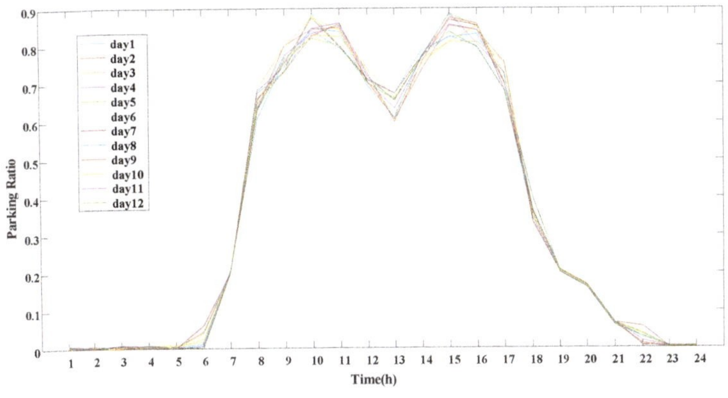

Figure 10. Typical parking ratio curve of office type at different days.

Figure 11 shows typical parking ratio curves of a commercial type building. The parking characteristics of commercial format are related to many factors, such as pedestrian flow, position and development of the UC. At the early opening period of the UC, the parking demand is low because of the low business ratio. With the increase of business, the demand for parking spaces also increases. Overall, the parking ratio during opening hours is much higher than at other times. The parking ratio is at a very low level in other times, especially in the early morning.

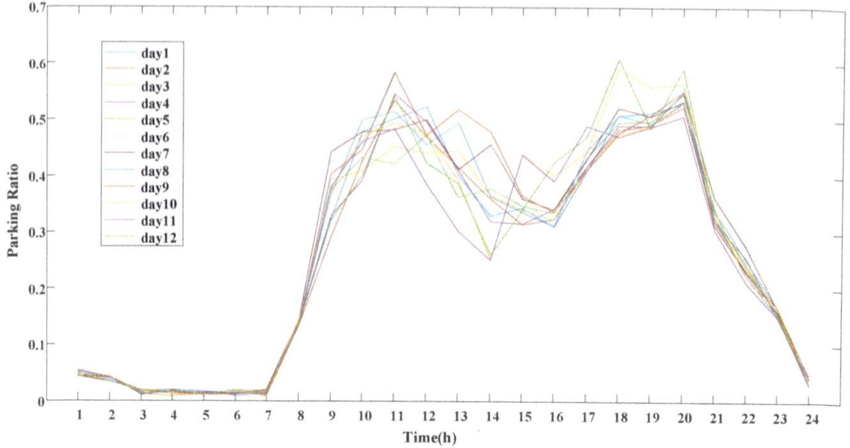

Figure 11. Typical parking ratio curves of commercial type buildings at different days.

We can observe that there is a strong correlation between the load characteristics and parking characteristics of office business and commercial business type buildings. The calculation and analysis of the load ratio and its corresponding parking ratio are carried out. The correlation coefficients of different dates are obtained in Table 3. There is a strong positive correlation between the load ratio and parking ratio of office type and commercial type buildings. Although the correlation coefficients of residential type buildings are low, the parking ratio of residential type buildings remains at a high level. Therefore, our research takes advantage of EVs as mobile energy storage terminals, and the correlation between parking ratio and load ratio of an UC, and presents the strategy of peak-shaving during the peak load time of the UC.

Table 3. Correlation between the load ratio and parking ratio of different function types.

Date	Commercial	Office	Residential
Day1	0.93026	0.960493	−0.32142
Day2	0.958525	0.95422	−0.31743
Day3	0.951995	0.95055	−0.27929
Day4	0.937235	0.953497	−0.14804
Day5	0.95543	0.953404	−0.32844
Day6	0.949204	0.957675	−0.16634
Day7	0.953815	0.950329	−0.34091
Day8	0.943993	0.951152	−0.16013
Day9	0.955598	0.95061	−0.27576
Day10	0.932579	0.953988	−0.27335
Day11	0.928337	0.944777	−0.32217
Day12	0.931294	0.946525	−0.31062

3. Math Description and Control Strategy

According to the analysis presented in the previous section, the market entry of large-scale UCs basically breaks the upper limit of the original urban grid planning. This will cause power shortages and equipment failures in some areas. The construction speed of large-scale UCs is usually very fast. Even if the grid projects that need supporting construction are synchronously set up to carry out preliminary work, they cannot be completed and put into operation when a large UC is put into use. It is often necessary to adopt emergency distribution network projects to complete the temporary transition of the power supply scheme. Large-scale UCs have high electricity consumption and large daily load peak-valley differences. If they blindly invest in power grid construction according to the maximum load requirements, it will inevitably lead to unnecessary waste of resources. In order to meet the multi-functional needs of the UC, we should support the construction and use of large garages to meet the parking demands of consumers. With the development of EVs, more and more EVs will replace petroleum fuel vehicles. Random charging of these EVs will also have a huge impact on the power grid, greatly increasing the risk of power grid accidents. In our research, the advantages of EVs as mobile energy storage terminals are put forward. According to the load and parking characteristics of the UC, a large UC garage can be used to control the charging of EVs in an orderly way, making them participate in the UC load control and solving the potential safety problems caused by the disorderly charging of EVs. The large daily load peak-valley differences of the UC cause the problem of excessive distribution network equipment resources [40], as well as the problem of lagging behind the planning of the power grid due to the large number of UCs.

In order to control the task conveniently and reduce the operation time and size of the necessary control center, our research proposes a two-tier real-time control strategy based on the EV user layer and the grid load decision-making layer. The lower layer divides the EVs into groups according to the load characteristics and parking characteristics of the UC, and then integrates the state information of the EVs in each group. We calculate the maximum electricity that can participate in the regulation in each regulation interval of each group. Due to the randomness of the behavior of EVs, the dynamic callable power of the first regulation period is updated continuously according to the changes in the number of EVs. The integration state information of each group calculated by the user layer will be send to the grid layer. The power grid decision-making level determines the power size of each group by formulating a dynamic target load curve according to the dynamic callable EV power of each group in each period. Thus, the load of the UC can be reduced to fill valleys and peaks. The upper decision is fed back to the lower. The lower level determines the charge power of each vehicle according to the power required by each group.

Aiming at EVs participating in load regulation of UC, our research puts forward the following main ideas:

(a) The users' willingness to participate in UC load control will be affected by battery status, periods of initial parking and other factors. Therefore, UC managers can specify price strategies to attract more users to participate in load control. Many scholars have studied the price strategy of V2G, and our research does not elaborate on this topic. When the user agrees to participate in load control, the load control center will collect the relevant data of the EVs to formulate a control strategy.

(b) The start parking period of EVs is random, and the parking requirement in the UC is large. If the new parked EV data is imported when each vehicle stops, the EV charging strategy will be changing constantly, which makes the strategy infeasible. Therefore, our research divides the peak period into equal time intervals. At the end of the last adjustment interval $(K - 1)$, the control strategy of the next interval (K) is worked out. When EVs are parking in the T-interval, they will not participate in the load regulation of the K-interval. The EVs parking in T-interval will participate in the strategy formulation of the next RI $(K + 1)$. This method can avoid the strategy from changing too frequently.

(c)   According to the analysis of different function types of UC, there are some typical parking characteristics of business type buildings such as short parking periods, high turnover ratio and large parking quantities. Although the total battery capacity which can participate in the strategy formulation is very large, most of them can't be controlled flexibly. They can only participate in the load control in their short parking period. As for office buildings, most of the parking periods are as long as the working hours, and the turnover ratio is low. That means they can be controlled flexibly. The parking ratio in residential type buildings is always at a high level. Therefore, the parking EVs of residential type can be controlled flexibly, so we use a large number of EVs in commercial type buildings for inflexible control, and EVs in office and residential type buildings for further flexible control.

(d)   The number of parking EVs in the UC is large. In order to reduce the computational load of the control center, our research clusters the EVs in the UC according to the previous analysis. Parking EVs in commercial areas is mainly for consumers, whose initial parking time is random. We take them as a group. In order to ensure the availability if a suitable numbers of EVs that can participate in load regulation flexibly, we formulate preferential policies to attract and sign agreements with users with long parking time and low frequency in office type and residential type buildings. According to the parking period, those EVs in the UC are divided into office EV group and residential EV group. According to the parking characteristics, the EVs are divided into groups, and the group is taken as a whole to participate in the formulation of the strategy and get the total discharge power of each group in each RI. Then, the discharge of each group is allocated to each EV.

3.1. Regulation Interval And EV Model

According to Figure 8 and previous study on the load characteristics of UC, the peak load regulation period will be 9:00–22:00, so the peak load of the UC lasts 13 hours a day. The scheduling cycle is divided into several regulation intervals Δt on average. Therefore, the number of RI in the peak load hour is:

$$K = \frac{13}{\Delta t} \tag{8}$$

The discharge strategy does not change in each regulation period. If Δt is too small, the strategy changes quickly. The computation will be so complex that there is no enough time to work out the strategy. If Δt is too long, the uncertainties of EV will increase during the regulation period, and less EV satisfy the Δt parking period conditions [41]. The number of EVs that can participate in dispatching decreases, which affects the total battery capacity provided for the UC.

When the EV parks in the garage, the control center will collect the state information represented by Γ

$$\Gamma = [ID_{EV} \; T_{arrive} \; T_{leave} \; SOC_{arrive} \; SOC_{leave} \; L] \tag{9}$$

where ID_{EV} is the car number of an EV; T_{arrive} is the time when the car arrives at the UC; T_{leave} is the time for leaving the UC; SOC_{arrive} is remaining SOC of EVs parking at k-interval; SOC_{leave} is the SOC when a user leaves the garage; L represents which functional areas the EV belong and whether the car participates in discharge control.

When the EV is parked in the garage, the garage dispatching system will collect the status information of the vehicle. At the same time, the user's willingness to participate in load regulation is sought [42]. Users willing to participate in load regulation will be required to input the earliest time they plan to leave the garage and the minimum target power when the user leaves.

According to the state information of the car, the parking period $T_{park}(n)$ and the maximum capacity $SOC_{con}(n)$ providing for UC of the n-th EV can be obtained by using Equations (10) and (11):

$$T_{park}(n) = T_{leave} - T_{arrive} \tag{10}$$

$$SOC_{con}(n) = SOC_{arrive} - SOC_{leave} \tag{11}$$

As shown in Figure 12. $T_{avai}(n)$ is the regulation interval which the n-th EV can participate in. It can be calculated as follows:

$$T_{avai}(n) = T_{leave} - T_{arrive} \tag{12}$$

where, T_{leave} means the beginning time of the control period interval not greater than T_{leave}, and T_{arrive} means the end time of the control period interval not less than T_{arrive}.

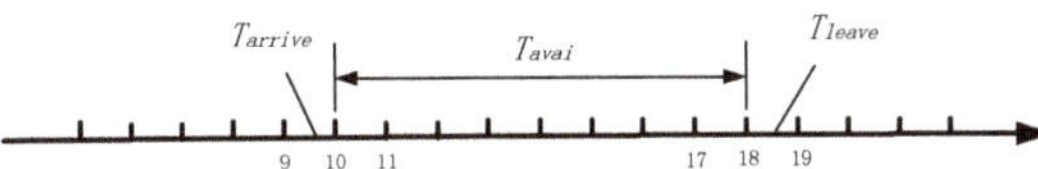

Figure 12. The regulation interval that the EV can participate in.

The garage parking information matrix of the k-th interval Λ^k can be obtained from the parking state information of vehicles and for the further application:

$$\Lambda^k = \begin{bmatrix} \Gamma_1^1 & \Gamma_2^1 & \cdots \\ \Gamma_1^2 & \ddots & \\ \vdots & & \Gamma_i^k \end{bmatrix} \tag{13}$$

3.2. Clustering of EVs

EVs are divided into three groups based on the analysis of Section 2:

(1) The first type is the EVs parked in residential type of UC buildings and contracted with the control center. The vehicles will park in the garage most of the time, which always keeps a high parking ratio in the garage. These cars are parked in garages most of the day and have shorter travel times plus shorter distances. They can be observed as a dynamic balanced whole part. They can participate in load regulation during the peak load period. By signing a contract with the owner, the control center can get the total amount of electricity of EVs in advance and take them as a whole part. We define those EVs as Group A.

The control center gets the target SOC of each user set by the contract and calculates the controllable electricity. Then, the maximum total controllable electricity of Group A in a day is obtained by adding up that of each EV. Considering that some EVs will leave temporarily during peak load period, we define SA_{total} as the controllable electricity of group A participating in regulation in one day which can be obtained by multiplying a coefficient on the total electricity:

$$SA_{total} = \eta \cdot \sum C(n) \cdot (SOCA_{start}(n) - SOCA_{target}(n)) \tag{14}$$

where, $C(n)$ is the battery capacity of the n-th EV; $SOCA_{start}(n)$ is the n-th EV's SOC of starting parking and $SOCA_{target}(n)$ is the target SOC set by the EV owner.

(2) The second type is the EVs parked in office type buildings and contracted with the control center. The parking period of these contracted EVs is the same as the working hours. Therefore, the time for these vehicles to participate in load regulation is working hours. These vehicles are classified as Group B. In the paper, the regulation period of Group B is 9:00–18:00. Before 9:00 a.m., the cars will be parked in the garage. We define SB_{total} as the total electricity involved in the regulation according to the target SOC set by the users:

$$SB_{total} = \eta \cdot \sum C(n) \cdot (SOCB_{start}(n) - SOCB_{target}(n)) \tag{15}$$

(5) The third type is the EVs parked in commercial type building garages. The starting parking time and parking period of these vehicles are related to the purpose of consumers going to the UC. The number of cars parked is large, but we cannot estimate the number of cars parked in each RI accurately. When the vehicle stops in the (K-1)-interval, the control center will inquire about the owners' willingness to participate in load regulation. For the vehicles involved in the regulation, the electricity of each vehicle in K-interval is calculated according to Equations (8) and (11), and we define SC_{total}^k as the total controllable electricity of Group C in the k-th RI which can be obtained by adding up the electricity of each vehicle parked in the k-th RI:

$$SC_{total}^k = \eta \cdot \sum C(n) \cdot (SOCC_{start}(n) - SOCC_{target}(n)) \tag{16}$$

3.3. Objective Function And Constraints

The electricity of group C has been determined in each RI and cannot be controlled. The controllable total electric quantity of group A and group B during peak load periods need to be determined as well. We define that P^k represents the conventional load after subtracting the discharge power in the k-th interval. P_A^k and P_B^k represent the total discharge capacity of group A and group B in the k-th interval. They can be decided by the objective function Equation (17) which minimizes the mean square deviation of load [43].

$$F = \min \frac{1}{22/\Delta t - k} \sum_{k}^{K} \left(P^k - P_B^k - P_A^k \right)^2 \tag{17}$$

The discharge power of each EV participating in load regulation should be within the threshold power allowed in the garage. The total amount of discharge electricity in each RI of each vehicle is not greater than the total controllable electricity. At the same time, the electric quantity of each vehicle in each RI is not greater than the total electricity, that is:

$$\begin{cases} PA^k(n), PB^k(n) \le P_{max} \\ \sum_{k=1+9/\Delta t}^{22/\Delta t} \Delta t \cdot PA^k(n) \le SA(n) \\ \sum_{n=1}^{N_A} \sum_{k=1+9/\Delta t}^{22/\Delta t} \Delta t \cdot PA^k(n) \le SA_{total} \\ \sum_{k=1+9/\Delta t}^{18/\Delta t} \Delta t \cdot PB^k(n) \le SB(n) \\ \sum_{n=1}^{N_B} \sum_{k=1+9/\Delta t}^{18/\Delta t} \Delta t \cdot PB^k(n) \le SB_{total} \end{cases} \tag{18}$$

where P_{max} is the threshold power allowed in the garage; N_A is the number of EVs in Group A; $PA^k(n)$ is n-th EV's discharge power in the k-th RI of Group A; $SA(n)$ is the n-th EV's total controllable electricity in Group A; N_B is the number of EVs in Group B; $PB^k(n)$ is the n-th EV's discharge power in the k-th RI of Group B and $SB(n)$ is the n-th EV's total controllable electricity in Group B.

3.4. Discharge Power of Each EV

After getting the total discharge electricity of Group A and Group B in each RI, the discharge electricity will be allocated to each EV. When the EV's power is less than the average value of discharge, the vehicle will discharge all the electricity during the RI. The discharge power of the other EVs is the average power of the vehicles excluded from the EVs whose power is less than the average value [44]. Taking group A as an example, the average discharge power PA_{ave}^k in the k-th RI is:

$$PA_{ave}^k = \frac{SA_{total}^k}{\Delta t \cdot N_A^k} \tag{19}$$

where SA_{total}^k is the total electricity in Group A that will discharge in the k-th RI, N_A^k is the number of Group A EVs in the k-th RI.

We define that $SA_{total}^k(n)$ is the controllable electricity of the n-th EV in the k-th RI and $PA^k(n)$ is the discharge power of the n-th EV in the k-th RI.

If:

$$SA_{con}^k(n) < \Delta t \cdot PA_{ave}^k \tag{20}$$

then:

$$PA^k(n) = \frac{SA_{con}^k(n)}{\Delta t} \tag{21}$$

otherwise:

$$PA^k(n) = \frac{SA_{total}^k - \sum SA_{con}^{k\prime}(n)}{\Delta t \cdot (N_A^k - N_A^{k\prime})} \tag{22}$$

where $SA_{con}^{k\prime}(n)$ is the controllable electricity of Group A in the k-th RI excluding the EVs whose power is less than PA_{ave}^k and $N_A^{k\prime}$ is the number of EVs whose power is less than PA_{ave}^k.

The discharge power allocation of Group B ($PB^k(n)$) is the same as that of Group A, so it will not be discussed repeatedly in our research.

In order to maximize the available electricity in Group C to participate in the regulation, the EVs in Group C will feed back to the UC with the maximum discharge power until the EV leaves or the remaining SOC reaches the target.

We define $PC^k(n)$ as the discharge power of the n-th EV in group C in the k-th RI. It can be calculated by the following expression:

$$PC^k(n) = \frac{SC_{con}^k(n)}{\Delta t} \tag{23}$$

If:

$$PC^k(n) \geq P_{\max} \tag{24}$$

then, the discharge power is decided by $P_{\max}$:

$$PC^k(n) = P_{\max} \tag{25}$$

Based on the above mathematical description, the load regulation strategy can be briefly summarized as follows. Firstly, we divide peak load period into several regulation intervals Δt on average, and get the garage parking information matrix Λ^k, and then, we calculate the parking period $T_{park}(n)$ and the maximum capacity $SOC_{con}(n)$ of each EV. Secondly, EVs are divided into three groups and the total controllable electricity SA_{total} and SB_{total} and uncontrollable electricity SC_{total}^k in each RI can be calculated by Λ^k. Thirdly, we set the total discharge power of group A and group B in each RI (P_A^k and P_B^k) as the decision variables, and then, set the objective function as Equation (17) and constraints as Equation (18). Finally, calculate the discharge powers $PA^k(n)$, $PB^k(n)$ and $PC^k(n)$ of each EV in the different groups in each RI.

4. Case Analysis

4.1. Case Statement

Our study chooses an UC as the research object, which integrates the functions of commercial retail, business offices, hotel catering, comprehensive entertainment facilities and apartment housing as a whole thing. It covers an area of 360,000 square meters and consists of a commercial building, four residential buildings and an office building. The parking space can hold 2000 vehicles, and 500 vehicles can be charged and discharged here now. The typical conventional daily load of the UC is shown as Figure 13.

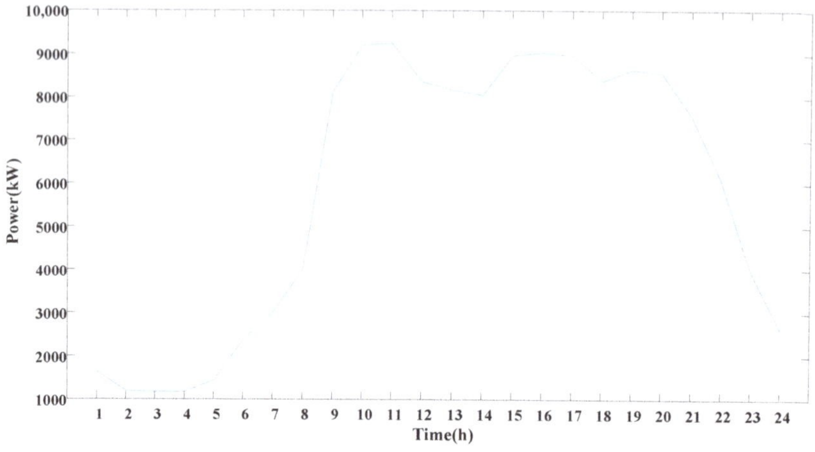

Figure 13. Typical conventional grid load of a UC.

At present, the proportion of EVs is still very low compared with traditional vehicles. We assume that 50 office EVs agree to establish a contract to participate in the load regulation during working hours and 50 household EVs participate as well. The number of EVs entering the commercial type area is 4000 a day. The parking characteristics of Group C are randomly generated by using the typical parking ratio fitting curve of commercial type buildings analyzed in Section 2. Suppose the initial SOC of each EV starting discharging in the garage meets a normal distribution N (0.65, 0.05).

4.2. Influence of The Permeability

The scale of Group C is large, so the permeability of EVs is an important factor affecting the efficiency of the strategy. We assuming that the length of the RI is set to be 30 min, and we change the different permeabilities for Group C. When there is no group A and group B involved in the regulation, the impact of EVs participating in UC load regulation with different peak load permeabilities is shown in Figure 14. It can be observed from the figure that peak load decreases with the increase of permeability of EVs. EVs in Group C can relieve the peak load pressure. However, group C loads cannot be regulated. As shown in Figure 14, load reduction is obvious in the evening when the parking time of group C is concentrated. However, the effect of load regulation is poor at noon when the number of vehicles in group C is small. It is the reason why peak load decreases about to 9.43% when the permeability reaches 100%.

The effects of load regulation are obviously improved as shown in Figure 15 when group A and group B EVs participate. Group A and Group B EVs can discharge more power if the load is high and discharge less or no power if the power load is low. When the permeability reaches 100%, the peak load decreases by 14.25%, and the peak load becomes more stable.

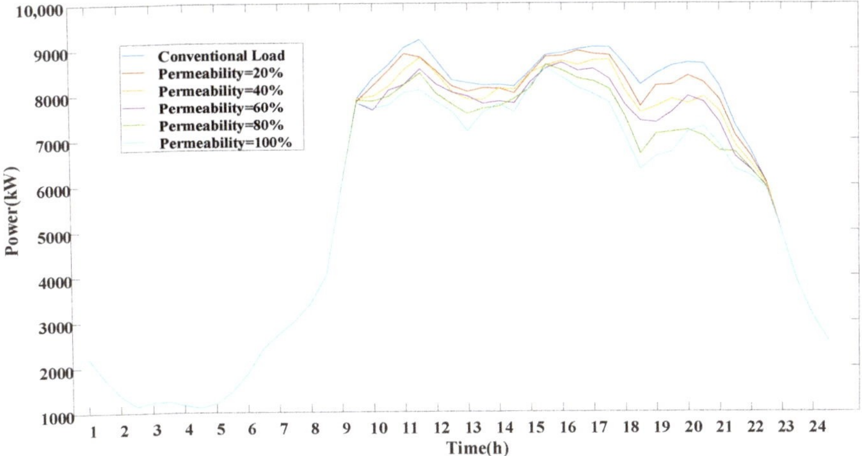

Figure 14. The load when only group C EVs participate in load regulation with different permeabilities.

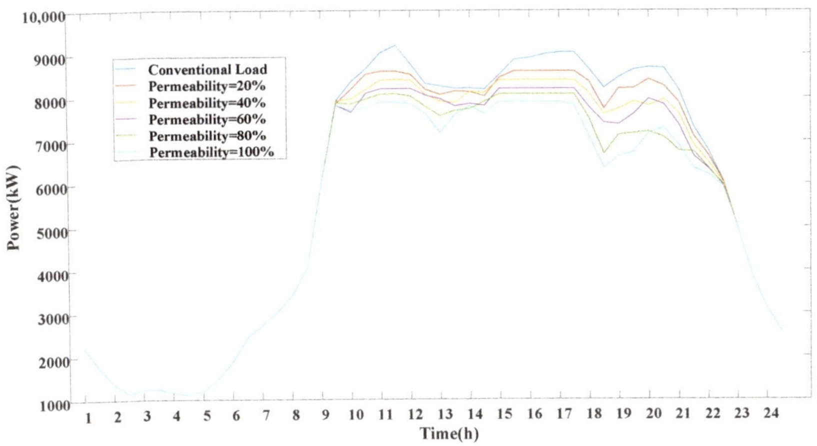

Figure 15. The load when group A,B and C EVs participate in load regulation with different permeabilities.

The load of UCs can be reduced effectively by increasing the permeability of EVs, although the permeability of EVs is still very low at present, and the peak load reduction is limited. However, with the increase of EVs' popularity and permeability, this strategy will play an important role in improving the load situation of UCs. It is very effective to reduce the peak load of UC by this strategy.

4.3. Influence of The Regulation Interval

Another influence on the effect of load regulation strategy is the length of RI. If the RI is too long, the parking period in the garage of EVs cannot meet the length of RI. The number of EVs of Group C participating in load regulation will be reduced greatly. If the length of RI is too short, the frequency of the strategy formulation will increase, which will increase the operational burden of the control center.

Assuming that the permeability of Group C is 40%, we set the RI as 15, 30 and 60 min, and explore the influence of different interval lengths when only group C EVs participate in load balancing. The simulation results are shown in Figure 16. When the RI is too long (60 min), the number of EVs that can satisfy the RI is very small, and it has little effect on load regulation. When the regulation time is shortened to 15 min, load regulation strategy works well. The peak load decreases about 9.44%, which is same as the situation (RI = 30 min, permeability = 40%, without Group A and B EVs). The shorter interval makes more EVs participate in the discharge, so that more electricity from Group C EVs can participate in the regulation.

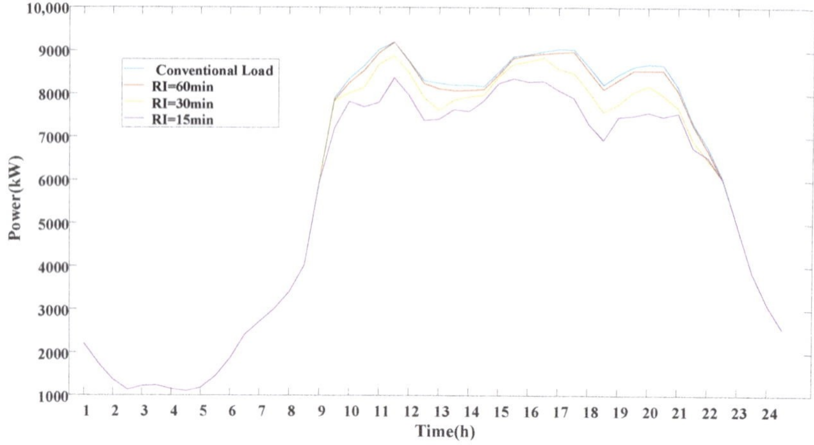

Figure 16. The load when only group C EVs participate in load regulation with different RIs.

As shown in Figure 17, when group A and group B EVs participate in load regulation, the peak load is further reduced by 14.6%. The numbers of EVs of group A and B that participate in load regulation is relatively small compared with Group C ones. However, they still play an important role in peak load reduction and load fluctuation reduction. From the simulation results, this strategy can effectively reduce the peak load. Therefore, in the construction of an UC, less money will be spent in the transformation of the local distribution system and the operation of the local power grid will be safer.

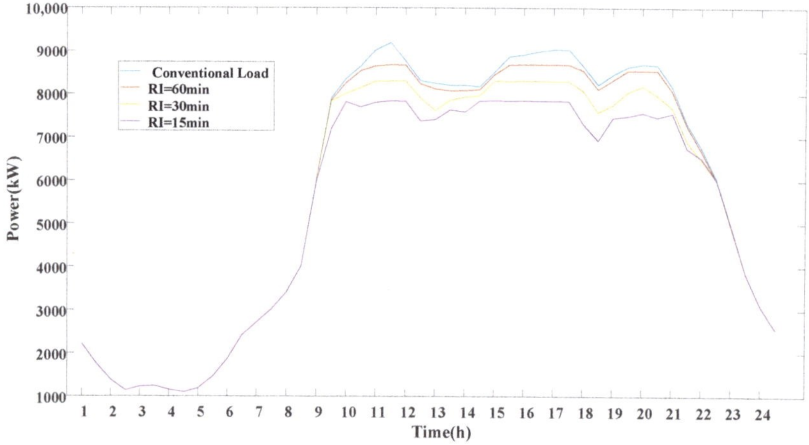

Figure 17. The load group when A,B and C EVs participate in load regulation with different RIs.

However, when the length of the interval is short enough, the number of EVs that can participate in the discharge will not change anymore. Shortening the interval length will no longer optimize the load curve regulation of the UC, but will increase the operational load of the control center. Therefore, it is important to adopt an appropriate RI length in any load regulation strategy. The actual position and operational location of UC are different, so the parking characteristics of cars are different, which means the best RI length is different. It is necessary to determine the optimum RI length by analyzing the actual parking characteristics.

5. Conclusions

During the construction of an UC, due to the huge power consumption during peak periods, the capacity of power distribution equipment is high, and it is often necessary to transform the nearby power facilities, which will increase construction and maintenance costs. The popularity of EVs can improve the environment, but the stochastic large-scale charging of cars may pose a threat to the regional grid [45]. In view of the above problems, a strategy for EVs participating in load regulation of a UC was proposed. The stochastic charging load of EVs was modeled according to the users' habits. It was verified that the peaks of charging load and conventional loads will be superimposed. This situation may have a bad influence on the operation of the power grid. Because it is difficult to put forward a load regulation strategy if an UC is regarded as a whole, we divide UCs into different functional areas. The relationship between parking ratio and load ratio was obtained after dividing UCs into different functional areas. There are large numbers of vehicles in the commercial functional areas with high mobility (Group C), and vehicles in office and functional areas with low mobility (Group A and B). Then, we formulate the load regulation strategy: we divide the peak load period into several equal RIs, and establish a the parking information database. The EVs in group C of each RI will feed back to UC with maximum discharging power. The objective function was established to minimize load variance during the peak period of the UC, and the discharge power of Group A and B EVS of each RI was calculated by an objective function. Finally, the case analysis proved that the strategy in our research can effectively reduce the load of the UC during the peak load period. The results show that with the increase of the permeability, the number of EVs participating in load regulation also increases, which means more electricity from EVs is fed back to the UC to make the peak load smaller. Shortening the RI can make more vehicles from group C participate in load regulation and also improve the effect of load regulation. In the future, the strategy of charging EVs in UCs during valley load period will be studied to reduce the peak-valley load difference of the UC and improve the energy utilization structure of the UC. Considering that EVs can be used as mobile power sources, the characteristics of vehicles parked in the UC are analyzed in depth.

Author Contributions: All authors contributed equally to conceptualization, methodology development, software implementation, validation, formal analysis, investigation, writing-original draft preparation, writing-review and editing, visualization, and supervision. All authors have read and agreed to the published version of the manuscript.

Funding: This research was funded by the Chongqing Science and Technology Commission of China under Project No. cstc2018jcyjA3148; graduate scientific research and innovation foundation of Chongqing, China under Grant No. CYS1807; graduate research and innovation foundation of Chongqing, China under Grant No. CYB18009.

Conflicts of Interest: The authors declare no conflict of interest.

References

1. Li, G.; Zhang, X.-P. Modeling of Plug-in Hybrid Electric Vehicle Charging Demand in Probabilistic Power Flow Calculations. *IEEE Trans. Smart Grid* **2012**, *3*, 492–499. [CrossRef]
2. Shafiee, S.; Fotuhi-Firuzabad, M.; Rastegar, M. Investigating the Impacts of Plug-in Hybrid Electric Vehicles on Power Distribution Systems. *IEEE Trans. Smart Grid* **2013**, *4*, 1351–1360. [CrossRef]
3. Shafiei, E.; Thorkelsson, H.; Asgeirsson, E.I.; Davíðsdóttir, B.; Raberto, M.; Stefansson, H. An agent-based modeling approach to predict the evolution of market share of electric vehicles: A case study from Iceland. *Technol. Forecast. Soc. Chang.* **2012**, *79*, 1638–1653. [CrossRef]

4. Clement-Nyns, K.; Haesen, E.; Driesen, J. The Impact of Charging Plug-In Hybrid Electric Vehicles on a Residential Distribution Grid. *IEEE Trans. Power Syst.* **2009**, *25*, 371–380. [CrossRef]

5. Amini, M.H.; Kargarian, A.; Karabasoglu, O. ARIMA-based decoupled time series forecasting of electric vehicle charging demand for stochastic power system operation. *Electr. Power Syst. Res.* **2016**, *140*, 378–390. [CrossRef]

6. Kevin, M.; Verschueren, T.; Haerick, W. Optimizing smart energy control strategies for plug-in hybrid electric vehicle charging. In Proceedings of the Network Operations & Management Symposium Workshops, Osaka, Japan, 19–23 April 2010.

7. Zhong, J.; He, L.; Li, C.; Cao, Y.; Wang, J.; Fang, B.; Zeng, L.; Xiao, G. Coordinated control for large-scale EV charging facilities and energy storage devices participating in frequency regulation. *Appl. Energy* **2014**, *123*, 253–262. [CrossRef]

8. Sun, B.; Huang, Z.; Tan, X.; Tsang, D.H.K. Optimal Scheduling for Electric Vehicle Charging with Discrete Charging Levels in Distribution Grid. *IEEE Trans. Smart Grid* **2016**, *9*, 624–634. [CrossRef]

9. Donadee, J.; Ilic, M.D. Stochastic Optimization of Grid to Vehicle Frequency Regulation Capacity Bids. *IEEE Trans. Smart Grid* **2014**, *5*, 1061–1069. [CrossRef]

10. Pavić, I.; Capuder, T.; Kuzle, I. Value of flexible electric vehicles in providing spinning reserve services. *Appl. Energy* **2015**, *157*, 60–74. [CrossRef]

11. Chenye, W.; Hamed, M.R. PEV-based combined frequency and voltage regulation for smart grid. In Proceedings of the IEEE Pes Innovative Smart Grid Technologies IEEE Computer Society, Washington, DC, USA, 16–20 January 2012.

12. Zhou, C.; Qian, K.; Allan, M.; Zhou, W. Modeling of the Cost of EV Battery Wear Due to V2G Application in Power Systems. *IEEE Trans. Energy Convers.* **2011**, *26*, 1041–1050. [CrossRef]

13. Asensio, M.; Contreras, J. Stochastic Unit Commitment in Isolated Systems with Renewable Penetration under CVaR Assessment. *IEEE Trans. Smart Grid* **2015**, *7*, 1356–1367. [CrossRef]

14. Psarros, G.N.; Nanou, S.; Papaefthymiou, S.V.; Papathanassiou, S.A. Generation scheduling in non-interconnected islands with high RES penetration. *Renew. Energy* **2018**, *115*, 338–352. [CrossRef]

15. Thomas, D.; Deblecker, O.; Ioakimidis, C.S. Optimal design and techno-economic analysis of an autonomous small isolated microgrid aiming at high RES penetration. *Energy* **2016**, *116*, 364–379. [CrossRef]

16. Muñoz, E.R.; Razeghi, G.; Zhang, L.; Jabbari, F. Electric vehicle charging algorithms for coordination of the grid and distribution transformer levels. *Energy* **2016**, *113*, 930–942. [CrossRef]

17. Wang, L.; Sharkh, S.; Chipperfield, A.; Chipperfield, A.J. Optimal coordination of vehicle-to-grid batteries and renewable generators in a distribution system. *Energy* **2016**, *113*, 1250–1264. [CrossRef]

18. Huang, S.; Yang, J.; Li, S. Black-Scholes option pricing strategy and risk-averse coordination for designing vehicle-to-grid reserve contracts. *Energy* **2017**, *137*, 325–335. [CrossRef]

19. Graabak, I.; Wu, Q.; Warland, L.; Liu, Z. Optimal planning of the Nordic transmission system with 100% electric vehicle penetration of passenger cars by 2050. *Energy* **2016**, *107*, 648–660. [CrossRef]

20. Hui, L.; Yun, L.; Feng, L. Dynamic economic/emission dispatch including PEVs for peak shaving and valley filling. *IEEE Trans. Ind. Electron* **2018**, *66*, 2880–2890.

21. Mozafar, M.R.; Amini, M.H.; Moradi, M.H. Innovative appraisement of smart grid operation considering large-scale integration of electric vehicles enabling V2G and G2V systems. *Electr. Power Syst. Res.* **2018**, *154*, 245–256. [CrossRef]

22. Lance, N.; Regina, M.C. A cost benefit analysis of a V2G-capable electric school bus compared to a traditional diesel school bus. *Appl. Energy* **2014**, *126*, 246–255.

23. Zhao, Y.; Noori, M.; Tatari, O. Vehicle to Grid regulation services of electric delivery trucks: Economic and environmental benefit analysis. *Appl. Energy* **2016**, *170*, 161–175. [CrossRef]

24. Zhang, L.; Li, Y. Optimal Management for Parking-Lot Electric Vehicle Charging by Two-Stage Approximate Dynamic Programming. *IEEE Trans. Smart Grid* **2015**, *8*, 1722–1730. [CrossRef]

25. Lund, H.; Kempton, W. Integration of renewable energy into the transport and electricity sectors through V2G. *Energy Policy* **2008**, *36*, 3578–3587. [CrossRef]

26. Locment, F.; Sechilariu, M.; Forgez, C. Electric vehicle charging system with PV Grid-connected configuration. In Proceedings of the Vehicle Power & Propulsion Conference, Chicago, IL, USA, 6–9 September 2012.

27. Göransson, L.; Karlsson, S.; Johnsson, F. Integration of plug-in hybrid electric vehicles in a regional wind-thermal power system. *Energy Policy* **2010**, *38*, 5482–5492. [CrossRef]

Energies **2020**, *13*, 2939

28. Kou, P.; Liang, D.; Gao, L.; Gao, F. Stochastic Coordination of Plug-In Electric Vehicles and Wind Turbines in Microgrid: A Model Predictive Control Approach. *IEEE Trans. Smart Grid* **2016**, *7*, 1537–1551. [CrossRef]

29. Chukwu, U.C.; Mahajan, S.M. Real-Time Management of Power Systems with V2G Facility for Smart-Grid Applications. *IEEE Trans. Sustain. Energy* **2013**, *5*, 558–566. [CrossRef]

30. Alam, M.J.E.; Muttaqi, K.M.; Sutanto, D. Effective Utilization of Available PEV Battery Capacity for Mitigation of Solar PV Impact and Grid Support with Integrated V2G Functionality. *IEEE Trans. Smart Grid* **2015**, *7*, 1562–1571. [CrossRef]

31. Vachirasricirikul, S.; Ngamroo, I. Robust LFC in a Smart Grid with Wind Power Penetration by Coordinated V2G Control and Frequency Controller. *IEEE Trans. Smart Grid* **2014**, *5*, 371–380. [CrossRef]

32. Khodayar, M.E.; Wu, L.; Shahidehpour, M. Hourly Coordination of Electric Vehicle Operation and Volatile Wind Power Generation in SCUC. *IEEE Trans. Smart Grid* **2012**, *3*, 1271–1279. [CrossRef]

33. Fettinger, N.S.; Ten, C.W.; Chigan, C. Minimizing residential distribution system operating costs by intelligently scheduling plug-in hybrid electric vehicle charging. In Proceedings of the Transportation Electrification Conference and Expo (ITEC), Michigan, MI, USA, 18–20 June 2012.

34. Yang, H.; Chung, C.Y.; Zhao, J. Application of plug-in electric vehicles to frequency regulation based on distributed signal acquisition via limited communication. *IEEE Trans. Power Syst.* **2012**, *28*, 1017–1026. [CrossRef]

35. Sortomme, E.; El-Sharkawi, M.A. Optimal Scheduling of Vehicle-to-Grid Energy and Ancillary Services. *IEEE Trans. Smart Grid* **2011**, *3*, 351–359. [CrossRef]

36. Nezamoddini, N.; Wang, Y. Risk management and participation planning of electric vehicles in smart grids for demand response. *Energy* **2016**, *116*, 836–850. [CrossRef]

37. Sales ranking of EVs in 2019. Available online: https://www.sohu.com/a/371183113_205282 (accessed on 1 February 2020).

38. Gong, L.; Cao, W.; Zhao, J. Load modeling method for EV charging stations based on trip chain. In Proceedings of the 2017 IEEE Conference on Energy Internet and Energy System Integration, Beijing, China, 26–28 November 2017.

39. Weckx, S.; Driesen, J. Load Balancing With EV Chargers and PV Inverters in Unbalanced Distribution Grids. *IEEE Trans. Sustain. Energy* **2015**, *6*, 635–643. [CrossRef]

40. Wang, Y.; Huang, Y.; Wang, Y.; Yu, H.; Li, R.; Song, S. Energy Management for Smart Multi-Energy Complementary Micro-Grid in the Presence of Demand Response. *Energies* **2018**, *11*, 974. [CrossRef]

41. Badawy, M.O.; Sozer, Y. Power Flow Management of a Grid Tied PV-Battery System for Electric Vehicles Charging. *IEEE Trans. Ind. Appl.* **2017**, *53*, 1347–1357. [CrossRef]

42. Xie, D.; Chu, H.; Gu, C.; Li, F.; Zhang, Y. A Novel Dispatching Control Strategy for EVs Intelligent Integrated Stations. *IEEE Trans. Smart Grid* **2015**, *21*, 1. [CrossRef]

43. Dubey, A.; Santoso, S. Electric Vehicle Charging on Residential Distribution Systems: Impacts and Mitigations. *IEEE Access* **2015**, *3*, 1871–1893. [CrossRef]

44. Binetti, G.; Davoudi, A.; Naso, D.; Turchiano, B.; Lewis, F.L. Scalable Real-Time Electric Vehicles Charging With Discrete Charging Rates. *IEEE Trans. Smart Grid* **2015**, *6*, 2211–2220. [CrossRef]

45. Beaude, O.; Lasaulce, S.; Hennebel, M.; Mohand-Kaci, I. Reducing the Impact of EV Charging Operations on the Distribution Network. *IEEE Trans. Smart Grid* **2016**, *7*, 2666–2679. [CrossRef]

Review

Energy Management Strategies for Hybrid Electric Vehicles: Review, Classification, Comparison, and Outlook

Fengqi Zhang [1,*], **Lihua Wang** [1], **Serdar Coskun** [2], **Hui Pang** [1], **Yahui Cui** [1] and **Junqiang Xi** [3]

[1] School of Mechanical and Precision Instrument Engineering, Xi'an University of Technology, Xi'an 710048, China; 1170211016@stu.xaut.edu.cn (L.W.); huipang@163.com (H.P.); cyhxut@xaut.edu.cn (Y.C.)

[2] Department of Mechanical Engineering, Tarsus University, Tarsus, Mersin 33400, Turkey; serdarcoskun@tarsus.edu.tr

[3] School of Mechanical Engineering, Beijing Institute of Technology, Beijing 100081, China; xijunqiang@bit.edu.cn

* Correspondence: zfqdy@126.com

Received: 29 April 2020; Accepted: 28 June 2020; Published: 30 June 2020

Abstract: Hybrid Electric Vehicles (HEVs) have been proven to be a promising solution to environmental pollution and fuel savings. The benefit of the solution is generally realized as the amount of fuel consumption saved, which by itself represents a challenge to develop the right energy management strategies (EMSs) for HEVs. Moreover, meeting the design requirements are essential for optimal power distribution at the price of conflicting objectives. To this end, a significant number of EMSs have been proposed in the literature, which require a categorization method to better classify the design and control contributions, with an emphasis on fuel economy, providing power demand, and real-time applicability. The presented review targets two main headlines: (a) offline EMSs wherein global optimization-based EMSs and rule-based EMSs are presented; and (b) online EMSs, under which instantaneous optimization-based EMSs, predictive EMSs, and learning-based EMSs are put forward. Numerous methods are introduced, given the main focus on the presented scheme, and the basic principle of each approach is elaborated and compared along with its advantages and disadvantages in all aspects. In this sequel, a comprehensive literature review is provided. Finally, research gaps requiring more attention are identified and future important trends are discussed from different perspectives. The main contributions of this work are twofold. Firstly, state-of-the-art methods are introduced under a unified framework for the first time, with an extensive overview of existing EMSs for HEVs. Secondly, this paper aims to guide researchers and scholars to better choose the right EMS method to fill in the gaps for the development of future-generation HEVs.

Keywords: Hybrid Electric Vehicles (HEVs); energy management strategies (EMSs); driving cycle prediction; optimization

---

1. Introduction

Hybrid Electric Vehicles (HEVs) are composed of different types of energy sources and power converters, which generally refer to vehicles consisting of an internal combustion engine (ICE) with an electric motor. HEVs seem to be the most economically viable solution so far and probably for the upcoming decades. The general goal to develop HEVs is to reduce fuel consumption and emissions while ensuring drivers' power demands by investigating the appropriate energy management strategies (EMSs). Energy management aims to obtain an optimal power split in view of complex driving conditions, as well as to minimize fuel consumption and emissions. It is commonly acknowledged that improvements in the fuel economy of HEVs, and thus the consequent reduction in emissions, depend crucially on their energy management strategies (EMSs) [1]. The complex configuration and behavior of multi-source hybrid energy systems introduce challenges to the performance of EMSs. Regardless of the topology of the powertrain, the EMS aim is to instantaneously manage the power flows from the energy converters to achieve the control objectives [2]. The optimal control algorithms employed under a given driving cycle are therefore the representative research outline in the field of energy management strategies.

Various EMSs for HEVs have been conducted in recent years. In the existing literature reviews, a number of classifications for the energy management strategy are reported [3–7]. Generally, EMSs can be divided into three categories: rule-based EMSs, local optimization-based EMSs, and global optimization-based EMSs [3]. An overview of EMSs for plug-in Hybrid Electric Vehicles is presented in [4]. The classification of energy management, such as rule-based control strategies and optimization-based control strategies, are introduced according to their mathematical models and the approach commonly used. In [5], EMSs are divided into two categories as rule-based and optimization-based methods for parallel Hybrid Electric Vehicles, and the pros and cons of each approach are compared. Finally, some real-time implementation issues are discussed from different aspects (e.g., computational burden and optimality). The different classifications for hybrid vehicles focusing on hydraulic drives is introduced and discussed in [6]. Different kinds of approaches like offline and online strategies are classified and compared. As intelligent transportation system (ITS) technology has emerged and machine learning methods have been widely used, some new EMSs have been developed to improve the performance requirements (e.g., adaptability and real-time implementation). However, there is still a need for a comprehensive review of the EMSs to better elucidate the state-of-the-art approaches and potential future research directions. To this end, the present review, different from the aforementioned review papers in EMSs, proposes a comprehensive hierarchical classification scheme for the first time. In the first category, offline EMSs are presented based on the level of driving information under global optimization-based EMSs and rule-based EMSs. In the second category, online EMSs are layered as instantaneous optimization-based EMSs, predictive EMSs, and learning-based EMSs. Since the presented scheme covers various approaches in terms of targeted solution objectives, optimality, and real-time implementation, an important number of literature studies are extensively overviewed. The principle of each approach along with its pros and cons are illustrated and compared within the design and operational characterization of the proposed scheme. Finally, a good number of emerging innovative EMSs and recent literature that have not been covered in previous review papers are summarized and important future trends for HEVs are highlighted. This study is intended to serve as a comprehensive reference for researchers in the field of development and optimization of EMSs.

The remainder of the paper is organized as follows. Different powertrain topologies of Hybrid Electric Vehicles are briefly discussed and compared in Section 2. In Section 3, a hierarchical classification scheme of EMSs is presented. In the following Sections 4 and 5, offline and online EMSs categories are stated in more detail. Each approach is elaborated and compared according to its principles, as well as pros and cons. Some important future trends of EMSs are discussed in Section 6.

2. The Powertrain Topologies of Hybrid Electric Vehicles

It is well known that there are mainly three kinds of topologies for Hybrid Electric Vehicles: series, parallel, and power-split. A series hybrid powertrain is regarded as a simple extension of a battery-powered electric vehicle that is propelled only by motor. The engine drives a generator, producing electrical power, which can be summed to the electrical power coming from the energy storage system and then transmitted, via an electric bus, to the electric motor(s) driving the wheels [8]. In principle, the advantage of the series hybrid powertrain is that only electrical connections between the main power conversion devices are required. Thus, vehicle packaging and design are simplified. Meanwhile, the engine that is completely off the wheels offers great freedom in selecting speed and load, thus allowing the engine to operate at a high-efficiency region. On the other hand, the series hybrid powertrain requires two energy conversions (i.e., from mechanical to electrical in the generator, and from electrical to mechanical in the motor), which result in a loss of efficiency, even when there is a direct mechanical connection between the engine and the wheels in the existing configuration. As a result, in some cases, a series hybrid electric vehicle consumes more fuel than a traditional vehicle, especially in highway driving. Furthermore, one of the two electromechanical energy converters must be sized to meet the maximum power demand of the vehicle, as it is the primary source of propulsion [2,8]. The series topology is shown in Figure 1.

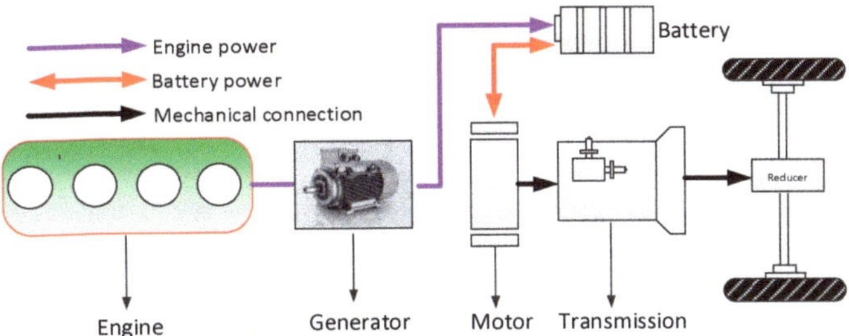

Figure 1. Series hybrid electric vehicle.

As for the parallel topology, the engine is connected to the powertrain by a mechanical coupling device while the motor propels the vehicle. Either engine or motor could propel the vehicle according to different load conditions, which makes it possible to greatly increase the fuel economy. The motor provides the power when the vehicle operates at lower speed to reduce fuel consumption. Thus, this configuration is capable of maintaining a higher efficiency and better fuel economy. The power summation is mechanical rather than electrical, and the engine and the electric machines (one or more) are connected with a gear set, a chain, or a belt; thus, their torques are summed and transmitted to the wheels [8]. In this configuration, there is no need to size one of the two electromechanical energy converters to meet the maximum power demand for parallel hybrid powertrain; however, unless it is significant oversize, the electric motors have less power than those used in a series hybrid powertrain (since not all the mechanical power goes through them), thus reducing the possibility of regenerative braking. Meanwhile, the engine operating conditions cannot be regulated as freely as in a series hybrid powertrain, since the engine speed is mechanically related (via the transmission system) to the vehicular velocity [2,8]. The parallel topology is illustrated in Figure 2.

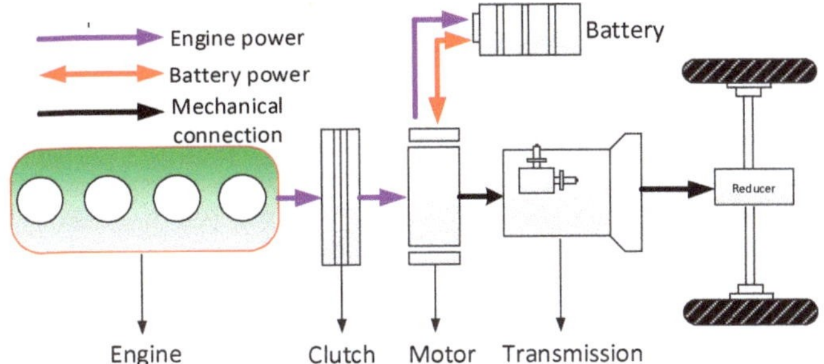

Figure 2. Parallel hybrid electric vehicle.

As for the power-split topology, the most important improvement is the ability to operate as either a series or parallel topology, which provides more operation modes to substantially improve the overall efficiency under complex driving conditions. Although the series path is generally avoided because it is less efficient, the main feature of this design is that the engine, generator, and motor speed are decoupled, allowing additional freedom in control. The engine and two electric machines are connected to a power split device (usually a planetary gear set), so that the power from the engine and the electric machines can be merged through both a mechanical and an electrical path, allowing series and parallel operations [8].Compared to the parallel hybrid powertrain, the power-split architecture is the most flexible and represents a higher control ability on the engine operating conditions while adopting the double energy conversion, which is typical of a series operation only in a small portion of the total power demand, thus decreasing overall losses [2,8]. The power-split topology is presented in Figure 3.

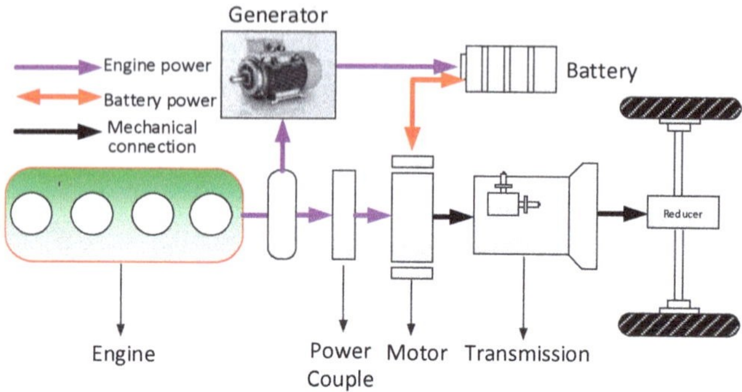

Figure 3. Power-split hybrid electric vehicle.

3. The Classification of EMSs

In this paper, we propose a new hierarchical classification scheme of EMSs for all kinds of Hybrid Electric Vehicles via two main headlines: (1) offline EMSs are categorized according to the information level of the driving conditions utilized, including global optimization based-EMSs and rule-based EMSs; and (2) online EMSs are represented as instantaneous optimization-based EMSs, predictive EMSs, and learning-based EMSs. The classification of the EMSs is illustrated in Figure 4. It is noted that a flexible EMS can include a mixture of various techniques (offline and online) to form an integrated EMS for improving the fuel economy and performance. Thus, in this paper, these combinations with other techniques may be included while providing a particular EMS classification. For offline EMSs, two categorizations are illustrated: the global optimization-based and rule-based EMSs. The main goal of global optimization-based EMSs is to achieve a global optimal power split under a given driving cycle and provide modified online EMSs. They are not directly applicable in real-time control due to their computational complexity and the requirement of a priori knowledge of the entire driving cycle. However, it can be used as a benchmark to adjust the control parameters. Typical methods, such as dynamic programming, can implement global optimization over given driving cycles, but it cannot be directly employed in a real vehicle. Therefore, this method can be used to evaluate the performance of other optimization methods to extract the control rules. Rule-based EMSs are considered as an offline method since the rules are derived from pre-production tests. Rule-based EMSs are based on pre-defining a series of control rules to determine the power split while it cannot achieve optimal allocation of power as compared to offline globally optimized energy management. Online EMSs, however, are based on local optimization and causal with the potential of being applied in real-time control. Among these strategies, Instantaneous optimization EMSs can minimize the instantaneous fuel consumption at each instant without a priori knowledge of the entire driving cycle and only obtain local optimal results. The instantaneous optimization-based EMSs are (1) the equivalent consumption minimization strategy (ECMS); (2) adaptive-ECMS (A-ECMS); and (3) robust control (RC). As a fundamental method, ECMS can be used for real-time implementation due to its adjustability, which is related to the equivalent factor (EF). It is realized that the performance of

ECMS is closely tied to the equivalent factor. The next question on how to select an appropriate equivalent factor remains a key issue for ECMS. Therefore, different methods are proposed to adjust the equivalent factor online and split the power on the basis of ECMS, for example A-ECMS. Next, the discussion continues for the predictive EMSs, whereby the main idea is to optimize the power split based on the predicted velocity over a certain horizon. The future power demand over the horizon is calculated via the traffic information received through ITS and GPS. As the intelligent transportation system technologies are increasingly utilized in traffic management systems, useful information of the preceding vehicle through communication channels among the vehicles lead to an implementation of predictive control that distributes the power by maximizing the fuel economy over a certain time window. Thus, the driving cycle prediction is significant for predictive EMSs. As a common solution method, model predictive control (MPC), which depends on the accuracy of a vehicle model for prediction, can be implementable in predictive control for HEVs. Learning-based EMSs mainly update the control parameters by training data to improve the adaptability to the changing driving conditions.

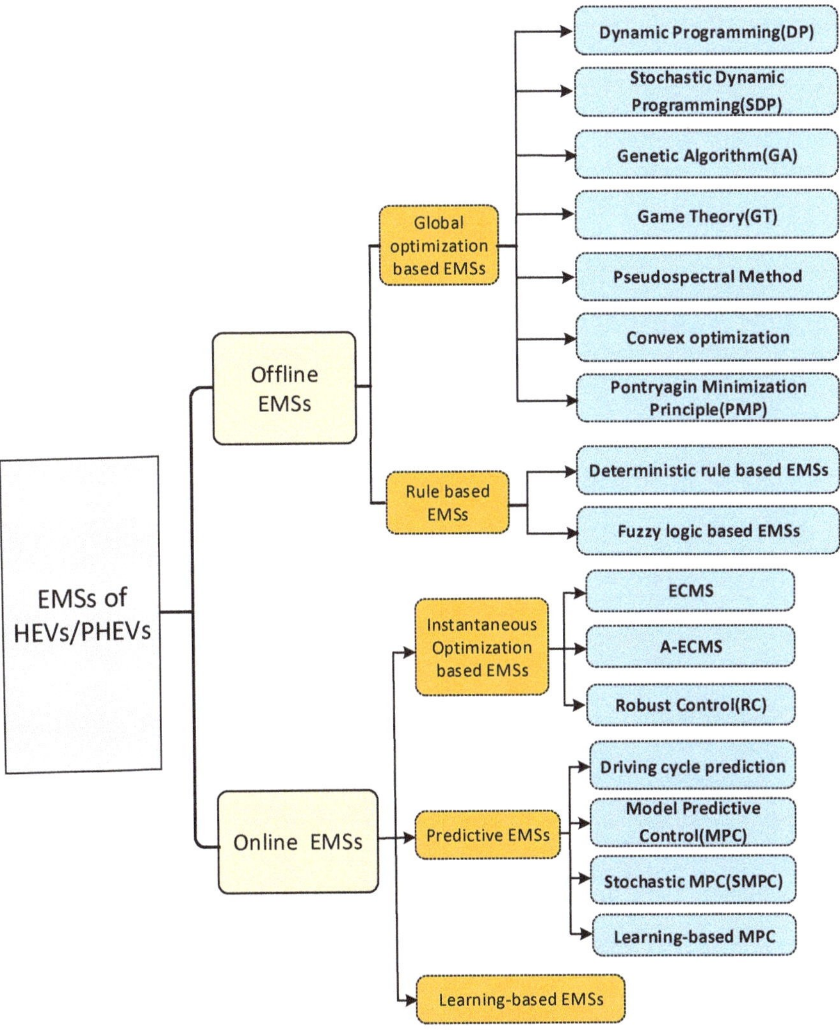

Figure 4. The proposed categorization of the energy management strategies.

4. Offline EMSs

4.1. Global Optimization-Based EMSs

These types of methods are non-causal and seek global optimal solutions since they need a prior knowledge of the typical driving cycle. Because of the non-casual solution, they cannot directly be employed in real-time problems; however, non-causal optimal solutions can be obtained offline under a given driving cycle, which can provide a benchmark for other algorithms or modified online EMSs. Thus, as a benchmark, these methods can be adopted to obtain globally optimal results under a specific driving cycle. The commonly methods, such as dynamic programming (DP), stochastic dynamic programming (SDP), genetic algorithm (GA), game theory (GT), robust control (RC), pseudospectral method, and convex optimization, are illustrated and compared in this section. To clearly illustrate the pros and cons of each approach, a comparison of different approaches is shown in Table 1. The "computational complexity" requires low computational burden to score well since this is desirable for fast operation and efficiency. The "adaptability" refers to the flexibility of the EMSs adapted in different driving cycles. It scores well when the control parameters are easy to adjust to different driving cycles for fuel economy. The SDP can provide the best adaptability in comparison with other methods. The "priori knowledge of driving cycle" denotes the amount of driving future information required for calibration and formulation. For these methods, the DP requires the most a priori knowledge of the future information of the driving cycle and obtains the best fuel economy.

Table 1. Comparison of different approaches.

Approaches	Main Advantages	Main Disadvantages	Literature
DP	• achieves global optimal results • benchmark for other EMSs	• less adaptability to changeable driving cycles • highest computational complexity (3-level) • prior knowledge of entire driving cycle	[9–22]
SDP	• more adaptability • achieves near-optimal fuel economy	• highest computational complexity (3-level) • requires driving cycle database	[23–28]
GA	• global optimality • good global search performance	• higher computational complexity (2-level) • less adaptability	[29–33]
GT	• trade off among conflicting objectives • consider driver behaviors in EMSs	• highest computational complexity (3-level) • poor adaptability	[34–45]
Pseudospectral method	• global optimality • more accurate numerical computation	• higher computational complexity (2-level) • requires analytic expressions for vehicle models	[46–49]
Convex optimization	• fast computation • easy to implement	• requires convex models • limited applications	[50–54]
PMP	• achieve near-optimal results • lower computational burden	• complex mathematical models • require co-state estimation	[55–68]

Note: The computational complexity of other algorithms refers to the computation time compared to dynamic programming (DP). The smaller of the level represents less computation burden compared to DP.

4.1.1. Dynamic Programming (DP)

Dynamic programming, as an offline optimization approach, can realize a global optimal solution for a given driving cycle; however, it cannot directly be used in a real vehicle EMS because it is impossible to know the future driving conditions (speed, road slope as well as traffic dynamics). DP also suffers from considerable computing time for solving the optimal problem of the backward duration of the trip from the future state to find the initial control input in a feasible region. Especially, the computation burden increases as the dimension of the system states raise. However, as a benchmark, it can be used to determine the operating conditions that yield a globally optimal fuel consumption, which is then further used to evaluate the performance of other energy management algorithms and extract some heuristic rules. Moreover, it can be employed to obtain an optimal solution over a prediction horizon for model predictive control, such as in [9].

The basic principle of DP is illustrated as shown in Figure 5. The optimal process is formulated as to find the best cost function from A to F. Firstly, the feasible region is discretized and cast into the grid to calculate all possible paths from A to F. Then, starting from F and proceeding backwards, the best path is computed from F to E at time *t*. Similarly, the global optimal solution is calculated step-by-step starting from E and to an ending at the initial state. The shortest path is A-B-J-H-E-F and the minimum cost is 1.2 + 0.6 + 0.7 + 0.6 + 0.8 = 3.9. The general optimal objective function is defined as follows

$$J = \sum_{k=0}^{N-1} [L(x(k), u(k))] + G(x(N)) = \sum_{k=0}^{N-1} [\dot{m}_{fuel}(k) + \mu \cdot \text{NO}_x(k) + \alpha \cdot PM(k)] + \beta(SOC(N) - SOC_f)^2 \quad (1)$$

where N is the driving cycle time; L is the cost function, including fuel consumption; NO_x is emissions, etc.; and G is the constraint of the state-of-charge (SOC) and gear shifting.

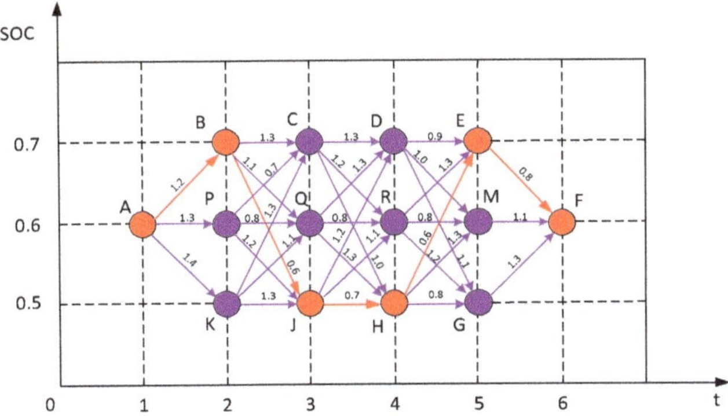

Figure 5. The principle of dynamic programming.

Guzzella et al. [10] put forward an energy management strategy with DP for parallel Hybrid Electric Vehicles. Dynamic programming is used to design an optimal gear shift strategy in [11]. A cost function representing a combination of fuel consumption and emissions over a driving cycle is defined to sustain battery SOC. The optimal gear shifting schedule that can be implemented to a real vehicle is extracted from DP by splitting the power between the engine and the motor. To reduce the computational complexity and implement easily, Patil et al. [12] proposed a novel dynamic programming that is calculated by a backward simulation model for a series hybrid electric vehicle. This approach evaluates state constraints before choosing the optimal paths rather than using penalty functions, which can avoid the requirement interpolation by considering transitions to only the finely discretized nodes of the state space.

A novel dynamic programming, on the basis of machine learning, is proposed in [13]. An EMS for a power-split HEV with an on-line trained neural network is developed to predict traffic congestion and road types. DP is adopted to split the power between engine and motor over a specific driving cycle. The neural network is utilized to predict the traffic conditions and road type with vehicle historical data. This is called an on-line intelligent energy management strategy by combining a machine self-learning algorithm and dynamic programming. In [14], a gear shifting strategy and a power allocation strategy for a hydraulic hybrid vehicle were obtained by dynamic programming, which is utilized in a real-time controller by extracting the control rules. Simulation results show that fuel savings can be improved by 47% over a conventional vehicle. Kutter et al. [15] combined dynamic programming with an equivalent consumption minimization strategy to solve the conflict between global optimality and real-time capability, which is performed by an independent calculation of the main control parameters using dynamic programming, and the power split is optimized online by the ECMS. In [16], a weighted, improved dynamic programming technique is proposed to allocate the power for a hybrid fuel cell vehicle, proving that it converges faster than the traditional dynamic programming methods that suffer from a dimensionality problem. Simulation results reveal that, when compared to the rule-based EMSs, lower costs and a lower hydrogen consumption are achieved using the weighted, improved dynamic programming. To improve the computation efficiency, Zhuang et al. [17] extracted a mode shift map for a multi-mode hybrid powertrain with the DP optimal results using the support vector machine. This can be combined with ECMS to implement real-time control. More works can be viewed in related studies [18–21].

It is well known that DP is a numerical method to solve a dynamic optimal control problem. However, it may lead to optimization inaccuracy when the continuous states are implemented in a discrete framework. To address this issue, Berkel et al. [22] proposed a new implementation method by extending the discrete method by storing the quantization residual after the nearest neighbor, rounding of the continuous state at each node. This can avoid the implementation difficulty of the interpolation method and the inaccuracy of the discrete method.

4.1.2. Stochastic Dynamic Programming (SDP)

Although DP is regarded as a useful tool to obtain a global optimal solution, it is impossible to know exactly the whole driving cycle conditions (speed, road slope, etc.) in advance. To address this issue, stochastic dynamic programming is proposed by researchers. The basic principle of stochastic dynamic programming is that assuming that the sequence of values can be modeled using Markov chain power, the state transition matrix map of the future driver's power demand is generated to estimate the driver's power demand. The power sequence demand is calculated by discretizing the historical driving data at a certain step, and the determination of the current power demand is made in terms of the vehicular speed. The maximum likelihood estimation method is utilized to obtain the state transition probability from the current state to the next one by distributing the total power using discrete dynamic programming. It has been successfully applied as a promising approach for obtaining a quasi-optimal policy that is implementable on-line and in real time, since only historical driving data is needed without a priori knowledge of the driving cycle. However, there are differences between the power demand using the Markov chain model and actual driver power demand, leading to poor adaptability to different driving cycles because of the complexity and randomness of the actual driving cycles. Moreover, the computation process for solving the SDP is still time consuming due to the policy iterations. The future discounted costs are chosen based on the mathematical expediency, leading to difficulties in validation on engineering applications.

The state transition probability is described as in Formula (2).

$$p_{i,j,a} = P\{P_{dem}(k+1) = i | P_{dem}(k) = j, v(k) = a\} \tag{2}$$

where I and j is the power demand at state $k+1$ and k, respectively, and a is the velocity at state k.

In [23], a stochastic dynamic programming algorithm for a power-split hybrid vehicle is proposed, which is performed by establishing drive power sequence demand over different driving cycles based on the Markov process to obtain a state transfer matrix of the driver's power demand. The optimal problem is formulated to maximize the fuel and electricity economy as the objective function in a constraint domain, on the condition of the torque of the engine and motor, as well as the battery charging and discharging power. The energy price was introduced into the objective function. The simulation results are compared with that of the charge-depleting and charge sustainability (CD-CS) strategy in terms of fuel consumption, engine control principle, engine start-stop control, and energy price.

Researchers mainly focus on minimizing the fuel consumption by using SDP. To incorporate the drivability, Opila et al. [24] formulated a stochastic dynamic programming to gain a trade-off between fuel economy and drivability, including engine start–stop and gear shifting time. The driving cycle is modeled by the Markov process considering driver power demand as a stochastic process. The simulation results in FTP and NEDC demonstrate that fuel savings of the proposed EMSs improve by 11%. The influence of engine start–stop and gear shifting time on fuel economy is also investigated and compared with baseline EMSs. In [25], an optimal energy management for a series hybrid electric vehicle is presented on the basis of SDP and considering the fuel consumption and emissions. However, the computational burden is intractable for SDP due to the large state space in this problem. Thus, a new neurodynamic programming (NDP) is proposed to solve the issue. Finally, an SDP controller and NDP controller are compared with a baseline one, indicating that both SDP and NDP can achieve significant fuel economy compared to rule-based EMSs. References [26–28] can be referred to for more information on the subject.

4.1.3. Genetic Algorithm (GA)

The genetic algorithm (GA) in evolutionary computing has become one of the most popular algorithms among modern optimization algorithms due to its good global search performance and low algorithm complexity [29]. As a random search method, the genetic algorithm is performed by global searching to converge to an optimal solution based on the law of biological evolution. These advantages are well suited to optimizing the rules, parameters, or evaluation criteria in EMS for better performance [29]. The optimization problem is solved by simulating biological phenomena, such as genetic variation. GA can be applied in EMSs to obtain global optimal solutions; however, the computational load is heavy, especially for more variables due to the repeated searches, and can be regarded as an offline optimization method, which guides researchers to select the optimal parameters (e.g., engine size and battery size) for an HEV. Zhou et al. [30] obtained the optimal parameters by GA and analyzed the energy management for fuel cell Hybrid Electric Vehicles. Figure 6 shows the basic flow of the genetic algorithm. The main steps to implement the GA are as follows:

(1) Initial population: Select an initial population in a feasible solution domain.
(2) Genetic operation: A new population is generated by the selection, crossover and variation of the initial population to converge to the global optimal solution.
(3) Decide if the population meets the ending criteria, referring to the iterations of the intelligent optimal algorithm.

Piccolo et al. [31] put forward an energy management strategy using the genetic algorithm to implement global optimization, which can be performed by adjusting the control parameters to minimize fuel consumption and emissions. This method can obtain the global optimal solution and yield better robustness; however, the computational complexity is higher than the other EMSs. To improve the optimal performance of a genetic algorithm, Liu et al. [32] proposed a hybrid genetic algorithm for a series hybrid electric vehicle, with faster convergence and better adaptability compared to the traditional GA that performs the global search randomly. The proposed algorithm can acquire fast convergence to a global solution using the quadratic programming algorithm. In addition, the GA is combined with other algorithms to address the energy management optimization problem.

In [33], an energy management strategy is proposed based on fuzzy logic and genetic algorithm optimization. The membership function of a fuzzy logic controller is optimized using the genetic algorithm. The simulation results show that the presented EMSs are clearly capable of improving the fuel economy and reducing the gas emissions as compared to the deterministic rules without adjustment by the GA.

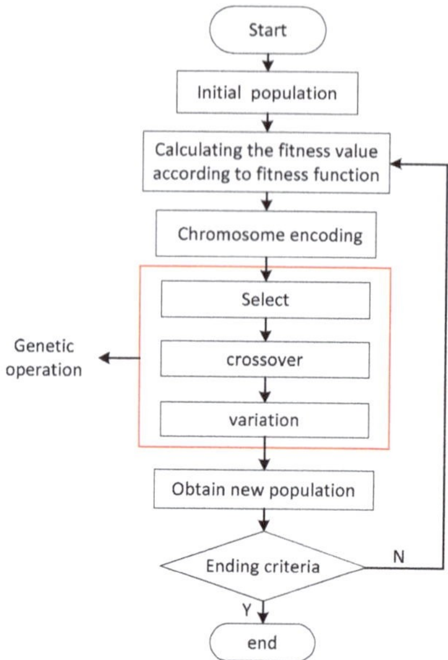

Figure 6. The flow of the genetic algorithm.

4.1.4. Game Theory (GT)

As a branch of operational research, game theory is commonly used in multi-subject optimization problems by taking into account the forecast and actual behavior of individuals in a game. In the 1950s, cooperative game theory enjoyed its peak and non-cooperative game theory began to develop [34]. During this time, a legendary figure, John F. Nash, deserves special mention for his two essays in 1950 [35] and 1951 [36], firstly using rigorous mathematical language and then simple words to accurately define the Nash Equilibrium, which was a significant milestone in game theory history. The basic idea of game theory is to determine, through formal reasoning alone, what strategies the players ought to choose in order to pursue their own interests rationally, and what outcomes will result if they do so [37]. In recent years, game theory-based EMSs, which are sensitive to the variations in vehicle parameters, have been developed.

Gielniak et al. [38] proposed an integrated system approach based on game theory for automotive electrical power and energy management systems. The objective of the players is to maximize their payoff that is a function of vehicle performance and powertrain efficiency. Yin et al. [39] formulated the energy management problem as a non-cooperative current control game. The Nash equilibrium is analytically derived as a balanced solution that compromises the different preferences of the independent devices. Dexireit et al. [40,41] designed a controller for a parallel hybrid electric vehicle using game theory with the objectives of fuel economy and emission. First, the vehicle operating conditions and the powertrain are viewed as two players in a finite-horizon non-cooperative game. A cost function of this game is formed by weighting the fuel consumption, NOx emissions, and the deviation of the battery SOC from the setpoint, as well as the deviation from the vehicle operating conditions. The policy is established as a function of wheel speed, torque, and battery SOC to decide the control mode of the engine, motors, and battery. Compared to traditional EMSs, this control policy is independent of the time and driving cycle. Therefore, it can achieve better performance under different driving cycles. Test results validate that the game theory controller substantially outperforms the baseline controller under NEDC. Xu et al. [42] proposed a game-theoretic energy management strategy with velocity prediction for a hybrid electric vehicle. A recurrent neural network structure was realized to predict the future velocities and Nash equilibrium of game-theoretic energy management, and was implemented through the best response functions. Chen et al. [43] developed a game-theoretic approach for solving the complete vehicle energy management problem of a hybrid heavy-duty truck with a high-voltage battery and an electric refrigerated semi-trailer. The solution concept is based on a two-level single-leader multi-follower game model. The game-theoretic approach presented the optimal performance in the simulation. Chen et al. [44] introduced an adaptive game-theoretic approach for solving the complete vehicle energy management problem of a hybrid heavy-duty truck with a high-voltage battery and an electric refrigerated semi-trailer. The proposed method enhances the game-theoretic approach, such that the strategy is able to adapt to real driving behavior. The fuel reduction results are compared and the adaptive game-theoretic approach shows improved and more robust performance over different drive-cycles compared to the non-adaptive one. A game-theoretic solution concept for solving the complete vehicle energy management (CVEM) of a hybrid heavy-duty truck can be found in [45].

4.1.5. Pseudospectral Method

Pseudospectral method, also known as the discrete variable representation method [46], is a direct numerical algorithm for optimal control problems. In the energy management problem, the optimal control theory is utilized to optimize the energy distribution. The pseudo spectral method can be used as a direct numerical method to obtain the optimal energy distribution. The continuous energy management optimization problem can be solved by discretizing and transforming it into a nonlinear programming problem. Hu et al. [47] proposed a double objective charging optimization strategy for two kinds of lithium-ion batteries, by considering the influence of battery charging time and charging energy loss on HEV energy management. A multi-objective optimal charging control problem was constructed, and then solved by using the Radua pseudospectral method. Zhou et al. [48] utilized the pseudospectral method to solve an HEV energy management problem and optimized the energy management and co-state trajectory simultaneously. The results showed that the computation efficiency of the pseudospectral method is higher than that of DP, while the optimization performance is close to DP. Wu et al. [49] developed a hierarchical EMS with the pseudospectral method for Hybrid Electric Vehicles, which incorporates velocity planning, with a tradeoff between fuel consumption and path tracking accuracy.

4.1.6. Convex Optimization

As an optimization algorithm, convex optimization is utilized for solving convex problems [3], whose objective function and constraints are convex. In convex optimization problems, the results of local optimization and global optimization are consistent, which greatly simplifies the solution process [50]. As compared to other global optimization algorithms, it is easy to obtain optimal solutions with a higher computation efficiency. The optimization of HEV energy management can be regarded as a nonlinear programming problem, which can be transformed into a semi convex problem by using a convex optimization method that offers a simplified calculation process and better optimization effect. Murgovski et al. [51] presented an EMS with convex optimization for a plug-in hybrid electric bus. The influence of battery size, gearshift, and engine on/off on energy management was investigated by transforming these problems into semi convex problems with a convex optimization method. In addition, the optimal results obtained from the convex optimization were compared with dynamic programming. Nafisi et al. [52] considered the influence of the power grid on the energy management of plug-in HEVs, and proposed a two-level optimization method based on convex optimization to reduce the energy loss. However, the disadvantage of convex optimization is that the objective function and inequality constraint must be convex [53], and it yields limited applications. Especially for a parallel HEV, the gearshift strategy should be devised separately, instead of optimizing the gearshift and power split simultaneously, such as in [54].

4.1.7. Pontryagin's Minimum Principle (PMP)

PMP is an analytical optimization method to solve optimal control problems to provide a necessary condition. PMP transforms a global optimization problem into an instantaneous Hamiltonian optimization problem, derived from DP through a variational approach. Thus, an optimal solution can be obtained by minimizing the instantaneous Hamiltonian that includes fuel consumption and battery SOC. Similar to ECMS, an optimal co-state is a key factor that needs to be determined appropriately. A shooting method is commonly adopted to calculate the optimal co-state λ, for example in [55]. More works can be found in [56–59]. The form of instantaneous optimization shown in PMP makes it possible to implement real-time control. The basic principle is generally formulated as Equations (3)–(7). It is obvious that a differentiable objective function is required for deriving the optimal solution; however, it is difficult to obtain a continuous Hamiltonian for Hybrid Electric Vehicles, especially for a parallel HEV. To this end, a simplified PMP is proposed in [60] to avoid the adaptation mechanism of the co-state for real-time applications. The main drawback of the control concept is that the PMP-based EMS will not guarantee optimality if no information regarding the future driving condition is provided [61].

The augmented cost function for a general problem can be given as Equation (3):

$$Q = \varphi(x_f, t_f, v) + \int_{t_0}^{t_f} L(x, u, t) dt \tag{3}$$

where $L(x, u, t)$ is the cost function and $\varphi(x_f, t_f, v)$ is presented as Equation (4):

$$\varphi(x_f, t_f, v) = \varphi(x_f, t_f) + v^T \psi(x_f, t_f) \tag{4}$$

The state dynamic is described as Equation (5) and $x(t_0) = x_0$ is also satisfied:

$$\dot{x} = g(x, u, t) \tag{5}$$

Thus, the Hamiltonian function can be formulated as Equation (6):

$$H(x, u, t) = L(x, u, t) + \lambda^T g(x, u, t) \tag{6}$$

where λ^T is the co-state. Given the problem settings in Equations (3)–(6) and assuming the problem is convex, the necessary condition that minimize Equation (3) are given as Equation (7).

$$\begin{cases} \dot{\lambda} = -H_x^T \text{ and } \varphi_{x_f} = \lambda^T(t_f) \\ \dot{x} = g(x, u, t) \text{ and } x(t_0) = x_0 \\ H_u = 0 \\ (\varphi_t + H)\big|_{t_f} = 0 \\ \psi(x_f, t_f) = 0 \end{cases} \tag{7}$$

In [62], three kinds of EMSs, namely DP, PMP, and ECMS, are conducted and compared. By comparing ECMS and PMP, it is found that they are similar in terms of equivalent factor and co-state. The author suggested that the ECMS becomes the implementation of the optimal solution of PMP, which also obtains results close to the DP optimal solution, with an improvement in comparison to the traditional ECMS. To adjust the control parameters, adaptive PMP is proposed using the total trip length and the average cycle speed in [63]. The results demonstrate that improvement in fuel consumption can reach 20% compared to an on-board controller. Kim et al. [64] proposed an EMS-based on PMP considering the battery efficiency of the plug-in Hybrid Electric Vehicles (PHEVs) and derived an additional condition for the inequality state constraints. The results prove that the PMP can achieve similar performance to the global optimal results obtained by DP. In [65], PMP is introduced by solving the Hamiltonian function to find the battery current command, and the simulated annealing algorithm is used to calculate the engine-on power and the maximum current coefficient. The simulation results demonstrate that the proposed algorithm can reduce the fuel consumption as compared with charge-depleting and charge-sustaining EMS. Although PMP is utilized to solve the optimal control problems for the energy management by simplifying the engine fuel map, engine on/off control is not considered. To address this issue, the approximate PMP is proposed in [66]. A piecewise linear approximation to fit the fuel rate map for a plug-in HEV has been developed based on PMP to avoid distortion in the fuel map. The results show that the engine state switching frequency is reduced by 43.40% with engine on/off optimal EMSs.

Previous works mainly focus on the determination of an optimal co-state with future driving cycles or a prior knowledge of the driving cycles, such as [67]. Kim et al. [61] presented an adaptive energy management strategy with PMP by analyzing the past driving patterns and updating the control parameters with an assumption that vehicles operate under repeated driving conditions (e.g., commuting buses). In real conditions, the driving cycle is affected by numerous factors, for example, driver behaviors and traffic conditions. To this end, Park et al. [68] investigated a PMP-based energy management strategy for plug-in HEVs incorporating the driver's characteristics to improve the adaptability of PMP.

4.2. Rule-Based EMSs

Generally, rule-based EMSs can be performed by predefining the logical rules according to the HEV system characteristics and operation mode. The rules are determined based on the battery SOC, driver power demand, and vehicle velocity through an "if–then" structure. Given these rules, the power split can be performed to meet the driver power demand and maintain the SOC at a certain range. Instead of a prior knowledge of the driving cycle, this method mainly depends on logical rules and local constraints. The control parameters cannot be tuned due to a lack of future information on the driving cycle, making it less adaptable to varying driving conditions. The typical methods, like deterministic rule-based control and fuzzy rule-based methods, are introduced in the following sequel.

4.2.1. Deterministic Rule-Based EMSs

In this method, based on the engine map and motor efficiency map, a series of logical rules are predefined to split the power between the engine and motor, considering the efficiency of the motor and engine and battery SOC simultaneously. The control rules are easy to implement on-line by a look-up table due to its simplicity. Thus, it is widely utilized in the commercial application of vehicle controllers. The rules are commonly devised based on specific driving cycles (e.g., ECE). However, the varying traffic conditions make it less adaptable to different driving cycles. Peng et al. [69] present a rule-based EMSs for a parallel hybrid electric vehicle. Thus, conventional rule-based power management is not optimal for real driving cycles since a unique approach to design the logical rules does not exist. In most cases, this depends on the engineer's experiences and driving cycles. In the following subsections, rule-based strategies, including on/off and power follower EMSs, are discussed in more detail.

(1) on/off EMSs

As for this strategy, a battery SOC is always maintained between its preset minimum and maximum thresholds by turning the engine on/off. The basic control rules are as follows:

① The engine starts to work at the highest efficiency region or sub-optimal emissions area and supplies constant power when the battery SOC is lower than the preset minimum threshold.

A portion of the engine power is provided to the motor to satisfy the power requirement while the rest is used on charging the battery.

② The engine is shut off when the battery SOC increases to the pre-set maximum threshold and only the battery provides the driving power.

In some cases, a surge of instantaneous power may be supplied from the battery with this EMSs, which makes battery charge and discharge period shorter and the engine start–stop frequently. The main advantage is that the average efficiency of the engine is higher and the battery charging and discharging period became shorter, but leads to negative effects, such as more power loss due to frequent engine start–stop, less total energy efficiency, and shorter battery life [70]. Although this method is simple relative to the optimal EMSs, it cannot satisfy the vehicle power demand at all operating conditions.

(2) The power follower EMSs

Based on the battery SOC and vehicle load, the output engine power as well as the moment to start or shut off the engine are determined to satisfy the driver power demand. The control rules are as follows:

① If the power demand is less than the maximum engine power at its operating speed, the operation point is adjusted to work at the minimum output power line.

② If the battery SOC is higher than the preset minimum value and lower than maximum value while driver power demand is less than the battery capacity and greater than the maximum engine power at the operating speed, the engine operates at the maximum output power line and the rest of the power demand is supplied by the battery.

③ If only the battery SOC is higher than the preset maximum value and able to satisfy the power demand, the engine should be shut off.

The main advantage of this strategy is that it can reduce the frequency of battery charging and discharging and lower the system energy loss to extend the battery life. This method yields better adjustability for engine output power to the power demand, but the engine operation region becomes wider to lower the overall efficiency.

The rule-based EMSs is easy to implement on-line; however, it is not optimal and cannot guarantee the optimality for different driving cycles. It is also not capable of adjusting the control parameters to achieve the best fuel economy due to the complexity of the driving conditions.

4.2.2. Fuzzy Logic-Based EMSs

Fuzzy logic control theory is composed of fuzzy set theory and fuzzy logic. The former is an extension of TRUE and FALSE (1 and 0) set theory and the latter is an extension of conventional logic in how the system determines the output [71]. Fuzzy relations depend largely on the similarity or the degree of similarity between data sets, and fuzzy reasoning is represented by the IF–THEN format, giving birth to some popular reasoning approaches, for example, the Mamdani method [72] and Takagi–Sugeno method [73]. Fuzzy logic-based EMSs have been conducted throughout the years in the literature [74–77]. Fuzzy logic-based EMSs aim to split the power with fuzzy rules. In this method, the fuzzy logic rules are usually developed according to the driver power demand and SOC. A fuzzy logic controller consists of a set of linguistic rules and each of them includes one antecedent and two consequents. Looking into a hybrid system as a nonlinear and time-varying plant, fuzzy logic controllers are adjustable to implement in real-time with sub-optimal control by a set of fuzzy logical rules. Moreover, it is important to devise a membership function in optimizing the power split. Thus, GA is adopted to optimize the membership function in the reference [78]. Some other forms of modified fuzzy logic-based EMSs can be referred to in [79–81].

In [82], a fuzzy logic controller (FLC) for parallel Hybrid Electric Vehicles is designed. In [83], a multi-input fuzzy logic controller for a power-split hybrid vehicle is presented and compared to rule-based EMSs in terms of fuel economy and emissions. Given the desired driver torque, vehicle speed, and battery SOC, the power is distributed using the FLC method. This method achieves a better fuel economy with good adjustability compared to the conventional rule-based EMSs. Lee et al. [84] presented a fuzzy logic-based energy management strategy to minimize the NO_x emissions while meeting the driver power demand. The proposed fuzzy logic controller uses an electrical motor speed as well as an acceleration pedal stroke as the control inputs. It is claimed that the proposed fuzzy logic controller could reduce about 20% of the NO_x emissions compared with the conventional vehicle. However, the main challenge of this method is that it cannot guarantee the SOC charge-sustainability of the battery. To address this problem, Lee et al. [85] proposed a more sophisticated fuzzy logic controller that includes a power balance controller and a driver's intention predictor for the energy management. Baumann et al. [86] developed an inclusive fuzzy logic controller based on road load estimation to compensate for the difference between the actual engine torque and the required torque. To enhance the adaptability of the fuzzy-based EMS, Tian et al. [87] presented an EMS for a plug-in hybrid electric bus using adaptive fuzzy logic-based with an optimal SOC reference generated by a neural network and followed by a fuzzy logic controller.

In principle, the fuzzy rule-based EMSs can be utilized to adjust the control parameters to a limited extend by predefining a set of fuzzy rules. However, this approach yields less adaptability due to the difficulty in selecting a proper membership function based on different inputs.

5. Online EMSs

Online EMSs are causal and local optimization-based since they generally do not require a priori knowledge of the whole driving cycle. They can be implemented in real-time with a limited computational burden by converting the global optimization problem of off-line EMSs into an instantaneous optimization problem. Due to less computational effort, on-line EMSs yields the potential of being implemented in real-time control problems. Three categories are included, namely instantaneous optimization-based EMSs, predictive EMSs, and learning-based EMSs. The instantaneous optimization-based EMSs determines the power split with optimal algorithm utilizing the current driving cycle information while the predictive EMSs mainly employ future information to optimize the power split. Furthermore, the instantaneous optimization EMSs mainly focus on determining the optimal power split by minimizing the performance indexes (e.g., fuel economy, emissions, and drivability) at each instant. In the following subsection, these EMSs are extensively reviewed and important headlines are highlighted.

5.1. Instantaneous Optimization-Based EMSs

This kind of approach is to optimize the power split by minimizing the instantaneous fuel consumption and other performances (e.g., emissions and drivability) at each instant. These EMSs can achieve the best performance at each instant without a priori knowledge of the driving cycle and it is easy to implement in real-time. Instead of predefining the logical rules, instantaneous optimization EMSs mainly focus on optimization and implementation on-line, resulting in better fuel economy and adjustability compared to simple rule-based EMSs. However, only local optimal results can be obtained instead of global optimization as is possible in offline EMSs.

Due to its reasonable computation burden and no requirements of previewed knowledge, these are capable of being applied to a real-time controller and achieving approximate optimal results in comparison with DP. In recent years, many researchers focus on instantaneous optimization EMSs, including equivalent consumption minimization strategies (ECMS), adaptive-ECMS, Pontryagin's minimum principle (PMP), and robust control. In the following section, these are introduced and discussed in more detail.

3.1.1. Equivalent Consumption Minimization Strategy (ECMS)

The main idea of ECMS is that the power is distributed by minimizing the instantaneous equivalent fuel consumption at each instant by converting the electricity consumption into the equivalent fuel consumption. In contrast to other EMSs, the control variable in ECMS is the equivalent factor (EF), which is defined as the relation between the energy consumption of the secondary power source and power requirement. The equivalent factor plays a significant role in improving the fuel economy. Thus, selecting a suitable equivalent factor according to different driving cycles is a key issue. For this method, it is easy to implement for real-time control, achieving sub-optimal results without prior knowledge of the driving cycle. The standard ECMS generally adopts a constant optimal EF obtained from an iterative method; however, it cannot adapt to the varying driving conditions. Thus, other forms of ECMSs are proposed, such as adaptive ECMS [88,89], telemetry ECMS [90], predictive ECMS [91], ANFIS-based ECMS [92], artificial neural network-enhanced ECMS [93], and a driving-style based ECMS [94]. Since fuel consumption is the main design objective, two key issues need to be considered for ECMS implementation. One is the drivability, in that the optimal torque usually jumps frequently at each instant without incorporating engine or motor response time, which may lead to oscillation of the powertrain. Another is the computation efficiency, in that it cannot directly be utilized in a real vehicle controller although yielding a lower computational burden compared to DP. Instead, it can be implemented online in a look-up table. Additionally, it is more challenging to adjust the EF in real driving cycles.

The basic principle of an ECMS is illustrated in Figure 7, which is depicted for a parallel HEV. The energy flow when the battery is discharging is shown in Figure 7a. In this state, the electric motor supplies mechanical power. The route of the red dots is concerned with the return of the used instantaneous electrical energy in the future, which means that the used electricity is converted into equivalent consumption. The energy flow when the battery charging is shown in Figure 7b. In this state, the engine supplies the mechanical power. The mechanical energy is received and converted into electrical energy by the motor, and then is stored in the battery. The red dotted route is related to the use of this electrical energy for generating mechanical power in the future. This part of the mechanical energy will not have to be generated by the engine, which is considered as fuel-saving. The power split is then determined by minimizing the equivalent fuel consumption.

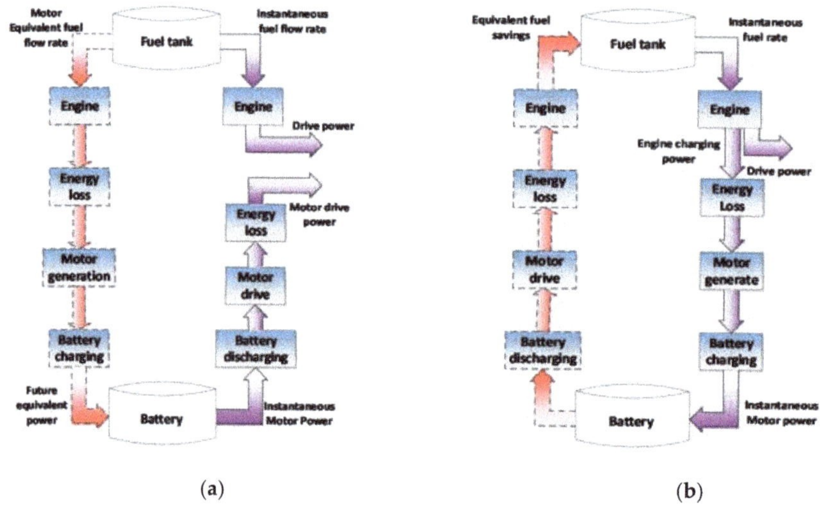

Figure 7. The energy path during charge and discharge in a parallel hybrid electric vehicle (HEV): (**a**) battery discharging; (**b**) battery charging [95].

The equivalent fuel consumption rate is given as Equation (8).

$$\dot{m}_{eqv} = \dot{m}_f + \dot{m}_e = \dot{m}_f + \frac{s}{Q_{lhv}} P_e \tag{8}$$

where $\dot{m}_f$ is the engine instantaneous fuel consumption; s is the equivalent factor; P_e is the motor power, where the value is negative when braking and the value is positive when driving; and Q_{lhv} is the fuel lower heating value.

As an instantaneous optimization method, an equivalent consumption minimization strategy (ECMS) is firstly introduced by [95] and an instantaneous optimization algorithm is supplemented (see [96–98]). Nüesch et al. [99] proposed an approach that minimizes the fuel consumption using ECMS for a diesel hybrid electric vehicle while tracking a given reference trajectory for both battery SOC and NO$_x$ emissions adjusted by a PI controller. By hardware-in-the-loop (HIL) experiments, the proposed method not only improves the fuel economy but also implements feedback regulation of the SOC and NO$_x$ emissions. Gao et al. [100] introduced an ECMS for series Hybrid Electric Vehicles in comparison with on/off EMSs and power follower strategy. The on/off EMS mainly optimizes the operation region of the engine while the power follower EMS optimizes the operation region of the battery charging and discharging. The main objective of the ECMS is to implement system optimization in terms of battery and engine efficiency, which can achieve better fuel economy.

To ensure battery SOC charge-sustainability and keep the EMSs simple to implement, Skugor et al. [101] proposed an energy management strategy for a power-split hybrid electric vehicle, integrating rule-based EMS and ECMS to optimize the fuel economy. One-dimensional directional search-based and two-dimensional directional search-based instantaneous ECMSs were analyzed, in which the former was performed in two variants, corresponding to the engine maximum torque target line and constant-power target line, while the latter gave special attention to the offline optimization of the target region size. The simulation results indicate that the optimization solution of the rule-based + ECMS is close to that of dynamic programming under an HWFET (Highway Fuel Economy Test) cycle.

In [102], ECMS is deployed to solve the optimization problem for a hybrid system of fuel cells and batteries, obtaining suitable energy management of the hybrid system by minimizing the hydrogen consumption. In [103], Park et al. applied ECMS for the power distribution between the engine and the motor of Hybrid Electric Vehicles. To find the optimal equivalent factor for a certain driving cycle, a parameter optimization method based on a model applying a genetic algorithm was studied. The results represent a promising improvement in fuel economy and the optimal equivalent factor is considered as a good initial value for vehicle calibration.

5.1.2. Adaptive Equivalent Consumption Minimization Strategy (A-ECMS)

As explained previously, the performance of an ECMS for real-time control is closely related to the equivalent factor. Therefore, how to tune the equivalent factor is essential to improve the performance of energy management strategies. The equivalent factor is generally decided by the future power requirement and the current SOC as well. To achieve this goal, A-ECMS is proposed by refreshing the control parameters according to the future power demand and current one. The basic principle of an A-ECMS is that the equivalent factor is regulated accordingly by the current SOC, predicting the velocity and driver's power demand in real-time, keeping the SOC in a certain range and minimizing the fuel consumption. The PI adaptor is commonly adopted in [104]; however, the PI parameters needs to be adjusted appropriately. Thus, a fuzzy logic-based PI adaptor is proposed in [105] to adapt to the changing driving conditions. Furthermore, incorporating the uncertainty of the driving cycles and future information from ITS are utilized in adjusting the EF. The typical structure of an A-ECMS with ITS is illustrated in Figure 8. With the GPS/ITS and feedback information, the future power demand is estimated over a certain horizon. The equivalent factor is estimated and tuned online to maintain the prescribed SOC by the adaptor. The A-ECMS can be implemented in real-time control without a priori knowledge of the driving cycle.

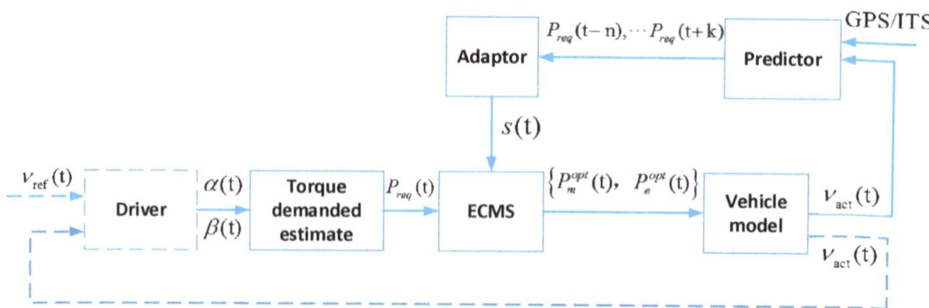

Figure 8. Structure of the adaptive equivalent consumption minimization strategy (A-ECMS). Note that $s(t)$ is the equivalent factor, and $\alpha(t)$ and $\beta(t)$ is the acceleration and deceleration pedal, respectively.

In [106], Musardo et al. put forward an A-ECMS by estimating the equivalent factor based upon different road loads to update the control parameters, which minimize fuel consumption and maintain the battery SOC at a certain range. Sezer et al. [107] introduced a novel ECMS for series Hybrid Electric Vehicles considering the efficiency of the engine, battery, and generator to gain the combined fuel consumption and emissions cost map, which optimizes the engine-generator set and ensures battery charge sustainability. Sciarretta et al. [97] proposed a new approach for redefining an equivalent factor according to the coefficient of charging and discharging of the battery, which presents great robustness and reduces the fuel consumption by 30% in comparison to the traditional approaches.

Other approaches to estimate the equivalent factor by combining the ECMS with other optimization algorithms are proposed in [108–110]. In [108], Zhang et al. proposed two kinds of methods, such as DP and backward ECMS to estimate the equivalence factor and adopted a backward ECMS sweeping over the estimated future velocity and exacting the future 3-D terrain information to adjust the parameters of the ECMS. In [109], Kim et al. developed a method based on Pontryagin's minimum principle (PMP) to calculate the optimal equivalence factor. He et al. [110] presented an energy management strategy that combines rule-based strategy and ECMS for fuel cell vehicles to reduce the hydrogen consumption.

The development of Intelligent Transportation Systems offers a promising way to predict the velocity and estimate the equivalent factor for an ECMS. The velocity and position of each vehicle as well as the traffic information in front of a target vehicle can be provided through vehicle-to-vehicle communication (V2V) and vehicle-to-infrastructure communication (V2I) with a DSRC protocol to make it possible to adjust the equivalent factor according to the updated predicted velocity. Serrao et al. [111] indicate that PMP can be shown as the underlying optimization principle for ECMS, but online implementation is unfeasible due to its iterations in finding the initial value of the dynamic equivalent factor for charge-sustaining (CS) operation. Mohd et al. [112] proposed a velocity prediction method combining a car-following model and cell-transmission model (CTM) based on Inter-Vehicle Communication (IVC) and Vehicle-Infrastructure Integration (VII). A computationally efficient CS HEV powertrain optimization strategy was then analytically derived based on the PMP and CS condition to adjust the co-state according to the predicted velocity to evaluate the performance of the proposed strategy. Zhang et al. [113] proposed an adaptive ECMS on the basis of velocity prediction through V2V and V2I communications to improve the robustness of the ECMS and maintain a good SOC charge sustainability. To incorporate the future information into the EF adaptor, Sun et al. [114] developed an adaptive-ECMS to improve the fuel economy by updating the EF periodically, with the predicted velocity obtained from neural networks.

On the basis of recognizing the driving pattern, the equivalent factor can also be assessed. The driving cycle pattern can be identified by the previous driving pattern. Thus, the equivalent factor is adjusted adaptively based on driving pattern recognition. In [115,116], Gurkaynak et al. put forward an energy management strategy using ECMS for a parallel hybrid electric vehicle to obtain sub-optimal results. The optimal performance is related to the vehicle model and equivalent factor, which is updated by driving cycle identification using a neural network algorithm. Simulation results demonstrate that ECMS can obtain approximate optimization results in comparison to DP. In [98], a novel method is developed to calculate the equivalent factor determined by the change rate of the SOC in ECMS without a priori knowledge of the entire driving cycle. The robustness and adjustability are demonstrated through different driving cycles compared with that of estimating the equivalent factor under specific driving cycles. To catch energy-saving opportunities, Rezaei et al. [117] proposed a novel energy management based on an adaptive equivalent consumption minimization strategy for series Hybrid Electric Vehicles by determining a range for the optimal EF of ECMS. Most of the literature ignore the vehicle lateral dynamic in devising the EMS; to this end, Li et al. [118] developed an energy management strategy considering the vehicle lateral dynamic with an adaptive ECMS.

5.1.3. Robust Control

Robust control is a branch of control theory whose approach to controller design explicitly deals with uncertainty. The robust control method is utilized in designs for them to function properly, provided that there are uncertain parameters and that disturbances exist within some forms (parametric or structural) [119]. As for this method, the energy management is formulated as an optimal problem, represented by the state–space equation. The HEV model generally needs to be simplified to devise a closed-loop system, which can be stable and of strong anti-jamming ability by designing state–feedback gain matrices, as well as achieving sub-optimal results with higher computational complexity.

In [120], to overcome the presence of parameter uncertainty in the optimal problem, an optimal-heuristic EMS is presented. The solution is real-time implementable since it is based on a discrete-time description of the system and the optimal solution can be analytically found. In [121], a robust energy management strategy for a fuel cell hybrid vehicle is proposed to solve the sensitivity issue regarding driving cycle uncertainty. This approach improves the robustness of the energy management strategy against driving cycle variations while minimizing the H_2 consumption. Pisu et al. [122] discussed three kinds of energy management approaches for a parallel hybrid electric vehicle, namely rule-based EMSs, an adaptive equivalent consumption minimization strategy (A-ECMS), and H_∞ control, compared with DP that presents the disadvantages of computational complexity and requiring a priori knowledge of the driving cycle. The rule-based EMSs that is of lower computational burden is easy to implement by pre-defining a series of control rules, dependent on the brake and accelerator pedal angle, battery SOC, and the torque demand. The A-ECMS is implemented by establishing an optimization cost function, which takes into account electricity consumption, fuel consumption, and NO_x emissions, and adds a penalty function on an equivalent factor. The state–feedback H_∞ control method aims at minimizing fuel consumption by computing a control gain matrix. Simulation results show that an A-ECMS achieves a similar performance in comparison with DP. For the A-ECMS, the optimal control can be calculated offline and stored in the controller as a look-up table to reduce the computational load, whereas the dynamic characteristics of the components is neglected.

In principle, although the robust control method can provide dynamic optimization to adjust the control parameters, it only can achieve sub-optimal solutions because of its simplification of the models.

To clearly show the pros and cons of each method, a comparison of different approaches is summarized in Table 2.

Table 2. Comparison of different approaches.

Approaches	Main Advantages	Main Disadvantages	Literature
ECMS	• easy to implement • on-line implementation	• less adaptability • obtain local optimal results	[88–103]
A-ECMS	• on-line implementation • more adaptability	• complex EF adaptor • obtain local optimal results	[104–118]
RC	• robustness with uncertainty parameters • more adaptability	• high computational complexity • higher vehicle model complexity	[119–122]

Note: The adaptability refers to the flexibility of the energy management strategies (EMSs) adapted in different driving cycles.

5.2. Predictive EMSs

The main purpose of predictive EMSs is to optimize the power split utilizing predictive information related to the uncertainty and disturbance of a driving cycle. This strategy requires future driving cycle information (e.g., future velocity) that can be predicted with available information (e.g., road conditions and traffic conditions). Thus, to a large extent, the performance of this strategy depends on the power reference provided at each prediction horizon. In other words, it is mainly based on the predicted velocity on a flat road without considering road slope. Therefore, it is significant to predict the vehicular velocity accurately in implementing such approach. Generally, it is impossible to predict the whole cycle accurately. Alternatively, it should be partially predicted if only a small part of the upcoming trip is considered [123]. In addition, the factors affecting the prediction accuracy include driver behavior, road condition, dynamic traffic conditions, preceding vehicles, etc. Inaccurate prediction may worsen an EMS's performance. Therefore, in order to improve the prediction accuracy, more surrounding information needs to be effectively considered. The optimal control input is obtainable by minimizing the performance indexes (e.g., fuel consumption and emissions) over a certain horizon, and this approach is in real-time implementation to adapt to the changing driving conditions. In view of this, researchers increasingly adopt predictive EMSs to improve the fuel economy. Model predictive control (MPC) is commonly employed to implement predictive energy management. Apart from this approach, predictive ECMS [124] can also be performed.

The general cost function of the predictive EMSs is commonly formulated as Equation (9) and the constraint is given as Equation (10). The optimal problem can be solved by minimizing Equation (9) under the constraint Equation (10).

$$J = \int_{k}^{k+H_p} [(\dot{m}_f(u(t))^2 + \lambda F(t))]dt \tag{9}$$

$$\begin{cases} SOC_{min} \leq SOC \leq SOC_{max} \\ w_{e\_min} \leq w_e \leq w_{e\_max} \\ w_{m\_min} \leq w_m \leq w_{m\_max} \\ P_{m\_min} \leq P_m \leq P_{m\_max} \\ P_{e\_min} \leq P_e \leq P_{e\_max} \end{cases} \tag{10}$$

where J is the cost function; H_p is the prediction horizon; $\dot{m}_f(u(t))$ is the fuel consumption; $u(t)$ is the control input (e.g., engine torque, motor torque, and gearshift); $F(t)$ is the other performance factors, such as emissions and drivability, etc.; λ is the penalty coefficient; $SOC_{min}(t)$ and $SOC_{max}(t)$ are the minimum SOC and maximum SOC, respectively, with SOC being the state of charge; $w_{e\_min}(t)$ and $w_{e\_max}(t)$ are the minimum and maximum speed of the engine; $w_{m\_min}(t)$ and $w_{m\_max}(t)$ are the minimum and maximum speed of the motor, respectively. $P_{m\_min}(t)$ and $P_{m\_max}(t)$ are the minimum and maximum power of the motor; $P_{e\_min}(t)$ and $P_{e\_max}(t)$ are the minimum and maximum power of the engine; $P_m(t)$ and $P_e(t)$ are the power of the motor and engine, respectively; and $w_m(t)$ and $w_e(t)$ are the speed of the motor and engine, respectively.

In the following subsection, typical prediction techniques as well as predictive EMSs are elaborated.

5.2.1. The Driving Cycle Prediction Approach

It is important to predict the driving cycle for EMSs, especially for predictive EMSs. The main challenge of EMSs is that the power split is conducted under a given standard driving cycle, which cannot achieve the best fuel economy due to the uncertainty of the driving cycles. Especially for the city condition, many uncertain factors exist, such as traffic congestion and driving habits. Thus, it is very important to predict the driving cycle for the energy management of HEVs. In this section, typical prediction methods are introduced. More predictive techniques can be advised in [125].

A. Driving pattern recognition

The driving database can be obtained by dividing the standard driving cycles into several segments to extract the feature parameters, including velocity, acceleration, and deceleration. The whole driving cycle can then be constructed by comparing the current driving pattern with all the past databases to find a match. At present, this approach is widely used in recognizing driving patterns. However, the identified driving cycle may be different from the actual one due to the complexity and uncertainty of the real driving cycle. For this approach, the fuzzy recognition method as well as artificial neural network are commonly adopted.

Langari et al. [126] proposed an intelligent energy management strategy for a parallel hybrid electric vehicle on the basis of the driving pattern identification. They utilize vehicle static information (e.g., velocity and acceleration) to improve fuel economy in different driving conditions. The cycle characteristic parameters, such as maximum speed, minimum speed, acceleration, and deceleration, are used to recognize the driving cycle. Wu et al. [127] proposed a learning vector quantization (LVQ) algorithm by extracting the driving condition parameters to recognize the driving pattern, which can be integrated into a fuzzy torque distribution controller for improved adaptability. Simulation results demonstrate that this method enhances the fuel economy more effectively than that of without driving cycle recognition. Murphey et al. [128] also extracted the driving characteristic parameters from standard driving cycles by dividing them into several segments and classifying historical data for different roadway types. The collected data can be trained with a neural network (NN) to identify the type of driving cycle. Finally, the current driving cycle can be identified according to the input parameters from the NN. Simulation results show that the EMS with driving identification can significantly improve the fuel economy.

Driving pattern recognition is usually adopted in optimization for a city bus due to its relatively fixed route. Zhu et al. [129] proposed a dynamic optimization method based on driving cycle self-learning in view of a relatively fixed route for a series of hybrid city buses. The velocity and mileage for certain routes are accumulated by an on-board information unit. The database server receives the data through GPRS and extracts the kinematics segment, and then the clustering approach is used to construct the entire driving cycle. Finally, dynamic programming is utilized to optimize the control parameters and load them into a hybrid controller unit. Bender et al. [130] presented an energy management strategy for hybrid hydraulic vehicles based on driving cycle prediction in terms of the repetitive operation characteristics for a city bus. The velocity and acceleration are captured by a GPS and on-board unit. After data processing and filtering, current driving data, including the beginning and ending position of the interval, is extracted to obtain the speed–position profiles as a history database. In the following, the current velocity profile can be predicted by comparing the start and stop position of each new interval with that of the acceleration process included in the history database if the set threshold value is satisfied. Finally, DP was implemented according to the predicted velocity profile to evaluate the effect of prediction error on fuel economy. The results show that fuel savings increased by 5% with the recognition of the driving cycle.

B. Traffic flow modeling

The velocity can also be obtained by modeling a driving cycle approximately with the help of a traffic flow model in the field of transportation. Due to the relationship between vehicle speed and traffic flow, the velocity is estimated using a mathematical model as well as a probability method with historical traffic data (e.g., traffic volume, speed, and occupancy). The traffic flow models (e.g., macroscopic or microscopic models) are utilized to predict the velocity, only reflecting the regular characteristics approximately and neglecting other factors. Furthermore, it is difficult to accurately represent the actual cycle conditions because of the uncertainty of the actual driving conditions.

In [131], a piecewise modeling approach is proposed to obtain an entire driving cycle assuming that velocity and acceleration are kept constant at different intervals. The velocity is also given by analyzing the historical data. Meanwhile, considering the influence of road slope on fuel economy, an energy management strategy based on Multi-Information Integrated Trip modeling is developed. The influence of interval length on fuel economy is analyzed, indicating that a long interval length leads to less computational time and a worse fuel economy using dynamic programming. In [132], an optimal EMS is introduced based on a traffic flow model called the gas-kinetic model for highways. Simulation results show that the gas-kinetic model can reflect the dynamic characteristic of the actual driving conditions and improve the fuel economy under different driving cycles.

C. Driving cycle prediction based on an Intelligent Transportation System (ITS)

An Intelligent Transportation System (ITS) aims to provide innovative services related to traffic management and enables various users to be better informed about traffic conditions and having a safe trip. The ITS does not only offer traffic information for an energy management strategy but also provides a promising way to enhance the road traffic safety via intelligent vehicle technology. One way to do so, the vehicular velocity can be predicted over a certain horizon by accumulating real-time traffic data (e.g., traffic condition, signal phase and timing, and road grade) with Global Position System (GPS), Geographic Information System (GIS), vehicle-to-vehicle communication, an on-board units. The predictive EMSs can be then be implemented considering this future information. The corresponding performances of the EMSs can be highly improved, since multi-source information from ITS, GPS, and GIS could be combined for reducing the uncertainty of future driving conditions to further improve the prediction accuracy [125].

To improve the prediction accuracy, He et al. [133] presented a driving cycle prediction method with real-time traffic data from the communication between a vehicle and infrastructure (V2I) to predict the velocity using neural networks. The predicted velocity is sent to a vehicle's on-board unit to calculate the driver's power demand. The influence of prediction error, penetration rate, and window size on fuel economy are also analyzed. Simulation results show that fuel savings can increase by 14% under the UDDS (Urban Dynamometer Driving Schedule) and the average velocity prediction error with V2I communication is 13.2%. Considering the influence of road terrain on an energy management strategy, Zhang et al. [134] proposed a new strategy for solving the problem of not achieving the best fuel economy with traditional energy management due to a lack of information about the upcoming driving cycle. With future road terrain determined by Geographic Information System (GIS), the optimal results using DP and ECMS in comparison to rule-based EMSs are analyzed for the case of having a terrain preview and no preview. The results show that fuel savings with a terrain preview increase from 1% to 4% and enhance the longevity of the battery on the uphill. Fu et al. [135] proposed a real-time optimal energy management strategy based on driving cycle prediction, utilizing information attainable from Intelligent Transportation Systems (ITS). The effect of prediction error on the optimal results is also analyzed. The results of using model predictive control (MPC) and A-ECMS are compared, respectively, which is based on different prediction errors using standard and actual driving cycles as a prediction benchmark. The results indicate that a small deviation in the final SOC and fuel economy are introduced when the prediction error is small. Thus, it is important to investigate the influence of prediction error on fuel economy due to sensor precision and delay of ITS. Gong et al. [136–138] modeled a driving cycle using real-time traffic data from ITS, GIS, and GPS. Two kinds of models are introduced, utilizing historical driving data and only real-time driving data. The accumulated historical data is classified into urban, highway, and countryside conditions. The characteristic parameters (e.g., maximum acceleration, maximum speed limit, and average waiting time) of each segment are extracted to generate an approximate driving cycle. Simulation results under the two kinds of models were analyzed, indicating that the EMSs with historical data modeling is better than those without them. The fuel economy of the proposed energy management algorithm is better than the rule-based EMS.

D. Driving cycle prediction using artificial intelligence

Machine learning is the science of having computers to act without being explicitly programmed, which can be used to build smart robots (for perception and control), text understanding (web search and anti-spam), computer vision, medical informatics, audio, database mining, and other areas. As a machine learning algorithm, Artificial Neural Network (ANN) is deployed in classification, prediction, pattern recognition, and clustering. The application of ANN to predict driving and handling behaviors [139], city power load [140], and traffic flows [141] have demonstrated its strong capability in predicting nonlinear dynamic behaviors. In [142], three kinds of prediction methods, including exponentially varying, the Markov process, and ANN, are compared. The prediction is performed over each receding horizon and the predicted velocity is utilized for energy optimization of a power-split HEV. The results show that the ANN-based velocity predictor yields the best performance for predictive energy management. In [143], considering the vehicle-to-vehicle communication (V2V) and vehicle-to-infrastructure communication (V2I) information, a Bayesian Network approach is presented to predict the velocity by assuming a stochastic model of the velocity of the preceding vehicle. The results demonstrate that the prediction yields a higher accuracy within a certain horizon.

5.2.2. Model Predictive Control (MPC)

Model predictive control (MPC) describes the development of tractable algorithms for uncertain, stochastic, and constrained systems. As a mathematical method, model predictive control aims to optimize a future system output by calculating the system input trajectory [144]. The main idea of MPC is that the future control output is predicted by an online optimization according to historical information, as well as by future input and output. The principle diagram of an MPC is shown in Figure 9. Upon the error between the reference and the predictive output, the control sequence can be obtained by combining the historical input, historical output, and predictive input. However, it requires a higher computational burden if the vehicle model is complex. It can be used as a real-time energy management strategy when the computational load is decreased.

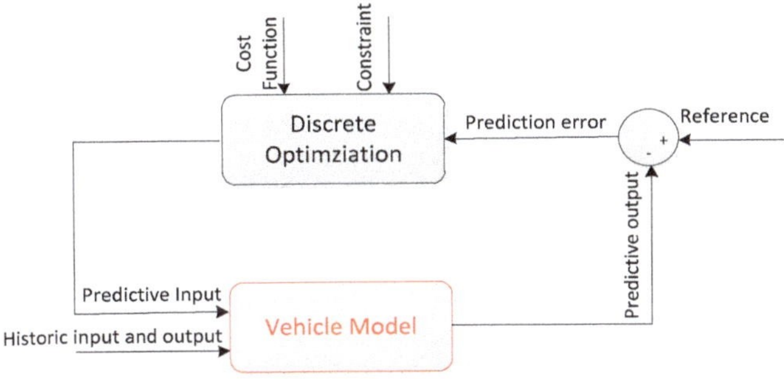

Figure 9. The principle of the model predictive control.

In recent years, MPC has been widely adopted in EMSs. The purpose of MPC-based EMSs is to optimize the power split over a prediction horizon and update the control input, by transforming the global optimization problem into a local optimization for the whole driving condition. Compared with other EMSs, MPC is a rolling horizon optimization method based on system prediction information. The main advantages of an MPC are that it can deal with constraints explicitly, i.e., state variables, input, and output constraints. In addition, the constraints can be formulated as quadratic or nonlinear programming problems, by predicting the system dynamic behavior [145]. The traditional optimization algorithm cannot effectively deal with the impact of the uncertainty of the future working conditions on vehicle performance. In view of this, MPC adopts local optimization, rolling optimization and feedback correction to solve this problem efficiently. To this end, some forms of MPC has been developed in optimizing the power split, such as hybrid MPC [146], distributed MPC [147,148], variable horizon MPC [149], adaptive MPC [150], and tube-based MPC [151].

Generally, linear MPC and nonlinear MPC are commonly formulated in optimizing the power split. Borhan et al. [115] presented an MPC-based energy management strategy for the power split of a hybrid electric vehicle. Energy management is a constrained nonlinear optimal problem. The MPC is utilized to split the power between the engine and motor to regulate the engine operating point at each sample time. Simulation results of the nonlinear MPC show a noticeable improvement in the fuel economy with respect to linear time-varying MPC. In [152], the power management based on nonlinear MPC with an adaptive prediction time horizon is proposed. An MPC-based control algorithm based on load profile prediction is proposed. In this approach, results show that the MPC-based solution yields better performance for total energy consumption in comparison to the conventional approach and strongly depends on the performance of the prediction algorithm. If the predicted velocity could match well with the measured velocity, then the time horizon increases, and vice versa. Thus, it is significant to decrease the computational load to improve the performance of energy management.

In [153], an integrated predictive power management controller is studied. A model-based control approach for a plug-in HEV is proposed to minimize the overall CO_2 emissions. The energy management is formulated as a global optimization problem and cast into a local problem by applying Pontryagin's Minimum Principle. Simulation is conducted to calibrate the control parameters (e.g., environmental factors, vehicle usage condition, and geographic scenarios) and investigate their influence on the fuel economy. Results show that the sensitivity of proposed EMSs on the driving cycle is not significant. To design a torque controller for a parallel hybrid electric vehicle, He et al. [154] developed a torque demand control approach based on MPC. The torque distribution controller has the function of the torque split, torque demand, torque compensation, and torque limit. The engine torque controller is designed based on a nonlinear mean-value model with MPC. The difference between the demand torque and the actual engine torque is compensated for by motor torque because of the nonlinear and lag of the engine torque response. The motor torque controller is developed based on a linear MPC and the transient torque load of a hybrid powertrain is estimated with a PI observer.

5.2.3. Stochastic Model Predictive Control (SMPC)

The common MPC generally utilizes the predicted velocity provided by an exponential estimation or neural network, which have been well studied in [155,156]. This method is based on the standard driving cycles without considering the uncertainty of the driving cycles. Thus, it yields less adaptability to the changing driving cycles. To this end, the stochastic model predictive control (SMPC) is proposed in [157–159], which mainly employs the predicted velocity by a Markov chain and an MPC in optimizing the power split. In this method, the distribution of the driver's future power demand can be obtained by a Markov chain and the MPC is then adopted to obtain the optimal power split. A linear optimization method is utilized for solving the SMPC with a lower computational burden, which can be regarded as an online EMS. Stochastic model predictive control (SMPC) accounts for model uncertainties and disturbances based on their probabilistic description [160].

In [161], a new model predictive control for a series hybrid electric vehicle is proposed based upon the Markov chain process. In this method, the driver power demand is modeled as a Markov chain to represent the driver future power request. All possible distribution of power demand in the next step with all possible Markov states at each time step are generated iteratively. In the following, SMPC is used to split the power between the engine and the motor over a distribution of future power demand given the current one at each sample time. Simulation is conducted under standard driving cycles to obtain a better fuel economy, compared to other deterministic approaches. The advantage of SMPC is that its optimization is feasible in real-time control with respect to SDP. Xie et al. [162] proposed a model predictive energy management for plug-in HEVs based on Pontryagin's Minimum Principle. The design utilizes a Markov chain model to predict the velocity and achieves a higher computation efficiency. Most of the literature works do not consider the battery aging in devising EMSs, especially for a plug-in HEV. For this purpose, Chen et al. [163] developed a nonlinear model predictive control for a power-split HEV considering battery aging. Better battery aging performance is achieved compared with that without considering battery aging while obtaining a similar fuel economy performance.

5.2.4. Learning-Based SMPC

Learning-based SMPC aims to integrate an MPC with machine learning algorithms to improve the performance of the MPC controller in a data-driven way. In contrast to SMPC, which assumes that the driver's power demand can be modeled offline based on the Markov chain, learning-based SMPC can update the Markov chain by online learning, which allows adjusting to variations in the driver behavior with minimal computational effort in real-time control. Therefore, it can dynamically adapt to the changing driving behaviors, such as environmental changes and varying traffic conditions. This approach is more realistic than SMPC in terms of capturing driver actions as well as driving styles. In addition, a traditional MPC generally assumes that the vehicle model parameters are time-invariant; however, actual vehicle models are usually time-varying, such as the vehicle load and battery life that change with time, for construction vehicles or hybrid electric buses. Thus, to devise a robust MPC-based energy management, a new learning-based model predictive control should be developed by adjusting the model parameters adaptively and updating the system model dynamically with online learning, which can capture the dynamic characteristics of the control objectives.

In [164], the driver's power demand Markov chain model is updated by online learning, which can be reconfigured in real-time for accommodating the changes in driver behavior. To capture the driver behaviors, online learning of the Markov chain is introduced to tackle the uncertainty that arises from the environment around the vehicle. By updating the Markov model, the controller can adapt to the changes in driver behavior with less computational effort. Learning-based SMPC is adopted to determine the power split of a series hybrid electric vehicle, where the driver model predicts the future power request that relates to the driving style and driving cycle. Simulation results for standard and real-world driving cycles show that learning-based SMPC improves the performance of classical MPC with the learned pattern of driver behavior.

5.3. Learning-Based EMSs

The learning-based EMSs aim to update the control parameters of EMSs online by interacting with the environment to adapt to the various traffic conditions. They generally employ massive historical and real-time driving-related data to obtain the optimal solution. For this method, the precise model data is not required. Reinforcement learning (RL) and machine learning are commonly used to devise such EMSs. This method can capture the dynamic traffic conditions and yield to potential real-time applications. In [165], the concept of EMSs based on learning is introduced to combine the optimal EMSs with the learning method to enhance the robustness of the EMSs. A predictive energy management strategy for a parallel HEV is designed by means of a reinforcement learning approach. Similar work can also be found in [166–169]. Moreover, an overview of reinforcement learning-based EMSs can be found in [170]. A reinforcement learning system is composed of two items: a learning agent and an environment where the learning agent interacts continuously with the environment. The state of the environment can be observed at each instant for the learning agent. The learning agent then selects an action, which is subsequently inputted to the environment. The reward associated with the transition is calculated and fed back to the learning agent while the environment shifts to a new state because of the action. Together with each state transition, the agent can receive an immediate reward to produce a control policy that represents the current state to the best control action for that state. At each instant, the agent makes a decision based on its control policy. Finally, the optimal policy can lead the learning agent executing the best series of actions to maximize the cumulated reward over time, which can be learned after satisfactory training. The advantage is that the design is a model-free control and provides more adaptability compared other EMSs. However, these require more driving-related data for training.

To obtain a trade-off between optimal fuel savings and real-time performance, Qi et al [171] proposed a reinforcement learning-based real-time EMS for PHEVs by learning the optimal decisions from historical driving cycles. In [172], a deep reinforcement learning-based PHEV EMS is devised to autonomously learn the optimal behaviors from its own historical driving cycles to adapt to the changes in driving conditions. Most of the works in the literature ignore battery health in devising learning-based EMSs; to this end, in [173], a reinforcement learning-based real-time energy management is developed for PHEV by considering the battery health. To further achieve higher computation efficiency, Sun et al. [174] developed a reinforcement-learning-based EMS by combining the ECMS for fuel cell Hybrid Electric Vehicles.

Apart from the learning-based EMSs, the distributed optimization (DO) approach was recently proposed to solve the complete vehicle energy management problem. Romijn et al. [175] proposed a distributed optimization (DO) approach for a hybrid truck with a refrigerated semi-trailer, an air supply system, an alternator, a dc–dc converter, a low-voltage battery, and a climate control system. A dual decomposition is firstly applied to the optimal control problem such that the problem related to each subsystem can be solved separately. Then, an Alternating Direction Method of Multipliers method is used to efficiently solve the optimal control problem for every subsystem in the vehicle. Simulation results show that the fuel consumption can be reduced up to 0.52% by including auxiliaries in the energy management problem, assuming that the auxiliaries are continuously controlled. The computation time is reduced by a factor of 64 up to 1825, compared with solving a centralized convex optimization problem.

6. Conclusion and Future Trends

The EMSs of Hybrid Electric Vehicles have been extensively studied and compared. The offline EMSs aim to minimize fuel consumption globally. Although they cannot be directly implemented in a real vehicle, they provide a benchmark for other energy management strategies and obtaining modified online EMSs. The online EMSs are relatively easy to implement in a real vehicle due to a lower computational burden and no prior knowledge of the whole driving cycle, while achieving similar performance (e.g., fuel economy) as compared to the offline EMSs. The instantaneous optimization-based EMSs are a promising way to compromise real-time implementation and fuel consumption minimization. Driving cycle prediction is important in predictive EMSs. As the ITS technology is increasingly developed, the predictive EMSs are capable of better adjustability and represent a better performance compared to other EMSs. Although different EMSs have been conducted in recent years, offering remarkable solutions, some important future trends need to be further considered.

6.1. The Predictive EMSs Considering Dynamic Traffic Conditions with ITS

The main challenge of current EMSs is that solutions are generally devised under specific driving cycles, which bring about the impossibility to attain optimal results in a real cycle. Despite the global optimization that obtains optimal results in theory, it is hard to implement in real-time vehicle controllers because of computational complexity. Predicting the driving cycle is an effective way for real-time optimization-based energy management strategies considering dynamic traffic conditions and future information. In this design, the real-time traffic data can be dynamically obtained with intelligent transportation system (e.g., vehicle-to-vehicle communication technology).

If the driving cycle can be predicted as accurate as possible by taking into consideration traffic congestion and road slope, the EMSs can be effectively performed. Therefore, incorporating dynamic traffic conditions into EMSs and investigating predictive EMSs based on driving cycle prediction are possible future trends.

6.2. Real-Time EMSs Incorporating Components Response and Accurate Vehicle Models

Currently, most of the EMSs aim to minimize the fuel consumption and emissions to obtain global results. Although these EMSs can provide a benchmark for researchers, it is more challenging to implement in real-time. The vehicle controller not only determines the power split but also is responsible for acquiring data, monitoring the operation state, and diagnosing the fault required for a high real-time performance. This is essential for vehicle prompt response once the control command has been received. The computational complexity of global optimization is acceptable for simulation, whereas it is impossible to update control parameters for real-time application. The adopted numerical optimization methods for simplified vehicle models can reduce the computational complexity, which, on the other hand, become less attractive if the nonlinear characteristic of the vehicle model is considered. Considering the nonlinear characteristics of the vehicle model results in obtaining a higher accuracy for the optimization results.

Rule-based EMSs are commonly performed in real vehicles. Because the other optimization algorithms are hard to implement due to their computational complexity, they utilize simplified vehicle models, which lead to unexpected energy management results in practice. Consequently, how to simplify the vehicle model and reduce the computational complexity to ensure the real-time performance of the optimization algorithm will be an urgent problem that needs to be solved in the near future.

6.3. Multi-Objectives EMSs Incorporating Battery Aging and Drivability

It is well known that a hybrid electric vehicle is a complex and nonlinear system composed of many components: engine, motor, and transmission, which are highly coupled. Multiple performance indexes (e.g., drivability and fuel economy) are influenced by each other. Thus, it is important to trade-off different performance objectives since these are conflicted with each other in a variety of operating mode switches. In addition, battery aging will also affect energy efficiency and fuel economy.

Currently, the energy management strategy mainly aims at minimizing fuel economy and emissions while neglecting other performance factors involving battery life as well as drivability. Thus, how to incorporate these performance indexes to implement them in an integrated optimization is a key issue.

6.4. Adaptive EMSs Considering Driver Characteristics and More Influential Factors

To the best of our knowledge, most of the EMSs are demonstrated by simulation over a specific driving cycle. However, the actual driving condition is complex and diverse; for instance, traffic congestion in cities, highways, urban, and suburban areas. In addition, driving behaviors (e.g., driving styles) is another important factor in the driving cycle. Different drivers may take different actions toward the same situation, leading to uncertainty in driving cycles. The optimal results of the EMSs are strongly dependent on the driving cycle while it is hard to adapt to various driving conditions using existing EMSs. Therefore, to develop adaptive EMSs may be a promising solution for the HEVs.

6.5. Multi-Dimension EMSs Including Route Planning and Velocity Planning

It is well known that the performance of EMSs is related to the vehicular velocity and traffic conditions. The changing traffic conditions make it challenging to implement a high energy-efficiency-oriented energy management strategy. This is due to the uncertainty of the vehicle route and velocity affected by traffic conditions. Moreover, different routes present a distinct traffic condition, even if the vehicle operates on the same route since the traffic condition may be diverse. All these factors bring uncertainty and disturbance for optimizing EMSs. Traditional EMSs mainly consider fuel-to-powertrain optimization instead of combining economic route planning and optimal velocity planning. Thus, how to integrate powertrain optimization, route, and velocity planning to further improve the energy efficiency is a key challenge.

Author Contributions: Wrote the original manuscript, F.Z.; edited and proofread the manuscript, L.W. and S.C.; reviewed and provided good suggestions, H.P., Y.C. and J.X. All authors have read and agreed to the published version of the manuscript.

Funding: This research was funded by National Natural Science Foundation of China (Grant No.51905419) and Natural Science Basic Research Program of Shaanxi (Grant No.2019JQ-503), and the Fundamental Research Fund for the Central Universities of China (grant No. 300102229514 and No.300102229502).

Conflicts of Interest: The authors declare no conflict of interest.

References

1. Zhang, F.Q.; Hu, X.S.; Langari, R.; Cao, D.P. Energy management strategies of connected hevs and phevs: Recent progress and outlook. *Prog. Energy Combust. Sci.* **2019**, *73*, 235–256. [CrossRef]
2. Onori, S.; Serrao, L.; Rizzoni, G. *Hybrid Electric Vehicles: Energy Management Strategies*; Springer: Berlin/Heidelberg, Germany, 2016.
3. Wang, Q.; You, S.; Li, L.; Yang, C. Survey on energy management strategy for plug-in hybrid electric vehicles. *J. Mech. Eng.* **2017**, *53*, 1–19. [CrossRef]
4. Wirasingha, S.G.; Emadi, A. Classification and review of control strategies for plug-in hybrid electric vehicles. *IEEE Trans. Veh. Technol.* **2011**, *60*, 111–122. [CrossRef]
5. Salmasi, F.R. Control strategies for hybrid electric vehicles: Evolution, classification, comparison, and future trends. *IEEE Trans. Veh. Technol.* **2007**, *56*, 2393–2404. [CrossRef]
6. Karbaschian, M.; Söffker, D. Review and comparison of power management approaches for hybrid vehicles with focus on hydraulic drives. *Energies* **2014**, *7*, 3512–3536. [CrossRef]
7. Tran, D.-D.; Vafaeipour, M.; Baghdadi, M.E.; Barrero, R.; Hegazy, O. Thorough state-of-the-art analysis of electric and hybrid vehicle powertrains: Topologies and integrated energy management strategies. *Renew. Sustain. Energy Rev.* **2019**, *119*, 109596. [CrossRef]
8. Miller, J.M. *Propulsion Systems for Hybrid Vehicles*; The Institution of Electrical Engineers: London, UK, 2004; Volume 45.
9. He, H.; Zhang, J.; Li, G. Model predictive control for energy management of a plug-in hybrid electric bus. *Energy Procedia* **2016**, *88*, 901–907. [CrossRef]
10. Donitz, C.; Vasile, I.; Onder, C.; Guzzella, L. Dynamic programming for hybrid pneumatic vehicles. In Proceedings of the 2009 American Control Conference, St. Louis, MO, USA, 10–12 June 2009; IEEE: New York, NY, USA, 2009; pp. 3956–3963.
11. Lin, C.C.; Peng, H.; Grizzle, J.W.; Kang, J.M. Power management strategy for a parallel hybrid electric truck. *IEEE Trans. Control Syst. Technol.* **2003**, *11*, 839–849.
12. Patil, R.M.; Filipi, Z.; Fathy, H.K. Comparison of supervisory control strategies for series plug-in hybrid electric vehicle powertrains through dynamic programming. *IEEE Trans. Control Syst. Technol.* **2014**, *22*, 502–509. [CrossRef]
13. Murphey, Y.L.; Park, J.; Kiliaris, L.; Kuang, M.L.; Abul Masrur, M.; Phillips, A.M.; Wang, Q. Intelligent hybrid vehicle power control-part ii: Online intelligent energy management. *IEEE Trans. Veh. Technol.* **2013**, *62*, 69–79. [CrossRef]
14. Wu, B.; Lin, C.C.; Filipi, Z.; Peng, H.; Assanis, D. Optimal power management for a hydraulic hybrid delivery truck. *Veh. Syst. Dyn.* **2004**, *42*, 23–40. [CrossRef]

15. Kutter, S.; Bäker, B. Predictive online control for hybrids: Resolving the conflict between global optimality, robustness and real-time capability. In Proceedings of the 2010 IEEE Vehicle Power and Propulsion Conference, Lille, France, 1–3 September 2010; pp. 1–7.

16. Fares, D.; Chedid, R.; Panik, F.; Karaki, S.; Jabr, R. Dynamic programming technique for optimizing fuel cell hybrid vehicles. *Int. J. Hydrog. Energy* **2015**, *40*, 7777–7790. [CrossRef]

17. Zhuang, W.; Zhang, X.; Li, D.; Wang, L.; Yin, G.J.A.E. Mode shift map design and integrated energy management control of a multi-mode hybrid electric vehicle. *Appl. Energy* **2017**, *204*, 476–488. [CrossRef]

18. Peng, J.; He, H.; Xiong, R. Rule based energy management strategy for a series–parallel plug-in hybrid electric bus optimized by dynamic programming. *Appl. Energy* **2017**, *185*, 1633–1643. [CrossRef]

19. Yang, Y.; Hu, X.; Pei, H.; Peng, Z. Comparison of power-split and parallel hybrid powertrain architectures with a single electric machine: Dynamic programming approach. *Appl. Energy* **2016**, *168*, 683–690. [CrossRef]

20. Liu, B.; Li, L.; Wang, X.; Cheng, S. Hybrid electric vehicle downshifting strategy based on stochastic dynamic programming during regenerative braking process. *IEEE Trans. Veh. Technol.* **2018**, *67*, 4716–4727. [CrossRef]

21. Liu, J.; Chen, Y.; Zhan, J.; Shang, F. Heuristic dynamic programming based online energy management strategy for plug-in hybrid electric vehicles. *IEEE Trans. Veh. Technol.* **2019**, *68*, 4479–4493. [CrossRef]

22. Van Berkel, K.; de Jager, B.; Hofman, T.; Steinbuch, M. Implementation of dynamic programming for optimal control problems with continuous states. *IEEE Trans. Control Syst. Technol.* **2015**, *23*, 1172–1179. [CrossRef]

23. Moura, S.J.; Fathy, H.K.; Callaway, D.S.; Stein, J.L. A stochastic optimal control approach for power management in plug-in hybrid electric vehicles. *IEEE Trans. Control Syst. Technol.* **2011**, *19*, 545–555. [CrossRef]

24. Opila, D.F.; Wang, X.Y.; McGee, R.; Gillespie, R.B.; Cook, J.A.; Grizzle, J.W. An energy management controller to optimally trade off fuel economy and drivability for hybrid vehicles. *IEEE Trans. Control Syst. Technol.* **2012**, *20*, 1490–1505. [CrossRef]

25. Johri, R.; Filipi, Z. Optimal energy management of a series hybrid vehicle with combined fuel economy and low-emission objectives. *Proc. Inst. Mech. Eng. Part D J. Automob. Eng.* **2014**, *228*, 1424–1439. [CrossRef]

26. Zou, Y.; Kong, Z.; Liu, T.; Liu, D. A real-time markov chain driver model for tracked vehicles and its validation: Its adaptability via stochastic dynamic programming. *IEEE Trans. Veh. Technol.* **2016**, *66*, 3571–3582. [CrossRef]

27. Xu, F.; Jiao, X.; Sasaki, M.; Wang, Y. Energy management optimization in consideration of battery deterioration for commuter plug-in hybrid electric vehicle. In Proceedings of the 2016 55th Annual Conference of the Society of Instrument and Control Engineers of Japan (SICE), Tsukuba, Japan, 20–23 September 2016; pp. 218–222.

28. Du, Y.; Zhao, Y.; Wang, Q.; Zhang, Y.; Xia, H. Trip-oriented stochastic optimal energy management strategy for plug-in hybrid electric bus. *Energy* **2016**, *115*, 1259–1271. [CrossRef]

29. Lü, X.; Wu, Y.; Lian, J.; Zhang, Y.; Chen, C.; Wang, P.; Meng, L. Energy management of hybrid electric vehicles: A review of energy optimization of fuel cell hybrid power system based on genetic algorithm. *Energy Convers. Manag.* **2020**, *205*, 112474. [CrossRef]

30. Zhou, S.; Wen, Z.; Zhi, X.; Jin, J.; Zhou, S. *Genetic Algorithm-Based Parameter Optimization of Energy Management Strategy and Its Analysis for Fuel Cell Hybrid Electric Vehicles*; 0148-7191; SAE Technical Paper: New York, NY, USA, 2019.

31. Piccolo, A.; Ippolito, L.; Galdi, V.Z.; Vaccaro, A. Optimisation of energy flow management in hybrid electric vehicles via genetic algorithms. In Proceedings of the 2001 IEEE/ASME International Conference on Advanced Intelligent Mechatronics. Proceedings (Cat. No.01TH8556), Como, Italy, 8–12 July 2001; IEEE: New York, NY, USA, 2001; pp. 434–439.

32. Xudong, L.; Yanping, W.; Jianmin, D. Optimal sizing of a series hybrid electric vehicle using a hybrid genetic algorithm. In Proceedings of the 2007 IEEE International Conference on Automation and Logistics, Jinan, China, 18–21 August 2007; pp. 1125–1129.

33. Zhang, Y.; Meng, D.; Zhou, M.; Lu, D. Management strategy based on genetic algorithm optimization for phev. *Int. J. Control Autom.* **2014**, *7*, 399–408.

34. Zhang, H.; Su, Y.; Peng, L.; Yao, D. A review of game theory applications in transportation analysis. In Proceedings of the 2010 International Conference on Computer and Information Application, Tianjin, China, 3–5 December 2010; pp. 152–157.

35. Nash, J.F. Equilibrium points in n-person games. *Proc. Natl. Acad. Sci. USA* **1950**, *36*, 48–49. [CrossRef]

36. Nash, J. Non-cooperative games. *Ann. Math.* **1951**, *54*, 286–295. [CrossRef]
37. Colman, A.M. *Game Theory and Its Applications: In the Social and Biological Sciences*; Psychology Press: London, UK, 2013.
38. Gielniak, M.J.; Shen, Z.J. Power management strategy based on game theory for fuel cell hybrid electric vehicles. In Proceedings of the IEEE 60th Vehicular Technology Conference (VTC2004-Fall 2004), Los Angeles, CA, USA, 26–29 September 2004; pp. 4422–4426.
39. Yin, H.; Zhao, C.; Li, M.; Ma, C.; Chow, M.-Y. A game theory approach to energy management of an engine–generator/battery/ultracapacitor hybrid energy system. *IEEE Trans. Ind. Electron.* **2016**, *63*, 4266–4277. [CrossRef]
40. Dextreit, C.; Kolmanovsky, I.V. Game theory controller for hybrid electric vehicles. *IEEE Trans. Control Syst. Technol.* **2014**, *22*, 652–663. [CrossRef]
41. Dextreit, C.; Assadian, F.; Kolmanovsky, I.; Mahtani, J.; Burnham, K. *Hybrid Electric Vehicle Energy Management Using Game Theory*; 0148-7191; SAE Technical Paper: New York, NY, USA, 2008.
42. Xu, J.; Alsabbagh, A.; Yan, D.; Ma, C. Game-theoretic energy management with velocity prediction in hybrid electric vehicle. In Proceedings of the 2019 IEEE 28th International Symposium on Industrial Electronics (ISIE), Vancouver, BC, Canada, 12–14 June 2019; pp. 1084–1089.
43. Chen, H.; Kessels, J.; Donkers, M.; Weiland, S. Game-theoretic approach for complete vehicle energy management. In Proceedings of the 2014 IEEE Vehicle Power and Propulsion Conference (VPPC), Coimbra, Portugal, 27–30 October 2014; pp. 1–6.
44. Chen, H.; Kessels, J.T.; Weiland, S. Online adaptive approach for a game-theoretic strategy for complete vehicle energy management. In Proceedings of the 2015 European Control Conference (ECC), Linz, Austria, 15–17 July 2015; pp. 135–141.
45. Chen, H. Game-Theoretic Solution Concept for Complete Vehicle Energy Management. Ph.D. Thesis, Technische Universiteit Eindhoven, Eindhoven, The Netherlands, 2016.
46. Orszag, S.A. Comparison of pseudospectral and spectral approximation. *Stud. Appl. Math.* **1972**, *51*, 253–259. [CrossRef]
47. Hu, X.; Li, S.; Peng, H.; Sun, F. Charging time and loss optimization for linmc and lifepo 4 batteries based on equivalent circuit models. *J. Power Sources* **2013**, *239*, 449–457. [CrossRef]
48. Zhou, W.; Zhang, C.; Li, J.; Fathy, H.K. A pseudospectral strategy for optimal power management in series hybrid electric powertrains. *IEEE Trans. Veh. Technol.* **2016**, *65*, 4813–4825. [CrossRef]
49. Wu, J.; Zou, Y.; Zhang, X.; Du, G.; Du, G.; Zou, R. A hierarchical energy management for hybrid electric tracked vehicle considering velocity planning with pseudospectral method. *IEEE Trans. Transp. Electrif.* **2020**, *6*, 703–716. [CrossRef]
50. Martinez, C.M.; Hu, X.; Cao, D.; Velenis, E.; Gao, B.; Wellers, M. Energy management in plug-in hybrid electric vehicles: Recent progress and a connected vehicles perspective. *IEEE Trans. Veh. Technol.* **2016**, *66*, 4534–4549. [CrossRef]
51. Murgovski, N.; Johannesson, L.; Sjöberg, J.; Egardt, B. Component sizing of a plug-in hybrid electric powertrain via convex optimization. *Mechatronics* **2012**, *22*, 106–120. [CrossRef]
52. Nafisi, H.; Agah, S.M.M.; Abyaneh, H.A.; Abedi, M. Two-stage optimization method for energy loss minimization in microgrid based on smart power management scheme of phevs. *IEEE Trans. Smart Grid* **2015**, *7*, 1268–1276. [CrossRef]
53. Boyd, S.; Boyd, S.P.; Vandenberghe, L. *Convex Optimization*; Cambridge university press: Cambridge, UK, 2004.
54. Nüesch, T.; Elbert, P.; Flankl, M.; Onder, C.; Guzzella, L. Convex optimization for the energy management of hybrid electric vehicles considering engine start and gearshift costs. *Energies* **2014**, *7*, 834–856. [CrossRef]
55. Xie, S.; Li, H.; Xin, Z.; Liu, T.; Wei, L. A pontryagin minimum principle-based adaptive equivalent consumption minimum strategy for a plug-in hybrid electric bus on a fixed route. *Energies* **2017**, *10*, 1379. [CrossRef]
56. Kang, C.; Song, C.; Cha, S. A costate estimation for pontryagin's minimum principle by machine learning. In Proceedings of the 2018 IEEE Vehicle Power and Propulsion Conference (VPPC), Chicago, IL, USA, 27–30 August 2018; pp. 1–5.
57. Zhang, J.; Zheng, C.; Cha, S.W.; Duan, S. Co-state variable determination in pontryagin's minimum principle for energy management of hybrid vehicles. *Int. J. Precis. Eng. Manuf.* **2016**, *17*, 1215–1222. [CrossRef]

58. Li, X.; Wang, Y.; Yang, D.; Chen, Z. Adaptive energy management strategy for fuel cell/battery hybrid vehicles using pontryagin's minimal principle. *J. Power Sources* **2019**, *440*, 227105. [CrossRef]

59. Ghasemi, M.; Song, X. A computationally efficient optimal power management for power split hybrid vehicle based on pontryagin's minimum principle. In Proceedings of the ASME 2017 Dynamic Systems and Control Conference, Tysons, VA, USA, 11–13 October 2017.

60. Nguyen, B.-H.; German, R.; Trovão, J.P.F.; Bouscayrol, A. Real-time energy management of battery/supercapacitor electric vehicles based on an adaptation of pontryagin's minimum principle. *IEEE Trans. Veh. Technol.* **2018**, *68*, 203–212. [CrossRef]

61. Kim, N.; Jeong, J.; Zheng, C. Adaptive energy management strategy for plug-in hybrid electric vehicles with pontryagin's minimum principle based on daily driving patterns. *Int. J. Precis. Eng. Manuf. Green Technol.* **2019**, *6*, 539–548. [CrossRef]

62. Serrao, L.; Onori, S.; Rizzoni, G. A comparative analysis of energy management strategies for hybrid electric vehicles. *J. Dyn. Syst. Meas. Control* **2011**, *133*, 031012. [CrossRef]

63. Onori, S.; Tribioli, L. Adaptive pontryagin's minimum principle supervisory controller design for the plug-in hybrid gm chevrolet volt. *Appl. Energy* **2015**, *147*, 224–234. [CrossRef]

64. Kim, N.; Rousseau, A.; Lee, D. A jump condition of pmp-based control for phevs. *J. Power Sources* **2011**, *196*, 10380–10386. [CrossRef]

65. Chen, Z.; Mi, C.C.; Xia, B.; You, C.W. Energy management of power-split plug-in hybrid electric vehicles based on simulated annealing and pontryagin's minimum principle. *J. Power Sources* **2014**, *272*, 160–168. [CrossRef]

66. Hou, C.; Ouyang, M.G.; Xu, L.F.; Wang, H.W. Approximate pontryagin's minimum principle applied to the energy management of plug-in hybrid electric vehicles. *Appl. Energy* **2014**, *115*, 174–189. [CrossRef]

67. Zhu, M.; Wu, X.; Xu, M. *Adaptive Energy Management Strategy for Hybrid Vehicles Based on Pontryagin's Minimum Principle*; 0148-7191; SAE Technical Paper: New York, NY, USA, 2020.

68. Park, K.; Son, H.; Bae, K.; Kim, Y.; Kim, H.; Yun, J.; Kim, H. Optimal control of plug-in hybrid electric vehicle based on pontryagin's minimum principle considering driver's characteristic. In Proceedings of the International Conference on Vehicle Technology and Intelligent Transport Systems, Porto, Portugal, 22–24 April 2017; pp. 151–156.

69. Jinming, L.; Huei, P. Modeling and control of a power-split hybrid vehicle. *IEEE Trans. Control Syst. Technol.* **2008**, *16*, 1242–1251. [CrossRef]

70. Ehsani, M.; Gao, Y.; Longo, S.; Ebrahimi, K. *Modern Electric, Hybrid Electric, and Fuel Cell Vehicles: Fundamentals, Theory, and Design*; CRC press: Boca Raton, FL, USA, 2004.

71. Liu, W. *Introduction to Hybrid Vehicle System Modeling and Control*; John Wiley & Sons: Hoboken, NJ, USA, 2013.

72. Mamdani, E.H. Application of fuzzy algorithms for control of simple dynamic plant. *Inst. Electr. Eng.* **1974**, *121*, 1585–1588. [CrossRef]

73. Takagi, T.; Sugeno, M. Fuzzy identification of systems and its applications to modeling and control. *IEEE Trans. Syst. ManCybern.* **1985**, *SMC-15*, 116–132. [CrossRef]

74. Syed, F.U.; Kuang, M.L.; Smith, M.; Okubo, S.; Ying, H. Fuzzy gain-scheduling proportional–integral control for improving engine power and speed behavior in a hybrid electric vehicle. *IEEE Trans. Veh. Technol.* **2008**, *58*, 69–84. [CrossRef]

75. Denis, N.; Dubois, M.R.; Desrochers, A. Fuzzy-based blended control for the energy management of a parallel plug-in hybrid electric vehicle. *IET Intell. Transp. Syst.* **2014**, *9*, 30–37. [CrossRef]

76. Dawei, M.; Yu, Z.; Meilan, Z.; Risha, N. Intelligent fuzzy energy management research for a uniaxial parallel hybrid electric vehicle. *Comput. Electr. Eng.* **2017**, *58*, 447–464. [CrossRef]

77. Li, S.G.; Sharkh, S.; Walsh, F.C.; Zhang, C.-N. Energy and battery management of a plug-in series hybrid electric vehicle using fuzzy logic. *IEEE Trans. Veh. Technol.* **2011**, *60*, 3571–3585. [CrossRef]

78. Yu, H.; Tarsitano, D.; Hu, X.; Cheli, F. Real time energy management strategy for a fast charging electric urban bus powered by hybrid energy storage system. *Energy* **2016**, *112*, 322–331. [CrossRef]

79. Li, J.; Zhou, Q.; Williams, H.; Xu, H. Back-to-back competitive learning mechanism for fuzzy logic based supervisory control system of hybrid electric vehicles. *IEEE Trans. Ind. Electron.* **2019**, *67*, 8900–8909. [CrossRef]

80. Ma, K.; Wang, Z.; Liu, H.; Yu, H.; Wei, C. Numerical investigation on fuzzy logic control energy management strategy of parallel hybrid electric vehicle. *Energy Procedia* **2019**, *158*, 2643–2648. [CrossRef]

81. Li, J.; Zhou, Q.; He, Y.; Williams, H.; Xu, H. Driver-identified supervisory control system of hybrid electric vehicles based on spectrum-guided fuzzy feature extraction. *IEEE Trans. Fuzzy Syst.* **2020**. [CrossRef]

82. Salman, M.; Schouten, N.J.; Kheir, N.A. Control strategies for parallel hybrid vehicles. In Proceedings of the 2000 American Control Conference. ACC (IEEE Cat. No.00CH36334), Chicago, IL, USA, 28–30 June 2000; Volume 521, pp. 524–528.

83. Montazeri-Gh, M.; Mahmoodi-k, M. Development a new power management strategy for power split hybrid electric vehicles. *Transp. Res. Part D Transp. Environ.* **2015**, *37*, 79–96. [CrossRef]

84. Hyeoun-Dong, L.; Seung-Ki, S. Fuzzy-logic-based torque control strategy for parallel-type hybrid electric vehicle. *Ieee Trans. Ind. Electron.* **1998**, *45*, 625–632. [CrossRef]

85. Hyeoun-Dong, L.; Euh-Suh, K.; Seung-Ki, S.; Joohn-Sheok, K.; Kamiya, M.; Ikeda, H.; Shinohara, S.; Yoshida, H. Torque control strategy for a parallel-hybrid vehicle using fuzzy logic. *Ind. Appl. Mag.* **2000**, *6*, 33–38. [CrossRef]

86. Baumann, B.M.; Washington, G.; Glenn, B.C.; Rizzoni, G. Mechatronic design and control of hybrid electric vehicles. *IEEE/ASME Trans. Mechatron.* **2000**, *5*, 58–72. [CrossRef]

87. Tian, H.; Wang, X.; Lu, Z.; Huang, Y.; Tian, G. Adaptive fuzzy logic energy management strategy based on reasonable soc reference curve for online control of plug-in hybrid electric city bus. *IEEE Trans. Intell. Transp. Syst.* **2017**, *19*, 1607–1617. [CrossRef]

88. Onori, S.; Serrao, L. On adaptive-ecms strategies for hybrid electric vehicles. In Proceedings of the International Scientific Conference on Hybrid and Electric Vehicles, Malmaison, France, 6–7 December 2011.

89. Zeng, Y.; Cai, Y.; Kou, G.; Gao, W.; Qin, D. Energy management for plug-in hybrid electric vehicle based on adaptive simplified-ecms. *Sustainability* **2018**, *10*, 2060. [CrossRef]

90. Geng, B.; Mills, J.K.; Sun, D. Energy management control of microturbine-powered plug-in hybrid electric vehicles using the telemetry equivalent consumption minimization strategy. *IEEE Trans. Veh. Technol.* **2011**, *60*, 4238–4248. [CrossRef]

91. Han, J.; Kum, D.; Park, Y. Synthesis of predictive equivalent consumption minimization strategy for hybrid electric vehicles based on closed-form solution of optimal equivalence factor. *IEEE Trans. Veh. Technol.* **2017**, *66*, 5604–5616. [CrossRef]

92. Tian, X.; He, R.; Sun, X.; Cai, Y.; Xu, Y. An anfis-based ecms for energy optimization of parallel hybrid electric bus. *IEEE Trans. Veh. Technol.* **2019**, *69*, 1473–1483. [CrossRef]

93. Xie, S.; Hu, X.; Qi, S.; Lang, K. An artificial neural network-enhanced energy management strategy for plug-in hybrid electric vehicles. *Energy* **2018**, *163*, 837–848. [CrossRef]

94. Yang, S.; Wang, W.; Zhang, F.; Hu, Y.; Xi, J. Driving-style-oriented adaptive equivalent consumption minimization strategies for hevs. *IEEE Trans. Veh. Technol.* **2018**, *67*, 9249–9261. [CrossRef]

95. Paganelli, G.; Delprat, S.; Guerra, T.M.; Rimaux, J.; Santin, J.J. Equivalent consumption minimization strategy for parallel hybrid powertrains. In Proceedings of the Vehicular Technology Conference (VTC Spring 2002), Birmingham, AL, USA, 6–9 May 2002.

96. Kleimaier, A.; Schroder, D. An approach for the online optimized control of a hybrid powertrain. In Proceedings of the 7th International Workshop on Advanced Motion Control. Proceedings (Cat. No.02TH8623), Maribor, Slovenia, 3–5 July 2002; pp. 215–220.

97. Sciarretta, A.; Back, M.; Guzzella, L. Optimal control of parallel hybrid electric vehicles. *IEEE Trans. Control Syst. Technol.* **2004**, *12*, 352–363. [CrossRef]

98. Khodabakhshian, M.; Feng, L.; Wikander, J. Improving fuel economy and robustness of an improved ecms method. In Proceedings of the 2013 10th IEEE International Conference on Control and Automation (ICCA), Hangzhou, China, 12–14 June 2013; pp. 598–603.

99. Nüesch, T.; Cerofolini, A.; Mancini, G.; Cavina, N.; Onder, C.; Guzzella, L. Equivalent consumption minimization strategy for the control of real driving nox emissions of a diesel hybrid electric vehicle. *Energies* **2014**, *7*, 3148–3178. [CrossRef]

100. Gao, J.P.; Zhu, G.M.G.; Strangas, E.G.; Sun, F.C. Equivalent fuel consumption optimal control of a series hybrid electric vehicle. *Proc. Inst. Mech. Eng. Part D J. Automob. Eng.* **2009**, *223*, 1003–1018. [CrossRef]

101. Skugor, B.; Deur, J.; Cipek, M.; Pavkovic, D. Design of a power-split hybrid electric vehicle control system utilizing a rule-based controller and an equivalent consumption minimization strategy. *Proc. Inst. Mech. Eng. Part D J. Automob. Eng.* **2014**, *228*, 631–648. [CrossRef]

102. Torreglosa, J.P.; Jurado, F.; García, P.; Fernández, L.M. Hybrid fuel cell and battery tramway control based on an equivalent consumption minimization strategy. *Control Eng. Pract.* **2011**, *19*, 1182–1194. [CrossRef]

103. Park, J.; Park, J.H. Development of equivalent fuel consumption minimization strategy for hybrid electric vehicles. *Int. J. Automot. Technol.* **2012**, *13*, 835–843. [CrossRef]

104. Sun, C.; He, H.; Sun, F. The role of velocity forecasting in adaptive-ecms for hybrid electric vehicles. *Energy Procedia* **2015**, *75*, 1907–1912. [CrossRef]

105. Zhang, F.; Liu, H.; Hu, Y.; Xi, J. A supervisory control algorithm of hybrid electric vehicle based on adaptive equivalent consumption minimization strategy with fuzzy pi. *Energies* **2016**, *9*, 919. [CrossRef]

106. Musardo, C.; Rizzoni, G.; Staccia, B. A-ecms: An adaptive algorithm for hybrid electric vehicle energy management. In Proceedings of the 2005 44th IEEE Conference on Decision and Control and 2005 European Control Conference (CDC-ECC '05), Seville, Spain, 12–15 December 2005; pp. 1816–1823.

107. Sezer, V.; Gokasan, M.; Bogosyan, S. A novel ecms and combined cost map approach for high-efficiency series hybrid electric vehicles. *IEEE Trans. Veh. Technol.* **2011**, *60*, 3557–3570. [CrossRef]

108. Chen, Z.; Vahidi, A. Route preview in energy management of plug-in hybrid vehicles. *Control Syst. Technol. IEEE Trans.* **2012**, *20*, 546–553. [CrossRef]

109. Kim, N.W.; Lee, D.H.; Zheng, C.; Shin, C.; Seo, H.; Cha, S.W. Realization of pmp-based control for hybrid electric vehicles in a backwards-looking simulation. *Int. J. Automot. Technol.* **2014**, *15*, 625–635. [CrossRef]

110. Hemi, H.; Ghouili, J.; Cheriti, A. A real time energy management for electrical vehicle using combination of rule-based and ecms. In Proceedings of the 2013 IEEE Electrical Power & Energy Conference, Halifax, NS, Canada, 21–23 August 2013; pp. 1–6.

111. Serrao, L.; Onori, S.; Rizzoni, G. Ecms as a realization of pontryagin's minimum principle for hev control. In Proceedings of the 2009 American Control Conference, St. Louis, MO, USA, 10–12 June 2009; pp. 3964–3969.

112. Mohd Zulkefli, M.A.; Zheng, J.; Sun, Z.; Liu, H.X. Hybrid powertrain optimization with trajectory prediction based on inter-vehicle-communication and vehicle-infrastructure-integration. *Transp. Res. Part C* **2014**, *45*, 41–63. [CrossRef]

113. Zhang, F.; Xi, J.; Langari, R. Real-time energy management strategy based on velocity forecasts using v2v and v2i communications. *IEEE Trans. Intell. Transp. Syst.* **2017**, *18*, 416–430. [CrossRef]

114. Sun, C.; Sun, F.; He, H. Investigating adaptive-ecms with velocity forecast ability for hybrid electric vehicles. *Appl. Energy* **2017**, *185*, 1644–1653. [CrossRef]

115. Borhan, H.; Vahidi, A.; Phillips, A.M.; Kuang, M.L.; Kolmanovsky, I.V.; Di Cairano, S. Mpc-based energy management of a power-split hybrid electric vehicle. *IEEE Trans. Control Syst. Technol.* **2012**, *20*, 593–603. [CrossRef]

116. Gurkaynak, Y. *Neural Adaptive Control Stategy for Hybird Electric Vehciles with Parallel Powertrain*; Illinois Institute of Technology: Chicago, IL, USA, 2011.

117. Rezaei, A.; Burl, J.; Solouk, A.; Zhou, B.; Rezaei, M.; Shahbakhti, M. Catch energy saving opportunity (ceso), an instantaneous optimal energy management strategy for series hybrid electric vehicles. *Appl. Energy* **2017**, *208*, 655–665. [CrossRef]

118. Li, L.; Coskun, S.; Zhang, F.; Langari, R.; Xi, J. Energy management of hybrid electric vehicle using vehicle lateral dynamic in velocity prediction. *IEEE Trans. Veh. Technol.* **2019**, *68*, 3279–3293. [CrossRef]

119. Robust Control. Available online: https://en.wikipedia.org/wiki/Robust_control (accessed on 30 June 2020).

120. Morales-Morales, J.; Cervantes, I.; Cano-Castillo, U. On the design of robust energy management strategies for fchev. *IEEE Trans. Veh. Technol.* **2015**, *64*, 1716–1728. [CrossRef]

121. Motapon, S.N.; Dessaint, L.A.; Al-Haddad, K. A robust h2-consumption-minimization-based energy management strategy for a fuel cell hybrid emergency power system of more electric aircraft. *IEEE Trans. Ind. Electron.* **2014**, *61*, 6148–6156. [CrossRef]

122. Pisu, P.; Rizzoni, G. A comparative study of supervisory control strategies for hybrid electric vehicles. *IEEE Trans. Control Syst. Technol.* **2007**, *15*, 506–518. [CrossRef]

123. Karbowski, D.; Kim, N.; Rousseau, A. Route-based online energy management of a phev and sensitity to trip prediction. In Proceedings of the 2014 IEEE Vehicle Power and Propulsion Conference (VPPC), Coimbra, Portugal, 27–30 October 2014; pp. 1–6.

124. Vadamalu, R.; Beidl, C.; Barth, S.; Rass, F. Multi-objective predictive energy management framework for hybrid electric powertrains: An online optimization approach 27th aachen colloquium. In Proceedings of the 27th Aachen Colloquium Automobile and Engine Technology, Aachen, Germany, 8–10 October 2018.

125. Zhou, Y.; Ravey, A.; Péra, M.-C. A survey on driving prediction techniques for predictive energy management of plug-in hybrid electric vehicles. *J. Power Sources* **2019**, *412*, 480–495. [CrossRef]

126. Langari, R.; Jong-Seob, W. Intelligent energy management agent for a parallel hybrid vehicle-part i: System architecture and design of the driving situation identification process. *IEEE Trans. Veh. Technol.* **2005**, *54*, 925–934. [CrossRef]

127. Wu, J.; Zhang, C.H.; Cui, N.X. Fuzzy energy management strategy for a hybrid electric vehicle based on driving cycle recognition. *Int. J. Automot. Technol.* **2012**, *13*, 1159–1167. [CrossRef]

128. Murphey, Y.L.; Zhihang, C.; Kiliaris, L.; Jungme, P.; Ming, K.; Masrur, A.; Phillips, A. Neural learning of driving environment prediction for vehicle power management. In Proceedings of the IEEE International Joint Conference on Neural Networks, Hong Kong, China, 1–8 June 2008; pp. 3755–3761.

129. Zhu, D.W.; Hui, X.; Yin, Y.; Song, Z.B. The dynamic optimization of control strategy for hybtion city bus based on driving condition self-learing. *J. Mech. Eng.* **2010**, *46*, 33–38. [CrossRef]

130. Bender, F.A.; Kaszynski, M.; Sawodny, O. Drive cycle prediction and energy management optimization for hybrid hydraulic vehicles. *IEEE Trans. Veh. Technol.* **2013**, *62*, 3581–3592. [CrossRef]

131. Yang, B.; Yaoyu, L.; Qiuming, G.; Zhong-Ren, P. Multi-information integrated trip specific optimal power management for plug-in hybrid electric vehicles. In Proceedings of the 2009 American Control Conference, St. Louis, MO, USA, 10–12 June 2009; pp. 4607–4612.

132. Qiuming, G.; Yaoyu, L.; Zhong-Ren, P. Trip based optimal power management of plug-in hybrid electric vehicles using gas-kinetic traffic flow model. In Proceedings of the American Control Conference, Seattle, WA, USA, 11–13 June 2008; pp. 3225–3230.

133. He, Y. *Vehicle-Infrastructure Integration Enabled Plug-in Hybrid Electric Vehicles for Energy Management*; Clemson University: Clemson, SC, USA, 2013.

134. Zhang, C.; Vahidi, A.; Pisu, P.; Li, X.; Tennant, K. Role of terrain preview in energy management of hybrid electric vehicles. *IEEE Trans. Veh. Technol.* **2010**, *59*, 1139. [CrossRef]

135. Fu, L.; Ozguner, U.; Tulpule, P.; Marano, V. Real-time energy management and sensitivity study for hybrid electric vehicles. In Proceedings of the American Control Conference (ACC), San Francisco, CA, USA, 29 June–1 July 2011; pp. 2113–2118.

136. Gong, Q.; Li, Y.; Peng, Z.R. Optimal power management of plug-in hev with intelligent transportation system. In Proceedings of the 2007 IEEE/ASME international conference on advanced intelligent mechatronics, Zurich, Switzerland, 4–7 September 2007.

137. Gong, Q.; Li, Y.; Peng, Z.R. Optimal power management of plug-in hybrid electric vehicles with trip modeling. In Proceedings of the ASME International Mechanical Engineering Congress and Exposition, IMECE 2007, Seattle, WA, USA, 11–15 November 2007; American Society of Mechanical Engineers: New York, NY, USA, 2008; Volume 16, pp. 53–62.

138. Qiuming, G.; Yaoyu, L.; Zhong-Ren, P. Trip-based optimal power management of plug-in hybrid electric vehicles. *IEEE Trans. Veh. Technol.* **2008**, *57*, 3393–3401. [CrossRef]

139. Lin, Y.; Tang, P.; Zhang, W.J.; Yu, Q. Artificial neural network modelling of driver handling behaviour in a driver-vehicle-environment system. *Int. J. Veh. Des.* **2005**, *37*, 24–45. [CrossRef]

140. Santos, P.J.; Martins, A.G.; Pires, A.J. Designing the input vector to ann-based models for short-term load forecast in electricity distribution systems. *Int. J. Electr. Power Energy Syst.* **2007**, *29*, 338–347. [CrossRef]

141. Vlahogianni, E.; Golias, J.C.; Karlaftis, M.G. Short-term traffic forecasting: Overview of objectives and methods. *Transp. Rev.* **2004**, *24*, 533–557. [CrossRef]

142. Sun, C.; Hu, X.S.; Moura, S.J.; Sun, F.C. Velocity predictors for predictive energy management in hybrid electric vehicles. *IEEE Trans. Control Syst. Technol.* **2014**, *23*, 1197–1204.

143. Moser, D.; Waschl, H.; Schmied, R.; Efendic, H.; del Re, L. Short term prediction of a vehicle's velocity trajectory using its. *SAE Int. J. Passeng. Cars Electron. Electr. Syst.* **2015**, *8*, 364–370. [CrossRef]

144. Kouvaritakis, B.; Cannon, M. *Model Predictive Control*; Springer International Publishing: Cham, Switzerland, 2016.

145. Huang, Y.; Wang, H.; Khajepour, A.; He, H.; Ji, J. Model predictive control power management strategies for hevs: A review. *J. Power Sources* **2017**, *341*, 91–106. [CrossRef]

146. Li, G.; Goerges, D. Hybrid modeling and predictive control of the power split and gear shift in hybrid electric vehicles. In Proceedings of the 2017 IEEE Vehicle Power and Propulsion Conference (VPPC), Belfort, France, 11–14 December 2017; pp. 1–6.

147. Joševski, M.; Abel, D. Distributed predictive control approach for fuel efficient gear shifting in hybrid electric vehicles. In Proceedings of the 2016 European Control Conference (ECC), Aalborg, Denmark, 29 June–1 July 2016; pp. 2366–2373.

148. Joševski, M.; Abel, D. Gear shifting and engine on/off optimal control in hybrid electric vehicles using partial outer convexification. In Proceedings of the 2016 IEEE Conference on Control Applications (CCA), Buenos Aires, Argentina, 19–22 September 2016; pp. 562–568.

149. Cao, J.; Peng, J.; He, H. Research on model prediction energy management strategy with variable horizon. *Energy Procedia* **2017**, *105*, 3565–3570. [CrossRef]

150. Zhou, F.; Xiao, F.; Chang, C.; Shao, Y.; Song, C. Adaptive model predictive control-based energy management for semi-active hybrid energy storage systems on electric vehicles. *Energies* **2017**, *10*, 1063. [CrossRef]

151. Joševski, M.; Abel, D. Tube-based mpc for the energy management of hybrid electric vehicles with non-parametric driving profile prediction. In Proceedings of the 2016 American Control Conference (ACC), Boston, MA, USA, 6–8 July 2016; pp. 623–630.

152. Marx, M.; Soffker, D. Optimization of the powerflow control of a hybrid electric powertrain including load profile prediction. In Proceedings of the 2012 IEEE Vehicle Power and Propulsion Conference (VPPC), Seoul, Korea, 9–12 October 2012; pp. 395–400.

153. Stockar, S.; Marano, V.; Canova, M.; Rizzoni, G.; Guzzella, L. Energy-optimal control of plug-in hybrid electric vehicles for real-world driving cycles. *IEEE Trans. Veh. Technol.* **2011**, *60*, 2949–2962. [CrossRef]

154. He, L.; Shen, T.; Yu, L.; Feng, N.; Song, J. A model-predictive-control-based torque demand control approach for parallel hybrid powertrains. *IEEE Trans. Veh. Technol.* **2013**, *62*, 1041–1052. [CrossRef]

155. Zhang, S.; Xiong, R.; Sun, F. Model predictive control for power management in a plug-in hybrid electric vehicle with a hybrid energy storage system. *Appl. Energy* **2017**, *185*, 1654–1662. [CrossRef]

156. Xiang, C.; Ding, F.; Wang, W.; He, W. Energy management of a dual-mode power-split hybrid electric vehicle based on velocity prediction and nonlinear model predictive control. *Appl. Energy* **2017**, *189*, 640–653. [CrossRef]

157. Li, L.; You, S.; Yang, C.; Yan, B.; Song, J.; Chen, Z. Driving-behavior-aware stochastic model predictive control for plug-in hybrid electric buses. *Appl. Energy* **2016**, *162*, 868–879. [CrossRef]

158. Xie, S.; He, H.; Peng, J. An energy management strategy based on stochastic model predictive control for plug-in hybrid electric buses. *Appl. Energy* **2017**, *196*, 279–288. [CrossRef]

159. Xie, S.; Hu, X.; Xin, Z.; Li, L. Time-efficient stochastic model predictive energy management for a plug-in hybrid electric bus with an adaptive reference state-of-charge advisory. *IEEE Trans. Veh. Technol.* **2018**, *67*, 5671–5682. [CrossRef]

160. Mesbah, A.; Kolmanovsky, I.V.; Di Cairano, S. Stochastic model predictive control. In *Handbook of Model Predictive Control*; Springer: Berlin/Heidelberg, Germany, 2019; pp. 75–97.

161. Ripaccioli, G.; Bernardini, D.; Di Cairano, S.; Bemporad, A.; Kolmanovsky, I.V. A stochastic model predictive control approach for series hybrid electric vehicle power management. In Proceedings of the 2010 American Control Conference (ACC), Baltimore, MD, USA, 30 June–2 July 2010; pp. 5844–5849.

162. Xie, S.; Hu, X.; Xin, Z.; Brighton, J. Pontryagin's minimum principle based model predictive control of energy management for a plug-in hybrid electric bus. *Appl. Energy* **2019**, *236*, 893–905. [CrossRef]

163. Cheng, M.; Chen, B. Nonlinear model predictive control of a power-split hybrid electric vehicle with consideration of battery aging. *J. Dyn. Syst. Meas. Control* **2019**, *141*, 81008. [CrossRef]

164. Di Cairano, S.; Bernardini, D.; Bemporad, A.; Kolmanovsky, I.V. Stochastic mpc with learning for driver-predictive vehicle control and its application to hev energy management. *IEEE Trans. Control Syst. Technol.* **2014**, *22*, 1018–1031. [CrossRef]

165. Liu, T.; Hu, X.; Li, S.E.; Cao, D. Reinforcement learning optimized look-ahead energy management of a parallel hybrid electric vehicle. *IEEE/ASME Trans. Mechatron.* **2017**, *22*, 1497–1507. [CrossRef]

166. Xu, B.; Malmir, F.; Rathod, D.; Filipi, Z. *Real-Time Reinforcement Learning Optimized Energy Management for a 48v Mild Hybrid Electric Vehicle*; SAE Technical Paper: New York, NY, USA, 2019. [CrossRef]

167. Lian, R.; Peng, J.; Wu, Y.; Tan, H.; Zhang, H. Rule-interposing deep reinforcement learning based energy management strategy for power-split hybrid electric vehicle. *Energy* **2020**, *197*, 117297. [CrossRef]

168. Wang, P.; Northrop, W. *Reinforcement Learning Based Energy Management of Plug-in Hybrid Electric Vehicles for Commuter Route*; 0148-7191; SAE Technical Paper: New York, NY, USA, 2020.

169. Han, X.; He, H.; Wu, J.; Peng, J.; Li, Y. Energy management based on reinforcement learning with double deep q-learning for a hybrid electric tracked vehicle. *Appl. Energy* **2019**, *254*, 113708. [CrossRef]
170. Hu, X.; Liu, T.; Qi, X.; Barth, M. Reinforcement learning for hybrid and plug-in hybrid electric vehicle energy management: Recent advances and prospects. *IEEE Ind. Electron. Mag.* **2019**, *13*, 16–25. [CrossRef]
171. Qi, X.; Wu, G.; Boriboonsomsin, K.; Barth, M.J.; Gonder, J. Data-driven reinforcement learning–based real-time energy management system for plug-in hybrid electric vehicles. *Transp. Res. Rec.* **2016**, *2572*, 1–8. [CrossRef]
172. Qi, X.; Luo, Y.; Wu, G.; Boriboonsomsin, K.; Barth, M.J. Deep reinforcement learning-based vehicle energy efficiency autonomous learning system. In Proceedings of the 2017 IEEE Intelligent Vehicles Symposium (IV), Los Angeles, CA, USA, 11–14 June 2017; pp. 1228–1233.
173. Xiong, R.; Cao, J.; Yu, Q. Reinforcement learning-based real-time power management for hybrid energy storage system in the plug-in hybrid electric vehicle. *Appl. Energy* **2018**, *211*, 538–548. [CrossRef]
174. Sun, H.; Fu, Z.; Tao, F.; Zhu, L.; Si, P. Data-driven reinforcement-learning-based hierarchical energy management strategy for fuel cell/battery/ultracapacitor hybrid electric vehicles. *J. Power Sources* **2020**, *455*, 227964. [CrossRef]
175. Romijn, T.C.J.; Donkers, M.; Kessels, J.T.; Weiland, S. A distributed optimization approach for complete vehicle energy management. *IEEE Trans. Control Syst. Technol.* **2018**, *27*, 964–980. [CrossRef]